CONTEMPORARY ERGONOMICS 1995

CONTEMPORARY ERGONOMICS 1995

Proceedings of the Annual Conference
of the Ergonomics Society
University of Kent at Canterbury
4–6 April 1995

Edited by

S.A. Robertson
University College London

Taylor & Francis
Publishers since 1798

UK   Taylor & Francis Ltd, 4 John Street, London WC1N 2ET

USA   Taylor & Francis Inc., 1900 Frost Road, Suite 101, Bristol, PA 19007-1598

A catalogue record for this book is available from the
British Library.

ISBN 0-7484-0328-0

Printed in Great Britain by Burgess Science Press,
Basingstoke, on paper which has a specified pH value
on final paper manufacture of not less than 7.5 and is
therefore 'acid free'.

Contents

Donald Broadbent
Memorial Address

A FORESEEABLE RISK OF INJURY

Stephen Pheasant

Consulting Ergonomist
81 Arlington Road
Southgate
London N14 5BA

This paper is concerned with the mechanisms and processes of
musculoskeletal injury at work. It is written from the standpoint of a
consulting ergonomist practising in the area of personal injury
litigation. The interaction between ergonomic and personal risk
factors is discussed. This special case of keyboard injury then raised.
It concludes with a discussion of psychosocial risk and an exploration
of the concept of an occupational neurosis as a compensatable injury.

Introduction

Ergonomics is the science of work: of the people who do it and the ways in
which it is done; the tools they use the places they work in and the psychosocial
features of their working lives. Ergonomists speak of fitting the job to the worker;
and of fitting the product (system, environment etc) to the user. What constitutes a
good fit? The classical answer to this question is to say that it is one which
maximizes *both* the productivity *and* the safety and well being of the workforce.
This sounds good. The *Realpolitik* of working life is often very different.
Work is an economic process. In the final analysis, working enterprises
exist for the production of goods and services which are offered for sale in the market
place at competitive prices. The enterprise hires labour. The labour force sell their
time, effort and skills (again in a competitive market) in exchange for tokens
wherewith to purchase the outputs of the production process. And thus is the cycle
perpetuated. The up-front goods and services which are produced, are not the sole
outputs of the process however. It also has by-products, some of which are
desirable and some of which are not. For the individual working person, the
satisfaction of a job well done or the sense of purpose which work may bring to his
life, stand on the plus side. Work may also, all to often however, be physically
injurious to the worker - and, although these things are perhaps more difficult to
measure, psychologically injurious as well. Injury to the working person is a by-
product of the production process. Personal injury entails personal loss. The
question therefore becomes - how shall such losses be distributed? Or in plainer
words - who pays? Shall the losses (financial and otherwise) be borne solely by the
injured party; or by the community at large? Or shall they be passed back to the
enterprise - perhaps such a way that these costs of production are reflected in the final

market price of the product in question?

Work entails effort. Effort leads to fatigue. Fatigue causes discomfort. Fatigue is a normal physiological state which is reversible with rest. With rest, the everyday aches and pains associated with fatigue, or with mild overexertion of the muscles at work, may be expected to resolve uneventfully. If rest and recovery are inadequate however, there comes a point at which something more serious supervenes. We have crossed the boundary which divides normal physiology from pathophysiology. The boundary is an indistinct one. Subjectively, the aches and pains of everyday life merge imperceptibly into the pain which signifies incipient or actual physical injury. People vary in their pain thresholds. In part, this is probably due to differences in neurophysiology; and in part to differences in stoicism and in the psychological construction we place on pain. The stoic is more inclined to believe that he can "work his way through the pain". The belief is often erroneous. The symptoms become more severe and more unremitting - and the pathophysiological states which underlie these symptoms become increasingly difficult to reverse. Secondary and tertiary pathologies arise. Chronic pain is physiologically quite different from acute pain; and it is quite different in its meaning for the person concerned. Physical injury - and the chronic pain and limitations on activity which stem from it - can have profoundly deleterious consequences for the individual's psychological health and well being. When treatment is ineffective and rehabilitation unsuccessful, anxiety, depression and a sense of helplessness and hopelessness supervene. The injured person gets trapped in a downward spiral of pain and inactivity, reinforced by his anxiety and depression in which the whole of his life becomes coloured with negativity and despair.

This is the natural history of musculoskeletal injury at work. (Or one amongst a number of possible histories.) Injury is a process rather than an event. We cannot comprehend the injury in its acute manifestations unless we take it in the contexts of both its antecedents and precursors and its long term consequences for the injured person. The purpose of this paper is to explore the injury process. It is written from the standpoint of a consulting ergonomist, whose professional practice lies almost entirely in the area of personal injury litigation. The author is instructed on behalf of both plaintiffs and defendants - the former more often than the latter.

About half of the cases I deal with are back injuries, resulting from lifting and handling (or some biomechanically equivalent activity). Of these, a certain proportion are ones in which the history of the person's condition strongly suggests that it was caused by the cumulative effects of a period of physically stressful activity - which resulted in an overuse injury to the anatomical structures in question - rather than a discrete event such as an accident or a single episode of overexertion. The occupational groups involved are diverse. The back injuries currently on my files range from chorus girls to coal miners - via nurses, factory hands, farm workers, shop assistants and academics. Back trouble is no respecter of persons; and no walk of life holds a monopoly on unsafe systems of work.

Most of the remainder of my cases involve people who are suffering from conditions which we could generically call "repetitive strain injuries (RSI)" or work related upper limb disorders (WRULD)". The two terms are, as far as I am able to discern, entirely synonymous - and far too much hot air has been expended in debating these matters and I for one have lost interest in what people call them. Some of the conditions falling into this overall category are relatively well defined and localized clinical entities: involving the traumatic inflammation of the soft tissue structures of the muscle tendon unit (eg as tenosynovitis, peritendinitis, epicondylitis etc); or the well known peripheral entrapment neuropathies (eg carpal tunnel syndrome etc). Many people with "RSI" however suffer from something very much more obscure, which is not described in the ordinary medical textbooks and for which these books have no name.   The two subgroups are sometimes called the Type I and Type II forms of RSI/WRULD respectively.

In this author's experience, the discrete and clinically well defined forms of RSI/WRULD (Type I) tend to occur in people who do repetitive (or otherwise hand intensive) blue collar jobs in industry - like people who work in poultry processing factories or on motor car assembly lines etc.   Whereas the obscure and innominate forms of RSI/WRULD (Type II) are more characteristic of keyboard users.   I would not claim this to be infallibly the case - and I have certainly encountered exceptions (which are often interesting in themselves) - but there is a strong tendency in this direction.   This leads me to believe that there may be fundamentally different causative mechanisms involved.   It may also be that an injury which, in the first instance, is manifest as a discrete condition; may, as it progresses to chronicity, develop into something more obscure.   In the view of the "RSI sceptics", who claim with some justification to represent the consensus of the medical and surgical establishments, the obscure and innominate forms of RSI/WRULD do not exist. People who claim to suffer from such conditions are deluded.   They are "eggshell personalities who need to get a grip on themselves".   In essence these conditions are neurotic disorders which become manifest as arm pain, when certain emotionally vulnerable individuals are placed in certain occupational situations - often ones involving keyboard work.   To date, the consensus view of the RSI sceptics has prevailed in the English courts.   So far, the sceptics have carried all before them. The author does not share their view.

What is an Injury?

This is a surprisingly difficult question to answer.   The Oxford Dictionary provides the following definition: Injury - harm, damage; wrongful action or treatment.   In its everyday usage, the word has a range of meanings.   Of these, it is possible to distinguish (among others) "medical" meanings and "ethical" meanings.

In the medical context the word "injury" is narrowly used to refer to a physical lesion, in which structural damage to tissue has resulted from the action of some external agent - typically mechanical impact or overload, heat, cold etc.   The word "trauma" is synonymous.   By analogy we speak also of a psychological injury (or trauma).   States of dysfunction without structural damage occupy something of a disputed territory - in that they would seem to fall somewhere between the narrow and literal medical meaning of structural damage and the broader usage of the term, which may be applied by analogy.

In the ethical context, an injury is damage or harm resulting from a wrongful act; or, more narrowly, the wrongful act itself.

The etymology of the word injury is an interesting one.   It derives from the latin *injuria*; which is the feminine substantive form of the adjective *injurius*, meaning wrongful or unjust (in-*jus*, right).   The "ethical" meaning of the word injury is thus the original one; and the "medical" usage is one that it has acquired with time.   The nearest latin equivalent to the "medical" meaning would probably be *vulnus* (which we could also render as "wound" - from which we derive the present day adjective vulnerable.

We refer to a person who has been injured as a victim.   This word also has an interesting history.   It derives from the Latin *victima* - meaning a sacrificial victim.   Having been imported into English by early translators of the Bible its use gradually grew to include both the "victims" of wrongful acts and people who had suffered as a result of blind mischance and were thus "victims of circumstance".   In other words it has lost both its technical meaning to do with religious observances and its ethical meaning - and now merely means a person who suffers.   The meanings of words are apt to drift with the years and as they are imported from one language into another.

The diversity of meanings of the word "injury" can lead us into difficulties in the medicolegal context. If we choose to define injury sufficiently narrowly, or if we fail to make it clear just how we are using it, then we make it mean pretty much what we please. We can, for example, show that someone who suffers from one of the obscure and innominate forms of RSI/WRULD has not sustained an injury. Thus the RSI sceptics will argue that without a physical lesion there is no "injury"; that in the obscure and innominate forms of "RSI" there is pain and dysfunction without detectable signs of structural damage; and that it therefore follows that "RSI" is not an "injury". Much the same argument is posed for chronic back trouble. But what does this achieve? Nothing. As an argument it is devoid of content and entirely contingent on an idiosyncratic usage of the word injury.

Under English Common Law, there is no liability without fault. The plaintiff must show that he or she is the victim of a wrongful act rather than merely a victim of circumstance. Or to put it another way, he must show, "on the balance of probabilities": first, that he has sustained an injury; secondly that this injury was caused by his work; and thirdly that it was negligently (ie wrongfully) caused (ie that he has sustained injury in the ethical sense). Negligence being a *tort* or civil wrong. For the injury to have been negligently (ie wrongfully) inflicted it must (*inter alia*) have been both reasonably foreseeable and avoidable by means which were reasonably practicable - but which the defendant failed to adopt.

What does the law regard as an injury? This again is not a simple question - not least because the legal usage, like the everyday usage, has drifted with the years. In the time honoured maxims of legal latin *injuria* is a legal wrong (or the violation of a right). Thus *volenti non fit injuria* means "he who voluntarily enters into a course of action cannot claim to have been wronged"; rather than "he who volunteers cannot complain if he gets hurt". We may likewise have an *injuria absque damno*, which is a legal wrong which does not lead to any loss or damage; or a *damnum absque injuria* which is loss or damage which does not stem from a legal wrong; and neither confers a right of action at law. The former corresponds to what we should nowadays call "negligence in the air" - that is negligence from which no harm results. Examples of the latter include what the lawyers call "true accidents" for which nobody is to blame; or work injuries which result from risks which would not have been foreseen by a reasonable person or which could not have been avoided by means which were reasonably practicable. Nowadays however, when we read the Particulars of Injury, as set out in the Statement of Claim of an action for personal injury, they refer to the harm which the Plaintiff alleges he has suffered rather than the wrongs which have been done to him. The latter will typically be found in the Particulars of Negligence.

There are formal definitions of injury under various items of statute law, but they are not really very helpful. For example in Section 3 of the Law Reform (Personal Injuries) Act (1948) we are told that the term "personal injury" can in principle include "any disease or any impairment of a person's physical or mental condition". This is enough to indicate that the law is prepared to use the term injury broadly rather than narrowly, but little more.

The Common Law concept of injury turns on the legal maxim De *Minimis Non Jurat Lex*: the law does not concern itself with trifles. In essence the Plaintiff first has to show that he or she is suffering from something more than the ordinary aches and pains of everyday life. These do not constitute an injury at law and no redress is offered for these. Thus a defence offered in many RSI cases is that the plaintiff has in reality suffered from nothing worse than the ordinary everyday discomforts which we experience at work - but, because of certain personality defects,

therefore the contractile strength of muscle typically increases more rapidly than the tensile strength of its soft tissue attachments and the other anatomical structures of the musculoskeletal system which are subject to loading during muscular exertion. This also may create a vulnerability to overexertion injury.

When considering the risks of work there may also be some advantage in distinguishing between *overuse* and *misuse*. One might for example argue that an injury resulting from the repetitive dynamic loading of tissue is one caused by overuse; whereas an injury resulting from prolonged static loading is caused by misuse. The distinction is perhaps a fine one.

Turning now to personal risk factors and vulnerability to injury. Fitness for work is part of this. Some people, by virtue of their strength, flexibility, co-ordination, etc, are capable of tolerating greater levels of physical stress in the working situation than others. There is good evidence for this in the case of occupational back injury and the chances are that it is also true for the WRULDs - although the subject has not been so well explored epidemiologically. We differ also in our anatomical construction. The short leg with a compensatory scoliosis or the anomalous thumb tendon may render us susceptible to back trouble or de Quervain's tenosynovitis if we become nurses or chicken factory workers. There are numerous constitutional risk factors which may render us more than averagely likely to develop carpal tunnel syndrome - either spontaneously or as a result of having been exposed to the risks of work. Or we may have hypermobile joints or more than averagely easily injured soft tissues or peripheral nerves. Superimposed over these differences are the effects of "normal wear and tear" an the "degenerative changes" contingent on normal ageing.

Perhaps some of us age more rapidly than others - in that because of some, as yet unidentified, biochemical or immunological factor, our soft tissues are more susceptible to such degenerative changes. Some tissue types may wear out more easily than others. This has been invoked as an explanation for both degenerative disc disease and conditions like tenosynovitis and tennis elbow. In either case, the evidence cited for the proposition is that people with problems at one site in the back or the upper limb (as the case may be) are more likely to have problems at other sites also. The converse explanation (and one which in this author's view is just as plausible, if not more so) is that heavy work may be injurious to any or all parts of the spine; and that any given type of hand-intensive work may likewise be injurious to a variety of anatomical structures in the upper limb. There is good epidemiological evidence that at any age (past thirty-something) people in heavy jobs are more likely to show objective signs of degenerative disc disease than people in light jobs (Hult 1995). The occupational and personal risk factors interact. It seems likely that the same would also be the case for the WRULDs.

Given that it may be shown, on the balance of probabilities, that the plaintiff is not so excessively vulnerable as to have developed the condition of which he complains spontaneously; then, in terms of liability at law, constitutional at-riskness is neither here nor there. First, there is no burden of proof on the plaintiff to show that the defendant's negligence was the sole cause of his injuries. It suffices for his case, if he can show that the negligence of which he complains, made a "significant" or "material" (i.e. non trivial) contribution (*McGhee v. National Coal Board*, 1973). Secondly, injuries stemming from an unforeseen or unforeseeable vulnerability on the part of the plaintiff are subsumed under the so-called "eggshell skull" rule. For this rule to apply however the causative acts or omissions of the defendant must in the first place be tortious - i.e. negligent. The plaintiff must therefore show that the working practices in question would have entailed a risk of injury, which was both significant and foreseeable, for a person of normal fitness.

What is normal fitness? Presumably it means something like being not-much-more-than-averagely unfit.   Or to put it another way, falling somewhere in the middle part of the range of constitutional at-riskness (were we able to measure this, which we can not).   How far into the lower tail of this distribution do we go? There is no ready answer to this question - neither can there be.

Keyboard Injury

Keyboard injury is a special case.   It is easy to see that if you spend your time hauling geriatrics in and out of bed and on and off of the lavatory, you are likely to "do your back".   It is likewise easy to see that if you spend your working day pulling the guts out of chickens, something or another in your hand or arm is likely to wear out.   But in keyboard work (particularly when using the light action electronic keyboard) wherein lies the potential for injury?   To answer this question I return first to the particular types of RSI/WRULD characteristically suffered by the keyboard worker and secondly to the distinction between overuse and misuse.

As we have noted above, keyboard workers tend, for the most part, to suffer from the obscure and innominate forms of "RSI", which -in the consensus of the medical establishment - do not exist.   But their lives can be blighted by this condition - to the extent that they sometimes attempt suicide, sometimes successfully.   So, notwithstanding the scepticism of the medical establishment, this is a condition to be taken very seriously indeed.

The condition (or conditions) in question is sometimes referred to a s "diffuse" RSI.   I do not like the use of the term "diffuse" in this context, in that it carries connotations of vagueness and insubstantiality.   "Disseminated" is a better description.   In this author's view, the disseminated overuse syndrome which affects the upper limbs of keyboard users (and others) is sufficiently constant in its clinical presentation to be regarded as an entity in its own right (Pheasant 1994 b, 1995).   Its characteristics are summarized in table 1.

Table 1. The Disseminated Overuse Syndrome

- ■   Characteristic of keyboard users.
- ■   Insidious in onset.   Difficult to reverse.
- ■   Multiple sites of tenderness in upper limb(s), shoulder girdle and cervico-dorsal region: first dorsal interosseous, medial and lateral epicondyle, upper trapezius, scalenes, sternomastoid, inter-scapular etc.
- ■   Signs of "adverse neural tension" often present.
- ■   Histochemical and ultra-structural changes in muscle tissue and the sensitization of nociceptive mechanisms have all been demonstrated beyond reasonable doubt.
- ■   Secondary or tertiary psychopathology very common.

How does this condition arise?   The probability, as the author currently understands it, is that the direct agent of injury is the prolonged static muscle loading which results from working at the keyboard.   In part this is localized to the particular muscle groups which are active in order to maintain the person's working posture at the keyboard -the tension in these particular muscles being superimposed over a more generalized muscle tension contingent on overall discomfort, the psychological stress of work and so on.   Electromyographic studies have shown that the levels of muscle tension which keyboard work entails (especially if the working posture is unsatisfactory, for example because of a poorly designed or incorrectly adjusted workstation) may be as much as 30% or more of the maximum capacity of the muscle groups in question (Onishi et al, 1982; Hagberg, 1982). Levels of contraction of this magnitude result in a partial occlusion of the bloodflow to the working muscle - leading to an "energy crisis" in the muscle tissue and the accumulation of metabolites. This leads in the first instance to a localized muscle fatigue. Lundervold's classic early electromyographic studies of typists showed that with the onset of fatigue, muscle activation spreads to groups which were initially quiet (Lundervold 1958). So the scene is set for a more generalized muscle fatigue to set in. The physiology of muscle fatigue is complex (Pheasant 1991). In essence it is contingent on a disturbance of the *internal milieu* of the muscle tissue. Dennet and Fry (1984) have documented biopsy changes in biopsy specimens, taken from the hand muscles of people suffering from "RSI", which are strongly suggestive of an adaptive response to such a disturbance - such as might enable the muscle to deal better with the energy crisis which results from prolonged static overload. The adaptation is imperfect however.

Increasingly profound degrees of muscle dysfunction supervene. The accumulation of noxious metabolic by-products initiates a neurological "cascade effect" leading to the activation of "trigger points" in the muscles and the sensitization of the nociceptive mechanisms, probably both peripherally and centrally (Gibson et al 1991, Helme et al 1992) - or what is sometimes referred to a "pain amplification". The condition thus becomes self maintaining - and exceedingly difficult to reverse clinically.

When cases of keyboard injury come to trial, the defence will generally be based on three propositions. First, that the condition in question is not an injury; second, that even if it is an injury it was not caused by the plaintiff's work; and thirdly that even if it was caused by her work there was no negligence on the part of her employers.

If the process of pathogenesis proposed above is broadly correct - in its essentials if not necessarily in its details - then it is difficult to see how one could say that the syndrome in question falls within the *de minimis* rule. This being so then "RSI" is an injury at law. The defence will argue however that the plaintiff has never suffered from anything more than ordinary fatigue and "the aches and pains of everyday life"; she has become neurotically pre-occupied with these; and mistakenly believes herself to have been injured. They will further argue that "RSI" is a condition which has symptoms (pain) but no objective physical signs - and that it cannot therefore be an "injury" in the medical sense of the word. To support these propositions they will seek to present the plaintiff as a neurotic hypochondriac, who is unable to cope with the ordinary stresses of work and who expresses her inner distress in the form of physical pain - because for one reason or another she cannot express her emotions in the ordinary way. Or to put it more technically, she is suffering from a somatoform pain disorder or a form of conversion hysteria (Lucire 1986). This then sets the scene for the second and third propositions: that "RSI" is not caused by work (since it is a neurosis) and that there is no negligence (since the plaintiff is neurotically over-reacting to the ordinary physical and mental pressures of working life).

The cornerstone of this argument is removed, if it can be shown that the syndrome in question has an underlying organic pathology. In the author's view, the available scientific evidence is sufficient to show that it does - if not "beyond all reasonable doubt", then certainly "on the balance of probabilities" - which is the degree of certainty which the civil courts require. The evidence indicates that the syndrome in question has both a somatic pathology (Dennet and Fry, 1984); and a neuropathology (Gibson et al 1991, Helme et al 1992). It is abundantly clear however that it also has a psychopathology.

Anyone who has had dealings with people who suffer from chronic "RSI" will tell you that they are very often (but not always) severely emotionally disturbed - to an extent which, if the emotional disturbance could be shown to be secondary to their physical injuries, would fall outwith the limits of the *de minimis* rule.   But are they disturbed because of these injuries or were they disturbed people in the first place? Such epidemiology as is available points to the former conclusion - at least at the statistical level (Spence 1990).

This is not to say however that individual RSI sufferers may not have latent psychological vulnerabilities prior to the development of their condition; and it is not to say that these vulnerabilities were not contributory factors in its onset; but which of us can lay claim to being wholly sound and integrated personalities; and why is it that we do not get RSI unless we are exposed to the stresses of certain kinds of work.

Psychosocial Injury.

I have argued elsewhere that when musculoskeletal injuries fail to resolve as expected, it is because a chronic pain syndrome has supervened - this syndrome being a self-perpetuating condition contingent on physical re-injury, neurological sensitization, disuse and psychosocial reinforcement.   We could call this a post-injury syndrome (Pheasant 1994 a,b).   At law, such a syndrome would be seen as stemming from the original injury - and, if the original injury was negligently caused and the "chain of causation" was unbroken by any *novus actus inerveniens* - the defendant would be liable for the plaintiff's chronic disability. (See for example *Bird v Hussain*, 1993 for a recent judgement in this area.)

I am assured however that there is such a phenomenon as purely psychogenic pain.   It has been argued for example that this is "a masked form of depression" (Blumer and Heilbronn, 1982).   Although one could ask perhaps, since chronic pain and overt depressive illness so commonly go together, wherein lies the "masking"?

Let us suppose, purely for the purposes of argument, that we found ourselves dealing with a case of what was genuinely an "occupational neurosis": that is, a work related condition, which was devoid of organic pathology and existed solely "in the mind".   Or to put it slightly less strongly a condition which was essentially psychogenic notwithstanding that "mind" has a physical substrate (*pace* Descartes).

There is nothing particularly new about the notion of an occupational neurosis - the so called "occupational cramps" are conventionally thought of in these terms, but are nonetheless Prescribed Diseases for which Disablement Benefit is payable under the Social Security Act, 1975.

Our hypothetical occupational neurosis could be manifest in terms of chronic pain; emotional or behavioural disturbance; or some combination of these.

Let us further suppose that we could show causation - in that the psychological stresses of the person's working life could be shown (on the balance of probabilities) to have made a material contribution to this neurotic illness.   Is there here the possibility of a cause of action at law?

Munkman (1993) cites some interesting judgements in this area, although none are from the occupational context.   In one case, a plaintiff suffering from hysterical paralysis of his legs following an accident was awarded damages not far short of those he would have received if his legs had been amputated.   In another case, where two sisters were struck by a motor car and one of them died, the surviving sister was awarded damages for depressive illness brought on by feelings of "survivor guilt". One suspects however that these were cases in which the issue of negligence per se was never really in question.

On 16 November 1994, Mr Justice Coleman gave judgement for the plaintiff in the landmark case of *Walker v Northumberland County Council*.   This was the first occupational stress claim to come to trial in an English court.   The plaintiff, a social worker suffered two nervous breakdowns, as a result of being massively overloaded at work.   The plaintiff's claim was unusually well documented in that there were detailed records on file of frequent and repeated written requests for assistance in coping with his case load, which to all intents and purposes had gone ignored.   The defendants were not found liable for the first nervous breakdown on the grounds that it was not a reasonably foreseeable outcome of the circumstances in which the plaintiff found himself.   But they were found liable for the second breakdown, in that they had been "put on notice" by the first one - and judgement was given for the plaintiff.

The response of the press, somewhat predictably, was to speak of the "floodgates of litigation" being opened - and opinions offered on the subject verged at times on the intemperate.   An editorial in the *Daily Telegraph* (18 November 1994) concluded: "If the courts seem minded to offer further licence in this direction, then the Government should urgently seek legislative means to stop them".   Where does it go from here?   Let us speculate a little.

Adams (1992) has put together a set of case histories concerning bullying at work.   The principal feature which these have in common is that people who are in positions of authority in the workplace - and have what can only be described as warped personalities - have on an ongoing and continuous basis sought to undermine the emotional well being of their subordinates - to the extent that the victims of this process of psychological assault have become seriously emotionally disturbed as a result.   Under the legal concept of vicarious liability, the employer is liable for the negligent acts or omissions of his agents and servants - unless they can be shown to be "on a frolic of their own".   Perhaps the sadistic infliction of emotional suffering by a warped personality can be regarded as such.   But what if the organizational hierarchy deliberately condones this in the interest of motivating the workforce, or can be shown to have turned a blind eye?   How did the sadist reach the position of authority in the first place and why was he allowed to stay there?   What does this tell us about the nature of the organization in question?

Perhaps the ongoing stress to which the plaintiff is exposed does not stem from the psychopathology of her boss but merely from the boss's incompetence -   for example because he is himself bad at time management or he lacks the foresight to plan the work load of the department in advance and has developed a style of crisis management.   As a result, he is forced on a frequent and repeated basis, to impose unreasonable deadlines on his subordinates.   Does this sound at all familiar?   This is clearly bad management.   But how far does bad management have to go before it becomes negligently bad management?

We have traditionally thought of occupational stress solely as a problem for the individual in question - and sought to teach him to manage his stress better by learning more effective coping strategies.   We have seen it as a matter of stress-proofing the individual - as a matter of fitting the person to the job rather than fitting the job to the person.   To see occupational stress solely in these terms however is to miss a very important part of the picture.   Stress is also about the way that power is distributed in organizations. In many cases it is about the way that power is abused. Stress is not solely an issue in the domain of human psychology or psychophysiology - or, for that matter, the domain of the empirical sciences in general - it lies partly in the domain of ethics.   And this, as I see it, is equally true for ergonomics as a whole.

References

Adams, A. 1992 *Bullying at Work How to Overcome and Confront It.* (Virago, London).

Adams, MA and Hutton, WC. 1985 Gradual disc prolapse. *Spine*, 10, 524-531

Alcock v. Chief Constable of South Yorkshire Police. 1992 1 AC.

Bird v. Hussain, 6 July 1993, QBD, cited *Personal Injuries and Medical Negligence Newsletter*, Oct 1993.

Blumer, D and Heilbronn, M. 1982 Chronic pain as a variant of depressive disease: the pain-prone disorder. *Journal of Nervous and Mental Diseases*, 170, 381-406.

Gibson, SJ et al. 1991 Cerebral event-related responses induced by CO_2 laser stimulation in subjects suffering from cervico-brachial syndrome. *Pain*, 47, 173-182.

Hagberg, M 1982 cited in Grandjean, E. *Ergonomics of Computerized Offices* (Taylor and Francis, London) p 104.

Helme, RD et al. 1992 RSI revisited: evidence for psychological and physiological differences from an age, sex and occupation matched control group. *Australian and New Zealand Journal of Medicine*, 22, 23-29.

Hult, L. 1954 Cervical, dorsal and lumbar spinal syndromes. *Acta Physiologica Scandinavica*, Suppl 17.

Lucire, Y. 1986 Neurosis in the workplace. *Medical Journal of Australia*, 145, 323-340.

Lundervold, A. 1951. Electromyographic investigations during typewriting. *Ergonomics*, 1, 226-233.

McGhee v. National Coal Board. 1973 1 WLR 1.

McLoughlin v. O'Brian 1983 AC 410.

Munkman, J. 1993. *Damages for Personal Injuries and Death* (Butterworths, London).

Mullis, A. and Oliphant, K. 1993. *Torts* (Macmillan, London).

Onishi N. et al. 1982 Arm and shoulder muscle load in various keyboard operating jobs of women. *Journal of Human Ergology*, 89-97.

Pheasant, S. 1991 *Ergonomics, Work and Health* (Macmillan, London).

Pheasant, S 1994 a Musculoskeletal injury at work: natural history and risk factors. Ch 10 in Richardson, B. and Eastlake, A. (eds) *Physiotherapy in Occupational Health* (Butterworth Heinemann, London) pp 146-170.

Pheasant, S. 1994 b Repetitive strain injury: towards a clarification of the points at issue. *Journal of Personal Injury Litigation*, Sept 1994, 223-230.

Pheasant, S. 1995 *Bodyspace: Anthropometry, Ergonomics and the Design of Work*, second edition, London, Taylor and Francis.

Spence, S. 1990 Psychopathology amongst acute and chronic patients with occupationally related upper limb pain versus accident injuries of the upper limbs. *Australian Psychologist, 25, 293-305.*

Value definition 1
"Everyone realises the value of ergonomics."

Ergonomics was born in the Second World War and grew throughout the period of the "Cold War". Clearly the need for ergonomics, or as the American would then have it, Human Factors Engineering, was self evident. "Our" weapons must work effectively and be ready in time to counter "their" weapons whenever and wherever they were deployed. Budgets were large, "scientific method" was popular and confidence in scientists was high!

Ergonomists working in the military and related fields could point to their successes, or more likely, the failures of others, (often engineers) and obtain work. There was, of course, the constant gripe that the ergonomist became involved in the project too late and that critical (I mean to imply "wrong") decisions had already been taken. However, that could just be seen as establishing the alibi! Ergonomics was, obviously, a good thing but not everybody realised it! In the "white heat" of technological advancement there was nothing an ergonomist could not do! Ergonomics was, just "Common Sense" confounded by technical jargon, just as Bronowski (1951) argued all Science was!

This modest certainty still prevails in many academics, researchers and, to a lesser extent, practitioners! However, just because we know the value of ergonomics not everybody else has the confidence we do! Business is, for example, sceptical and must have the worth of the subject justified to it in the financial terms that it understands!

We know that ergonomics is not a theoretical subject, it is the application of knowledge and experience to a human activity system with the intention of minimising any undesirable factors. It is a holistic amalgam of all its root disciplines. We must seek to get these points across by demonstrating the subject's worth. It is, of course, good to have confidence in the product; now we must ensure that everybody else learns to have it too.

Value definition 2
"Ergonomists have different values from the organisations they serve."

Those at the announced birth of ergonomics, in 1949, (Murrell 1965) clearly could not be pure ergonomists as their original training had already occurred; Society at this time lacked knowledge of the appropriate terminology. Just as the first "Physicists" described themselves as "Natural Philosophers" or Mathematicians the founders of our discipline were to be found working under various titles. The cover of Alphonse Chapanis's 1959 classic (1965 printing) notes him to be a Professor of Psychology and suggest that the book should be filed under "Business, Economics". (Sadly not the last time that our discipline has found itself filed under economics!)

Ergonomics was created from an amalgam of other objects and disciplines at a time when University Departments or Research Groups did not emerge quickly. Thus the early ergonomists operated in Departments in which they qualified but undertaking tangential or novel activities. This would also be true in Industry where ergonomists might be found in Personnel Departments, Safety Offices, and Engineering, Design or Medical Groups. With some enlightened and meritorious exceptions the same is true today.

The mutli and interdisciplinary nature of the subject will always remain and many of those who could describe themselves as ergonomists still prefer to attach themselves to an area perceived to be more established. In the past year I have met Biomechanists, Heating Engineers, Health Chemists, Physiologists and Psychologists who have produced ergonomic reports. Many of these people could describe themselves as ergonomists but choose not to do so. When asked why they did not describe themselves as ergonomists, the most common reasons given were the title of their posts followed by a lack of understanding of ergonomics in Society. I experienced the reverse when I applied for a Lectureship in Ergonomics at another University only to find that the workload was virtually all physiology and biomechanics; apparently the "wrong" title was being used for internal political reasons!

I too must admit that I have, on occasions, avoided the inevitable confusion with "economics" or "work-study" and described myself as a Practical Philosopher! We must seek to remove these misunderstandings, often of our own making, and we must adopt the techniques of business to demonstrated the benefits to be obtained from using ergonomists. The values of the ergonomist concerning the well-being of people, the belief of solution obtained via inquiry must become recognised as important but this is unlikely to happen until our "eccentricities" are more widely known and our attributes financially justified. We must all work to over come our ignorance of the ways of business before we can overcome the comment so often heard when ergonomists meet; "why doesn't my boss understand me!"

Value definition 3
"Which do Product Designers value most, ergonomics or aesthetics?"

When faced with a decision between two products designed to undertake the same task does the potential purchaser select the one that looks good but might not work or the one that works but looks less good? To answer this question more information is required, perhaps most importantly what tests can be undertaken by the buyer? Do they have the option of a trial? Will they undertake realistic tests or just rely on intuition, initial perception, looks and the brand name?

Well this is, of course, the wrong question. How we get others to value ergonomics? How can we sell ergonomics and show that products incorporating "good" ergonomics work better than those that do not. Are we, as a professional society, even able to agree on what good ergonomics is?

Clearly if we are to encourage the general public to value ergonomics then we must emphasis when it has been incorporated and draw attention to the benefits that follow. These are early days but this is increasingly occurring. Some specific products, notably work seating, have been endorsed by respected ergonomics. These, and others, have been involved in particular products from the earliest design stages, however, these examples are rare and not without controversy. Consider, for example, the debate concerning the Maltron Keyboard and it's role in the prevention or cure of "RSIs" or, as is now generally preferred, "WRULDs". I can recall seeing versions of this keyboard at exhibitions in the early 1980s yet it appears that the definitive, accepted by all, research to validate the design has yet to be undertaken. As recently as the November 1994 issue of "The Ergonomist" the originator of the keyboard is suggesting that an appropriate organisation should seek funds to undertake the experiment! (This cycle of the debate started in October 1994 issue.)

Of course, not all of those claiming that their products are "ergonomic" are concerned about the niceties of justification or attribution. For example, a pen allegedly based on "the ergonomic perfect shape of an egg" may be fine if you want to lay it but why should it be easy to write with? How can we ensure that ergonomics is appropriately applied, understood and thus valued by the general public?

Perhaps one way forward is to join with the Professional Societies of others involved in the design and production of products to fight for individual credits. For example, an

Architect will display a board on a site and a typical Film, might credit 100 people from the most famous actor down to those that provided the catering or accountancy services. The Swedish company IKEA often acknowledges the Designer of the product but what about the others involved? Perhaps, the work of the identifiable ergonomist/ergonomics group will be valued and sought out by prospective purchasers. (I may not be able to tell which firm of accountants were used by watching the film but other film makers recognise the names and, I expect, seek out those successful in keeping the production within budget!) Good product ergonomics may not be obvious either but the lack of it could concern or injure the user.

The manufacturer and suppler might also welcome the labelling of products with details of those concerned with their production. The clear identification of responsibility could be important if recalls were required or a legal action started under the recent EEC driven Consumer Protection Regulations (SI 1994:2328). (This legislation would appear to offer another approach, for a plaintiff seeking to establish liability if they were injured while operating equipment supplied by others. I would expect to see, for example, cases concerning injured "keyboarders" before the Courts in due course The regulations do, however, offer a defence of "due diligence" and this could be expected to benefit from crediting those involved in the product's design and manufacture.)

Value definition 4
"It might not look best but it is the safest, ergonomics does have value."

Once we can define what constitutes good ergonomics and provide the information by which the purchaser can check a product or service we must ensure that it is recognised and acted upon. At present, even when a design includes some well thought out ergonomic features the marketing people tend to use it only as jargon. The benefits of the ergonomics must be shown to be self evident and not just regarded as some cynical Unique Selling Point (USP). Ergonomics must not be held in the same regard as the P45 fluoride formulation included in the stripes of the, apparently "ideal" toothpaste!

Virtually any marketeer will tell you that ergonomics can help sell products but that it is rarely a "contract clincher". For example, could you find office furniture designed to support computers that was not sold as "ergonomic"? I doubt it! In this market the term "ergonomic" has value but this value is not often quantified by ergonomists. The marketeers using the term may not understand our subject nor treat it with respect. Graphic equalisers, described as ergonomic, appear, for example, to be an essential part of "budget hi-fi" systems. Yet when you pay more the idea that you should modify the sound on the CD or watch the jumping green lights is rejected; the display and tone controls are removed; "Source Direct" is the desired requirement.

I know of a company manufacturing an "ergonomic" steam iron that had a simple three position steam switch; off, half and full rate. The iron did not sell well and studies suggested that one reason was that the irons of their competitors had greater adjustability. The solution was to change the switch, not the mechanism, into a knob and make it click ten times as it was rotated. A "flash" was added to the box - "Now with 10 position steam control". Sales increased dramatically! The replacement of a three position switch with a three position, ten click, rotary control would not be the normal advice of an ergonomist but to the marketing department this design change "added value" to the iron!

If we are to educate the general public then we must all do our part. Speak on Broadcast Media, write in the press and generally work to spread the word as wide as possible.

Value definition 5
"The value of the ergonomics contract to the University was a Lectureship."

Who are these "Experts" that might value ergonomics for others? In the case of a Domestic Product then the question about what something is worth is usually answered by the statement that it is worth what you can sell it for. This might appear glib but it is undoubtedly true. (Would you covet a cheap Porsche, buy low priced perfume or, without reservation, employ an "Expert" who only charges 1/5 of the market rate?) Can ergonomics be good if it is cheap, perhaps it is just applied common sense!

Perhaps "ergonomics" cannot yet be valued, for domestic products, because the market place is not aware enough of the benefits and the information required to identify those applying ergonomics is not displayed. However, we are generally selling not to the general public but to manufacturers, Product Designers, Government Agencies, etc. Perhaps we are too cheap!

Until the last decade most ergonomic advice available, other than in-house groupings, came from academics or University Departments/Research Groups. I fear that many of these, historically, did not charge commercial rates for their help. They also suffered from the perception that it was often the precision and rigour of the research work required for PhD opportunities that was for sale and not a near correct answer later that afternoon. Thus the valuation was undertaken according to different criteria by prospective client and consultant.

Recently two other groups have entered the market and are offering advice on ergonomic problems. This advice is often very cheap and appears to come from; "just finished" students testing the "freelance experience" and professionals in related disciplines with spare capacity. In the case of the latter I have been undercut by a local General Practitioner Practice (Doctor, Physiotherapist and Nurse) and a Marketing Agency seeking work for a Placement Student studying Personnel Management! Their "ergonomics" is unlikely to be comprehensive or of an appropriate standard but it was, apparently, cheap.

Discussions concerning the rates charged by those seeking ergonomics Consultancy work are, usually, candid. However those offering to work in the Health and Safety Field could have a listing in The Health & Safety at Work Journal. This publishes listings (Table 1 overleaf) under 28 categories of activity of which one is "ergonomics". The only other category that closely applies to our sphere of activity was "Manual Handling" as "Office & Buildings" was generally used by those offering environmental monitoring from an Occupational Health. The activities are classified under two headings; "activities most commonly carried out by the consultancy" or "other activities within a Consultancy's scope". There was growth (26%) in Consultancies offering ergonomics between the 1990/1 and 1994 editions at the same time as the number of Consultancies listed dropped by 33%.

The daily fees quoted as charged by the Consultancies are, in most cases given in wide bands (£150-£450 is typical) while others did not provide any details. In the 1994 list the mean daily rates quoted for all Consultancies excluding those offering ergonomics can be estimated as £421 (Bands 200-225 to 800-825) while the £383 (Bands 150-175 to 675-700) and £340 (Bands 150-175 to 625-650) are the estimates for those organisations offering ergonomics either as a main (n=16) or subsidiary (n=13) activity respectively and providing fee details. Table 2 (overleaf) shows how daily fees have changed during the 1990s over a period during which inflation was about 25% and suggests that competition is pushing down the fees charged for ergonomics and thus the perceived "value" is also falling!

Table 1. Summary of Consultancies listed. (1990/1 was a biannual listing).

	Totals (%)		"most commonly"		"other"	
	1990/1	1994	1990/1	1994	1990/1	1994
All Consultancies listed	227	151				
Those offering "Ergonomics"	23 (10%)	29 (19%)	14	14	9	15
Those offering "Manual Handling"	23 (10%)	54 (36%)	18	11	5	43

Table 2. Comparison of Mean Daily Rates(£) between 1990/1 & 1994.

	1990/1		1994		Change
Consultancies	£Mean	£SD	£Mean	£SD	£(%)
All Consultancies	319	116	421	157	102(32%)
"Most commonly" -ergs.	348	112	383	117	35(10%)
"Other" - ergonomics	387	102	340	108	-47(-12%)

The anecdotal evidence for the longer established Consultancies is not so bleak but if the image of Ergonomics as of high value/worth is to be maintained then the perception of falling fees and that ergonomics can be offered by non-ergonomists must be changed. At present, I suspect, Ergonomics Consultancy is too often seen as an additional revenue stream and not the main focus of activity. It must be realised that for the small to medium sized company the option of employing an ergonomist within the organisation does not exist. Thus as they learn the advantages of the incorporation of ergonomics into their structures, their products and services they will expect to discover pricing based upon utility and benefit gained, not just an hourly rate for, perhaps, an unspecified length of time. At the very least the latter approach leads to concerns that the job is being spread out and that the client is being "milked". Not a reputation we should wish to reinforce, especially for those of us working in Academic Institutions. This, hopefully erroneous, perception of academics is already well established among the general public!

As other areas of activity undertaken by ergonomists do not, as yet, appear to be such a free market as can be found in Health and Safety Consultancy the rates charged are not published. It will be interesting to see what, if any, difference the Ergonomics Society's new Consultancies Register will have to this aspect of value. To what extent, if any, will it monitor fee levels and the quality of service offered.

Value definition 6
"Manufactures should add value to their products by improving the ergonomics"

The possible objectives available in the marketing of products or services, can, according to the Marketing Guru (and no known relation) Michael Porter (1989) be reduced to only two true strategies; Quality ("Differentiation") leadership and Cost leadership. With the obvious exception of Not-For-Profit organisations it is the profitability of their products and/or services that drives the organisation and its Stakeholders.

It is, of course, possible for products and services to be targeted at different niches and thus a "mixed" ("focused") strategy can be applied. Thus a company might seek to market the "best" kettle at a number of different price points, £14.99, £19.99, etc. It is also true that a company will seek to maximise their market share by offering products/services targeted at different markets. IBM tried that with its "Ambra" products and Ford still maintains separation with Jaguar, lest people start to view the latter as an expensive Ford. The synergy between BMW and Range Rover is clear but the marketing problems of the reverse must also be considered; could the classic "Mini" devalue the BMW Marquee?

Thus it is only these two aspects (Cost or Differentiated ("Quality") Leadership) of competitive strategy that can benefit from the application or incorporation of ergonomics. However, this view highlights the two fundamental approaches, using ergonomics to improve the service or product offered or to reduce the costs that must be borne by the revenue stream concerned. The model Porter devised is shown in Figure 1.

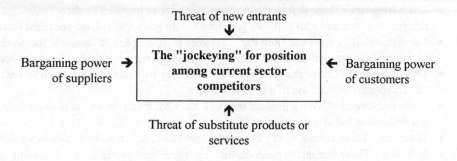

Figure 1. Michael Porter's Model of Competitive Advantage (1979).

Quality Leadership

I have mentioned earlier some difficulties in the incorporation of ergonomics into a product or service so that the customers' perception of quality within that item is enhanced. This is an area in which we must all work by requesting ergonomic evidence to support both our own purchases and those we have influence over. How often when considering a purchase do you discuss the ergonomics of the various products explicitly with the sales staff? For example, does your Purchasing Department specify the type of container that supplies come in or does it just specify the contents. I have recently helped a company with a manual material handling problem by training the purchasing department who then specified 25Kg kegs bound to pallets rather than the 50Kg drum that had been previously used. Including this specification in the tender documents did not, I understand, lead to an identifiable additional cost. I am also aware of a company that now specifies "Copier side" delivery for its paper and will not accept delivery elsewhere. In this case it is the delivery driver who brings the pallet truck and transports, in bulk, the paper by the lift rather than the office staff carrying the boxes by hand. (You will note that the Bargaining Power of Customers is an influence identified in Figure 1)

While it is often not easy for us to precisely quantify the financial benefit from the incorporation of ergonomics into a product or service that is really the role of others, the marketing Department for example. The ergonomist by considering "who benefits" and demonstrating marketable coherence with a "Quality Leadership Strategy" can give the "Marketeers" the guidance and confidence to sell ergonomics, surely something they will be better at than we are! This raises the question as to just who is the end/final customer, to

whom should we be selling the benefits of services or products incorporating ergonomics?
Thirteen "Customer Groups" can be identified, all of which should be considered during the
development, evaluation and modification of any product, service or system.

- Bystanders - These people have nothing to do with the product but are affected by it.
- Figurehead - The Company or Charity President, who undertakes no day-to-day
  activities for the organisation.
- Signature - A person who settles the account, on the advice of others.
- Owners - Those that make the rules by which the product or service must abide.
- Customer Representatives - Those individuals or groups, possibly self appointed, who
  decide what the actual customer may be exposed to and thus have the option of buying.
- Customers - Those that buy the product or service but do not, necessarily, use it
  regularly.
- Users (regular) - Those that us the product or service as intended by the
  designer/supplier and who will seek guidance before going outwith the specified limits.
- Users (normal) - Unlike the "Regular" customer these people will "misuse" the product
  or service but in a predicable and acceptable (although not necessarily advisable) way.
- Users (abusers) - Unlike the "Normal" customer they will exceed what could be regarded
  as "acceptable misuse" use of a product/service.
- Users (explorers) - Given a product they will see what it can do, unlike abusers they are
  not unthinking but investigating.
- Wreckers - These are not, in any real sense, users and are generally known as vandals!
- Installers - Those that install products may be "naive" or "professional" depending upon
  the complexity of the product and level of experience they have.
- Maintenance - Those that fix it when it has gone wrong. Depending on the product they
  too can be either "Naive" or "Professional".

The best inclusions of ergonomics will take into consideration all the above
"customers", encourage desirable/acceptable use and discourage/prevent the unacceptable.
In all cases we must seek to minimise the consequences of failure - "fail soft" design. (Any
approach can, of course, have undesirable emergent properties. If, for example, secondary
car safety had not been improved so dramatically over recent years then "Ram-Raiding"
could not happen. The occupants of the car would be so injured during the impact with the
wall or window that they would be unable to get out of the car and steal from the shop! I
wonder what the value is of the slogan - "Ergonomists Made-Ram-Raiding Possible"!)
The stages through which a design is developed from the initial idea or brief through
concept development and ultimately to production is well established. (Figure 2. overleaf)
Ergonomics can be appropriately applied at all stages and its incorporation must be
managed just as the whole design process should be. If ergonomics is not included within in
the design process the only opportunities for ergonomists will occur late in the process when
the design concept, and perhaps detail, will be fixed. Not only may this limit the options but
it is also likely to cause frustration to the individuals concerned. The alternative is to
delegate the ergonomics to the designers and rely on the fact that most will have been
introduced to the topic during their education. This can be successful but it must be
regarded as a high risk approach to delegate to somebody who will know some ergonomics
but prefer design! Faced with any tension between looks and usability the former will
generally succeed.

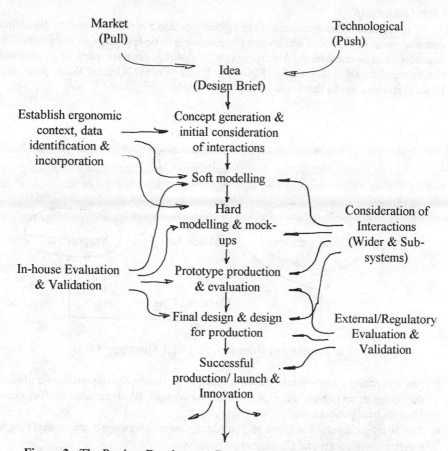

Figure 2. The Product Development Process and the role of ergonomics.

There is also the near fundamental problem that virtually all designers work virtually exclusively with visual representations of the product (2D or 3D and generated by hand or computer). Thus they have difficulty representing or communicating the sound to be made by the machine, the behaviour of the software interface or even the handling properties of the finished design. For example, the handling and pouring of a full kettle is not easy to test until the final stages of the design process when the "choice" materials and processes are used. By this time the company has invested considerable sums in tooling and is committed. This is one reason why the Designer sells the concept but engineers finish the project and why the management of the whole development process is so important.

There is an other aspect of "Quality Leadership" that need to be considered, the perception of Corporate Quality. For example, you may not know the detail or the product nor be able to test it fully but you do know that XYZ is a "good" name. You feel that you can safely buy their product, even thought it might cost a little more than a similar one made by the UVW corporation. The XYZ corporation, of course, know this and charge a premium for their products because of their name! Corporate Image will be discussed, below, in association with the "Cost Leadership Strategy"

Cost Leadership
In the case of the incorporation of ergonomics into a product or service the ultimate customer must be the focus while when ergonomics is to be applied to an organisation its structure must be considered. Mintzberg (1979) identifies six basic parts of an organisation that he represents diagramatically (Figure 3). Porter's (1985) "Genetic Value chain" would be an alternative model that could be also appropriate.

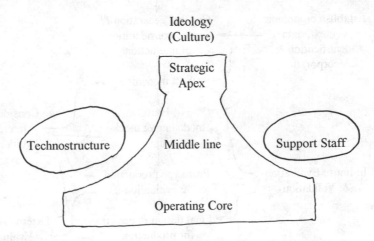

Figure 3. The Structure of Organisations (After Mintzberg 1979).

- *The Operating Core* - Where the basic work of producing the organisation's products or delivering its services is undertaken. Designers design, Workers assemble Televisions, Doctors treat patients, etc.
- *The Strategic Apex* - The home of Top Management, where goals are set and progress monitored with a general (Strategic) perspective.
- *The Middle Line* - The managers that operate directly between the Strategic Apex and the Operational Core. Often the level managing the introduction/implementation Ergonomics and the level most under pressure as "right-sizing" occurs.
- *The Technostructure* - The staff who design the systems by which progresses and the outputs of others are formally designed, monitored and controlled.
- *The Support Staff* - Those specialists that provide the support that the organisation needs outside of its direct operations. Public/Corporate relations, Canteen Staff, Legal Counsel, etc.
- *The Ideology (Culture)* - The halo of beliefs, traditions and folklore(experiences) that surrounds the whole organisation.

That ergonomics can benefit each of these above organisational areas is, to us, self evident, but we need to establish its value for the company concerned. I would suggest that each area should be evaluated task by task and initially allocated to one of three categories; no significant ergonomic problems or implications; significant ergonomic shortcomings already manifest in accidents, low production rates, slow induction/training speed, etc. and the wide category containing tasks with identifiable sub-optimal ergonomics with quantifiable risks and urgencies. In the latter category the order of attack is usually risk management with financial monitoring and control. From this the agreed Action Plan, incorporating ergonomics within the organisational, can be expected to emerge.

Tasks found to be acceptable may be left until changes are proposed in that structural area or those impinging upon it. At this time the initial "ergonomic audit" will need to be reworked. If major changes are proposed and the ergonomic implications are not readily identifiable then an allocation of funds against future work should be made. For example, the valuation of a project by discounting its lifetime cash flows is commonly undertaken. Different organisations will use this information in different ways both to compare projects competing for funding and to establish targets for the revenue stream. Net Present Value (NPV) and Internal Rate of Return (IRR) are commonly used and by increasing the Discount Rate employed (to cover the risk that additional ergonomics work will be required in the future) an underestimate of the project's profitability is produced. This will ensure that the project budget contains an allocation of funds for, as yet undefined, ergonomics work. Thus the financial measure selected is used for project comparisons in the usual way but modified to explicitly cover the risk (not a certainty) of future expenditure on ergonomics. Of course predictable expenditure must be specified and included in the calculations.

In the case of task with an identifiable risk the longer term solution will probably involve the same type of discounted cash flow costing as described above. However, with longer term projects the predicted expenditure might just not be enough. A more comprehensive argument will be required to help the valuation of the problem and thus the quantification of the ergonomics element.

Inevitability the first question is whether or not the problem is of sufficient size to risk the whole organisation. In the case of many multiple claims or a significant undesirable public event occurring this is not an unreasonable question to ask. What is the level of identifiable risk that the Board wishes to accept and what will it do to insure/hedge bigger risks? Figure 4 shows how such undesirable events might be initially considered.

	Major Accident/ Catastrophe	Catastrophic failure with significant impact
high numbers/ severe injuries (death)	(eg virtually any major accident not involving UK citizens!)	(eg BA DC10 aircraft crash, UK chemical plant explosion)
Number and severity of injuries per event (incident/ accident)		
low numbers/ slight injuries	"Counted" Accidents (eg road accidents, slipping/ tripping accidents in industry or Asset only accidents)	"Hyped" Accidents (eg minor road accident involving Royalty, a rock star or nuclear fuel)

low emotive incidents/ little engagement   highly emotive incidents or accidents/ high degree of concern

Emotive rating ("newsworthness"!)

Figure 4. "Hard" (severity) v "Soft" (emotive) classification of accident events.

The emotive axis will influence the Corporate Image and thus the marketing of the organisation's services or products while the actual consequences will impact directly upon costs, financial viability and thus, again, marketing considerations. Thus a "focused" strategy is referenced. Most hazards, thankfully, never become accidents and even fewer hurt or kill people. However, these events will happen and the risk should be estimated and managed in any organisation. (Reason's (1994) systemic approach to organisational error is most pertinent to this view.)

The costs of accidents have only been publically quantified infrequently. The fire on Piper Alpha (£125M (Financial Times 28.11.94)) is an example of a Catastrophic failure and HSE(1993) reports "Counted" examples. The HSE(1993) reported five cases studies where recorded accident rates and immediate direct costs were monitored:

- A Construction site of a major civil engineering company building a supermarket.
  87 accidents and £5833 loss per employee year representing 8.5% of the tender price.
- A Creamery manufacturing dairy products.
  11 accidents and £2886 loss per employee year representing 1.4% of operating costs
- A Transport company operating a fleet of tankers for the milk marketing board.
  14 accidents and £2446 loss per employee year representing 1/8% of operating costs & 37% of profits.
- A North Sea oil production platform.
  5 accidents and £17924 loss per employee year representing 14.2% of potential output.
- A NHS Hospital belonging to large metropolitan Health Board.
  7 (non medical accidents and £567 loss per employee year representing 5% of annual running costs)

The level of fines in the West Midlands region for criminal prosecutions under the Health and Safety at Work Etc. Act (1974) is typically between £1000 - £3500 per life and for other injuries about £750 per case (Bergman (1994)). An apparently extreme level of fine of £10,000 (plus £550 costs) was impose upon North Tyneside College after a student cut the tops off two fingers. (Newcastle Evening Chronicle 26.7.94).

In the case of civil cases the range of settlements is also large. The Mountenay (Hazzard) and others v Bernard Mathews) settlements ranged from £400 to £5000 for general damages while £79,000 was awarded for an Inland Revenue typist. (The Guardian 20.12.93) This latter settlement is the largest so far within England and Wales and the recent settlement of £72,000 against Clyde Shaw (a Motherwell Steelmaker) the largest in Scotland (Financial Times 3.9.94).

In all of these reports no details were given as to the proportion of the settlement that was covered by insurance, legal aid, etc. Legal costs and management time will be considerable more than these published sums and the long term Strategic and Corporate costs will be more difficult to quantify but may ultimately be a bigger threat to the business.

The full costs concerning an accident or other undesirable event can be listed under five headings that will generally be quantifiable by those concerned within the organisation and thus can be used to balance against the cost of the application of ergonomics. It must also be remembered that as British Industry tends to focus on "Core Businesses" then the effect of a particular failure or shortfall can be much greater than if a bigger range of revenue streams were operated. The degree to which these costs or liabilities are covered by insurance is, of course, also significant.

The job, production unit/line direct costs
- Lower output quantities that give rise to increased labour costs.
- Lower output quality involving re-working whenever it is identified, and the risk of customer dissatisfaction.
- Lower output quality and quantity due to "locum"/"stand in" operators replacing those "sick" and temporarily off work. This might show itself as an increase in the variance of quality, possibly subjective.
- Costs associated with the different production rates of injured or replacement workers as well as line inefficiency costs. This is especially important and obvious if Kanban, Just-in-Time (JIT), MRP, MRPII, etc. methods are used to closely monitor production and either require zero "buffer-stocks" or produce data for the stock value "Tied-up".
- Increased labour turnover resulting in significant recruitment and training costs as well as line inefficiency costs (including costs of agency staff to "cover").
- Cost due to damage to plant, work in progress, materials, etc.

The job, production unit/line indirect costs
- The cost of Injury claims ultimately causing increased employer liability insurance premiums. (Most policies were for "an indemnity unlimited in amount" rather than a specific agreed sum but this changed from 1 January 1995 when an ordinary, legally required, limit of £2M per event and £10M total became routine. "Top-up" insurance is obtainable but the premiums will be matched to the insured risk and it is unlikely that the basic insurance premiums will drop in line with the reduced cover provided.
- The cost of Injury claims leading to both direct (eg "sick pay") and indirect costs (eg management time used to deal with the situation) and ultimately increased insurance premiums. The impact of the financial costs of "sick pay" have, of course, increased since the responsibility for this largely passed from the State to the organization.
- Costs associated with additional pension payments/lump sum provision for people taking early retirement on "sickness" grounds.
- Reduction in morale and thus the ability to function effectively both at a factory floor and at a plant management level.
- Reduction of the organisation's public image locally affecting recruitment, esteem of staff (including Senior Managers/Directors), etc.
- Administrative costs associated with managing the accident or risk, etc.
- Poor motivation directly increases costs and limits the ability to respond to opportunities (eg staff are less likely to be enthusiastic about additional overtime working).
- Legal costs both for the company and, perhaps, the individual line/factory managers concerned to blamed for the accident.

Cost associated with the suppliers and customers of the organisation.
- The customers lose supplies or receive them late, perhaps after the main buying period (eg Christmas). The quality might be reduced resulting in additional repair/replacement costs. In the extreme the customer might have costs associated with the identification and contracting of alternative supplies. A supplier "rescuing" the original company's customer will probably seek to maximise their gain, perhaps by demanding long term contracts are signed.

- The suppliers might find themselves left with materials that are not required and which they are unable to sell elsewhere. This can lead to cash flow problems, especially for small companies. If sales are possible elsewhere then this might enhance the strength of the competition and make future trading more difficult.
- The confidence of the customer and supplier might be reduced for a considerable period of time, way beyond the time taken to settle the initial incident.

Wider company/corporation strategic costs
- The industry wide impact of a poor claims record concerning any significant company because of the suspicion that all plants operating in this field will be a similar risk.
- If poor management is clearly to blame then the Insurance Company might not fully settle the claim in full; Insurance companies only cover unexpected risk!
- Limitation of options on corporate identity coherence. Should companies link all operating units together and, if so, how closely.
- Reduction of the organisation's public image nationally effecting relationships with clients, financial institutions, etc. and thus the share price (shareholder wealth), rates on loans, bonds and derivatives, etc. These can be major "motivators" for senior staff.
- Concerns of major customers that would not wish to be publicly identified as associated with companies that injure or care little about their work-force; Brand and/or Corporate image congruency especially if members of the Investors In People(IIP) programme.
- The Wider effects of image damage influencing the behaviour of market makers, bankers, etc. This could ultimately influence share price (shareholder wealth) and perhaps rates on loans, bonds and derivatives, etc.
- Possible costs and embarrassment of any uninsured Board liability (civil and criminal) and perhaps, although unlikely, the ultimate DTI sanction of removal from the Board.

Wider Societal costs - not usually accepted as the organisation's responsibility
- Costs to those directly involved in the incident - the Dramatis Personae. These costs might be covered by the ultimate legal settlement but this can take several years and, without legal aid or Union backing, it may not be possible to pursue any contested claim.
- Costs to the Kith and Kin of those involved in the incident. These costs might be covered by legal settlement but this can take several years and is, by no means certain.
- Society as a whole will have cost that must be covered by all. These include DSS administrative costs, NHS costs (physical and psychological), loss of income tax, etc.

Conclusion

Ergonomics is of value to business, it does add value to products and services and can significantly reduce the costs associated with their supply. These costs include those associated with the general functioning of the organisation and those, largely uncontrolled costs associated with undesirable events. Thus ergonomists can support both strategies for gaining competitive advantage identified by Porter. However they must improve their skills of quantification and qualification so that the value they add is recognised by business.

Ergonomists are not, generally, trained in corporate financial management or even project appraisal/valuation but they are able to highlight the benefits to be obtained from, and show the costs associated with, their work. We must adopt models to remind ourselves of the need to consider the financial aspects of our work (Figure 5.) and delegate the detail to others The Ergonomist can, and should, support the organisation's managers to ensure that a full appraisal is undertaken of the value of ergonomics to the business.

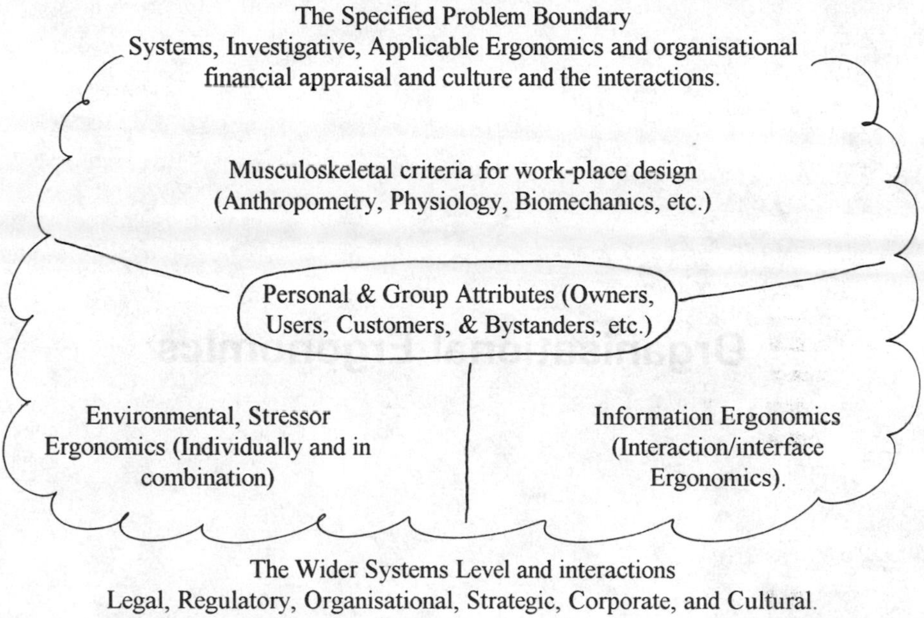

The Specified Problem Boundary
Systems, Investigative, Applicable Ergonomics and organisational
financial appraisal and culture and the interactions.

Musculoskeletal criteria for work-place design
(Anthropometry, Physiology, Biomechanics, etc.)

Personal & Group Attributes (Owners,
Users, Customers, & Bystanders, etc.)

Environmental, Stressor
Ergonomics (Individually and in
combination)

Information Ergonomics
(Interaction/interface
Ergonomics).

The Wider Systems Level and interactions
Legal, Regulatory, Organisational, Strategic, Corporate, and Cultural.

Figure 5. The Meta Ergonomics - Business Interaction Model.

References

Bergman, D., 1994, *The Perfect Crime?*, West Midlands Health & Safety Advice Centre.
Bronowski, J. 1951, *The Common Sense of Science*, (Heinemann Educational, London)
Chapanis, Alphonse, 1959, *Research Techniques in Human Engineering*, (The John Hopkins Press, Baltimore)
Health and Safety Executive, 1993, *The Costs of Accidents at work (HS(G)96)*, (HSE Books, London)
Mintzberg, Henry, 1979, *The Structuring of Organisations*, (Prentice-Hall, New York)
Murrell, K.F.H., 1965, *Ergonomics*, (Chapman and Hall, London)
Porter, Michael, 1979, *How Competitive Forces Shape Strategy*, Harvard Business Review, March/April.
Porter, Michael, 1985, *Competitive Advantage*, (Free Press, New York)
Reason, James, 1994, *Systems Approach to Organisational Error*, Paper presented to the 12th Triennial Congress of the International Ergonomics Association, Toronto(August15/9)
Sinclair, John. (Editor-in-Chief), 1992, *BBC English Dictionary*, (Harper Collins, London).
Legal
General Product Safety Regulation(SI1994:2328) established under the EEC Product Liability Directive (85/374)
Mountenay(Hazzard) & Others v Bernard Mathews (Norwich CC 4 May '93)IRS HSIB215

Organisational Ergonomics

USER PARTICIPATION IN SYSTEM DEVELOPMENT: THE IMPORTANCE OF ORGANISATIONAL FACTORS

C.M. Axtell, P.E. Waterson, C.W. Clegg

Institute of Work Psychology,
University of Sheffield
Sheffield , S10 2TN

In this paper we highlight the importance of organisational factors when conducting user participation in system development. The argument is illustrated with reference to a case study within a large organisation, where a system was being built in-house using an iterative and user-participative methodology. Organisational factors played a major role in influencing the practice of the methodology and so should be considered as part of the development process, rather than as extraneous to it. We end the paper by suggesting a way forward for the conduct of user participation in system development.

Introduction

The aim of this paper is to describe an innovative methodology for user participation in system development and analyse it within its organisational context. We aim to highlight the impact of organisational factors on the practice of the methodology and suggest a way forward for the conduct of user participation.

It is widely recognised that user participation in system development can improve system design processes and outcomes (e.g. Eason, 1989). Gould & Lewis (1985) recommend an early focus on users, early and continual user testing and integrated, iterative design. In practice participation of users can be interpreted differently by different people and can have different orientations and emphases; it is clearly not a unitary phenomenon. For example, user participation can be interpreted as the representation of users in meetings and committees, the interviewing of a sample of potential users, getting users to agree and sign off a set of system specifications or test a finished system for it's usability. Others recommend a more active involvement of users, such as cooperatively designing the system with developers (e.g. Bødker & Grønbæk, 1991).

However, detailed aspects of the method of involvement are not the only issues to focus on. The organisational setting in which user participation is attempted is also an important factor to consider (e.g. Grudin, 1991; Hornby & Clegg, 1992). Development of computer systems takes place in a wide variety of contexts, each with specific supports or pitfalls for the successful involvement of users. Grudin (1991) identifies several organisational factors which can hinder good computer interface design. Many of these have implications for user participation. For instance, there are challenges in the form of obtaining access to distant users. Also all members of the development team must be committed to the approach if user participation is to be effective. Another issue is the division of labour in organisations, where the different functions involved are separated and communications are often in written form, which means that opportunities for miscommunication and lack of coordination are high. This makes user participation more difficult. Managerial goals such as managing the development process and making effective decisions, can also hamper user involvement. For example, preferences for structured approaches which emphasise early design and getting it "right first time" can work counter to a more iterative user involvement approach.

We argue that organisational factors such as these are an integral part of user participation in development and need to be recognised as such rather than marginalised. We illustrate our argument with reference to a case study of user participation in a large organisation.

The Case Study

This study took place in the IT department of a large organisation in the UK., operating in a rapidly changing and highly political environment. The system being developed was to support the daily administrative activities of several thousand users, nation-wide. These activities varied slightly in different regions around the country.

As a result of dissatisfaction with existing approaches, a post-analysis, iterative methodology focusing on usability was developed within the organisation. Paramount in the process was the high degree of interaction and information flow between users and developers. There were two main components of the methodology (see Figure 1): (1) Cooperatives, where users and developers jointly designed prototypes of screens for individual sections of the system and navigation between them; (2) User Groups (UGs) typically of 4 to 5 people, who evaluated the prototypes before returning them to the Cooperatives for more work (e.g. changes and addition of data base links). This process was to go through several iterations before sections would be complete. Later evaluation stages incorporated role plays and scenarios that the UGs designed. Liaison between the users and developers was managed by a user support team. Although there was a project manager overseeing the development of the system, the different parties involved (users, developers, user support team, human factors specialists) had different reporting lines and accountabilities, with no formal overseeing manager in charge of the new methodology itself.

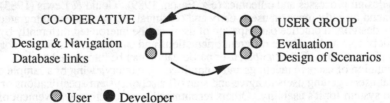

CO-OPERATIVE   USER GROUP

Design & Navigation   Evaluation
Database links   Design of Scenarios

User   Developer

Figure 1. Overview of The Methodology

Analysis was conducted using SSADM, and the new methodology followed on from this. Over time a number of developments took place such as a reduction in the level of contact between users and developers (i.e. no Cooperatives in operation), allowing developers more time to program. However, they had access to each other as and when required. A month or so after this, external consultants were introduced to conduct a Business Process Re-engineering (BPR) exercise and develop some additional functionality for the new system. They used a different methodology, with user involvement only in the requirements gathering and testing phase. Their promised completion date also brought forward the deadline for piloting the system. Personal contact between users and developers was reduced further during this phase as developers were engaged in programming in order to reach the deadline. During this time, feedback from UGs to developers was via written communication.

Methods

We employed a number of research methods and gathered data from a range of sources in order to gain a comprehensive overview of the design methodology and its use in context. The methods included interviews (with 12 managers, 18 users and 4 developers), questionnaires (responses from 67 users who had taken part in the methodology over it's course, and 6 developers), observation and video of the activities of users, and a tracer study which traced the flows of information between users and developers. The managers interviewed were from many different levels and functions, up to, and including, some senior managers, who had been involved in some way with the new methodology.

Findings

For reasons of space, we concentrate on a selection of the findings. One of the most striking findings overall, is that all the groups taking part (users, developers and managers) had positive attitudes towards the new methodology and the concept of involving users. All felt it was a better way of developing new systems than the usual method. There was agreement across the groups that the new methodology had led to the identification of errors that would have been costly to rectify later. Overall 81% of users agreed that they had a high level of influence on the development of the system. They also had positive attitudes towards the new computer system and felt a sense of ownership of it. Reports from the piloting phase of the project are also positive.

However, those involved identified issues which needed to be addressed relating to organisational factors as well as those internal to the methodology. Organisational factors fell into two categories; that of the more general organisational context (relating to the problem of capturing expertise from several thousand users, nation-wide and the political and changing environment) and that of it's relationship with other methods used in the organisation (SSADM, BPR). Our findings are therefore organised around the following issues, namely (1) The Organisational Context; (2) The Relationship with Other Methods; (3)The Internal Processes of the Methodology; and finally (4) The Inter-Relations between these issues.

(1) The Organisational Context:

The main issues here were the thousands of users working nation-wide, in different offices with local characteristics and demands, whose expertise needed to be captured and incorporated in the system. There was also the issue of the rapidly changing and political environment and the changing requirements ensuing from this. Issues here included recent (partially implemented) changes in the way work was carried out in users' offices, organisational policies, financial considerations, and the different interests and concerns of people in the organisation. Also some of the managers felt that senior management did not show enough active support for the new methodology. The organisation traditionally had a policy of using SSADM in development, and so changing to another methodology, which did not have a proven financial case as yet, was considered risky by some senior managers.

(2) The Relationship with other Methods:

One issue that arose here was the relationship between the new methodology and SSADM. Many managers commented that an outcome of the new methodology was a virtual re-writing of parts of the SSADM produced user specification. This meant going through lengthy change control procedures which disrupted the development process.

However, those with technical backgrounds still emphasised the need for detailed analysis and modelling (such as that which occurs with SSADM) to be used in conjunction with the new methodology. Some managers felt that users could have been more heavily involved when constructing the user specification, which might have helped alleviate the need for changes. 80% of users and 83% of developers agreed that this would have been an improvement in the methodology. Although not all of the specification could be changed, the advantage of the new methodology as it stood was being able to tackle many of the difficulties encountered before the system was developed. The question was therefore whether to and how to organise the relationship between a structured approach such as SSADM and the new methodology.

There was also the relationship between the new methodology and Business Process Re-engineering (BPR). There was a recognised need for BPR amongst the managers. Without this sort of exercise there was a danger of simply reproducing the existing functional organisation and it's current inefficiencies without challenging the methods of working that the system was to support. However, several managers, especially those working most closely with the new methodology were disappointed at the way at which it had been introduced (i.e. external consultants who went on to develop new functionality themselves rather than hand it back to internal specialists). There was also an issue of timing of the exercise, (i.e. near the end of the development process rather than at the beginning). Overall, there was a need to consider how the relationship between BPR and the new methodology should be managed.

(3) The Internal Processes of The Methodology:

There were many issues raised about the internal constituents and relationships within the methodology. For instance, there was the issue of selection of users. Many of the developers (83%) but fewer users (27%) felt that there should be more careful selection. There was also an issue of how many users should be involved. Some of the developers felt there should be fewer of them, but only a small minority of users felt the same. When the methodology first started there were too few users available for it to work the way it was intended. Users had to evaluate what they had just designed instead of handing it to someone else for an alternative view and there was also a less diverse group to evaluate it. However, some users said that there were also periods where they did not have enough to do, and so fewer of them may have been preferable at these times.

Many users felt they would benefit from longer placements and from being allowed to return again, as they felt they were "just getting the hang of it" when they had to leave. Users were involved for between one to five weeks each time, depending on how long their offices could release them for. 22% of respondents to the user questionnaire had been involved in the development of the system on more than one occasion. Some developers were concerned about the time it took new users to learn about the system and take an active part in development. So in this respect they agreed that longer placements for, or the recycling of users would be useful. However, all recognised the need to involve a variety of users because of the regional variations in the job. Developers, however, reported that sometimes they had to wait whilst users argued about what they wanted (particularly when they were from different regions) which disrupted development.

Another issue was what the role of users should be. Users felt they were in an advisory or evaluative role, bringing their business knowledge to the process. However, video analysis, of the activities of the user groups, showed that what they were doing was far more complex than that. They were: using their knowledge of the system and what would be technically possible, making detailed design decisions (such as with the on-line help system), checking usability, organising how the evaluation work should be carried out (at interview, some users felt they could have been more involved in this type of role) and preparing tasks (work based scenarios and entering data onto the database) to help with the evaluation. These activities are all important and go beyond straightforward usability testing which might normally be expected of user participation.

There were many issues about how the relationship between developers and users should be organised. When Cooperatives were in operation, many developers felt that users spent unnecessarily long periods of time in close contact with them, when they would have liked more time alone to program. However, when formal communications became reduced to written form only, many developers said they would have liked more contact with users. Over the methodology as a whole, more than 50% of users felt that feedback between users and developers was good. 84% of developers felt that they received good feedback from users, and 50% felt that users received good feedback from them. However, towards the end of the development period (i.e. during the involvement of external consultants), lack of direct feedback and the breakdown in communications was evident. The relationship between users and developers became uncoupled and asynchronous, making it both inefficient and ineffective. For instance from the tracer study, taking one day as an example, 37% of the 24 error reports that users completed were not formally raised with the developers. This was either due to them being duplicates of other issues raised or because the error could not be found or supported. Direct feedback about issues such as whether errors had already been raised or whether points being made were valid, would have been beneficial. Also screens were being designed without co-operation and immediate feedback from users. This meant that some had to be sent back to developers, due to easily preventable problems, such as abbreviations used on screens which did not make sense to users.

(4) Inter-Relation of Issues:

Our study revealed a number of issues to be addressed regarding the practical aspects of user participation. Some of these are, of course, intrinsic to the organisational context in which this work was undertaken. Any form of participative development, when there are thousands of users in offices all over the country, inevitably leads to problems of selection, getting volunteers and maintaining continuity of users. This relates to the problems of access to users that Grudin (1991) identifies. In this case constraints in local offices dictated the length of time users could be released and involved in the methodology. Another complication was the need to involve a wide variety of users, which meant that there had to be a fairly rapid turnover as having too many of them would cause a problem of not having enough to do. Also the issue of different working practices in different regions sometimes meant disagreements occurred, causing delays in the internal process.

Many mangers felt that more active support from senior managers was needed, particularly initially, as there was the problem of promoting the methodology to managers elsewhere in the organisation to attract people to take part. This had an impact on the number of users involved, which affected the internal workings of the methodology. The need for commitment from all concerned is an issue that Grudin also raises.

Other contextual factors affecting the methodology were decisions made by senior management, such as that to embark on a BPR exercise to challenge existing work practices and to bring in external consultants to do this. Here the need for BPR was considered more important than keeping user participation as it had been. The impact of managerial decision making is another point identified by Grudin.

Some of the issues to be addressed in the case-study arose from the problem of integrating this method of user participation with the methods for analysis (SSADM) and process redesign (BPR). This relates to the Grudin's point about preferred approaches for managing development. For example, the structure imposed by the SSADM user specification meant that there were certain specifications that users could not change and that the internal workings of the methodology were held up with lengthy change control procedures for those they could. The way the BPR exercise was undertaken using external consultants with their ways of working and promised deadline, meant that there was no longer any time (or desire from external consultants) to involve users to the same extent as they had been. Despite the efforts of those involved to incorporate some user involvement, the relationship between developers and users became uncoupled and asynchronised. In Grudin's terms, "division of labour" became more pronounced, leading to problems of communication and coordination.

It can be seen that many of the difficulties of user participation in this instance are due to wider organisational factors and can only be analysed and fully understood in their context.

Conclusions

The issues that we identify in the case study demonstrate that user participation cannot be seen in isolation, as it is embedded in connections with other ways of working and the larger organisational context, which all have an impact on it. This supports the work of other authors (e.g. Grudin, 1991). The role of organisational factors needs to be recognised and considered within the development process itself.

In this case a review of the development methodology should be undertaken, involving all relevant experts (including users) based on their collective experience with it, to address some of the issues outlined. One way forward for user participation in development would be to incorporate multiple forms of expertise (i.e. users, developers, senior managers, human factors and business process re-engineering specialists etc.) in a single project team, with one overseeing manager. This group would undertake all analysis, business process re-engineering, development and testing activities, allowing greater integration between the organisational issues and internal processes of development. The project team should be highly participative so that everyone, including users are involved in a broad range of activities. This conforms to Gould & Lewis's principle of integrated design under one management.

The study also points to the need to adopt a wider view of the role of users in large scale development projects more generally, rather than trying to prescribe a narrow, neatly defined role for them. After all, users in this case were already involved in and expressed an interest in more than just evaluating, advising or testing. This adaptability of users is a valuable resource that should be recognised. There is potential to expand their role to include participation in deciding how user participation could be conducted and in some of the more organisational decisions and processes which affect development, as well as in the design and evaluation activities themselves.

The case study has highlighted the importance of organisational factors and how they can impact on the user participation that takes place. Future research should aim to identify other issues such as these which need to be considered during the development process and investigate the success of projects which have attempted to incorporate them. In this way, through a variety of experiences, we can learn more about how to undertake user participation in system development successfully.

References:
Bødker, S. & Grønbæk, K. 1991, Design in action: from prototyping by demonstration to cooperative prototyping . In Greenbaum, J. & Kyng, M., (eds), *Design at Work: Cooperative design of computer systems.*(Lawrence Erlbaum Associates, Hillsdale, N.J)
Eason, K.D. 1989, Tools for participation: how managers and users can influence design. In K. Knight (ed.) *Participation in Systems Development* (Kogan Page, London)
Hornby, P. & Clegg, C. 1992, User participation in context: a case study in a UK bank, *Behaviour & Information Technology*, **11**,5, 293-307
Gould, J.D. & Lewis, C.H. 1985, Designing for usability: key principles and what designers think. *Communications of the ACM*, **28**, 3, 300-311
Grudin, J. 1991, Systematic sources of sub optimal interface design in large product development organisations, *Human Computer Interaction*, **6**, 147-196

THE EFFECTS OF A MANUFACTURING INITIATIVE ON EMPLOYEE JOBS AND STRAIN

Sharon K. Parker and Toby D. Wall

Institute of Work Psychology
University of Sheffield, Sheffield S10 2TN

Carol Myers

Occupational Psychology Branch
Employment Service, Sheffield S11 8JF

We present a study of the effects of a JIT-related initiative on shopfloor jobs and employee well-being. The focus is on: (a) the effect of employee participation in the implementation process on job characteristics and strain; and (b) the extent to which job properties within the JIT system affect strain. Results suggest that participation prevents increased work load and strain, but does not affect cognitive demands or autonomy. However, examining the relationship between job characteristics and strain for individuals working within the new system shows that autonomy is a key predictor of strain.

Introduction

Two themes are prominent in modern manufacturing. One is an increased emphasis on competitiveness, in pursuit of which companies are attempting to improve their responsiveness to customer demand and to enhance the quality of their products. However, the competitive advantage of responsiveness and quality may be nullified if it entails higher prices. Thus strenuous efforts are being made to control costs by reducing inventory and eliminating unnecessary work. To this end companies are adopting a variety of just-in-time inventory control (JIT) and related practices, such as that represented by the Toyota Production System (TPS).

The second theme focuses on stress at work, as reflected in the recent government white paper 'Health of the Nation'. CBI estimates that working days lost due to stress-related illness cost £5 billion per year; a figure that represents only the tip of the iceberg if other effects of strain, for example on performance and accidents, are also taken into account. Within manufacturing, the issue of stress may be especially important since it is among blue-collar employees that the poorest levels of mental health are typically found (Warr, 1992).

To many, these themes are linked by the belief that cost saving initiatives such as JIT and TPS affect employee strain. However, contradictory views have been expressed about the effects of such initiatives. For example, some commentators (e.g. Schonberger, 1986) argue that employees in such contexts will need to 'work smarter' and so will experience increased cognitive demands; while others (e.g. Delbridge and Turnbull, 1992) contend they will need to 'work harder', that is there will be increased work load. Similarly, some have suggested that modern initiatives provide the context for enriched, autonomous jobs (e.g. Lawler, 1992); whereas others argue that autonomy will be severely restricted, with "workers operating under clearly defined (and dictated) managerial guidelines, under constant surveillance, and solving managerially defined problems" (Delbridge and Turnbull, 1992; p. 68).

One constructive approach to dealing with these divergent views is to try to explain why there should be different outcomes. In this regard, it seems plausible to assume that the impact of JIT and related practices depends both upon choices made with regard to job content, and the implementation strategy adopted. The strategy explicitly recommended for some JIT initiatives is to involve employees in the change process. Such participation may bear on strain indirectly, by encouraging the design of jobs with less stressful properties, and more directly, by giving employees greater ownership of the outcome. This latter effect was illustrated in a simulation study by Seeborg (1978), which showed how jobs redesigned by subjects for themselves had positive effects on their reactions whereas the imposition of exactly the same changes upon others had a much less positive impact.

In the present study we explore these issues by taking advantage of an opportunity which presented itself. A company had applied TPS to one of its assembly teams and, because of the economic benefits achieved, was planning to extend the initiative to the remaining teams. In the first stage of implementation, employees were fully involved in the redesign process. Having learnt from this experience, however, management sought to save time in the second stage by minimising employee participation. This situation provided the opportunity to address two questions. First, taking a group level of analysis: does employee participation in the change process affect the characteristics of the resultant job (in terms of work load, cognitive demands and autonomy) and employee strain? Second, taking an individual level of analysis: to what extent do the characteristics of jobs within the system affect strain?

Method

Organisational Background

The study was conducted in the assembly area of a car-seat manufacturing company. The company had a few years earlier reorganised into cells, each dedicated to a separate customer and staffed by a team of 10- 20 people. Though the cells were assembling to slightly different specifications, the core tasks were very similar.

The focus of our investigation was on the application of TPS procedures, which represented the first phase of a JIT initiative. TPS involved closely examining work within each cell to systematically attack and remove 'waste' or 'non-value-added activities', and thus reduce lead times and hence costs. In principle, TPS is based on a philosophy of participation where local expertise is used to identify potential improvements. Thus all employees contribute to the redesign of the lay-out, tooling, tasks and operating procedures. Such participation had taken place for the first stage of implementation of TPS encompassing one team, which had been introduced prior to the start of the study. However, the participation element was neglected for the two teams for whom TPS was implemented during the course of the investigation. Instead, cell leaders designed solutions that they then introduced in their cells with minimal employee involvement. For example, assemblers in the second stage commented:

On assembly (TPS) has been put in with no thought for how it will work, only one assembly worker was on the TPS team. It did not seem to have a lot of planning and all the work benches were made too small. None of the assembly team on my shift were asked about how we felt it could be made better.

When (my work area) changed to TPS, suggestions made were ignored by the "cell leaders" setting it up; they always knew better. A lot of the changes they enforced have made the job harder and slower.

Research Design and Procedure

The study was conducted over a six month period. For the assemblers involved in the first stage of implementation of TPS, measures were taken 3 months (time 1) and 9 months (time 2) after the change. For the assemblers in the second stage of implementation, measures were taken at the same time, which in their case was three months before (time 1) and three months after (time 2) the introduction of TPS. This enabled three comparisons to be made: (i) between both groups three months after implementation, which would reflect any effects of participation on job characteristics and strain; (ii) between measures taken before and after the changes in the second group, indicative of the impact of TPS (without participation) compared with the previous way of working; and (iii) between measures taken after 3 and 9 months of implementation within first stage group, providing a partial 'non-equivalent control group' to ascertain whether other more general influences were affecting measures.

At each measurement occasion, questionnaires were administered by the researchers in small groups during work hours. Confidentiality was emphasised, and the response rate was 80%.

Measures

For each measure described, Cronbach's alpha coefficient (alpha) at time 1 was used as the index of internal consistency reliability (alpha at time 2 was very similar).

Work load was measured by a 6-item scale assessing the extent to which people report pressure resulting from the amount and pace of their work (e.g. To what extent: 'do you find your job physically demanding?'; and 'do you find yourself working faster than you would like to complete your work?') (alpha = 0 .87.)

Autonomy was assessed by combining Jackson, Wall, Martin and Davids' (1993) scales of Timing Control and Method Control, which were highly correlated. The 10 items cover the extent to which people have choice over both the timing (e.g. scheduling) and the methods of their work (alpha = 0.88).

Cognitive demands were assessed by combining Jackson *et al.'s* (1993) measures of Attentional Demand and Problem-solving Demand, which were also highly correlated. The resultant 9-item scale thus assesses the extent to which jobs involve monitoring as well as more active cognitive processing (e.g. such as solving problems with no obvious correct answer) (alpha = 0.80).

Strain was measured by the 12-item version of the General Health Questionnaire (GHQ; Goldberg, 1972). This assesses whether people have recently had problems (e.g. feeling unhappy, being unable to face up to problems) to a greater or lesser extent than usual (alpha = 0.80).

Results

Part 1: The effects of TPS, with and without participation, on jobs and strain

Our first set of analyses focus on the effect of participation in the implementation of TPS on job characteristics and strain. We compare scores for the first stage employees (with participation) with scores for the second stage employees (without participation) both after 3 months of TPS. The relevant means are shown in columns b and c in Table 1. The statistical test of the difference in means shows that the employees who did not participate had significantly higher work load than the participating employees, though not for either cognitive demands or autonomy. They also reported a significantly higher level of strain.

A consistent pattern of findings is evident when one compares scores from before and after the introduction of TPS (without participation). The relevant means are shown in columns a and b of Table 1. Again, there has been a significant increase in work load (but not in cognitive demand or autonomy) and also in strain. Finally, from the comparison over time for the first stage group (during which TPS continued), shown in columns c and d, it can be seen that there are no general, site-wide influences that could account for the observed effects.

Table 1. Means for Second Stage and First Stage at time 1 (t1) and time 2 (t2), with statistical tests comparing differences in means

	Second Stage (N=20)		First Stage (N=15)		Statistical tests[1]		
	Before (t1)	After 3 mths (t2)	After 3 mths (t1)	After 9 mths (t2)	b v c	a v b	c v d
Variable	a $\overline{X}$	b $\overline{X}$	c $\overline{X}$	d $\overline{X}$	F (33)	F (1, 33)	F (1, 33)
Work load	3.17	3.48	2.55	2.29	4.65*	5.46*	2.61
Cog demands	3.24	3.32	3.50	3.18	< 1	< 1	3.23
Autonomy	2.62	2.78	3.19	3.31	< 1	< 1	< 1
Strain	.79	.95	.65	.62	5.20*	4.27*	< 1

[1] For b v c, the test is for the significance of the difference between means of independent samples. For a v b and c v d, it is a repeated measures test.
* p<.05.

In summary, these findings are consistent with the view that the drive to reduce inventory and eliminate inefficient work practices can increase employee work load and strain. However, this only seems to be the case for the second stage of implementation where employees did not participate in the change process.

Part 2: Predictors of strain within the TPS system
The fact that TPS, when introduced without participation, seemed to affect work load but not cognitive demands or autonomy, does not mean that the latter two variables are unimportant in relation to strain. Rather, these factors were not systematically affected by the change in work practices. Taking into account the possibility that cognitive demands and autonomy have been increased in some cases but reduced in others, we examined the relationship of these variables with job strain.
An initial analysis of the zero-order correlations for the total sample at Time 2, showed work load and autonomy to be strongly related to strain, but no equivalent association with cognitive demands. We thus confined our attention to the first two of these variables in a regression analysis predicting strain. In the first step, work load and autonomy together were significant predictors of strain (R = .61, p < .001), as was each independently of the other (ß for work load = -.29. p < .05; and ß for autonomy = .45, p <.01). However, the interaction term, entered in the second step, was not significant. The direction of beta weights suggests that the greater the work load, the higher the strain; but the greater the autonomy, the less the strain.
A stronger analysis involves exploiting the repeated measures component to examine change in strain as a function of change in job characteristics. This was possible for the 38 people present on both measurement occasions and was carried out using hierarchical regression procedures. Change in strain (i.e. Time 2 - Time 1 strain) was the dependent variable. The first variable entered was strain at Time 1.

Following this, Time 1 levels of autonomy and work load were entered in the equation. Finally, change in autonomy and change in work load were entered as the last step. Table 2 shows the results. As can be seen, change is autonomy is the main predictor of change in strain (β = -.46, p <.01); the greater the increase in autonomy, the less likely an increase in strain. Changes in work load, however, were not independently predictive of change in strain (β = .21, ns).

Table 2. Results of hierarchical regression predicting change in strain (N = 38)

Step	Variable entered into regression	R	R^2 Change	Final Beta
1.	Strain, Time 1	.23	n/a	-.34*
2.	Autonomy, Time 1	.39	.10	-.40*
	Work load, Time 1			.26
3.	Change in autonomy	.62**	.23**	-.46**
	Change in work load			.21

** p < .01; * p < .05

Conclusions

In this paper we have considered two questions addressing the stress-related implications of introducing new manufacturing initiatives. The first concerned whether employee participation in the change process affects the characteristics of the resultant job and employee strain. Results suggested that this was the case. Where employees participated in the implementation of TPS, there was no evidence of detrimental effects to job content or psychological well-being, and even some indication of benefit. In contrast, where employees were not involved in the change, increased work load and strain were observed. Employee participation may thus function as an important contingency variable that helps to account for the varying effects of new manufacturing initiatives on jobs and people. From a practical perspective, these results suggest there is much to gain from involving all employees, not just those in the pilot phase, in the introduction of modern manufacturing practices. Such participation enables local expertise to be used to design less physically demanding jobs, and is also likely to increase employees' ownership of, and commitment to, the new ways of working.

However, in this study, regardless of whether or not people participated in the redesign process, the application of TPS did not have a group effect on shopfloor autonomy or cognitive demands. This JIT-related practice thus had neither the decrease in autonomy and skill use predicted by some commentators (e.g. Delbridge and Turnbull, 1992); nor the increased autonomy and greater use of cognitive ability suggested by others (e.g. Schonberger, 1986). The effects of new initiatives on shopfloor jobs might not be as dramatic or pervasive as predicted. Indeed, it is likely that, in addition to the implementation strategy adopted, the effects depend on organisational choices that are made in relation to job design, rather than entirely on intrinsic characteristics of the new practices. Such an observation has been made relation to Advanced Manufacturing Technology (e.g. Wall et al., 1990).

The second question addressed in this paper concerned whether the characteristics of jobs within the system affect strain. In this regard, autonomy was shown to be the most important factor: the greater the increase in autonomy, the less likely that strain increased. This finding is consistent with previous literature that has identified a lack of job control as a major work stressor (e.g. Warr, 1992), and suggests that enhancing autonomy might be the most powerful way of alleviating the potential stressful effects of new manufacturing initiatives. As suggested by Karasek and Theorell (1990), autonomy allows employees to more effectively manage the demands within their environment. It also provides employees with an on-going opportunity to participate in daily decision-making and change. Moreover, evidence exists to suggest that increased job control can enhance system performance, particularly when the environment is uncertain and complex (Wall et al, 1990).

There is some debate, however, as to whether increased job control at the individual level conflicts with JIT principles (Klein, 1991). For example, the interdependence between processes that JIT depends on can reduce individual control over the timing of their work (e.g. individuals cannot re-order their work as it will have an immediate effect on the next employee). Similarly, the emphasis on standardising processes means that employees have less autonomy over how to do things. One way of compensating for this potential decrease in individual autonomy is to devolve control to the group (Klein, 1991). For example, each employee may have to operate in a standard and consistent way, but the group can be given the freedom to change the methods of work as long as product specifications are met.

Clearly, further investigation is needed to understand the organisational choices that exist in relation to job design within modern manufacturing settings. This will require expanding the concept of autonomy beyond individual control over the timing and methods of the work to include, for example, group levels of autonomy, control over the group boundary, and control over target setting. At the same time, there is a need to develop a better understanding of the mechanisms by which increased job control alleviates strain. There is also a need to consider additional demands that are becoming salient within modern manufacturing, such as greater production responsibility, the presence of performance monitoring (e.g. through electronic surveillance) and higher peer pressure arising from team working.

In general, there is a need to carry out longitudinal studies that either capitalise on existing change such as this study or monitor the effects of interventions deliberately targeted at reducing strain. It is only through such studies that we will understand the effects of new manufacturing initiatives on employee job characteristics and strain, and the extent to which contingency factors (such as participation and organisational choices about job design) influence these effects.

References

Delbridge, R. and Turnbull, P. (1992). Human resource maximisation: The management of labour under JIT manufacturing systems. In P. and P. Turnbull (eds) *Reassessing Human Resource Management.* (Sage, London) 56-73.

Goldberg, D. P. (1972). *The Detection of Psychiatric Illness By Questionnaire,* (Oxford University Press, Oxford)

Jackson, P. R., Wall, T. D., Martin, R., and Davids, K. (1993). New measures of job control, cognitive demand and production responsibility. Journal of Applied Psychology, **78,** 753 - 762.

Karasek, R., and Theorell, T. (1990). *Healthy Work: Stress, Productivity and the Reconstruction of Working Life.* (Basic Books, New York)

Klein, J. A. (1991). A re-examination of autonomy in light of new manufacturing practices. Human Relations, **44,** 21-38.

Lawler, E. E. (1992). *The Ultimate Advantage: Creating the High Involvement Organisation,* `(Jossey-Bass, San Francisco)

Schonberger, R.J. (1986). *World Class Manufacturing: The Lessons of Simplicity Applied.* (Free Press, New York)

Seeborg, I. S. (1978). The influence of employee participation. Journal of Applied Behavioural Science, **14,** 87-98.

Wall, T. D., Corbett, J. M., Martin, R., Clegg, C. W., and Jackson, P. R. (1990) Advanced manufacturing technology, work design and performance: A change study. Journal of Applied Psychology, **75,** 691-697.

Warr, P. P. (1992). Job features and excessive stress. In R. Jenkins and N, Coney (eds) *Prevention of Mental Health at Work,* (HMSO, London) 40-49.

ASSESSING THE SAFETY CULTURE INFLUENCE ON SYSTEM SAFETY

Richard Kennedy

Industrial Ergonomics Group
Department of Manufacturing & Mechanical Engineering
University of Birmingham
Birmingham B15 2TT

The expansion of complex socio-technical systems in many spheres of society has lead to a broadening of the 'human error' consideration in risk assessments. Risk assessments now need to incorporate managerial and organisational factors. The concept of Safety Culture is suggested as a way in which these factors may be evaluated and some of the different methods for Safety Culture assessment are described. Safety Culture assessment approaches are only in their early stages of development and would benefit from greater modelling of Safety Culture influences. By drawing on approaches used in Human Reliability Assessment (HRA), Safety Culture Failure Mechanisms (SCFMs) are suggested as a way in which the mechanisms by which failures occur are addressed, as well as its indicators or influencing factors.

1. Introduction

The past few decades has seen the development of increasingly sophisticated socio-technical systems. Such systems have been developed for a diverse range of applications such as generating the worlds energy, providing transport or merely to satisfy scientific curiosity.

The larger, more complex and tightly coupled the systems components become, the greater the potential for control difficulties (Perrow, 1984). Therefore the 'safety barriers' of these type of systems can potentially be breached by the adverse combination of a number of systems components. These system components range from the technical, reliability and individual operator elements to the wider social, organisational and managerial aspects.

All technological innovations have a risk potential, but as the world has become reliant on some of this technology a degree of risk is deemed an acceptable cost for the benefits derived from operating the technology. Although the risk involved in operating certain technologies (e.g. nuclear power and chemical plants) cannot be completely eliminated, it can be managed.

2. Risk Management and Human Reliability Assessment

The 'risk management' approach emphasises preventative measures to achieve system safety and thus avoid accidents. Regulations and safety criteria [stipulating the control level for the risk] are put in place at the different stages of the system life cycle. A safety management system is also designed and implemented to ensure safe operation of the technology within this legislation and criteria.

An important part of the risk management approach is a Probabilistic Safety Assessment (PSA) This attempts to assess the negative affects on the system due to all types of failure (i.e. human, hardware, software and environmental failure). As engineering approaches to safety and reliability have become more advanced the focus in terms of 'risk management' has increasingly moved towards human error. Human Reliability Assessment (HRA) is the process by which possible human errors are identified in the system, their error impact assessed and the potential for their occurrence reduced. Recently a British Standard for HRA has been introduced (BS5760, 1994) which reflects the increased importance of incorporating human error into overall risk assessment.

HRA has been reasonably successful at identifying, quantifying and reducing execution errors (slips and lapses), and has become better at certain cognitive errors. It has however been less successful at addressing other forms of human error. In particular management and organisational failures have not been incorporated within the error assessment boundaries of HRA.

It has only been relatively recently that risk assessment practise has moved 'upstream' in an attempt to include the actions of designers, managers and other decision makers [as well as individual operators] in the overall risk evaluation. There has been recognition of the increased needed to evaluate the more 'subtle organisational pre-conditions to failure' (Pidgeon and O'Leary, 1994). The most recent developments in understanding and assessing these 'subtle organisational pre-conditions to failure' has been with the arrival of the concept of "Safety Culture".

3. Safety Culture

The term "Safety Culture" has its origins from the enquiries into the Chernobyl nuclear accident where human errors and violations of procedures were referred to as poor Safety Culture (OECD 1987). The term has been more thoroughly defined by ACSNI (1993) who state that *" the safety culture of an organisation is the product of the individual and group values, attitudes, perceptions, competencies, and patterns of behaviour that determine the commitment to, and the style and proficiency of, an organisation's health and safety management."*
" Organisations with a positive safety culture are characterised by communications founded on mutual trust, by shared perceptions of the importance of safety and by confidence in the efficacy of preventative measures."
Attitudes are the fundamental part of Safety Culture but, as the ACSNI definition expresses, it is also made up of other inter-related elements such as values,

perceptions and behaviours. These elements will combine to give a much larger Safety Culture effect. Safety Culture is therefore described a 'gestalt' type of construct whereby the whole is greater than its parts (ACSNI, 1993). Thus it should not be viewed purely in terms of individual attitudes and beliefs about safety and risk.

4. Safety Culture Failure

Some recent accidents in large socio-technical systems have illustrated the consequences of Safety Culture 'failing'. The official enquiries into Bhopal, Chernobyl, Challenger, Herald of Free Enterprise, Kings Cross, Clapham and Piper Alpha all cite examples of errors committed at management level combined with a mixture of other general organisational failings. These failures include the competency and training of certain personnel; inadequate communication and fault reporting systems; maintenance of equipment not carried out; safety equipment deemed too expensive an outlay and so on.

These type of management and organisational failures are indicative of 'poor' Safety Culture. The management of safety is part of the overall organisational Safety Culture and it is for this reason that the above accidents are regarded as examples of where the Safety Culture 'failed.' However, learning the Safety Culture lessons from these past disasters has been a slow process. The root causes of the accidents cited above are similar, but a realisation of the relationship in their causal error elements has only recently been made.

The importance of Safety Culture failings in the etiology of major socio-technical system accidents emphasises the need for the development of approaches to assess Safety Culture as part of the overall risk assessment process. The risk assessment of a system from a Safety Culture perspective would need to assess the set of beliefs, norms, attitudes and social and technical practises which are concerned with minimising the exposure of individuals, within and beyond an organisation, to conditions considered dangerous or injurious (Turner et al 1989). HRA does not achieve this type of risk assessment, nor does it actually purport to being able to do so. Instead approaches to assessing Safety Culture and safety management have tended to have been developed independently of HRA.

5. Methods for Assessing Safety Culture

The techniques for assessing Safety Culture can be distinguished into three general categories of approach;

5.1. Frameworks for Organisational Safety Performance
In this type of approach the general [Safety Culture] health of the organisation is assessed by the presence or absence of indicators of safety performance. The frameworks may start their assessment at different points to each other in the organisational structure, but are similar in that they can potentially show where Safety Culture problems arise and how the root causes of these problems may be addressed. Examples of such frameworks include TRIPOD (Hudson et al, 1991); OSTI (Komaki et al, 1986); NOMAC (Haber et al, 1991); WPAM (Apostolakis, 1992); and STATAS (Hurst and Ratcliffe 1994).

5.2. Audit Tools for Organisational Safety Performance

The audit approach is similar to the framework approach as it uses performance indicators, which are often organised into groups. The scores on sub-sets of safety performance areas are usually weighted and then translated into an overall total safety index rating. This type of approach could therefore be described as being 'correlational' in nature and is a more structured examination of managerial and operational procedures than the framework approach. Examples of audit tools include ISRS (Loss Control Institute, 1988); MANAGER (Pitblado et al, 1990); MORT (Johnson, 1980); and ASCOT (IAEA, 1992).

5.3. Safety Culture Prompt Lists

An example of a prompt list to assess the adequacy of Safety Culture is contained in ACSNI (1993) and a list of Safety Culture indicators has also been developed by INSAG (1991). These approaches utilise a simple question and answer procedure and are the most basic way of getting a handle on the Safety Culture of the organisation.

6. Adequacy of Safety Culture Assessment Methods

Many human errors can be assessed both qualitatively and quantitatively via a combination of HRA and Ergonomics. However for the phenomenon of Safety Culture, the theory and techniques to assess such failures is very much in its infancy. There is little to suggest that the currently available approaches to Safety Culture assessment have been based on the psychology of organisational behaviour (or for example attitudes to risk), nor do they appear to be based on a model of safety management. Instead a series of performance indicators or quasi-empirical checklists have been developed with which to enter the organisation and, via audit and interviews, assess the overall health of the system.

These approaches have not always been validated. Furthermore, they focus on indicators of Safety Culture performance which may or may not correlate with the actual safety performance for the particular organisation being assessed. This is counter to the philosophies of HRA and Ergonomics which rely on detailed modelling of a situation to determine what can go wrong. The approaches adopted by HRA and Ergonomics can address the mechanisms by which failures occur, as well as indicators or influencing factors.

7. Safety Culture Failure Mechanisms

Safety Culture assessment methods could draw upon a parallel approach used in HRA and develop Safety Culture Failure Mechanisms. In HRA there are error modes, the external manifestation of what happened or is predicted to happen (e.g. close wrong valve). The Safety Culture equivalent might be, for example, failure to transmit information across a shift boundary. In HRA there are Performance Shaping Factors (PSFs), such as training, and the Safety Culture equivalent might be, for example, workload near the shift handover period. HRA also uses Psychological Error Mechanisms (PEMs), such as memory failure, which explains what happens inside the head of the operator (the internal failure mechanism), and this is in particular what is apparently missing from current Safety Culture applied research.

Such mechanisms might entail, for example, shift team cohesiveness, or inter-team rivalry. Many of the checklist approaches aimed at evaluating Safety Culture point indirectly and sometimes directly to these mechanisms (as well as the equivalent of Safety Culture PSFs), but have not been filtered into a taxonomy of Safety Culture Failure Mechanisms (SCFMs).

Safety Culture failures can begin to be modelled after SCFMs are derived. A detailed operational knowledge of the safety management structure of the organisation, including informal safety management aspects, needs to be derived and then the SCFMs can be applied to the tasks of safety management. The output is a set of task or safety management area vulnerabilities. These vulnerabilities are not vague concerns about safety management practices, but are specific definable problem areas that can be tackled and rectified by management, based on an understanding of why the problem exists.

8. Concluding Comments

As engineering and HRA approaches to risk management have improved, safety in modern day socio-technical systems is now largely dependent on the Safety Culture of the organisation. Indeed Joksimovich (1992) states for nuclear utilities that "it is perceived that the Safety Culture is now probably the most effective barrier against releases of radioactivity."

Approaches are currently being developed which attempt to measure the effects and 'adequacy' of Safety Culture in large socio-technical systems. This paper has briefly described these approaches and suggested some of the research needs for the development of Safety Culture evaluation tools. One of the research avenues is the development of Safety Culture Failure Mechanisms (SCFMs) which can potentially address the mechanisms by which Safety Culture failures occur, as well as indicators or influencing factors. Such an approach is considered necessary if the phenomenon of Safety Culture is to lose its intangibility and become a more measurable concept.

9. References

ACSNI (1993) *Organising for Safety: Third Report*. Advisory Committee on the Safety of Nuclear Installations. London HMSO.

Apostolakis, G., Okrent, D., Gruksy, O., Wu., J.S., Adam, R., Davoudian, K. and Xiong, Y. (1992) Inclusion of organisational factors into probabilistic safety assessments of nuclear power plants. *IEEE Fifth Conference on Human Factors and Power Plants*. Monterey, 7-11 June.

BS5760 (1994) *Reliability of Systems Equipment and Components. Part 2: Guide to the Assessment of Reliability*. British Standards Institute. London.

Haber, S.B., O'Brien, J.N., Metlay, D.S., and Crouch, D.A. (1991) *Influence of Organisational Factors on Performance Reliability: Overview and Detailed Methodology*. NUREG/CR-5538.

Hudson, P.T.W., Groeneweg, J., Reason, J.T., Wagenaar, W.A., Van der Meenan, R.J.W., and Visser, J.P. (1991) Application of TRIPOD to measure latent errors in North Sea gas platforms: validity of failure state profiles. Paper presented at the *SPE*

Conference on Health and Safety and Environment in Oil and Gas Exploration and Production. The Hague, 11-14 Nov.

IAEA (1992) *ASCOT Guidelines: Guidelines for Self-Assessment of Safety Culture and for Conducting a Review*. Assessment of Safety Culture in Organisations Team, Vienna.

IAEA (1991) *Safety Culture: A Report by the International Nuclear Safety Advisory Group*. IAEA No.75-INSAG-4, Vienna.

International Loss Control Institute (1988) *The International Safety Rating System (ISRS), 5th Edition*. Institute Publishing, Loganville, Georgia, USA.

Johnson, W.G. (1980) *MORT: Safety Assurance System*. Marcel Dekker Inc, New York.

Joksimovich, V. (1992) Safety Culture in nuclear utility operations. *IEEE Fifth Conference on Human Factors and Power Plants*. Monterey, 7-11 June.

Komaki, J.L., Zlotnick, S. and Jensen, M. (1986) Development of an Operant Based Taxonomy for an Observational Index of Supervisory Behaviour. Journal of Applied Psychology, 71, **2,** 260-269.

OECD Nuclear Agency (1987) *Chernobyl and the Safety of Nuclear Reactors in OECD Countries*. Organisation for Economic Cooperation and Development, Paris.

Perrow, C. (1984) *Normal Accidents*. Basic Books, New York.

Pidgeon, N.F. and O'Leary, M. (1994) Organisational Safety Culture: implications for aviation maintenance. In N. Johnston, N. McDonald and R. Fuller (eds) *Aviation Psychology in Practise*. Avebury Technical, Ashgate Publishing Ltd, Hants.

Pitbaldo, R.M., Williams, J.C., and Slater, D.H. (1990) Quantitative assessment of process safety programs. Plant Operations Progress, 9, **3**, 169-175.

Hurst, N.W. and Ratcliffe, K. (1994) Development and application of a structured audit technique for the assessment of safety management systems (STATAS). In *Hazards XII European Advances in Process Safety*. Institution of Chemical Engineers, Rugby.

Turner, B.A., Pidgeon, N., Blockley, D.I., and Toft B. (1989) Safety Culture: it's importance in future risk management. Position paper for the *Second World Bank workshop on safety control and risk management*. Karlstadt, Sweden.

General Ergonomics

AN ENGINEERING PSYCHOLOGY APPROACH TO CONCEPTUAL DESIGN

A Joint Paper by I.S.MacLeod & A.McClumpha*

Aerosystems International
West Hendford
Yeovil
Somerset BA20 2AL UK

* DRA(F) CHS
Farnborough
Hampshire,
GU14 6SZ UK

Several complex aircraft systems have recently been developed that failed to meet expected performance requirements. In all cases the performance shortfall could be partly attributed to aircrew related problems inherent in the design. Recognition of these problems has prompted a search for improved design methods that could be effectively applied throughout the human-machine system design life-cycle. An engineering psychology method has recently been evolved through a UK programme developing a complex system incorporating air and ground elements. The human role in the control and direction of human-machine system complexity is emphasised.

Introduction

This study was prompted by a MoD appreciation of the existence of severe operating problems recently encountered with newly designed complex Human-Machine Systems (HMSs) developed elsewhere in the world. The problems stemmed from a poor understanding of the functionality required at workstations and an inappropriate distribution of tasks both within and between components of the system. This resulted in excessive operator workload and inappropriate information presentation which led to reduced operator efficiency and situational awareness.

Awareness of the current problems associated with complex HMS caused DRA(F) CHS concern that in the development of 'new' and complex HMSs there is an absence of agreed human performance methods which can help to determine the system critical issues. For example, critical issues associated with the numbers, skills, and needs of the operators of complex HMS. Importantly, though existing methods help to define physical and limited cognitive activities when the system already exists, for new and complex HMSs both methods and techniques are sparse for the prescription of cognitively critical activities .

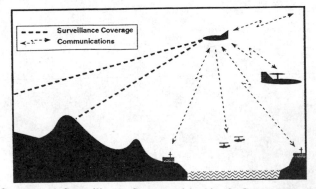

A Conceptual Surveillance System with Air & Ground Elements

Study Objectives & Utility

In response to a requirement from the UK MOD for a Human Engineering (HE) study concerning a new and complex surveillance system, comprising both air and ground elements, Aerosystems International undertook the work during 1994. MOD technical supervision to the study was provided by DRA(F) CHS.

The objective of the study was given by the Statement of Work as:

" .. to identify the human tasks that are critical to the efficient and effective operation of *[new system]* missions. The study shall also provide the means, through methods and models, to prescribe mission and task allocation of human activities within and between the components of the system."

The utility of the outputs from the study was to:

1) aid MoD in their initial advice to contractors on the Project Definition (PD) phase of the new system design;

2) assist MoD's later consideration of the PD studies concerning the approaches taken to new system design.

Traditional HE approaches to design have been shown to have drawbacks in their application to complex HMS. To meet the requirements of this study the development of a new approach to HMS design for 'new' and complex interactive system was required. This new approach will be briefly discussed by this paper.

The Study Problem Area

HMSs are becoming increasingly complex and technology driven, engineered to be capable of actions and computer based handling of data at rates beyond the capabilities of the human to assimilate and comprehend. Regardless of this complexity, the human is still required to direct and control overall system performance as this is still beyond the capabilities of engineered systems.

In relation to the increasing complexity of HMSs, the human's role has changed from a traditional systems one of physically based effort and control, to a role that is increasingly one of thinking rather than doing, directing rather than physically controlling. Moreover, the human control and direction within complex systems has become increasingly important and critical.

What is left to the human by traditional system design are the truly complex cognitive and judgmental tasks. But the system development process, instead of putting the primary emphasis on the characterisation and aiding of these key tasks, often concentrates instead on the design of the material parts of the system. The human is left to deal, as best they may, with a burgeoning problem of HMS control, direction, and the management of machine outputs and inputs.

The traditional approach to system design has typically been to specify the system in engineering terms as functions and then to allocate these system functions to human or machine. Such specification could cater for human physical inputs or controls but could only cover human cognitive activities in the broadest of terms, for example as duties or roles. The human was considered to be sufficiently capable of adaptation to cater for all functions that could not be assigned to the machine.

Traditional design methods cannot be used to adequately appreciate and cater for the changing human role within complex HMSs. Changing human roles require an associated change in the designer's appreciation of complex HMSs.

Designers Needs

One of the failures of human factors and engineering psychology has been that the outputs have often been too nebulous to add to any design specification. The HMS designer requires a tangible specification of the requirements of the system on which to create the design.

In addition, many of the 'human' disciplines have presented their outputs in purely deterministic engineering and quantitative terms to aid the acceptance of the output by engineers and designers. Frequently, however, the qualitative nature of the outputs does not support a quantitative form of presentation. Nor is human behaviour purely deterministic in nature. The basic argument of the study was that the specified functionality for a complex HMS must encompass both human cognition related and engineering related forms of functions.

This paper suggests that answers to the following issues will help in requirements definition for HMSs:

1) Definition of human and machine functional requirements for the HMS.
2) Knowledge of expected system performance.
3) Knowledge of formal system operating procedures or plans.
4) Knowledge of the expected competencies of the intended system operators (e.g. frames / spans of human control and responsibilities).
5) Knowledge of the expected performance limitations of the intended system operators (e.g. considerations on task, context & modes / states of human control).
6) Knowledge of the optimum forms of system control considering the first four points and the optimum forms of control for the human at the Human-Machine Interface (HMI).

Design Requirements based on Full Functional HMS Specification

The remainder of the paper will discuss the above issues. The full functional specification of a complex HMS requires that the human functionality and engineering functionality are both specified from the onset of design. This functionality specification should be progressively updated throughout the design of the HMS.

Adopted updates should be as a result of multi disciplinary trade-offs during design. These trade-offs must include consideration on the human cognitive functionality applicable to the HMS and the required human control and direction of the system. Therefore, part of the trade-off must be a consideration of the Human Machine Interface (HMI) using the extant knowledge on required human cognitive based HMS functionality, engineering functionality and man-machine system requirements related to human direction and control of the system.

Methods & Procedures

Any design of a complex HMS requires rules and measurement criteria on which to progress system design. HE inputs to design must be a result of the utilisation of appropriate and effective methods and procedures. Methods must not only be suited to the stage of the design life-cycle, they must have a clear and pertinent purpose. The methods must also produce outputs that can be useful to the progression of the HMS design, or produce outputs that can be usefully adopted by other methods at later stages of that design. HE methods must also be capable of utilising inputs of information from diverse sources such as analyses of previous HMS, pertinent studies or from Subject Matter Experts (SMEs).

This study used a variety of methods to capture the cognitive requirements. Three core methods to elicit and depict the SMEs knowledge were used - Goals Means Task Analysis (GMTA - Hollnagel 1993), Concept Mapping (Gowin & Novak 1984) and Task Features Elaboration (TFE). GMTA provide the formal task analysis and depiction of high level HMS tasks. The GMTA was mainly depicted in a diagrammatic form. Concept Mapping was used to glean the SME knowledge necessary to assist the conduct of the GMTA. TFE was used to demonstrate that ancillary task analysis methods are needed to the GMTA, to capture the information on cognition requirements needed to support design.

The study used versions of system flow diagrams to initially depict the MOD specified system functionality and to bridge any gaps in the specified functionality and the associated information flows required by the system. With the aid of the knowledge gleaned from a small band of highly skilled SMEs the GMTA covered the many diverse parts of the new system and the system co-ordination and communication requirements.

An important applied method of knowledge elicitation used questionnaires and group discussions to compile a conceptual mission story-line or account of possible mission events. This exercise provided the means of capturing the normal tasks that the system operators might perform, under normal situations and contexts, as well as the perturbations that might occur from the operating environment.

Using the tasks as depicted by the GMTA, plus the ancillary task analyses and the compiled story-line, an analysis was made of critical areas of system work and operator tasks. This analysis was performed from an engineering standpoint, and also from the standpoint of operators' work including their cognitive decision-making and control functions within the system.

Figure 2 below is a diagrammatic representation of a conceptual form of the man-machine system under discussion.

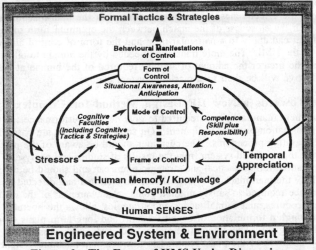

Figure 2 - The Form of HMS Under Discussion
This figure is pertinent to the subsequent discussion

System Interactions in the Human-Machine System (HMS).

Environmental Interactions

HMSs do not exist in a vacuum and must perform their roles in some form of operating environment. The new mission system will have an operating environment that is both air and land based. The human controllers of the system will mainly appreciate the system environment through an understanding of the performance of the system and the system sensors. The environment will also be appreciated directly by the operators through a use of their senses independant of the HMS.

Consideration of the system environment, system performance, and human expectations, will generate stressors and temporal appreciation affecting the operator's interpretation of the system's status in its world, their situational awareness and their inputs to system control and direction in that world.

Control Interactions

Control of the HMS will be related to operators' knowledge of short and long term plans of the mission (formal tactics and strategies), researched and compiled prior to the mission, and their competence in applying their cognitive faculties and mental skills to the direction of the system towards the achievement of perceived mission goals.

Part of this fit of plans is achieved by effective operation of the system through equipment operating procedures; part is achieved through control of the system by translating and applying human competence and cognitive facilities through HMI related activities; part is achieved through a maintenance of situation awareness, attention and the skilled anticipation of future events.

Maintenance of Effectiveness

The effectiveness of the human operation of the HMI, and through it system control, depends on what human cognitive mode (Hollnagel 1993) or state can be maintained in support of the use of cognitive faculties. These faculties support the required human cognitive tactics and strategies (Macleod & Taylor 1994) and, also, the effective utilisation of the competence of the operator within their available frame or span of control.

Moreover, the effectiveness of the human control of the man-machine system will depend on the closeness of the match between the optimum form of the control that the human can apply to a particular task, and the form of control allowed by the design of the HMI. The more translation needed by the human to allow HMS control, and the greater the amount of activity required of the human at the HMI, the less effective will be their man-machine system control.

Guidance Towards a New HE Design Method for Complex Systems

The equipment manufacturer is required to give design, performance and reliability information on their equipment. On such information are based the equipment operating procedures. The customer should be aware of the performance required from the specified system from the results of studies and analyses. Based on these latter results, formal tactical and strategic plans can and should be devised, indicating how the man-machine system should be operated and directed to achieve the performance required to satisfy mission goals. An awareness of the equipment operating procedures, and formal tactics and strategies, allows the operator to apply their skill through a formulation of cognitive short and long term plans (cognitive tactics and strategies). These cognitive tactics and strategies represent the operator's conception of the plans required to achieve man-machine system goals.

HMS Direction & Control

It is emphasised that only an operator can direct a man-machine system towards its given goals. Too often formal tactics and strategies are devised late in the design of a man-machine system and may represent a method of equating problems in the design. By explicitly approaching how effectively a system should and might be used by the operators, at as early a design stage as possible, it is suggested that the whole design process and final product must benefit.

Control of the system will be related to the operator's knowledge of formal system related plans, as formulated prior to the HMS's mission, and the competent usage of that knowledge through the individual operator's Frame of Control. The Frame of Control encompasses the cognitive faculties and mental skills needed for the performance of plans and direction of the system towards the achievement of mission goals. Moreover, operator performance of effective HMS control and direction implies that the operator can maintain an appropriate cognitive Mode of Control to support the required utilisation of cognitive tactics and strategies. To be effective, these latter plans should be related to the operator's knowledge of earlier and preplanned mission plans and goals (formal tactics and strategies).

As argued earlier, the operator is influenced by the environment, stressors and their appreciation of time pressures. At times these influences may make it difficult or impossible for the operator to maintain the Modes of Control related to the progression of normal tactics and strategies. Absence of plan supporting modes means that the human has to quickly adopt alternative, and sometimes undesirable, modes of control (Klein 1993). These modes of control include instinctive behaviour, behaviour dictated by innate human heuristics or, in extreme circumstances, panic. However, some of these latter modes of control can be effective in certain circumstances depending on the operator Frame of Control currently available, the recency of their work practice, their repertoire of skills and their training and experience in similar circumstances.

Further, the effectiveness of the human control of the man-machine system will depend on the closeness of the match between the optimum Form of Control that the human can sensibly apply to a particular task, and the effectiveness of the form of control allowed through the man-machine interface. The more translation needed by the human to effect control, and the greater the activity required of the human at the HCI, the less effective will be their control of the man-machine system.

Conclusion

The above discussion has suggested an avenue towards the determination of both cognitive and engineering functionality for complex HMS design. This avenue requires a consideration of various forms of short and long term plans (tactics and strategies). The need to determine a required level of operator competence, and the importance of work situation and context has been stressed. We have been able to identify costs and benefits of a variety of methods to a real system. Their efficacy must be judged over time through their continued application and improvement. It is beyond the scope of this paper to present procedures to assist in the consideration of Modes, Frames and Forms of Control within HMS design. Such procedures are outlined in the final report to the [new system] study.

Further, only by recruiting in a rigorous and scientific way the existing knowledge and skills of SMEs, through appropriate and explicit methods designed to capture the cognitive aspects of system use, can it be expected that the performance of an HMS will meet the desired goals. The use of SMEs should not be a once only exercise and a progressive refinement of their contribution should be made throughout the design life-cycle. Finally, if the current problems associated with specifying complex HMS are not addressed then the customer should be prepared to face the consequences in terms of costly options of redesign or the use of additional HMS operators.

References & Definitions

Gowin, D.B., & Novak. J.D. (1984) *Learning to Learn*, New York, Cambridge University Press.

Hollnagel, E. (1993) *Human Reliability Analysis: Context and Control*, Academic Press, London.

MacLeod, I.S. & Taylor R.M. (1994) 'Does Human Cognition Allow Human Factors (HF) Certification of Advanced Aircrew Systems?' in Wise, J.A., Hopkin, V.D. & Garland, D.J. (Eds) *Human Factors Certification of Advanced Aviation Technologies*, Embry-Riddle Aeronautical University Press, Florida.

Klein, G., (1993) *Naturalistic Decision Making: Implications for Design* CSERIAC, Wright-Patterson AFB, Ohio.

Final Report to [New System] Study, undertaken under UK MOD(PE) contract SLS 1b/132. (AeI 1539K/1/TR.1-2 dated December 1994).

Function: A function can be defined as a broad category of activity performed by an HMS.

Engineering Psychology: The application of psychology to the problems of HMS design.

BOREDOM AND EXPERT ERROR

Martyn B.A. Dyer-Smith & David A. Wesson

Centre for Business Research
Newcastle Business School
Newcastle, NE1 8ST

The bored have poor work records, and are prone to accidents and absenteeism. They also show a host of unfortunate behaviours from delinquency to substance abuse. We developed a computer task to test the predictions of an Inertial Resource Allocation Model of boredom. The procedure involves simple mental puzzles solved against a background of a continuous vigilance task. Some individuals had no problem adjusting to minor puzzle changes. Others experienced a significant shock effect. Individual variability correlated, as predicted by the theory, with work performance and absenteeism.

Introduction

Novice error in complex systems is easily understood. Expert error is more problematic. Fitts and Jones (1961) drew attention to the role of habit-interference in aircraft crashes, and Hendrick (1983) measured such effects. In reversed control stick conditions in a simulator (i.e., when a right hand turn is being made the stick is moved to the left) he found that the errors of experienced pilots were between twice and four times higher than those of novices. Experts may be more susceptible to attentional failures, and these may also be symptoms of habit-interference. The expert's skill is practised and by definition less stressed under normal conditions than that of novices. Task involvement may readily fall to a level insufficient for effective sustained attention. In a word, the expert may become bored.

Boredom rather than fatigue may explain why pilots or truck drivers doze off on longhauls. We know this happens and indeed it may be commonplace. In an inflight study of pilot behaviour (reported in Kiernan 1994) 5 out of the 9 pilots observed slept at their controls at some point, even though they knew they were being monitored by researchers. Boredom and fatigue are related, but not identical issues. Fatigue has been well researched. Here we will confine ourselves to the less well understood issue of boredom.

Familiarity, skill and system reliability ease the smooth performance of a standard task, but may reduce the ability to switch to novel tasks (as Hendrick' experiments eloquently demonstrated). This inflexibility, we argue, is a key symptom of boredom.

The cost of boredom

The bored are a liability both to themselves, and to others. Bored employees experience lapses in attention and fall asleep. They take longer to notice and correct errors, and have more accidents (Cox 1980; O'Hanlon 1981). The bored generally exhibit dysfunctional behaviours from poorer academic work to depression (Brissett and Snow 1993).

Symptoms of boredom

Overall the symptoms of boredom bear a striking similarity to those of low-level depression. Whereas a mentally healthy person "is often viewed as having an interest in, and engaging with, the environment," (Warr 1990) the depressed and the bored are both disengaged. This disengagement [we believe] signals an inefficiency in processes of mental resource or capacity allocation. The emotion of boredom may be the feeling that emerges in consciousness when allocated mental capacity exceeds that necessary to meet current environmental demands. This explanation can parsimoniously account for the most disturbing features of the work behaviour of the bored. Average performance level be unaffected by boredom (Hopkin 1990), but overall it may become increasing variable.

Modelling boredom

The basic limited capacity model has been effectively extended into the field of emotion (Keinan, Friedland & Arad 1991; Ellis & Ashbrook 1988, 1989). Given predictable stimuli and unchanging responses, task performance quickly requires diminishing resources as responses become automated. The tendency is always to reduce our resource allocation to the lowest level suggested as appropriate by feedback. But in many circumstances we cannot reduce resource allocation in this way. We are often required to pay more attention to tasks than feedback suggests they are worth. We believe that boredom arises directly from such 'forced' mismatches. An excess of allocation over demand is felt as boredom.

Many conditions (e.g., nature of the work, individual compulsion to strive, guilt about indolence) may prevent a requisite allocation of resource, and in general appraisal and adaptation cannot be instantaneous. If the mental system has some inertia, then temporary over-runs of allocated capacity are inevitable.

Such an Inertial Resource Allocation Model (IRAM) can also explain why boredom-induced errors take the form they do. If, as the system winds down, a load is placed on it by the demands of some novel occurrence, the direction of adaptation must be reversed. For a period, as the system strives to overcome its own inertia and wind up again, it will necessarily lag behind the demands of the situation. At any time, the instantaneously available capacity may be insufficient to meet demand and apparent errors of judgement occur. These errors would not occur if the system were up-to-speed, and the individual expectant.

The implications of the IRAM are that persons who are not adept at making appropriate mental resource allocations are also those who will suffer most from the subjective effects of boredom. These, as everyone can testify, are uncomfortable and even painful. The natural response to adverse situations that cause discomfort or pain is avoidance. The avoidance can take the form of various physical or mental disengagement strategies, ranging from daydreaming to absenteeism.

Research Hypotheses

We investigated these implications of the IRAM in experiments with two working populations, data entry clerks (n=108) and seafaring watchkeepers (n=35). These populations were selected on grounds of particular theoretical interest, as their work can be located at opposite extremes of a notional continuum of boredom (Klapp 1986). Monitoring of any sort, whether in a power station control-room, or ship's bridge, entails the passage of large periods of time when nothing noteworthy happens. By contrast data-entry involves continuous pressure of work which is both repetitive and generally uninteresting.

The major research hypotheses were:
(1) That the time taken to allocate mental resource to a changing task will vary individually.
(2) Those who adjust most efficiently will show an improved all round work record in both attendance and performance on work normally regarded as boring.

Overview of Experimental Task

A model boring situation was designed with features that were varied within three experiments. The ability to actually do the core task - perhaps dependent on mental ability - was of no particular interest. It was the ability to rapidly reallocate resource and engage with variants of that task - more akin to mental agility - that was the focus. The experimental task (SWAT, see footnote) runs on a standard PC with colour monitor. Subjects respond to questions with presses of a key. The three time series experiments involved dual tasks.- a vigilance task and a cognitive task.

The vigilance task (shape recognition) was designed to capture attention. A target shape is present at the bottom of the computer screen throughout. Shapes appear singly on the main screen in a seemingly random manner (both in time and position). Some are target shapes; the majority are dummies. Response time is measured and is important in discounting fatigue as cause of variability (fatigue, boredom and stress are often confounded).

The cognitive tasks embedded in the procedure are all inferential. They require the subject to deduce the logic of the position of presented shapes and answer the question "Is the shape in the CORRECT place?". The difficulty of this question may be varied by using alternative forms (e.g. "Is the shape in the WRONG place?") and by changing the feedback (e.g. "You are RIGHT" or "It is in the WRONG place"). The need to cognitively shift gears (Louis and Sutton 1991) is thus manipulated and a small mental load imposed despite the core task remaining the same. The effect of this 'shock' is measurable as an increase in response time.

Four alternate question/answer formats were presented in a time-series ABCDA arrangement, the last test always being a repetition of the first (i.e. A). In each part there are fixed number of repetitions of one form of question before the form changes. In this manner expectation is built up (i.e. that the next question will be the same as the previous one) before it is confounded by a change. The shock of that change is measured directly in 10 milli-secs.

The subjects were trained on the procedure in the early part of a session before experimental measure were made.

Calculated Variables

The output from the procedure yields these variables for analysis (others were recorded but are not relevant here).

REFLEX TIME - The mean of the time in 10 milli-secs between the 'correct' shape appearing on screen and the subject's response (hitting a key or button) during the main part of the Test.

THINK TIME - The mean of the time in 10 milli-secs between the appearance of questions about 'correct quadrant' and the subject's response (answering the question) during the main part of the Test.

THINK TIME VARIABILITY - The variability in THINK TIME for the subject across the main part of the Test.

Work performance data were extracted in encoded form (scaled 1-7) from the last Staff Appraisal reports of the clerical subjects by their Personnel Department staff. No comparable data was available for seafarers.

Results

The behaviours under test were remarkably stable over three different experiments conducted three months apart. The mean Kendall's Coefficient of Concordance (W) of all computed variables across experiments was .66, p ≤ 003 (W =1 would signal a perfect rank correlation). Over the duration of an individual experiment (c15mins) REFLEX TIME change was effectively nil (-0.002 secs). Fatigue may thus be discounted as an operant cause for the differences in performance.

The pattern of THINK TIME VARIABILITY was found to be individual. The distribution of all such scores was normal and figure 1 shows the record of two individual subjects results (S1 & S2) taken from the opposite tails of that distribution. There were no significant group differences between seafarers and clerical staff performance nor any difference between sexes.

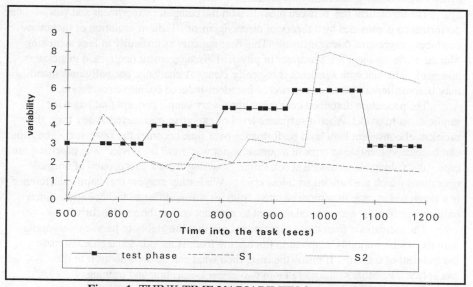

Figure 1. THINK TIME VARIABILITY for two subjects

The first and the last of the test phases were identical (see figure 1). The circumstances were thus created for comparison between early performance and late performance located approximately 15 minutes apart in time (a typical time-series experimental manoeuvre). As predicted, a general improvement in performance in task solving was compensated for by increasing variability. The effect is completely independent of both fatigue and inattention. The change in VARIABILITY was particular significant for the clerical staff (t = 3.20, p ≤. 001)

Individual THINK TIME and VARIABILITY correlated as predicted with parameters of job behaviour extracted from the clerical workers records.

Table 1. Correlations (Pearson's r) between experimental
and key work-performance variables

	OUTPUT	ACCURACY	ABSENCE
THINK	-.27*	-.35**	.17
VARIABILITY	-.19	-.39**	.26*

* p ≤ .05   ** p ≤ .01 (2-tailed)

If those who were unable or unwilling to complete the task are excluded, the correlations of experimental results to work performance data are markedly enhanced. The work data included two factors (accounting for 73% of the variance). Multiple regression of experimental variable on these factors yielded good predictions (to r =.5).

Discussion

The predicted effects of a repetitious task were found. There were marked differences between individuals on most experimental measures. The between-subject differences overwhelmed all between-group differences. Furthermore, the individual differences are related to performance and attendance records of clerical workers in exactly the manner predicted by the IRAM.

The theoretical link between measures on the computer experiment and work-related performance is provided by a theory of disengagement. Tedium, isolation or monotony produces progressive disengagement. This first appears as difficulty in task switching. Mental disengagement is a precursor of physical disengagement (expressed in sickness, absence). The link with accidents is logically clear. A challenge normally undemanding may become impossible after a period of boredom-induced cognitive regression.

The procedure described here measures a very simple concept that has wide implications in work. When a sufficient level of expertise is achieved in any task, the problem of consistent high level performance over time becomes the key issue. The expert can become vulnerable to error in a quite distinct way from the novice. The practised are especially vulnerable to sudden shock loadings arising from a background of normal operations which are handled with low effort. We further suggest that habit-interference is a much wider issue than mere negative transfer across different technologies. Such habit-interference is encountered moment to moment within the any repetitive task.

The indications from these experiments are that the ability to handle unengaging activity, whilst smoothly adapting to fluctuating load, is an individual characteristic independent of training. If this is the case, then what can be done about it in the workplace? Possible solutions fall into two classes; selection and training.

The selection issue is in principle straightforward. We propose that certain individuals with particularly low mental agility (or high inertia) are temperamentally unsuited for boring or repetitive work. On the other hand, there are others whose mental agility is high who are particularly well-suited. We believe we have generated a tool (SWAT) which is capable of making this discrimination reliably.

The training issue is more complex. The IRAM suggests that provision of stimulation alone for the bored or under-stimulated is probable futile. This proposition undermines great tranches of current social and workplace orthodoxies from provision of activities for bored teenagers to job-enrichment. The IRAM suggests that the educative aim should be to enhance efficient individual allocation of resource. There are two ways that this might be achieved. The first is the technique of meditation; the other the technology of biofeedback.

Becher and Kanter (1984) demonstrated that meditators show superior performance (detection without false alarms) on monitoring tasks than either non-meditators or novice meditators. The meditator, somehow, learns how to modulate their own thought processes. Biofeedback is a technological means to the same end.

Conclusion

We believe boredom is a key workplace and psycho-social issue. Mikulas and Vodanovich (1993) have arrived independently at the same conclusion from a completely different perspective from our own. Indeed they have gone on to propose that boredom as a state of being or consciousness should form the integrating principle of psychology. Certainly it is believed by many that "boredom is probably the most difficult and pervasive problem facing advanced industrial societies. And the more advanced the society, the more advanced the boredom" (The Economist quoted in Guest, Williams, and Dewe 1978). Perhaps in this age of the intelligent machine we should be educating more for indolence than for industry.

References

Bercher, D.B. and Kanter, D.R. 1984, Individual differences. In J.S. Warm, *Sustained attention in Human Performance,* Chap 5, 143-177, (John Wiley and Sons Ltd)

Brissett, D. and Snow, R.P. 1993, Boredom: where the future isn't, *Symbolic Interaction,* **16**(3), 237-256.

Cox, T. 1980, Repetitive Work. In C.L. Cooper and R. Payne (eds.), *Current concerns in occupational stress,* (John Wiley and Sons)

Ellis, H.C. and Ashbrook, P.W. 1988, Resource Allocation model of the effects of depressed mood states on memory. In K Fiedler, and J Forgas (eds.), *Affect, Cognition and social behaviour,* (Hogrefe)

Ellis, H.C. and Ashbrook, P.W. 1989, The "state" of mood and memory research: a selective review, *Journal of Social Behaviour and Personality,* **4**, 1-21.

Fitts, P.M. and Jones, R.H. 1961 Analysis of factors contributing to 460 'pilot error' experiences in operating aircraft controls. In H.W. Sinaiko (ed.), *Selected papers on Human Factors in the design and use of control systems,* (Dover, New York)

Guest, D, Williams, R. and Dewe, P. 1978, Job design and the psychology of boredom. *Proceeding of 19th Congress of Applied Psychology, Munich, West Germany.*

Hendrick, H.W. 1983 Pilot Performance under reversed control stick conditions, *Journal of Occupational Psychology,* **56**, 297-301.

Hopkin, V.D. 1990, The Human Factor Aspects of Single Manning, *The Journal of Navigation,* **43**(3), 346.

Keinan, G. Friedland, N. and Arad, L. 1991, Chunking and Integration: effects of stress on the structuring of information, *Cognition and Emotion,* **5**(2), 133-145.

Kiernan, V. 1994, Flying in the face of a pilot's need to sleep, *New Scientist,* **144**(1951) 9.

Klapp, O. 1986, *Overload and boredom,* (Greenwood Press, London)

Louis, M.R. and Sutton, R.I. 1991, Switching cognitive gears: from habits of mind to active thinking, *Human Relations,* **44**(1), 55-76.

Mikulas, W.L. and Vodanovich, S.J. 1993, The essence of boredom, *The Psychological Record,* **43**, 3-12.

O'Hanlon, J.F. 1981, Boredom: practical consequences and a theory, *Acta Psychologica,* **49**, 53-82.

Warr, P. 1990, The measurement of well-being and other aspects of mental health, *Journal of Occupational Psychology,* **63**, 193- 210.

* Footnote. The programme Smith & Wesson mental Agility Test (SWAT) is being further developed by ASE (part of NFER Nelson).

ECOLOGICAL ERGONOMICS:
UNDERSTANDING HUMAN ACTION IN CONTEXT

Neville Stanton

Department of Psychology
University of Southampton
Highfield
Southampton
SO17 1BJ

This paper describes a conceptual framework for organising ergonomics literature and data. It is argued that rather than tie the discipline to a cycle of testing, further advancement will come from theories of human performance with artefacts. Such theories might best be served by embracing the context within which performance occurs. This leads to the proposal for a contextual theory of human action.

Introduction

Ergonomics abounds with comparative studies that exhort the merits of one particular design over another, one particular environment over another, one particular training regime over another, and so on. Whilst these are undoubtedly useful studies in their own right, they answer questions to matters of current short-term concern, it is questionable whether they offer long-term advancements for ergonomics. Is the discipline to be chained to a cycle of testing?

Dowell & Long (1989) identified 4 deficiencies in the current status of the discipline, namely: poor integration, suspect efficacy, inefficient practices and lack of systematic progress. These problems do not bode well for advancement. By way of softening the blow, Dowell & Long suggest that the trial and error approach, characterised by the conception of Ergonomics as a craft, may be attributable to the relative youth of the discipline: genesis being dated as 12 July 1949 (Oborne, 1982).

The title 'Ecological Ergonomics' might strike the reader as a little odd: surely all ergonomics studies are inherently ecological? This is not the case. Generalising findings from one context to other contexts misses the essence of ecology: context has a direct bearing upon the phenomena being observed. In a different context, behaviour may be very different. Intuitively we know this to be true, which has led for calls for 'user-centred design' and 'field trials' in the development and evaluation of products.

The role of theory

From the arguments raised by Dowell & Long (1989), it might be reasonable to suppose that the discipline would benefit from more theory building by its disciples. Certainly, this would tackle some of the deficiencies they identified. The role of theory is to provide explanations that show causal linkages at work in the phenomenon we observe (Newell, 1990). In practice, theories are derived through the passage of time and considerable research effort. A life cycle of a theory might start with a conceptual framework. Conceptual schemes offer aid to theoretical advancement (Ryan, 1970), for example as a mechanism for classifying data and organising ideas. Analyses of this information may lead to prototheories, which are typically loosely formulated, but offer claims regarding causal sequences. As data is gathered and organised, the theory can be modified reiteratively in the light of new evidence. The process of generating hypotheses based upon the theory and testing these hypotheses and modifying the theory could go on *ad infinitum*. However, there are two points of exit. First, the evidence could become unequivicable turning the theory into a law. Second, the theory could be overthrown by a paradigm shift within the discipline which makes the theory untenable. However, in most cases, theories tend to exist on balance scales: the weight of evidence presented either tips in favour of, or against, the argument.

Models of human action

In an attempt to predict human behaviour with technological artefacts, theorists have tended to opt either for parsimony or for over-inclusiveness. An example of a parsimonious account is TOTE (Miller at al, 1960). Each TOTE unit contains four elements: Test-Operate-Test-Exit (see figure 1). TOTE offers a functional, goal-oriented, explanation of human behaviour. For example, consider the goal for an operator of maintaining a target state for a process variable. To ensure that the variable is kept at its target level (e.g. level of a liquid in a vat) the operator is required to conduct a TEST (e.g. compare the value of the variable against its target). If the variable is above or below the target state, the operator might OPERATE a valve to return the variable to its target state. After operating the valve the operator would be required to conduct another TEST, to check that the desired state had been reached. If the variable had been returned to the target value, the operator could EXIT that task. The original conception of TOTE appears to suggest a fractal-like relationship between a multiplicity of units.

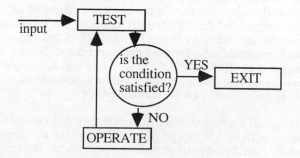

Figure 1. A TOTE unit

An example of over-inclusiveness is the model of Performance Shaping Factors (PSFs: Miller & Swain, 1987) where anything, and indeed everything, is thought to influence human behaviour. An example of such factors are illustrated in figure 2. PSFs include human-machine interface characteristics, task demands, tasl characteristics, instructions and procedures, stresses, environment, socio-technical aspects and individual factors.

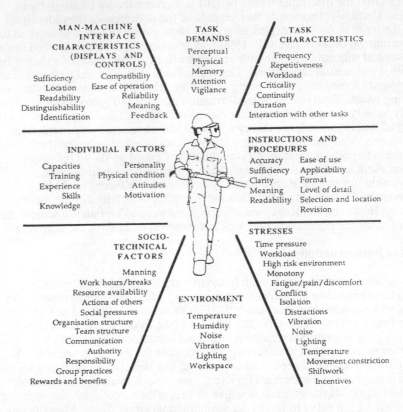

Figure 2. An illustration of PSFs (from Bellamy, 1990)

In the comparison of these two approaches, I am struck by the oversimplification of the first and the over-complexity of the second. However, I am drawn to parsimony. If we are to explain behaviour, we must look to the simplest description that accounts for the phenomenon observed. Recent research on PSFs suggests that most of the variance in behaviour can be accounted for by relatively few factors. In a study by Fujita (1993) 20% of the variation in operator performance was accounted for by training and experience. Similarly, in a study reported by Glendon et al (1994) 30% of the variance was accounted for by work pressure. These data suggest that a small pool of factors could be identified to account for most of the influence on human behaviour.

Carroll (1990) proposed the notion of a task-artefact cycle in order to explain the link between human activity, the design of artefacts and the situation in which the artefacts were used. Carroll suggests that these elements are interrelated: the task sets the requirement for the artefacts and the artefact, in situations of use, defines the task. He argues that the artefacts cannot be fully understood outside the situations within which they are used. These propositions lead me to believe that context plays a central role in the task-artefact cycle, see figure 3.

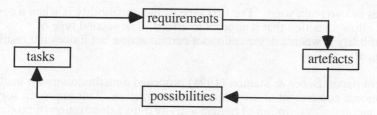

Figure 3. The task-artefact cycle (from Carroll, 1990)

Norman (1988) illustrates a cycle of human action that makes implicit reference to the context in which the activities are performed. He describes a seven stage model of action, comprising: perception, interpretation, evaluation, goal formation, intention, sequencing actions and execution of actions. Norman also draws the distinction between knowledge in the world and knowledge in the head. This is an important point, it suggests two types of context: external context and internal context.

The importance of context

At present there appears to be a rather lively debate between cognitive scientists and ecological psychologists regarding the explanation of human action. Norman (1993) offers a caricature of these two approaches. The symbolic cognition stance taken by cognitive science focuses exclusively upon the symbolic representations of the mind and the processing structures of the brain. Whereas the situated action stance taken by ecological psychology focuses almost exclusively upon the way in which structures of the world guide and constrain human behaviour. At either extreme end of these viewpoints, something is missing from the explanation.

Human *action* occurs as an *event* within a *context*. By understanding the relationship between these three entities (i.e. the relationships between human action, device states and context of the interaction), one might begin to propose a contextual theory of human action. In explaining behaviour, we often point to the context in which certain decisions were made, but their are two types of context: internal and external. Internal context (cf. knowledge in the head) refers to the knowledge, beliefs, experiences and motivations of the individual concerned. External context (cf. knowledge in the world) refers to the situational, temporal, informational, design and environmental characteristics present. To ignore either will result in an incomplete account.

An analysis of situated action

Gaver (1991) argues that information regarding the affordance of a device (i.e. the potential for action) is supplied not only by direct perception but also by exploration. Information about the operation of a device becomes apparent through interaction with that device. As the interaction progresses, new (previously unapparent) affordances are made available. Gaver calls these *sequential* affordances, when one action leads to a new affordance and so on. For example, a door handle has the affordance of grabbing, the grabbing leads to the affordance of turning and the turning leads to the affordance of pulling. Each action in turn leads to a new possibility that ultimately satisfies the goal of opening the door. Suchman (1987) refers to this type of activity as 'situated' planning. Errors may occur when there is an incompatibility between the affordance offered by a device and its' actual properties in two main ways. The first type of incompatibility is when a device does not allow an action that it appears to afford. The second type of incompatibility is when a device allows a certain action but it does not result in the desired goal.

In a recent paper, Baber & Stanton (1994) proposed a methodology for analysing the situational component of human-artefact interaction. This approach suggested that by mapping a description of human activity onto a description of product states one could begin to identify situations where the product would allow erroneous activity. The methodology makes the relationship between product design, situational use and errors explicit.

Contextual action theory

The value of an action performed, or a choice made, at one point in time often depends upon its relationship to the context within which it occurs. The coping model used in health psychology provides an analogous framework for a theory of contextual action. I am suggesting that it is necessary to consider how people cope with the operation of a device within a given context. This understanding will enable us to organise the existing research data to take the discipline beyond the last study to be published. Contextual Action Theory (CAT) would explain human action in terms of coping with technology within a context. To understand the theory, it is necessary to understand what enables people to cope, or fail to cope, with tasks.

In essence there are 5 main phases associated with contextual action. In the first phase the actual demands and actual resources are presented to the actor. The actual demands are made up of the design of the device, the tasks to be performed on the device, environmental constraints (e.g. time) and so on. Factors relating to external context have impact here. Actual resources are represented by the training, knowledge and experience of the actor with the device, and so on. Factors relating to internal context have impact here. The second phase consists of an appraisal of those demands and resources by the actor. It is noteworthy that actual demand and resources may differ from perceived demand and resources. A comparison of perceived demand and perceived resources occurs at the third phase. It is proposed that an imbalance at this stage would lead to some form of degradation. Possible pathways that the degradation might take would occur in the forth phase, these include emotional and behavioural responses or abandonment. Examples of emotional responses include decreases in user satisfaction and motivation. Examples of behavioural responses include increases in errors, increases in reaction time, increases in inefficient or inappropriate activities. The ultimate pathway would be for the user to abandon the interaction all together. The effects of these responses on the interaction with the device are appraised in the fifth phase, and fed back to the internal and external contexts.

It is possible to identify imbalance between actual (A) and perceived (P) demand (d) and actual (A) and perceived (P) resources (r). Hypothetically, imbalance would lead to some form of degradation in performance, as indicated by the use of pathways. Examples of the forms of imbalance are as follows:

$Ad>Pd$ or $Ad<Pd$
$Ar>Pr$ or $Ar<Pr$
$Pd>Pr$ or $Pd<Pr$

In addition to the above framework, CAT incorporates the *coping hypothesis*: people cope with technology. A basic premise of the notion of comparing demands with resources is that, if demands cannot be met by existing resources one way of compensating is to change the demands. Thus, a novice device user may interact in a different, less efficient, manner than an expert. The novice reduces the demands by making the device less demanding. An obvious example is the use of icons and menus in preference to function keys and commands in computing tasks. If this pathway was not sought we may indeed observe some form of degradation in performance. One might propose that a well designed interface helps novice users cope with a device, whereas experience helps expert users cope with a poorly designed interface.

Conclusion

In conclusion, I propose that it is vital for the growth of Ergonomics as a discipline that overall theories of human-artefact interaction are developed. CAT is one such approach that takes account of the context within which the interaction occurs.

References

Baber, C. & Stanton, N. A. 1994, Task analysis for error identification: a methodology for designing error-tolerant consumer products, *Ergonomics*, 37 (11), 1923-1941.

Bellamy, L. J. 1990, The quantification of human fallibility, *Journal of Health & Safety*, 6, 13-22.

Carroll, J. M. 1990, Infinite detail and emulation in an ontologically minimized HCI, In J. C. Chew & J. Whiteside (eds). *Proceedings of CHI '90*, (ACM, New York) 321-327.

Dowell, J. & Long, J. 1989, Towards a conception for an engineering discipline of human factors, *Ergonomics*, 32 (11), 1513-1535.

Fujita, J. 1993, What accounts for the operator performance? A simulator-based study of the influence of performance shaping factors on nuclear power plant operator performance, *Proceedings of Human Factors in Nuclear Safety* (IBC: London).

Gaver, W. W. 1991, Technological Affordances, *Proceedings of CHI '91* (ACM: New York).

Glendon, A. I.; Stanton, N. A. & Harrison, D. 1994, Factor analysing a performance shaping concepts questionnaire. In S. A. Robertson (ed), *Proceedings of the Ergonomics Society's Annual Conference* (Taylor & Francis, London) 340-345.

Ryan, A. 1970, *The Philosophy of the Social Sciences*, (Macmillan: London)

Miller, G. A.; Galanter, E. & Pribram, K. H. 1960, *Plans and the Structure of Behaviour*, (Holt, Rinehart & Winston: London).

Miller, D. W. & Swain, A. D. 1987, Human error and human reliability. In G. Salvendy (ed) *Handbook of Human Factors*, (Wiley: New York).

Newell, A. 1990, *Unified Theories of Cognition*, ()Harvard: Cambridge, Mass.).

Norman, D. A. 1988, *The Psychology of Everyday Things*, (Basic Books: New York).

Norman, D. A. 1993, Cognition in the head and in the world: an introduction to the special issue on situated action, *Cognitive Science*, 17, 1-6.

Oborne, D. J. 1982, *Ergonomics at Work*, (Wiley, Chichester).

CURSIVE TRANSCRIPTION ERRORS USING RESTRICTED DISPLAYS

Neil Morris   and   Dylan M. Jones

University of Wolverhampton   University of Wales,
School of Health Sciences   College of Cardiff
Psychology Division   School of Psychology
62-68 Lichfield Street   P.O. Box 901
Wolverhampton WV1 1DJ   Cardiff CF1 3YG

Portable, electronic information storage devices that can fit into the pocket
are becoming more common. They have some disadvantages relative to
larger computers. In particular, they have smaller screens presenting fewer
characters simultaneously and the lack of a printer may lead the user to
cursively transcribe information from the screen. This study examines the
effects of manipulating the time on screen and grammaticality of a seven
word display on cursive transcription errors and copying times.
Ungrammatical material took longer to transcribe but this was not modified
by time on screen. Both ungrammaticalness and brief presentation increased
copying errors. Recommendations for ameliorating these problems are
presented.

Introduction

Microchip technology has facilitated the development of information storage
systems that can be carried in the pocket. Such devices, for example electronic personal
organisers, are often only a little larger than a pocket calculator. While such devices have
many advantages, for example, they are much more portable than even a lap top computer,
they also have some inherent drawbacks. In particular, very small storage systems have of
necessity very small visual displays. Consequently, they tend to present relatively few
characters simultaneously on the screen. In addition, their extreme portability may exclude
the use of a printer to make a hard copy in most user environments. Thus if a hard copy is
required the user is likely to cursively transcribe the contents of the screen, especially if
there is likely to be any delay in using the information. These limitations are likely to lead
to errors in transcription and this paper examines some of the factors that are likely to lead
to such errors.

This study examines the speed and accuracy of transcription from restricted visual
displays by taking the simplest case where the transcriber has to provide a verbatim
transcript of the contents of the screen. There are likely to be two sources of error in this
task (assuming that there are no perceptual problems). First, there are likely to be short
term memory problems as the to-be-copied text must be represented in working memory
(Morris and Jones, 1988;1991) prior to transcription (see Baddeley, 1986 for a detailed
account of the nature of working memory). The extent to which working (short-term)
memory is required is likely to be a function of the amount of text presented and its
duration on screen. We manipulate memory load by varying the time on screen of the text
and thus avoid confounding effects when analysing transcription time. This is because
manipulating time on screen allows one to manipulate the memory load while still requiring
the same amount of material to be transcribed. The load is increased by reducing on screen
time so that the subject cannot opt to simply hold one or two words in working memory
when on screen time is brief; he or she must 'capture' all of the words in working memory

before the screen clears. With long on screen presentations the strategy of holding only one or two words becomes a viable option. Second, it is likely that users of restricted displays will maximize the amount of useful text that is displayed at any one time by representing material in a terse, non-redundant, note form. This will lead to violations of the rules of english grammar and reduce comprehensibility. It is known that material that is difficult to comprehend is much less memorable than simple, redundant material (Dooling and Lachman, 1971). We have therefore manipulated the grammatical properties of the texts and once again we have done this in a manner that does not violate the requirement that the same words must be recalled irrespective of the grammatical properties of the texts. This inevitably leads to a certain artificiality in the nature of the task but this is a necessary sacrifice in order that 'like can be compared with like'.

Thus this study requires subjects to transcribe, verbatim, single lines of text that are present, on screen, for between one and four seconds. With briefer presentations the burden on working memory is greater. In addition, the texts have been degraded semantically, as a result of a procedure that degrades the grammatical structure of the line of text, to examine the effects of the reduced redundancy that is normally a consequence of the predictable nature of english grammar. Finally it should be pointed out that verbatim transcription is very different from note taking. Note-taking usually requires the subject to impose some structure on the material and to be selective about what is transcribed. This is a very different activity to verbatim transciption. The subjects used in this study were required to copy, rapidly, material varying in meaningfulness, from restricted visual displays. They were not allowed to be selective in what they transcribed. The literature on note taking is discussed further in Jones, Morris and Quayle (1987).

Method

50 members of the Cardiff psychology department subject pool participated in this experiment. Both male and female subjects, aged 18-50 years, were used. They were run individually and paid £2.00 for a session lasting approximately 50 minutes.

Four, 42-word sentences were selected from Trevelyan's 'History of England'. The sentences were then grammatically degraded in a systematic manner. This was achieved by removing words and then reinserting them in the text in a transposed position. In this way various approximations to english prose were generated and these were designated as approximations 1-5.

Approximation 1: The four original texts with no transpositions.
Approximation 2: Every seventh word was deleted from the text. The six words so deleted were then randomly reinserted into the text in the six vacant positions.
Approximation 3: The texts were transformed in a manner exactly similar to that used to create approximation 2 except that every sixth word was deleted (a total of seven deletions) and randomly reinserted into the vacant positions.
Approximation 4: Every third word was randomly transposed (14 transpositions).
Approximation 5: This approximation was generated by deleting and then randomly reinserting every other word (21 transpositions).

Thus 5 versions of 4 different texts were generated and these versions varied in the degree to which they violated the rules of English grammar. This procedure was similar to that employed by Shannon (1949) to obtain word-order approximations. However the procedure used here differed in one crucial respect. *All the approximations of any one text contained exactly the same words.* Any differences between versions were purely grammatical and semantic and not the result of inserting new words. As the transcription time of the texts was a dependent variable this was a critical feature in the design.

There were 20 texts (5 levels of approximation for each of 4 source texts) each consisting of 42 words. Each of the 20 texts was then recorded onto a disc in files of 7 lines of 6 words. Subsequently these would be displayed on a VDU one line at a time by software developed for a BBC micro computer. The BBC micro computer also controlled the screen duration of each line of text. The duration of a given line was either 1, 2, 3 or 4 seconds. A pilot study indicated that subjects could indeed read any of the lines of text within one second but briefer durations created perceptual problems. The first line of any text was presented, across the centre of the screen, when the subject pressed the space bar. This also reset and started the computer's timing facility. The subject wrote down the line and then immdiately pressed the space bar again. This resulted in the presentation of the second line. The procedure continued in this manner until the subject had made seven bar presses. At this point a complete text had been presented one line at a time. Immediately after transcribing the seventh line the subject pressed the space bar again to indicate that he/she had finished and the total transcription time for all seven lines of text was displayed. Subjects were then given a fresh response sheet and pressing the space bar again instigated the presentation of the first line of the next text and reset and triggered the timing mechanism.

Before the transcription program was loaded a random number program was run to allocate each subject to the various condition orders. Firstly, a subject was given a text presentation order (e.g. Text 1, Text 3, Text 2, Text 4). Line duration (1-4 seconds) was then randomised between texts. Thus one subject might, for example, observe all the lines of Text 1 (and all five approximations) for 4 seconds. The order of Approximation to English (1-5) presentation within a text was also randomised. Thus all subjects experienced the same line duration throughout a given text regardless of its Approximation to English (e.g. subject one might observe text 1 first with its approximations in the order 1,5,4,2,3 but all these versions would have the same line duration). No subject observed two text sources for the same line duration.

In summary, then, the design of the experiment was such that the five approximations of any given text were presented to a subject in random order but *line presentation duration was invariant within a text and this applied to all five approximations*. Approximately equal numbers of subjects observed text 1, for example, at each presentation duration. Subjects had no control over line duration but they instigated presentation onset at a speed compatible with their writing speed. Subjects were allowed to rest between texts (i.e. when the transcription time was displayed). Every subject observed all text sources and was tested at all 4 presentation rates.

Subjects were provided with 20 response sheets which each had 7 numbered lines on them and details of text number, line duration and approximation to English. These were turned face down on completion. The procedure and the nature of the stimuli were fully explained to them at the beginning of the session.

Rapid, cursive transcription results in poor handwriting which can be difficult to interpret. To avoid experimenter bias an independent scorer, unaware of the purpose of the experiment, was employed. Transcription times were available and errors were also recorded. Error analysis consisted of scoring in terms of total errors, omissions (less than 6 words written on a line), intrusions (words not presented) and transpositions (responses recorded in inappropriate locations).

Results and Discussion

The data were analysed using analysis of variance with approximation to English as a within-subject factor and line duration as a between-subject factor. Comparisons were not made between texts as these could only be equated for number of words and other dimensions of text difficulty were not controlled (In fact similar patterns of results were forthcoming from the analyses of each text so the analyses of variance reported here all relate to Text 1). Error rates were very low and quite clearly not normally distributed. The error data was therefore transformed using log, + 1 before analysis. The means and standard deviations of the transcription times and the means of the raw data are presented in Tables 1 and 2 respectively.

Transcription time

This analysis revealed no effect of time on screen and no interaction. However there was a large effect of Approximation to English (p<0.001). Further analysis revealed that the significant effect could be accounted for by longer transciption times for

Table 1: Mean transcription time, in seconds, for each Approximation to English with on screen times of 1, 2 , 3 and 4 seconds. Standard deviations are shown in parentheses.

| Presentation rate (in seconds) | Approximation to English | | | | |
	1	2	3	4	5
1	100.70 (22.67)	100.25 (20.21)	102.07 (19.87)	102.83 (20.85)	104.30 (24.19)
2	100.06 (19.05)	104.16 (18.66)	105.03 (20.93)	108.80 (17.75)	110.32 (22.13)
3	97.78 (18.49)	102.06 (17.77)	101.54 (19.40)	108.64 (17.73)	109.78 (22.49)
4	102.45 (19.43)	107.75 (19.59)	103.50 (19.13)	112.53 (23.12)	116.20 (22.18)

Table 2: Total transcription errors, for each Approximation to English with on screen times of 1, 2 , 3 and 4 seconds. Figures are summed across all four texts and the figures in parentheses show the data represented as percentage errors. The standard deviations tend to be of similar magnitude to the raw score means.

| Presentation rate (in seconds) | Approximation to English | | | | |
	1	2	3	4	5
1	10.04 (5.98)	20.68 (12.31)	17.29 (10.29)	26.37 (15.70)	37.79 (22.49)
2	3.56 (2.12)	3.87 (2.30)	3.71 (2.21)	8.53 (5.08)	19.12 (11.38)
3	1.80 (1.07)	4.31 (2.58)	3.92 (2.33)	8.95 (5.33)	10.81 (6.43)
4	3.05 (1.81)	2.54 (1.01)	4.64 (2.76)	5.73 (3.41)	9.93 (5.91)

Approximations to English 4 and 5 and a similar but smaller trend for approximations 1, 2 and 3. However the interpretation of these results is tempered by the analysis of the error data.

Error scores

Because of the low frequency of errors analysis of variance was performed on the total errors. This analysis revealed main effects for both time on screen (p<0.001) and Approximation to English (p<0.001) but no interaction. As one might expect, there is a trend towards reduced errors with longer time on screen and also better performance with more grammatical texts. The results of the errors analyses were fairly clear cut. Most of the errors were omissions and these increased with brevity of display time and grammatical degradation. The fact that most errors were omissions is particularly interesting in that it suggests that the lengthening of transcription time with such degradation is marked. It takes longer to transcribe fewer words with low order Approximations to English. This is consistent with the view that generally semantic processing takes longer than 'shallower' processing (Craik and Lockhart, 1972).

The most interesting findings from these data are the large effects of line duration and the singularly large effect of approximation to English. The effect of line duration is to be expected. A brief display time requires the subject to commit more words to memory. However the effect of approximation to English suggests that rehearsal of this material is semantic rather than rote and that, in general, organisational factors in short-term memory are crucial to rapid cursive transcription of material with low redundancy.

Conclusion

What is abundantly clear is that copied material does not pass directly from the book or VDU to the notepad. Rather it has a brief sojourn in human short-term memory during which time it appears to be semantically processed to a deep level. However comprehensibility tends to improve recall of thematically relevant words only. One would expect that linguistic processing would encourage inference which may be undesireable when verbatim reproduction is necessary. If there is a mismatch between the transcriber's expectations of what will be presented and reality then the transcribers inferences may act as distractors. With prose this distraction effect necessarily trades off with the high level of redundancy in English prose which will, in general, improve performance by allowing one to make inferences from long term memory and thus relieving the burden on short term processes. With material that violates English grammar such inferences will be detrimental. It therefore seems to be the case that verbatim transcription may involve some form of linguistic processing which may be detrimental to performance when the material is not highly redundant.

With respect to the problem of simply loading up working memory, the recommendations for ameliorating this problem seem to be straightforward. Most of the errors attributable to this source in the present study are probably due to restricting the on screen time; this is unlikely to occur in many settings so it is sufficient to acknowledge that such a source of error is likely unless the transcriber opts to transcribe material in segments of a size that will not overload working memory. The problem of text redundancy and linguistic analysis is more serious. One could probably learn to avoid using inference but in many situations the use of inference is so advantageous that such avoidance is probaly not a useful skill to develop. A more satisfactory solution may be to deliberately introduce some redundancy into the stored material or, when this is not feasible, to develop a procedure (as with much radio communication) for double checking for all mismatches. A single transcription error in a message of low redundancy can radically change the meaning of the message and given the less than perfect performance of human memory systems, which seem to be important components in the act of transcription, it is worthwhile to develop procedures for reducing the likelihood that transcription errors will go undetected.

Acknowledgement

This research was supported by the Army Personnel Research Establishment, Farnborough. The views expressed in this paper are those of the authors.

References

Baddeley, A.D. (1986) *Working Memory*. Oxford: Oxford University Press.
Craik, F.I.M. and Lockhart, R.S.(1972) Levels of Processing: A Framework for Memory Research. *Journal of Verbal Learning and Verbal Behaviour*, **11**, 671-684.
Dooling, D.J. and Lachman, R. (1971) Effects of comprehension on retention of prose. *Journal of Experimental Psychology*, **88**, 216-222.
Jones, D.M., Morris, N. and Quayle, A.J. (1987). The psychology of briefing: A review and seven recommendations for improving performance. *Applied Ergonomics*, **18**, 335 - 339.
Morris, N. and Jones, D.M. (1988) The effect of relevant and irrelevant transcription on performance on a freightline simulation. In T. Megaw (ED.) *Contemporary Ergonomics, 1988*. (London: Taylor and Francis).
Morris, N. and Jones, D.M. (1991) Impaired transcription from VDUs in noisy environments. In E.J. Lovesey (Ed.) *Contemporary Ergonomics, 1991*. (London: Taylor and Francis).
Shannon, C.E. (1948) A mathematical theory of communication. *Bell Telephone System, Monograph B-1598*, Technical Publications.

TRACKING PERFORMANCE WITH
A HEAD-SLAVED POINTING INSTRUMENT WITH LAGS

Richard So and Michael Griffin

Human Factors Research Unit
Institute of Sound and Vibration Research
University of Southampton
Southampton SO17 1BJ

With a head-coupled 'virtual reality' system, head position is
continuously monitored and may be used to direct a pointing
instrument. Lags can occur when such instruments respond to a
directional input. When tracking a target, the position of a lagging
head-slaved instrument may be represented on the display as a reticle
(i.e. an aiming sight). Dual-axis tracking performance has been
studied with and without this reticle position feedback. Tracking
errors increased with increasing lags. The increase in the error was
statistically significant when the lags reached 67 ms. With lags up to
160 ms, no significant difference in tracking error was found with or
without the use of reticle position feedback.

Introduction

A head-coupled 'virtual reality' system usually consists of a helmet-mounted
display (HMD) and a helmet-pointing system. The pointing system measures the
orientation of the head and directs a device which produces optical images. The
images are presented on the HMD and may come from a camera controlled by the
head orientation or a computer which generates the appropriate view.

The measured head orientation can be used to direct a pointing instrument
and a 'head-slaved reticle' may be presented on the HMD as an aiming sight. A
potential application of such a system is in the field of tele-robotics. If lags occur
when the instrument follows the head position, there will be relative angular
displacements between the pointing angles of the instrument and the head. If these
displacements are ignored then the 'head-slaved reticle' can be presented at the
centre of the display and in-line with the head orientation. This condition will be
referred to as the condition with no 'position feedback'. Without 'position
feedback', the reticle remains at the centre of the display irrespective of any lag.

The relative angular displacement between the orientations of the instrument and the head may be displayed by offsetting the reticle from the centre of the display (Figure 1). This condition will be referred to as the condition with 'position feedback'. In this case, the pointing angle of the reticle truly reflects the orientation of the lagging instrument. When tracking a target, the 'head-slaved reticle' will appear to move in the opposite direction to the head movement when there are lags in the system. This type of lag has been referred to as 'reticle lag' (So and Griffin, 1993). In the context of this paper, the 'reticle lags' were pure time delays.

Hypotheses

With and without 'position feedback', it is hypothesised that tracking error with a lagging head-slaved instrument will increase with increasing duration of lag.

When 'position feedback' is used, the 'head-slaved reticle' represents the true position of the instrument and this information may enable the subjects to partially compensate for the lag. It is hypothesised that for the same lag duration, the use of 'position feedback' will improve tracking performance. With 'position feedback', subjects may change their tracking strategies according to the lag. It is hypothesised that there may be a carry-over effect when a subject is exposed to consecutive lags of different durations.

Experimental details

Method and design

Six male subjects participated in the experiment. Subjects were asked to move their heads to track a circular target (diameter one degree) with a 'head-slaved reticle'. Both the target and the reticle were presented on a helmet-mounted display. Images were focused at optical infinity on the display with a field-of-view of 17° x 17°. The position of an instrument was represented by a reticle which followed the head orientation with one of six imposed reticle lags: 0, 33, 67, 100, 133, 167 ms. The order of presentation of the six lags was arranged according to a 6 x 6 Latin square design adapted from Li (1964). For lags in consecutive runs, the order of occurrence was balanced. For example, among the six subjects, the lag of 67 ms was presented twice immediately before and twice immediately after the 0 ms lag. Subjects repeated a run with each lag three times consecutively. This formed 18 runs per session and there was one session per subject.

Although the study investigated the conditions with and without 'position feedback', some feedback was provided during the whole experiment. The reason was that, from the subjects' point of view, the 0 ms lag condition with 'position feedback' appeared the same as the six lag conditions without 'position feedback'. In these seven conditions, the reticle would always be presented at the centre of the display. In this study, instead of measuring the tracking errors with each of the six lags without 'position feedback', the errors were calculated using measurements obtained during the third run of the 0 ms lag condition with 'position feedback'.

The durations of the target motions were 120 seconds. The yaw axis target motion was generated by integrating a random time history once and then low-pass filtering at 0.4 Hz (24 dB/octave) and at 1 Hz (120 dB/octave). This filtered time

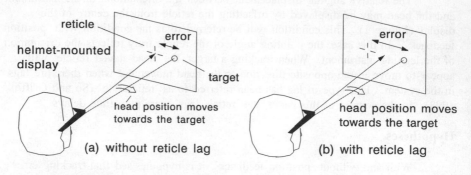

Figure 1.   Positions of a 'head-slaved reticle' on a helmet-mounted display
                 during a head movement towards a target.

history was presented in reverse order to form the pitch axis target motion. Six
combinations of the yaw axis and the pitch axis target motions were used. These
combinations were generated by starting the yaw axis target motions at different
instants along the original yaw axis target time history. The same target motion was
used for the three repetitions of each condition but was not used for any other
condition. The presentation of the six target motions was balanced over the six
subjects and six lags. The target spans in the yaw and pitch axes were ±40°
and ±30°, respectively.

        Subjects sat inside a dark environment so that the only visible stimuli came
from the helmet-mounted display. Four practice runs with different lags were given.
The four lags were in the order: (i) 0 ms, (ii) 133 ms, (iii) 67 ms and (iv) 167 ms.
The target motion used for the practice runs was similar to that used in the main
session but had a duration of 60 seconds and was not repeated in the main session.

Measurement

        A quasi-linear model of the tracking system is shown in Figure 2. The
system had a 24 ms inherent 'display lag'. This lag included the time delay in
measuring head position and displaying the target with the correct orientation to the
measured head position. Unlike 'reticle lag', 'display lag' does not affect the
relative displacement between the reticle and the head (So and Griffin, 1993).
Previous study has shown that tracking error is not significantly affected with an
additional display lag of less than 40 ms (So and Griffin, 1992). In this case, the
effect of the 24 ms inherent 'display lag' was assumed to be negligible. Two
dependent variables were used to assess the performance of subjects.

A) Mean radial error when 'position feedback' present

        This applied to all the experimental conditions. The displacement between
the reticle and the target was measured as the instantaneous radial error and the
mean radial head tracking error (MRE) was calculated with the following equation:

$$MRE = \frac{1}{N} \sum_{i=1}^{N} e(i) \qquad (1)$$

where $e(i)$ is the ith value of the radial error time history; N is the number of values in each time history.

B) Mean radial error when 'position feedback' not present
As discussed in the 'Method and design' section, this error was not a direct measurement, but was calculated. The angular displacement between the orientation of the lagging head-slaved instrument, *inst*(i), and the orientation of the target, $t(i)$, was calculated as the instantaneous radial error, $e(i)$:

$$\begin{aligned} e(i) &= t(i) - inst(i) \\ &= t(i) - h[i - (\text{reticle lag} / \text{T})] \end{aligned} \qquad (2)$$

where $h[i]$ is the ith value of the head position time history measured in the third run with 0 ms lag and 'position feedback'; T is the sampling period (16.7 ms).

The mean radial error (MRE) was calculated according to Equation (1).

Results and discussion

Effects of lag duration
Inspection of Figure 3 shows that as the lags increased, the mean radial tracking error increased. Wilcoxon matched-pairs signed ranks tests were performed to test the differences between the tracking error with the 0 ms lag and the tracking error with lags greater than 0 ms. For reticle lags greater than, or equal to, 67 ms, tracking performance was significantly degraded ($p<0.05$, all 3 runs, Wilcoxon).

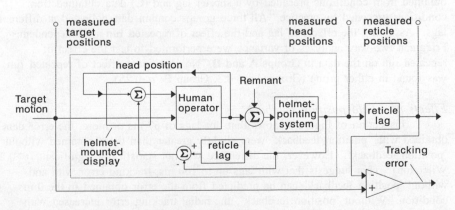

Figure 2.   A quasi-linear model of the tracking system used in the experiment (with 'position feedback').

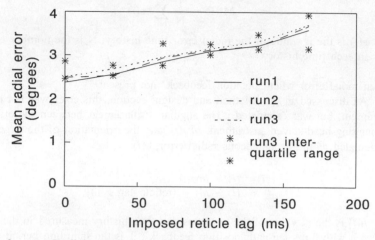

Figure 3.   Radial tracking error with different lags (median values of mean
                   radial error from 6 subjects).

Carry-over effects

During the experiment, subjects repeated each tracking run three times. For
each lag condition, Friedman two-way analyses of variance showed no significant
effect of run on radial tracking error (0 ms $p>0.05$; 33 ms $p>0.2$; 67 ms $p>0.7$;
100 ms $p>0.5$; 133 ms $p>0.6$; 167 ms $p>0.6$). This suggests that there was no
significant carry-over effect on tracking performance with consecutive runs having
different lags.

        Further analysis was carried out to separate the mean radial error data into
three groups: (A) data obtained from conditions preceded by a longer lag, (B) data
obtained from conditions preceded by a shorter lag and (C) data obtained from
conditions preceded by practice. All three groups contain data collected at different
lags. Assuming the effects of lag and the effect of repeated run are independent,
Friedman two-way analyses of variance were performed to test the effect of
repeated run on the data in Groups A and B. No significant effect of repeated run
was found in either group (Group A: $p>0.3$, Group B: $p>0.25$).

Effects of reticle position feedback

Inspection of Figure 4 shows that, for lags of 67 ms or more, the error data
obtained with 'position feedback' were slightly greater than those obtained without
'position feedback'. However, the difference was not statistically significant ($p>0.2$,
Wilcoxon). This suggests that with lags up to 160 ms, tracking error with and
without 'position feedback' can be predicted from the error obtained in the 0 ms
condition. Without 'position feedback', the radial tracking error increased with
increasing lags and the increase was significant for lags of 33 ms or more ($p<0.05$,
Wilcoxon).

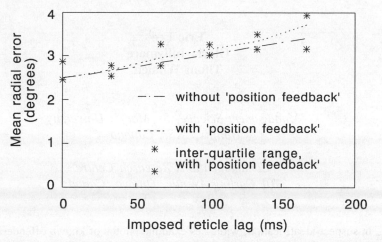

Figure 4.   Radial tracking error with different lags (with and without 'position feedback', median values of mean radial error from 6 subjects).

Conclusions

With position feedback, tracking performance with a lagging head-slaved instrument was degraded significantly with lags greater than, or equal to, 67 ms. Without the position feedback, the threshold lag was 33 ms. With lags up to 160 ms, no significant difference was found between tracking errors obtained with and without the use of position feedback.

With position feedback, repeating the tracking run had no significant effect on the tracking error. No evidence was found to support a carry-over effect in tracking error between consecutive runs with different lags.

Acknowledgement

This work was supported by Armstrong Aerospace Medical Research Laboratory through the Logicon Technical Services, Inc.

References

Li, C.C. 1964, *Introduction to experimental statistics*, McGraw-Hill Book Company. Library of Congress Catalog Card Number 63-23045, 195-196.
So, R.H.Y. and Griffin, M.J. 1992, Compensating lags in head-coupled displays using head position prediction and image deflection, *Journal of Aircraft*, **29**, (6), 1064-1068.
So, R.H.Y. and Griffin, M.J. 1993, Effect of lags on human performance with head-coupled simulators, Armstrong Laboratory, Air Force Material Command, Wright-Patterson Air Force Base, Ohio, Report AL/CF-TR-1993-0101. National Technical Information Service, 5285 Port Royal Road, Springfield, Virginia 22161, 40-42.

COMPUTERIZED FEATURE SYSTEMS FOR IDENTIFYING SUSPECTS: THE EFFECTS OF DELAY

Eric Lee[1]
Gloria Jollymore[1]
Thom Whalen[2]

[1]Management Science, St. Mary's University
Halifax, N.S., Canada

[2]Communications Research Centre
3701 Carling Avenue, Ottawa, Canada

In suspect identification, witnesses examine photos of known offenders in mugshot albums. The probability of correct identification deteriorates rapidly, however, as more are examined. Feature systems, which display mugshots in order of similarity to witness descriptions, increase success by reducing this number. Previous experimental tests failed to examine the effects on feature system performance of delays in eliciting witness descriptions, the use of live targets as suspects rather than photos, and the number of police raters/mugshot. In this study, 36 subjects searched 1000 mugshots for one of four different suspects, observed live, after delays of 1, 2, or 3 days. Delays had no effect, but two police raters/mugshot significantly improved performance. More raters are not required. Performance with live suspects was good (mean rank of targets = 20).

Introduction

In suspect identification, witnesses examine photos of known offenders in mugshot albums. The probability of correct identification deteriorates rapidly, however, as the number of mugshots examined increases (e.g., Ellis et al. 1989; Laughery et al. 1971, 1974; Lenorovitz and Laughery 1984). Feature approaches, where mugshots are displayed in order of similarity to a witness's description of a suspect, attempt to increase identification success by reducing this number (Ellis et al. 1989, Harmon 1971, Laughery et al. 1981, Lee and Whalen 1995a,b). In our computerized feature system, for example, both police raters and witnesses describe facial features of suspects on 107 subjective rating scales such as nose size: small 1 2 3 4 5 large. In experimental tests, the photo of target suspects is, for databases containing 1000 mugshots, on average among the first 10 or 20 photos examined (e.g., Ellis et al. 1989; Lee et al. 1993a, 1993b, 1995a, 1995b). This compares favourably with the 500 mugshots examined on average in album searches. Feature system users correctly identify more target suspects than do album users (e.g., Ellis et al. 1989). The most dramatic difference in performance, however, is in the rate of false alarms. Feature users consistently identify fewer incorrect people as the suspect. For album users, identification success declines precipitously as the number of photos examined by witnesses increase. Feature systems are highly tolerant of witness errors and omissions.

Previous experimental tests have failed, however, to examine the effects on feature system performance of delays in eliciting witness descriptions of suspects, the use of live targets as suspects rather than photos, and the number of police raters/mugshot. In those experiments, witnesses described target suspects immediately after seeing their photos. In practice, there is often a delay of several hours, or even several days, after an offense is committed before witnesses are required to describe suspects. Moreover, witnesses examined photos of target suspects, rather than viewing actual suspects (observed live as they would be during an actual crime). A third issue not addressed by previous studies is the question of how many police raters/mugshot are required for feature systems to work successfully. There is a strong, widely-held belief among researchers that since people vary considerably in their ratings of any given person's facial features, ratings from at least 10-12 police raters per suspect must be averaged to obtain an accurate estimate that can be used for comparison with witness feature descriptions (e.g., Ellis et al. 1989, Harmon 1971). We argue it is a myth, untested and unchallenged.

The primary purpose of this study is to evaluate empirically the effects of delay on performance of feature systems for identifying suspects. Many researchers argue that memory for faces deteriorates rapidly with delay between the time witnesses view a suspect and when they describe the suspect's facial features. Consequently, so the argument goes, feature system performance should similarly deteriorate rapidly with delay. In practice, delays of hours, days, and sometimes weeks elapse before police elicit witness descriptions of suspects. The implication is that all extant empirical tests of feature systems, because they all elicited witness descriptions of suspects with minimal delay, have yielded artificially inflated estimates of system performance.

The literature suggests otherwise. Many researchers have failed to find any difference in identification success between immediate testing and a two-day delay (Chance et al. 1975; Goldstein et al. 1971), 1-week delay (Laughery et al. 1974; Shepherd and Ellis 1973), 2-week delay (Deffenbacher et al. 1981), and up to a 3-month delay (Shepherd 1983). Shepherd (1983) did find a delay of 11 months impaired identification performance. Lindsay et al. (1994) and Deffenbacher et al. (1981) actually found significant increases in identification success after short delays. In contrast, only two studies have reported identification impairments for short delays (Krouse 1981, Davies et al. 1978).

The present study examines the effects of delay, live suspects, and number of police raters/suspect on the performance of feature systems.

Experiment

The Mugshot Database

The database consists of 1000 official mugshot photos of known offenders. (In contrast, Harmon and Ellis used photos of non-offenders.) Colour photos were taken under standard conditions -- frontal view of face from the shoulder up (90 x 125 mm prints). The suspects are all white males, aged 18-33 (99.5% are aged 18 to 27).

Each mugshot was coded on 107 facial features by one of 13 raters (6 males and 7 females in their early twenties). The raters received no training or instructions. Raters coded directly from the photo which was always available for inspection, as would be the case if police officers coded the mugshots. Coding time per mugshot was approximately five minutes. Each feature is coded on a 5-point Likert scale (e.g., narrow nose 1 2 3 4 5 broad nose).

The Feature Retrieval System

Both witnesses and police raters describe suspects using a printed questionnaire listing the 107 features. Similarity between witness and database descriptions is measured by a Euclidean metric, that is, we sum the squared deviations between witness and database feature descriptions and take the square root of the sum. In the Ellis system, similarity is measured by the number of feature matches. Preliminary research in our lab suggests the Euclidean metric minimizes retrieval rank relative to other metrics.

Method

The experimental design was a mixed ANOVA with two independent factors -- suspects (4) by time delay (1,2,3 days) -- and one repeated factor -- number of raters/suspect (1 to 5). The target suspects were four young white males, mean age = 24.0. The five raters included three men and two women with a mean age of 42.1 years. The 36 subject witnesses, 15 men and 21 women with a mean age of 29.3 years, included 15 full-time workers and 19 university students. All witnesses but one were white North Americans.

Raters described each target suspect on the 107 facial features. Suspects were rated live, and raters observed the suspects as much as they wished. None of the target suspects were judged to be unusually distinctive.

Subject witnesses were seated at a desk and asked to remember what the target suspect looked like. Suspects entered the room and stood for 10 sec approximately 4-6 feet from the witness. Witnesses were tested individually and observed a single suspect. One to three days later, witnesses returned to fill out a feature description questionnaire (107 items) on the suspect. Witnesses were randomly assigned to one of three delay conditions (1,2,3 day delay). Guessing was discouraged. Each target suspect was observed by three different witnesses at each time delay.

Synthetic experiments

Each target suspect was described by six different "police" or database raters (call them raters a, b, c, d, e, and f). For each of the six raters, or combination of raters, a separate database was constructed, resulting in a total of $2^6 - 1 = 63$ distinct databases (a, b, c, d, e, f, ab, ac, ..., abcdef). Each database was combined with our retrieval software to form 63 different mugshot retrieval systems, varying in the number of raters/mugshot ($\underline{r} = 1$ to 6). The "correct" database feature description of each suspect was the mean of the $\underline{r}$ raters in that particular database. These systems were then used to test the effect of number of database raters/mugshot on witness success in identifying suspects. If Ellis is correct, system performance should be poor for single-rater systems and should improve as the number of raters in the system increases.

Independent-groups and repeated-measures designs are the conventional methods for testing differences in performance among these 63 groups. Testing 63 independent groups would require the testing of many subjects and would, therefore, be prohibitively time consuming and expensive to conduct. On the other hand, conventional repeated-measures designs are precluded because they require the same target suspect to be examined by subject witnesses 63 consecutive times. Carry-over and practice effects confound the interpretation of this experimental design.

For use in this situation, we propose an alternative which we refer to as synthetic designs. They are a form of repeated-measures design. In conventional repeated-measures designs, all k treatments are administered to each subject. In contrast, in synthetic designs, only a single experimental treatment is administered empirically. In effect, testing of subject witnesses and testing of system performance are separated. Subject witnesses describe suspects by rating them on 107 5-point feature scales using printed forms. These same witness feature descriptions (i.e., a vector of 107 1's, 2's, ..., 5's) are then entered separately into each of the 63 suspect-identification systems (just as though the witness had entered them), and system performance (i.e., retrieval rank) determined for each. Our system responds with the rank order of the target suspect in the sequence which the witness would ordinarily examine the mugshots. (In this experiment, for security reasons, database mugshots are not presented to subject witnesses for examination.) Since witnesses describe all features before the system operates on them, witness feature descriptions are independent of the identification system used.

In effect, synthetic designs are a variation of the conventional repeated-measures design in which each subject serves in all conditions but is only tested once. Since subjects serve in all treatment conditions, and are thus necessarily correlated, the appropriate analysis is by repeated-measures analysis of variance. Since a subject's behaviour, in the present context a subject witness's feature description of a suspect, does not change from condition to condition, error variation is reduced. Synthetic designs offer several advantages over conventional designs: fewer subjects required, lower costs, and a complete absence of asymmetric transfer, range, carry-over, and practice effects which plague use of conventional repeated-measures designs.

The empirical testing of subjects is a cornerstone of system development and testing. In practice, however, the high financial and time costs incurred by empirical testing limit its use. Instead, system developers often rely on general guideline principles with minimal empirical testing. Synthetic designs, which significantly reduce the costs of empirical testing, offer an appealing alternative in some situations.

Results

The effect of delay on retrieval rank was not significant, $F(2,24) < 1$. However, the effect of number of raters, suspect, and raters by suspect were all significant, $F(4,96) = 50.78$, $F(3,24) = 11.53$, $F(12,96) = 12.62$, p's $< .001$, accounting for 27%, 19%, and 20% of total variation. All other effects were non-significant.

Table 1.
Effects of suspects and number of police raters/suspect on feature system performance (measured by mean retrieval rank of suspects)

No. raters per suspect	Retrieval rank	Suspect			
		A	B	C	D
1	78.4	22.0	46.2	47.9	197.4
2	20.1	2.4	10.1	10.5	57.1
3	10.3	1.3	5.1	5.1	29.8
4	7.2	1.2	3.6	3.4	20.6
5	5.4	1.1	3.0	2.2	15.4

    A Newman-Keuls multiple-comparison test of the number of raters effect
indicated a second rater/suspect significantly reduces retrieval rank. Addition of
further raters/suspect did not improve system performance further (see Table 1 --
means only). However, Table 1 shows that the effect of number of raters varies with
suspect. Performance is highly variable from suspect to suspect for single-rater
systems. These differences among suspects are eliminated, however, for two or more
raters/suspect. Simple main effects analysis confirms this interpretation with
significant differences among suspects only for single-rater systems, $F(1,120) =$
52.78, $p < .01$.

Discussion

    If computerized feature systems are to replace mugshot albums, they must
work with live suspects after realistic delays while requiring only a couple of
raters/suspect. In this study, subject witnesses searched a database of 1000 mugshots
for one of four different suspects, observed live, after delays of 1, 2, or 3 days, while
the number of police raters/database mugshot was systematically varied from one to
five. There was no effect of delay on system performance (as measured by the
number of photos examined by witnesses to find target suspects). There was,
however, a significant effect of number of police raters/mugshot. A second police
rater/mugshot significantly reduced the number of photos examined from an average
of 78.4 for single-rater systems to 20.1. More police raters/mugshot did not affect
performance further. Significantly, our feature system performed well with target
suspects seen live.

    Performance with live suspects was comparable to that in previous
experiments using photos as target suspects (Lee and Whalen 1995a,b). In those tests,
retrieval rank of target suspects averaged between 4 and 55 for single-rater systems,
depending on the experimental setting. Addition of a second rater/suspect consistently
improved system performance with retrieval rank averaging between 2 and 12.
Additional raters/suspect did not improve performance.

    Delay did not affect system performance. Many people find this surprising
because they expect memory for faces to deteriorate rapidly with time. The literature
on facial memory contradicts this intuitive theory (e.g., Laughery and Wogalter
1989).

    Many factors limit the generalizability of these results. Suspects did not act
out realistic crime scenarios, but simply appeared briefly and turned around 360
degrees. Factors such as realism, suspect movement, witness arousal, and lighting
conditions might affect system performance. However, system performance has been
good in preliminary experiments in our lab using realistic scenarios.

    Notwithstanding these limitations, the present results indicate that feature
systems are a promising alternative to the traditional album approach. Feature systems
work with live suspects, not just photos, and require only two police raters/suspect.
Most importantly, short delays of a couple of days have little effect on system
performance.

References

Chance, J.E., Goldstein, A.G., and McBride, L. 1975, Differential experience and
recognition memory for faces, *Journal of Social Psychology*, **97**, 243-253.

Davies, G.M., Ellis, H.D., and Shepherd, J.W. 1978, Face recognition accuracy as a function of of mode of representation, *Journal of Applied Psychology*, **63**, 180-187.

Deffenbacher, K., Carr, T.H. and Leu, J.R. 1981, Memory for words, pictures and faces: Retroactive interference, forgetting and reminiscence, *Journal of Experimental Psychology: Human Learning and Memory*, **7**, 299-305.

Ellis, H.D., Shepherd, J.W., Shepherd, J., Klin, R.H. and Davies, G.M. 1989, Identification from a computer-driven retrieval system compared with a traditional mug-shot album: A new tool for police investigations, *Ergonomics*, **32**, 167-177.

Goldstein, A.J., Harmon, L.D. and Lesk, A. 1971, Identification of human faces, *Proceedings of the IEEE*, **59**, 748-760.

Harmon, L.D. 1973, The recognition of faces, *Scientific American*, **229**, 70-82.

Krouse, F.L. 1981, Effects of pose, pose change, and decay on face recognition performance, *Journal of Applied Psychology*, **66**, 651-654.

Laughery, K.R. 1972, Photograph type and cross-racial factors in facial identification. In A. Zavala and J.J. Paley (eds.), *Personal appearance identification* (Charles Thomas, Illinois).

Laughery, K.R., Alexander, J.F. and Lane, A.B. 1971, Recognition of human faces: Effects of target exposure time, target position, pose position, and type of photograph, *Journal of Applied Psychology*, **51**, 477-483.

Laughery, K.R., Fessler, P.K., Lenorovitz, D.R. and Yorlick, D.A. 1974, Time delay and similarity effects in facial recognition, *Journal of Applied Psychology*, **59**, 490-496.

Laughery, K.R., Rhodes, B.T., and Batten, G.W., 1981, Computer-guided recognition and retrieval of facial images. In G. Davies, H. Ellis, and J. Shepherd (Eds.), *Perceiving and Remembering Faces*. (Academic Press, New York).

Laughery, K.R. and Wogalter, M.S. 1989, Forensic applications of facial memory research. In A.W. Young and H.D. Ellis (Eds.), *Handbook of Research on Face Processing* (Elsevier, North-Holland).

Lee, E.S. and Whalen, T. 1993a, Feature approaches to suspect identification: The effect of multiple raters on system performance, *Faces '93* (International Conference on Face Recognition), University of Wales, College of Cardiff, Cardiff, Wales, United Kingdom.

Lee, E.S. and Whalen, T. 1993b, Computer image retrieval by features: Suspect identification, *Proceedings of the 1993 Conference on Human Factors in Computing Systems INTERCHI'93* at Amsterdam, The Netherlands, (ACM, New York) 494-499.

Lee, E.S. and Whalen, T. 1995a, Computerized feature retrieval of images: Suspect identification, *Ergonomics*, in press.

Lee, E.S. and Whalen, T. 1995b, Feature approaches to suspect identification: The effect of multiple raters on system performance, *Ergonomics*, (accepted).

Lenorovitz, D.R. and Laughery, K.R. 1984, A witness-computer interactive system for searching mug files. In G.Wells and E. Loftus, *Eyewitness testimony*. (Cambridge University Press, New York).

Lindsay, R.C.L., Nosworthy, G.J., Martin, R., and Martynuck, C. 1994, Using mug shots to find suspects, *Journal of Applied Psychology*, **79**, 121-130.

Shepherd, J.W. and Ellis, H.D. 1973, The effect of attractiveness on recognition memory for faces, *American Journal of Psychology*, **86**, 627-633.

Wogalter, M.S. and Laughery, K.R. 1987, Face recognition: Effects of study to test maintenance and change of photographic mode and pose, *Applied Cognitive Psychology*, **1**, 241-253.

COULD HIGH HEEL COUNTERS IN RUNNING SHOES BE INJURIOUS TO SOME RUNNERS' ACHILLES TENDONS

Tay Wilson

Psychology Department
Laurentian University
Ramsey Lake Road
Sudbury, Ontario, Canada
P3E 2C6
tel (705) 675-1151
fax (705) 675-4823

An examination of heel counter height in a set of large (size 12-13) running shoes spanning 25 years and a set of small running shoes (size 8 female or 6 male) spanning 20 years revealed, for the large running shoes, a significant linear relationship with heel counter height increasing over time; but none for the smaller running shoes. The results are related to possible achilles injuries in big-footed runners, particularly those with a "sneaky pink panther" gait involving a low stride with the back foot staying on the ground as long as possible and a final vigourous drive off the back toe which torques the foot backwards and drives the achilles tendon against the heel counter.

In the last couple of decades there has been a considerable interest in the study of the foot, sponsored primarily by the shoe industry, in general, and the sports shoe industry, in particular. For instance, Falcao and D'Angelo (1992), in an attempt to aid the Brazilian shoe industry, have established the anthropometry of the feet of the Brazilian population as different from that represented by standard European data. Moreover, Hawes and Sovak (1994), sponsored by the sport shoe industry have developed a comprehensive set of variables describing the three dimensional nature of the foot along with normative data. Finally, Hawes, Heinemeyer, Sovak, and Tory (1994) have developed an approach to averaging digitized outline shapes of feet with the aim of assessing comfort of fit of shoes.

Buy any running magazine. Attractive colour pictures of shoes for sale appear on every second page. The dominant view is side-on giving predominance to the manufactures trademark logo and providing a visually pleasing silhouette topped at the back by a dramatically upsweeping heel counter. Cover up, with your fingers, the top of the heel counter in the picture. Without the high rising heel counter, the shoe would not look "quite right". It is well and good to have pretty pictures; but two questions are posed in this paper. First, are these heel counters, in large part, style driven? Second, could they be injurious to the achilles tendons of some types of runners?

Ron Hill, the great British distance runner, with twenty plus years of 160 km/week running, many championships, a world record 2:12 in the marathon is credited with several innovations in running gear. He is most famous for his invention of the string vest allowing maximum heat venting; but he is also credited with the invention of the "negative" or "dished down" heel counter, in the mid 1970's. This latter innovation was in response to the complaints of many runners that they were getting achilles tendon injury from the constant banging of the top of their heel counters on the tendon. Examination of current running shoes indicates that the design of the dished down heel counter appears to have been adopted by large numbers of shoe manufacturers. But has it really? What is needed is an examination of the height of the heel counters over time to see if the over-all height at the critical achilles tendon area has dropped.

Method

A set of 38 pairs of running shoes, size 12 and 13, purchased and used over the last 25 years by the author was measured for length of sole and height of heel counter. The length of sole was measured on the outside from the heel to the farthest forward point. The measurement did not include any wrap-up distance around the toe. The measurement of the heel counter was made inside the shoe from the lowest sink spot of the heel part of the sole to the top of the centre part of the heel counter which would be opposite the achilles tendon when the shoe was worn. This procedure was repeated for a similar set of 18 pairs of running shoes, size 6 male or 8 female, purchased and worn by a female runner over twenty years. Finally the same data was collected, at a local shoe store, for 6 pairs of large shoes and 6 pairs of small shoes from the 1994 stock. These were haphazardly selected by the store manager.

Results

In table 2 can be found, for 18 pairs of small running shoes, the year of manufacture, the shoe brand, size, sole length and heel counter height at the achilles tendon site. In table 2 can be found the similar data for 38 pairs of large running shoes.

For the large shoes (size 12-13) the correlations between the approximate year the shoe came out and the heel counter height was 0.65, the correlation between year and shoe length was 0.57 and the correlation between length and heel counter height was 0.63. The R^2 values (proportion of variation of one variable predictable from knowing values of the other variable) were 0.42, 0.32, and .40 respectively.

Linear regression was used to predict heel counter height from shoe year the least squares best fitting equation was counter height = 1.29 + 0.07 shoe year. The standard error of dependent variable estimate was 0.61 cm, small considering the average 7.9 for counter height. The standard error of the coefficient of shoe year (slope of the function) was 0.014 yielding a significant coefficient for shoe year (t = 5.43, df = 38, p < 0.01). A similar linear regression predicting counter height from sole length yielded a best fitting equation of counter height = -9.79 + 0.54 length.

The standard error of dependent variable estimate was 0.63 cm, again small considering the average 7.9 for counter height. The standard error of the coefficient of shoe year (slope of the function) was 0.11 yielding a significant coefficient for shoe length (t = 4.99, df = 38, p < 0.01).

For the small shoes (predominantly size 8 female) the correlations between the approximate year the shoe came out and the heel counter height was a minuscule -0.06, the correlation between year and shoe length was 0.31 and the correlation between length and heel counter height was a tiny 0.04. The R^2 values (proportion of variation of one variable predictable from knowing values of the other variable) were 0.004, 0.10, and .002 respectively. Linear regression was used to predict heel counter height from shoe year the least squares best fitting equation was counter height = 7.38 - 0.055 shoe year. The standard error of dependent variable estimate was 0.65 cm. The standard error of the coefficient of shoe year was 0.023 yielding a non-significant coefficient for shoe year (t = 0.25, df = 16, n.s.). A similar linear regression predicting counter height from sole length yielded a best fitting equation of counter height = -3.03 + 0.36 length. The standard error of dependent variable estimate was 0.61 cm, again small considering the average 6.9 for counter height. The standard error of the coefficient of shoe year was 0.27 yielding a non-significant coefficient for shoe length (t = 1.33, df = 16, n.s.).

Table 1. Year, brand, size, sole ln (cm) and heel counter ht (cm) for 18 prs of small shoes				
YR	BRAND	SIZE	LNG	COUNT
74	BARRET	8	26.6	7.1
78	NB BLUE	8	27.2	7.2
79	NKE YELL	8	27.8	7.7
79	NIKE TERRA	8.5	29	7.3
81	RBK GL160	6	27.5	6.7
83	NKE EQUAT	8.5	28.2	6.9
85	BKS TEMPO	6	27.7	5.8
85	RBK	8	27.7	6.4
87	BRKS SUPER	8	27.6	5.7
89	TURNTEK	8	28.3	7.5
90	NKE AIR	8	27.2	6.9
94	SAUC J500	8	27.6	6.4
94	NKE HUAR	8	28.2	7.1
94	ASC GEL122	8	27.8	7.3
94	RBK CLASS	8	27	6.1
94	SAUC J500	8	27.3	6.9
94	NKE ICAR	8	27.3	7.1
94	ASC LEGAR	8	28.1	8.1
M			27.67	6.9
S			0.56	0.63

YR	BRAND	SIZE	LNG	COUNT
	Table 2. Year, brand, size, sole ln (cm) and heel counter ht (cm) for 38 prs of large shoes			
70	ADIDA SL72	12.5	30.5	6.8
73	TIGR LEATH	12.5	31.5	6.5
73	TIGR LEATH	13	32	6.8
74	TIGERCUB	12	30.5	6.3
74	RBK STUD	12	30.5	7.1
77	NKE Y-WAF	13	32.5	8
82	BRKS GREY	13	32.2	6.3
82	TIGER	13	33	7.6
82	BRK SILVER	13	32.7	7.9
83	ASC LO TOE	13	32.8	7.9
83	ASC LO TOE	13	33	7.7
83	ASC LO TOE	13	32.8	7.7
83	BRKS CHAR	13	32.1	8.1
84	BRK SUPER	13	32.4	6.3
85	NB 557	13	34	7.7
85	NB 470	13	33.6	7.8
86	NKE RDRAC	13	33.2	8.7
86	RBK W-RD	13	33.7	7.8
86	NB 575	13	33	7.8
86	RBK W-TRN	13	33	7.8
88	TIGER GEL	13	33.6	7.5
90	ASC GELGN	13	32.5	8.8
91	NB 830	13	33.4	8.5
91	ASC GEL101	13	33.7	9
91	ADID ZX50	13	33.2	8.8
91	NKE AIR	13	33.3	8.5
92	NB 629	13	33.2	8.7
92	NB 91	13	33	8.4
92	NB 860	13	34.3	9.6
92	NB 860	13	33.7	7.9
92	BRK FLASH	13	33.2	8
93	BRK FLASH	13	32.8	8
93	NKE AALPHA	13	31.5	6.4
94	ASC SAGA	12	32.3	8.3
94	RBK	12	31.7	8.2
94	NKE PEG	13	33.5	8.6
94	RBK CLAS	12	31.7	8.3
94	NKE ICAR	13	33.9	8.3
94	NKE WINRN	13	33.2	8.4
94	NKE AIR	13	33.3	8.4
M			**32.74**	**7.87**
STD			**0.94**	**0.81**

Discussion

As can be seen from the results sections there is a clear linear relationship between shoe year and heel counter height accounting for 42% of the variation in the latter variable for the large shoes but no such linear relationship between these same variables for the smaller shoes. In order to assess some implications of these findings, it is worth noting that, over two decades, the female runner in question only had achilles discomfort with two pairs of shoes; in only one of these pairs was she required to trim the heel counter support down. However, the male runner, in question, had trouble with most of the shoes purchased in the eighties or later. For nearly all of these shoes drastic trimming of the top of the heel counter was necessary. In rough terms a heel counter of about 7.8 cm was as high as could be tolerated. With this information, it can be postulated that the runners most likely to be at risk due to high heel counters are those with large feet.

One further reason that large footed runners might experience tendon problems with high heel counters is that because of their relatively larger foot, they may torque their toes down and hence their heels up to the heel counter with relatively more force than a small footed runner who might even be more inclined to maintain their foot relatively un-torqued (rotated) in the drive-off and consequent rest phase.

There is one more variable that is perhaps of interest for further study of heel counters. That is, among older runners and some younger runners, there is a tendency to run in sort of an exaggerated walk or what I call the sneaky "Pink Panther" style; that is, by keeping the back leg on the ground as long as possible. This gait is very economical of energy particularly if a large foot is at the end of a long leg. The gait is further "sneaky" in that when watched from behind, it appears that the runner is not moving very fast and can be caught. Perhaps the most successful younger runner exhibiting the "sneaky pink panther" gait is Alberto Salazar who has won the New York marathon and this year won the difficult and prestigious Comrades marathon (53.75 miles or 86.55 km) from Durban to Pietermaritzburg in South Africa.

With this type of gait, there appears to be some additional stress on the calf muscle and associated tendons and any touching of parts of the heel counter might induce injury. Moreover, the runner, despite the low leg lift, tends to rock right over the toe of the back foot and drive off strongly from this toe. The consequent of this drive is a strong torque of the foot push the toe down and backwards and hence driving the achilles tendon hard against any heel counter that is too high.

References

Falcao, D.a and D'Angelo, M. 1992 Anthropometric Measurement of Brazilian Feet. In Contemporary Ergonomics, Praeger, London, pp 167-172.
Hawes, M. R. and Sovak, D. 1994. Quantitative Morphology of the human foot in a North American Population., Ergonomics vol. 37, no. 7, 1213-1226.
Hawes, M. R., Heinemeyer, Sovak D.and Tory, B. 1994. An approach to digitized Plantagram Curves, Ergonomics vol. 37, no. 7, 1227-30.

New Technology

The role of tacit skills
in jewellery manufacture

C. Baber and M. Saini

Industrial Ergonomics Group,
School of Manufacturing & Mechanical Engineering,
University of Birmingham,
Birmingham.
B15 2TT

**In this paper, the initial stages of research into
tacit skill in jewellery manufacture are described.
The aim of the work is to consider the interaction
between different types of technology and the
skills involved in the manufacture of jewellery. It
is proposed that visualisation and interpretation of
changing environmental feedback are essential to
the development and use of tacit skills in this
context, and that different forms of technology
interfere with different elements of the tacit skills.**

Introduction

New technology is becoming increasingly important in the manufacture of
jewellery; so much so that the British Jewellery Association has declared 1995
'new technology year'. Jewellery manufacturers across Europe are investing
heavily in processes and technologies which can increase product throughput and
reduce lead-times. However, for many people, jewellery manufacture still
maintains an aura of the archetypal craft industry. This raises the question of what
are the possible problems associated with using technology to either support or
replace craft skill ?

Jewellery manufacture is a scaled-down version of industrial manufacture.
Items of jewellery can be decomposed into discrete units, the units need to be cast
or otherwise produced and then assembled into a finished product. One might
feel that jewellery manufacture places additional emphasis on value-added aspects
of production, such as quality or hand finishing. However, not only is this
increasingly true of other spheres of manufacture, it is also not necessarily true of
all areas of jewellery manufacture (*vide*, the remarks of the ill-fated chair of major
UK jewellery manufacturer). One might feel that the reduction in scale of items in
the production process leads to a higher than average emphasis on manual skill,
and many aspects of jewellery production are still labour intensive. However, not
only have efforts been made to Taylorise a many of these tasks, in large-scale

manufacture, but also many of the tasks can, and have, been automated. In broad terms, our research interests lie in the interaction between automation and craft skill: can skill be extracted from craft workers and placed in technology, i.e., is it possible to fully automate craft skill, what are the consequences of partial automation ? It was decided to focus on jewellery manufacture for two reasons: there is a thriving jewellery quarter in Birmingham and jewellery manufacture is generally considered a craft.

Defining a role for 'New technology'

The introduction of new technology can be fraught with difficulty and applications may not fulfil their promise. One possible explanation of these problems is that too little consideration is given to the contribution that skilled manual workers make to the production process prior to the introduction of technology (Manwaring and Wood, 1985). Indeed, the introduction of new technology is often planned along strictly 'technocentric' lines, with the production process being decomposed into functional flow diagrams defined by quantitative data, such as operation times. While such an approach can produce objective descriptions of the process, and can be used to generate testable mathematical models to evaluate the possible effects of change, it loses sight of some other forms of information. By definition, the emphasis on 'hard' measurable time and motion data will not yield qualitative information such as the description of the intimate relationship between skilled workers and their tools and machines; a relationship which often adds a subtle degree of quality to the finished product (Littler, 1982).

This raises the question of how can we define the value-added benefits of manual skill in production. This is an issue of particular relevance in the manufacture of jewellery. Fully automated production has been possible for some time, although there seems to be consensus that such production is potentially limited in both scope and quality. There is a trade-off to be made between mass-production using total automation, and hand-made jewellery, and the determining factors must be quality and cost to the consumer.

It is proposed that very often jewellery manufacture relies on craft approaches to production, requiring flexibility and adaptability among its workers in order to manufacture short production runs of an enormous range of different products. In companies in which production is fully automated, manufacture tends to concentrate on specific items, e.g., rings. With the increasing concern over retention of market share, many UK jewellery manufacturers are investigating the potential benefits of technology.

We argue that, in this industrial context, it is important to understand the nature of craft skills involved in jewellery manufacture prior to the introduction of technology. However, as with many industries, it has been traditional to define skill in terms of social convention and job description (Scarborough and Corbett, 1992). If we combine the lack of precision of definition of the term 'skill' with rationalised, technocentric approaches to the introduction of technology, then it is highly probable that serious problems in quality and production will arise. Furthermore, there is an obvious concern that technology could lead to redundancy or deskilling.

Tacit Skill in Industrial Work

Assume that a manufacturing process has been designed to function as rationally as possible. Given a stable operating environment, there will be little or no need for human intervention. However, operating environments are rarely stable.

Consequently, there is a need to bridge the gap between the state of the operating environment and that of the ongoing process. It is the ability to monitor the state of the process and sense requisite changes which can be termed tacit skill. Myers and Davids (1993) suggest that tacit skill represents the difference between intellectual and experiential ways of knowing and acting. Tacit skill tends to be characterised by informal procedures and by a difficulty in articulating it's elements. Thus, while tacit skill represents an important dimension in many aspects of industrial work, the lack of precision excludes it from the rational, technocentric approaches to the design and introduction of new technology. Furthermore, as Hirschhorn and Mokray (1992) illustrate, the interaction between technology and tacit skill is often complex. For example, while a manually-operated lathe requires an operator to respond continuously to changes in machine state, a CNC machine requires the operator to predict and cater for future changes.

Manual Assembly Work Considered as Skilled
Research into the ergonomic aspects of manual assembly has been conducted since before the term ergonomics was coined. Studies in the munitions factories during WWI and subsequently in manufacturing industries led to a recognition that, rather than simply representing simplified, repetitive operations, assembly work could potentially exhibit characteristics of motor skills (Pear, 1924; Cox, 1934).

Research by Seymour (1967) showed that, in assembly work, grasp and position activities show more improvement than reach and move activities. Thus, one could propose that skilled assembly represents more efficient performance strategies, so that the movements within and between operations are smoother and less jerky for skilled than for nonskilled assemblers, that there will be less fumbling for skilled assemblers and simpler, consistent patterns of motion for fingers and thumbs. Even the simplest manual activity will require some form of mental activity, e.g. planning, initiation, control, termination, and checking (Crossman, 1964).

Baggett and Ehrenfeucht (1988) have shown how instructions based on typical conceptualisations of assembly tasks can facilitate performance; these were assumed to correspond to subassemblies of the finished product. The conceptualisations were assumed to be based on informational chunks, stored in subjects working memory. Baggett and Ehrenfeucht (1988) hypothesised "...that when a person is asked to build an object...the person conceptualises the object as a hierarchy of subassemblies and groups ...pieces according to these conceptual divisions." From this hypothesis, one might ask whether people develop and order the chunks prior to assembly, i.e., in the form of a physical model of the objects and a plan for assembly, or whether their chunking is more spontaneous.

Assembly requires the manipulation and joining of parts to form wholes. In order to achieve the apparently simple goal of placing a peg in a hole, a number of factors need to be considered, such as grasping the peg, determining the relative positions of peg and hole (aiming), moving the peg, inserting the peg and checking accuracy. All of these actions will have differing degrees of visual guidance, will require different levels of manual and finger dexterity, and call upon different levels of higher order cognitive activity, such as planning.

Studying Model-Making in Jewellery Manufacture
The manufacture of jewellery requires a number of processes, e.g., design ; model making; casting; stone-setting; polishing; electroplating. The processes

often interact with each, e.g., it is common for the model-makers to produce several versions of a design, and for the design to be modified as a result of the model. Similarly, it may be necessary for design and model making to take note the nature of the casting method to be used. In this study, attention was focused on the role of the model-maker.

The study employed verbal protocol to develop a description of the range of skills used, together with the planning and coordination of skilled work. Combining recent theories from skills literature with more traditional industrial ergonomics, we consider the role of tacit knowledge in the development and maintenance of skilled activity. On the basis of this analysis, we consider the pros and cons of automating craft skills.

The following extract is taken from the verbal protocol produced during the process shaping a shank for ring. The decomposition of the protocol into sections is used to illustrate the digression away from the main topic, i.e., elements of the task which need to be planned. The arrow indicates the topic flow, without digression.

"You are thinking about what you could be doing all the time

and also this shank is slightly more complicated
it could have had a simple shank and just come up
with what we call a spear point shank you joined on there

but this shank has got a wraparound
and divided or split around the collet,
so thats got to be split and wrapped
around the collet

you've got to have the right angle and finger size...

you know all the time you're sort of checking to see
that the angles are symmetric as it should be

once again it is still checking and checking
bearing in mind that the setter has got to let it
out a few thou more."

In this example, the respondent is describing a strategy by itemising problem stages. This strategy develops from a description of a simple task, i.e., produce a shank without split or curl, with more difficult elements added to it. In order to develop this strategy, it was necessary to visualise how previous approaches would need to be modified for the current problem. Indeed, in subsequent conversation, the model maker decided to sketch the various options in order to indicate their relative differences. In this instance, the jeweller appeared to work using a visual sketch, against which he compared subsequent development. The skilled craftsperson can be seen to develop a model or schema of the product based on visualisation of the product at specific stages during its production. These visualisation will be dependent upon previous experience of similar products, and are used to guide the planning of manual behaviour. This plan will then require further decomposition into several topics: begin with an intial visualisation of the product, compare current state of material with visualisation and select appropriate stragety to effect changes, choose tool to use, choose activity to perform using the chosen tool, chose appropriate fixture etc. For instance, in cutting a set of shapes from sheet-metal, another jeweller we interviewed was less concerned with saving space (as the remaining metal could be resold at face value) than with allowing one cut to finish with sufficient space and orientation to allow a subsequent cut to be easily performed. In this instance, the process required not only knowledge of the type of metal used, but also an ability to visualise the changing shape of the metal during cutting in order to plan appropriate cuts.

While there were several aspects of the work which involve visualisation and planning, there were other aspects of the work which seemed to relate to an interaction between experience and environmental cues. For instance, when heating a piece of metal, one of the jewellers we spoke to waited until it glowed 'cherry red', or, in another instance, he would use the change in sound from a machine as auditory feedback to indicate that a process had been completed. Thus, like many areas of skilled behaviour, we can begin to identify levels of control in the behaviour of skilled jewellers, such that an initial plan is decomposed into subplans and performance is continuously monitored using a range of feedback from the environment.

Interactions between technology and skill

Although there is interest in using technology in jewellery manufacture, there has been an almost continuous automation of some aspects of production. For example, while items could be polished by hand, they tend to be polished using a rotating wheel with a range of attachments, e.g., from hard stone to soft material. Let us further assume that polishing could be performed using a large drum, in which items are spun. In each case, the change from manual to increasing levels of automation in polishing alters the relationship between worker and product. While polishing by hand, the jeweller is free to modify the execution of polishing strokes and can easily manipulate the item. However, the process can be very time-consuming. The introduction of a rotating wheel could speed the process, while allowing the jeweller an opportunity to manipulate the item. However, the feedback received will change from a combination of kinesthetic and visual to visual (and possibly auditory or olfactory). This requires a modification of polishing behaviour to correspond to task requirements and to feedback from the environment. Finally, the use of a spinning drum requires someone to load and unload it, and provides no involvement in the polishing process. However, it does require items to be checked and, if necessary, touched-up after polishing. It strikes us that, while the first two approaches follow the current vogue for incorporating inspection into production, i.e., in which quality can be kept as part of the production process, the latter approach represents something of a retrograde step in which quality is inspected in after production.

To take another example, we have been considering model-making by hand. It is possible to produce models directly from computer-aided design (CAD) models, e.g., via stereolithography (Boon et al., 1992; Warnecke and Hueser, 1994). While this can radically shift the skill requirement from the use of hand-tools to the use of CAD, it retains the need to visualise the finished product. We are not certain of the interaction between visualisation and physical manipulation, but feel that CAD could remove this, possibly useful source of information. Alternatively, dies can be produced for stamping, using a spark-erosion machine. Sometimes the hard alloys needed for dies can not be fully shaped using spark erosion; this leads to the spark-eroder being used to define an initial shape which is finished by hand. Again the production process is altered, with manual skills being retained for the fine detail work, leaving coarse detail to be introduced via machine.

Conclusions

In many of the applications which we have considered, it seems to be vital to allow jewellers to retain an ability to interact with the item on which they are working in order to support the interaction between visualisation, planning and environmental feedback. Removing environmental feedback or the knowledge necessary to plan operations or the opportunity to compare item production with a visualisation is felt to have a negative impact on tacit skill. This, in turn, could have repercussions for product quality. However, we have also seen that some types of automation can require some adaptation of tacit skill and can lead to improvements in both performance and efficiency.

References

Baggett P.and Ehrenfeucht, A. (1988) Conceptualising in assembly tasks *Human Factors 30* 269 - 284

Boon, L., Ko, M., Gay, R., Leong, K. and Kai, C. (1992) Using computer-based tools and technology to improve jewellery design and manufacturing *International Journal of Computer Applications in Technology 5* 72 - 80

Crossman, E.R.F.W. (1964) Information processes in human skill *British Medical Bulletin 20* 32 -37

Hirschhorn, L. and Mokray, J. (1992) Automation and competency requirements in manufacturing: a case study In Ed. P. Adler *Technology and the Future of Work* Oxford: Oxford University Press

Littler, C. (1982) Taylorism, Fordism and job design In Ed. D. Knight, H. Wilmot and D. Collinson *Job Redesign: critical perspectives on the Labour process* Aldershot: Gower

Manwaring, T. and Wood, S. (1985) The ghost in the labour process In Ed. D. Knight, H. Wilmot and D. Collinson *Job Redesign: critical perspectives on the Labour process* Aldershot: Gower

Myers, C. and Davids, K. (1993) Tacit skill and performance at work *Applied Psychology: an international review 42* 117 - 137

Scarborough, H. and Corbett, J. (1992) *Technology and Organisation* London: Routledge

Seymour, W.D. (1967) *Industrial Skills* London: Pitman

Warnecke, H. and Hueser, M. (1994) Technologies of advanced manufacturing In Ed. W. Karwowski and G. Salvendy *Organisation and Management of Advanced Manufacturing* Chichester: Wiley

A THRESHOLD APPROACH TO COMPUTER-MEDIATED COMMUNICATION

Fraser Reid[*], **Linden Ball**[*,*], **Vlastimil Malinek**[†], **Clifford Stott**[†,††]
Andrew Morley[*] and **Jonathan Evans**[*]

[*]*Department of Psychology, University of Plymouth, Plymouth PL4 8AA*
[**]*Division of Psychology, University of Derby, Derby DE3 5GX*
[†]*MRC Applied Psychology Unit, 15 Chaucer Road, Cambridge, CB2 2EB*
[††]*Hewlett-Packard Laboratories, Filton Road, Bristol, BS12 6QZ*

We report on two experiments comparing synchronous, text-based computer messaging with face-to-face discussion for carrying out collaborative group tasks under high workload conditions. Substantial differences in the content of group discussions were found in both experiments. Computer mediation inhibited time critical messaging and promoted a normative style of group reasoning in which position statements and value judgements predominated over the exchange of hard facts. We argue that these results reflect the operation of a messaging threshold, in which the decision to send a computer message depends on the urgency and relevance of its content in relation to the costs associated with its effective communication.

There is now mounting evidence that computer-mediated communication (CMC) using conventional messaging systems (e.g. electronic mail, computer conferencing, etc.) has profound effects on group behaviour and decision making. At present, the single most widely held explanation for these effects is based on the claim that CMC systems filter out the social context cues present in normal face-to-face conversation (e.g. Kiesler & Sproull, 1992). The essence of the argument is that because communication is mainly by text, CMC messages are less able to convey the momentary nonverbal and paraverbal cues that regulate normal interaction and convey information about each person and their activities. In this paper, we focus on the central claim of the filter theory approach: that the absence of these cues shifts the group's attention to factual information and intellectual arguments (termed *informational influence*), and reduces the impact of conformity to group norms and preferences (*normative influence*).

Fairly good indirect evidence exists for this claim. Various experiments over the last ten years have shown that CMC leads to more equal participation among group members, and more extreme, unconventional, but less unanimous decisions, (Kiesler & Sproull, 1992). However, *direct* support for filter theory is much less compelling. Although CMC discussions are sometimes more formal and task oriented than face-to-face discussions, CMC is able to support expressively rich and playful interchanges, especially in well-established computer networks, and can enhance as well as suppress normative group behaviour (Spears & Lea, 1992).

The messaging threshold

The conclusion we draw from this is that CMC can support normative and informational influence equally effectively under certain conditions. In general, we can see no obvious reason why text-based CMC must *necessarily* be a poorer medium than speech for conveying social information. It may hinder expressively rich communication, but need not entirely prohibit it. We believe the key to understanding the psychological impact of CMC lies in the threshold costs to the sender of satisfactorily completing any conversational transaction, such as asking a question and obtaining an answer, making a proposal and having it accepted, and so on. It follows that whether or not a message is sent over a computer network depends more on the degree to which its urgency and relevance exceeds this messaging threshold, than one some fixed property of electronic media.

What are these transaction costs? We can suggest two at this stage. The first is simply the *effort* needed to process CMC messages. Sending a typewritten message is a particularly demanding cognitive task, and the resources needed to accomplish this task may at times be in short supply--especially if CMC is time-shared or alternated with other activities. A second kind of cost arises from the structural properties of ordinary conversation. When talking, speakers engage in sequences of rapid and closely coupled turns to establish that each has understood the other's utterances well enough for current purposes. This process is termed *grounding* (Clark & Brennan, 1993), and is thoroughly disrupted in CMC. It is costly to signal agreement, have a question answered, or repair a misunderstanding in computer-mediated conversations, and experienced CMC users adapt to this by composing longer, more carefully formulated messages, and by omitting certain kinds of conversational response altogether.

This suggested to us that the content of CMC messaging would turn out to be more sensitive to interactional and task conditions than suggested by filter theory. Under certain conditions, the processing and grounding costs of normative messaging might become prohibitive--for example, in dispersed groups working simultaneously on a joint report against a tight deadline. But when time constraints are absent, or when people use CMC for other reasons, normative messaging might become more prevalent.

Experiment 1: Collaborative report writing

We set out to test our threshold hypothesis by devising social and task conditions that would *reverse* the pattern of messaging predicted by filter theory. We reasoned that by increasing the cognitive load on subjects, only the most intense or persistent messages, normative or informational, would cross the messaging threshold. If at the same time social conditions favouring normative messaging were to be enhanced, an overall increase in normative relative to informational communication might result. To operationalise the first of these conditions a complex, realistic task with a significant cognitive workload was used. Undergraduate students, working in teams of four, were trained to use networked computers to prepare a sequence of collaborative reports over several extended work sessions. Students in eight of these teams (the *CMC teams*) were located in separate offices, and communicated with each other solely through by synchronous text-based messaging over the computer network. Students in six other teams (the *FTF teams*) worked alongside each other in a single office, able to see and speak to each other freely as they worked on their networked computers. FTF and CMC teams therefore differed solely in the communication bandwidth available to them. All FTF team interactions were video-taped.

The second set of conditions set out to increase the relevance and urgency given to normative communication by FTF and CMC teams. Recent experiments have shown that CMC can *strengthen* normative communication, provided team members identify with a shared group identity (Spears & Lea, 1992). In our experiment, we fostered group identification by having students form their own groups, devise team names and private passwords, team incentives, and by training them over several weeks to adopt a teamwork approach to the task.

Team interaction profiles

We used interaction process analysis (IPA) to detect the effects of these conditions on the normative and informational content of group interaction. The IPA system of categories (Bales, 1950) was adapted to the requirements of the present task, and was used to code the interactional content of all computer messages and the video recordings of face-to-face team interactions. A battery of measures was used to assess the performance of the teams and the quality of written reports they produced.

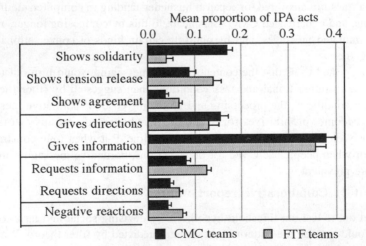

Figure 1. Team interaction profiles

To our surprise, we could find no effect of computer-mediation on the quality of the written reports: the only reliable difference was for CMC teams to take longer to produce them. However, very substantial differences between FTF and CMC teams were found in their *interaction profiles*. That is, teams differed in the patterning of their communication activity, irrespective of differences in their interaction base rates (see Figure 1: error bars depict standard errors of means). To begin with, tension release, agreement, and negative social-emotional reactions were all significantly rarer in CMC messaging. This result is predicted equally well by the filter and threshold hypotheses: each of these reactions is normally conveyed by nonverbal cues that are difficult to incorporate into a typewritten message. But two other results were far less consistent with filter theory. We found that CMC messages contained proportionally fewer task related requests, whilst gestures of solidarity-- including displays of camaraderie, affection, cooperation--were four times more prevalent in CMC messaging.

Taken together these results support the threshold hypothesis. Face-to-face, it doesn't cost much to produce an utterance, check that it has been understood, and wait for a reply: on a computer keyboard, it costs much more. As a result, *time critical messages*--those that have time-limited relevance, such as momentary emotional reactions, some kinds of urgent task request, brief acknowledgements, etc.--are rarer in CMC messaging. On the other hand, when group conditions heighten the relevance of emotional communication, expressive reactions may be important enough to warrant the effort of typing them out as a message. The emphasis by CMC teams on gestures of solidarity is clear evidence for CMC's capacity for normative messaging under certain conditions.

Experiment 2: Group decision making

Encouraged by these results, we focused attention in a second experiment on group decision making. This is a task in which normative and informational influence normally operate simultaneously, but to varying degrees depending on the decision problem. Some problems are factual in nature and can, in principle, be decided on the basis of an intellectual assessment of the evidence. Others are judgemental problems with no straightforwardly correct solution, and involve value-laden social, ethical, and aesthetic judgements (Kaplan, 1987). By using a *balanced* decision problem--one that allows both factual and judgemental argument--decision making groups would be free to adopt whichever style of reasoning they chose. We therefore constructed a decision problem, based on an authentic enquiry into a case of alleged child abuse, that was rich in factual and value arguments, and incorporated these into a set of hypertext case files installed on our network of computers.

Undergraduate students, again in groups of four, were first trained to use the networked computer system. This system gave each person unrestricted access to the case file assigned to them (but to no other case file), and joint access to a shared reference file. Each group sat as a panel of enquiry for a single three hour session, and decided what action to take in the case. However, decisions were not easily reached: case files were set up so that factual information was unevenly distributed among team members, and biased them towards contradictory views. As in our first experiment, ten *CMC panels* were located in separate offices, and communicated with each other solely by synchronous text-based messaging. An equal number of *FTF panels* worked alongside each on their networked computers. Again, all FTF panel interactions were videotaped. All panels completed a questionnaire assessing various aspects of the discussion process at the end of the experiment.

Styles of group reasoning

A category system devised by Martin Kaplan was applied to the argument content of all computer messages and to the video recordings of FTF panel discussions (Kaplan, 1987). This system allowed us to assess group reasoning styles by classifying discussion units into *informational arguments* (citations of facts from case files, and inferences from these facts), *normative arguments* (value assertions and statements of decision preferences), and *procedural arguments* (references to group procedures and computer materials).

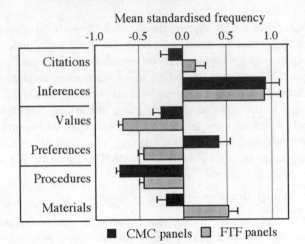

Figure 2. Group reasoning styles

We did not expect, nor did we find, a difference in panel decisions between experimental conditions, although CMC panels were predictably less able to reach unanimity than their face-to-face counterparts, and were less satisfied with the decisions they did reach. However, very substantial differences in the style of reasoning adopted by CMC and FTF panels did emerge (see Figure 2). First, we adjusted for differences in interaction base rates among panels by converting raw frequency counts in each coding category to standardised scores. What we found was that CMC discussions contained proportionally fewer citations and procedural arguments, but many more normatively-oriented value judgements and statements of decision preference. Only one argument category--inferences--showed no difference between experimental conditions.

What is clear is that CMC panels showed a marked preference for a *normative style of reasoning*, involving frequent position statements, appeals to values rather than facts, and a neglect of procedural matters. The result is that decisions were reached more often by vote than persuasion, and this was reflected in low levels of unanimity and decision satisfaction among CMC panels. This pattern strongly resembles the verdict-driven style of jury deliberation observed by Reid Hastie (Hastie, Penrod and Pennington, 1983).

Whilst clearly at odds with filter theory, this result is quite compatible with our threshold hypothesis. The transaction costs of typing computer messages accurately citing facts drawn from a case file will be enough to discourage CMC panels from focusing on factual information alone. However, they did not entirely abandon an informational style of reasoning. By summarising several case file entries in the form of abbreviated inferences and generalised conclusions, CMC panels continued to provide evidential support for their decision proposals. But what most distinguished CMC discussions were the frequent, low cost position statements, and the use of simple assertions of what is right or wrong, good or bad, natural or unnatural--a form of *ad hominem* argument frequently used in ordinary conversation when evidence is skimpy (Antaki, 1985). Although hard facts are available to CMC panels, the cost of reproducing them in the form of typewritten messages appeared to encourage this form of argumentation.

Reducing transaction costs

We believe these experiments illustrate the value of an approach to CMC based on the messaging threshold principle. Our results have implications for the design of CMC and other collaboration support systems, and stress the importance of minimising the transaction costs of communications media. Text-based media incur particularly high costs, and the advent of computerised voice mail and desktop video-conferencing may reduce these costs. However, we would argue that speech is richer than text only when the speaker has limited resources available to prepare a message. Furthermore, we have shown that the medium of communication is an important determinant of transaction costs, but alone it is not decisive. What is crucial is the capacity of the system to support complete conversational transactions, and it is against this criterion that new communications technologies must be judged.

References

Antaki, C. 1985, Ordinary explanation in conversation: Causal structures and their defence, European Journal of Social Psychology, 15, 213-230.

Bales, R. F. 1950, *Interaction Process Analysis: A Method for the Study of Small Groups*, (Addison-Wesley, Cambridge).

Clark, H. H. and Brennan, S. E. 1993, Grounding in communication. In L. B. Resnick, J. M. Levine and S. D. Teasley (eds.), *Perspectives on Socially Shared Cognition*, (American Psychological Association, Washington), 127-149.

Hastie, R., Penrod, S. D. and Pennington, N. 1983, *Inside the Jury*, (Harvard University Press, London).

Kaplan, M. F. 1987, The influencing process in group decision making. In C. Hendrick (ed.), *Review of Personality and Social Psychology, Volume 8; Group Processes*, (Sage, Beverly Hills) 189-212.

Kiesler, S. and Sproull, L. 1992, Group decision making and communication technology, *Organizational Behaviour and Human Decision Processes*, **52**, 96-123.

Spears, R. and Lea, M. 1992, Social influence and the influence of the 'social' in computer-mediated communication. In M. Lea (ed.), *Contexts of Computer-Mediated Communication*, (Harvester Wheatsheaf, London), 30-65.

COMPUTERS IN THE FIELD: NEW TECHNOLOGY AND THE VEGETABLE GROWER

Caroline Parker

HUSAT Research Institute
The Elms, Elms Grove
Loughborough, Leics LE11 1RG

This paper discusses the work of a MAFF funded project which is carrying out research into methods to effect faster technology transfer in horticulture by the use of computer based simulation models. There are three main user groups for the proposed system: biologists, consultants and growers. To date the project has completed the initial user requirements survey of the latter two groups and this paper presents the results of this work. Unlike many other industries, horticulture is just beginning to make use of computer technology. In addition, operations on the 'shop floor' are almost entirely controlled by that most unpredictable of variables, the weather. In spite of these factors research suggests that computer-based models would be of use to the industry and, in particular, to consultants and co-operative managers.

Introduction

Horticulture is very big business. Agriculture and horticulture together are the UKs second largest primary industry after North Sea oil. Largely because of this the industry has, since World War II, been supported by many government funded research centres. These institutes were established to provide basic research into food production and their efforts have doubled yields from most crops in the last 50 years.

Until recently the basic scientific knowledge generated by these centres was evaluated, tested and collated to provide free information to the industry by the government-funded Agricultural Development and Advisory Services (ADAS). However, in recent years the government has reduced its support, ADAS has been privatised and the research centres pressurised to provide much of their funding from industry. As a result, information and advice is now a marketable commodity and the three main user groups, biologists, consultants and growers, are constrained by the financial requirements of asking for, or providing, knowledge. This has resulted in information bottlenecks and a breakdown in the process of technology transfer.

This paper describes the work of the GRIME, 'GRaphical Integrated Modelling Environment', project which was established to try to improve one aspect of the information flow problem. The specific aim of the GRIME project is to research the linking of disparate biological models and to investigate methods which would allow these 'simulation' models to be disseminated for rapid use by the industry. One example is 'WELLN' (Greenwood, 1994), a piece of software which models the uptake of nitrogen by crops. It combines this information with knowledge about rainfall, soil type and crop requirements to predict the amount of fertiliser a grower needs to apply, thereby maximising yield and minimising water pollution.

GRIME is intended to be a complete system which will support the information flow from the biologist to the grower. Its design is illustrated in Figure 1 below.

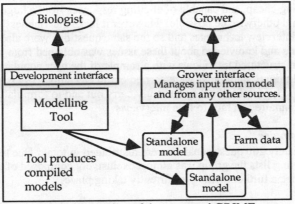

Figure 1: An illustration of the proposed GRIME system

There are two parts to the process: the development of models and their delivery to the grower. HUSAT's role in the project is to identify user requirements for both parts and to provide functional and interface specifications. Work to date has focused on the needs of the growers and this paper will concentrate on this aspect of the findings so far.

Method

User centred design philosophy suggests that certain knowledge about the domain is collected prior to the development of a system, i.e. who exactly will use the system, what will they use it for, what is the work context and environment in which it will be used and what is technically and logistically feasible (DTI, 1990). Based on this, the following were the areas identified to be addressed. Some indications of current thinking are also provided:

Is there a need for prediction/forecasting models in the grower/advisor population, or are the existing sources of information adequate? While there was some evidence for the need for simulation models no objective investigation had been conducted into its strength and user profile.

Do the technological and informational components required for prediction/forecasting models exist, and if not, is it likely these will become available in the near future? Earlier studies had suggested that the industry was not ready for computers. There was also concern that many simulation models needed detailed meteorological information which potential users would be unable to provide.

Where will prediction/forecasting models be of most use? Given the need to win user support at an early stage the project needed to discover the pests, diseases and nutrient requirements users considered most important.

What constraints do the users operate within? While some general information on the nature of the industry is published, much is locked behind commercial doors. The project had to identify the goals of the user groups, relevant external influences such as markets, and the pressures they exert. It also had to identify changes foreseen or in progress to judge their impact on the potential system.

What type of interface and functional requirements should be taken into consideration when building the system? The project needed to identify the particular user requirements for the system ie skill levels, use of language, units of measurement and existing methods.

The first three questions largely require yes/no answers, and it was felt they were capable of being incorporated into a postal questionnaire. This method has the advantage of being cheap, and capable of eliciting responses from a larger group of people than would otherwise be possible. However it lacks the 'richness' of data associated with interview techniques and so the same questions were also asked of a sample of growers and knowledge about these issues was obtained from background reading and from structured interviews with a sample of the user population.

Three preliminary interviews were carried out in December 1993, to identify the range of users, the types of tasks they currently carry out and to formulate questions for the postal questionnaire and face to face interviews.

Postal Questionnaire

Approximately 200 questionnaires were distributed in total. Due to the difficulty in obtaining mailing lists the group was almost exclusively composed of growers. To redress the balance a further mailing is currently taking place.

Interviews

Fifteen face to face interviews have been conducted to date, four with advisors, the rest growers. All sizes of company and a range of user 'roles' are represented.

Postal Questionnaire results

Over 40% of those contacted returned the questionnaire. Two thirds of these people own a computer, a large percentage of which are IBM or IBM clones capable of running Windows. They were generally used for accounting, word processing, farm/field records, stock records, management and presentation materials. Around a quarter also use their computers for staff/labour planning and general crop scheduling.

Almost three-quarters of those responding said they used the software themselves. Only 25% said the computer was used solely by others e.g. secretaries and office staff.

Half of the 26 respondents who said they did not currently own a computer are thinking of getting one in the future for management purposes, interest, integration with other technology or in response to parent company demands. Only 5 specifically said they would not be getting one.

Almost 2/3 of the sample population use meteorological data, mostly their own; although 44% also use the local Meteorological station. The types of data employed were very simple ie rainfall and min/max temperatures.

Over three quarters of the sample indicated that they would find useful some form of prediction or forecasting system concerned with pests particularly cabbage root fly, two thirds said they would like one concerned with spot producing diseases and a half one concerned with N,P & K nutrition. Other support systems received less support.

Interview results

Roles

Viewing the process from an information/product flow perspective there appears to be three main areas of activity illustrated in Table 1 below. The focus of the project is on the growing and initial preparation of the product and the key roles within it: the Nursery, the Grower, the Co-op and the Packer. The Nursery is concerned with the development of young plants, the Grower with mature plants. The Co-op is an umbrella organisation providing centralised agronomic and marketing services for a group of participating farms. The Packer is concerned with cleaning and packing the product for the various markets.

Table 1: Showing main roles in growing process

Stage	Area	Roles
1	Sources of information, advice and raw materials for feeding into the growing process.	Seedsmen Chemical Distributors Fertiliser Distributors Independent Agronomists Research Institutes
2	Growing and preparation of product s.	Nursery, Grower Co-op, Packer
3	Markets - sale of produce.	Retail, Wholesale, Processors, Supermarkets

There is usually a large overlap between these roles in the real world, some companies run nurseries and grow the crops, some grow and pack, some are independent, some are part of a Co-op, some are a mix of independent and Co-op. Each business is unique. Each takes information from wherever they can find it, depending on the price, although most employ some form of consultant agronomist.

The roles which feed into the growing/preparation process whilst remaining outside of it are those of Researcher, Advisor, Seedsman, Chemical Supplier and Fertiliser Supplier; again several of these may be combined in an existing company.

The third area is concerned with the final destination of the product, the markets, and these have been identified as Retail, Wholesale, Processor and Supermarket. These roles are unlikely to be combined.

Tasks

Tasks can be described under 5 main headings: Administration, Marketing, Glasshouse, Field and Packhouse. The Field and Glasshouse task areas are key to the project because they are the areas in which modelling/prediction systems will be of most use. They cover main task headings such as advanced scheduling, planting, harvesting, disease and pest control and spraying. Marketing is also an area of interest because fore-knowledge of potential peaks and troughs in production is important.

Goals

Figure 2 provides an initial description of the main goals and sub-goals to be achieved by the grower in order to be able to sell all of his/her product at the best price. They seem to decompose into two main areas: crop-based goals and market-based goals and to date the focus has been placed on the crop-based goals.

The person responsible for the production of the crop has to produce a high yield and therefore get the best return on the original investment; to ensure continuity of supply and thus avoid having to buy in additional product to meet contract demands; to ensure the crop meets the customer specifications ie is free from pests and diseases, is the correct and marketable size and is the specified colour and shape for the market niche.

The grower has to make decisions about the best way to achieve each goal and each decision is based on a variety of available information. The quality of this information impacts directly on the growers ability to achieve the key aim of selling all the product at the best market price. Much of the information is satisfactorily gained from past experience. The type of information which is more difficult to access is dynamic ie the level of pests/diseases currently in the crop, the level of nutrients in the soil, the size of the crop, the weather tomorrow; all of these have to be gleaned from human observation in the form of field walking and extrapolation from known facts and past experience.

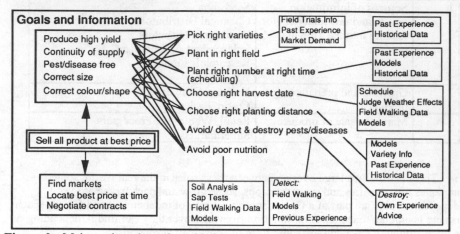

Figure 2: Main goals, sub-goals and information requirements for Brassica production.

Interface issues

Advisors and growers have different interface needs, however both would like an integrated method of storing information, the ability to make changes rapidly, to do 'what if' scenarios, and to adjust the system to suit their own particular needs. A short summary of the key points is given here and more information is provided in Parker (1994). Advisors works in different units to growers ie hectares & acres, as opposed to fields, plots, beds and rows. It is vital for any system which stores information to be able to work in at least the plot unit and preferably the bed unit as well. This is particularly true of on-line spray records as only one plot in a field may be sprayed.

Some form of visual representation of fields, often rough, is used by almost every grower with the possible exception of the very largest. People responsible for scheduling spraying and harvesting often use OS maps. They are also used to identify areas to be sprayed and are kept as part of spray records. Both advisors and growers use a term like 'period' to define a lump of time during which x amount of crop will be planted/harvested. The growing year is divided into these 'periods' and all scheduling documents refer to them. Whilst advisors may be familiar with and understand scientific terminology such as 'dry weight' the grower will not and will expect to be able to use familiar terms ie yield. It may also be necessary for some information to be encrypted or 'hidden' from the casual observer as the market is very competitive and paranoid and all users would like to be able to add their own comments to files.

Further detail on the results is presented in Parker (1994).

Discussion

The main difference between this industry and many others is the impact of an extremely unpredictable variable, the weather, on most decisions. The appearance of unusually dry, wet or cold weather can radically alter the approach required to successfully manage most tasks on the farm.

In spite of this the results of the studies so far suggest that a real need for prediction/forecasting models in the grower and advisor user groups. Marketing groups in the larger companies would also seem to be interested in a system to give a better forecast of the quantity of product they have to sell at a given time.

Although there is not currently a 100% take up of computer technology there are a considerable number in use and the trend is growing. It would seem therefore that the circumstances exist in which modelling information could be made available to growers and advisors as DOS/Windows based prediction/forecasting software.

The type of meteorological data to which the growers in particular have access is, however, rather limited. Models which rely on rainfall, min/max temperature and windspeed will be most successful.

Growers operate in an increasingly cut-throat world in a market which routinely overproduces by 40%. Twenty people run most of the vegetable producing businesses in the UK, all looking for contracts from around 5 main supermarket buyers. Supermarkets are the main driving force behind changes in the industry, they are seen as extremely valuable customers but unfair and immovable in their demands for quality and lower prices. As one grower put it "the only difference between a supermarket buyer and a terrorist is that you can negotiate with a terrorist". The issue of compatibility with existing and planned TQM systems to be set in place by the supermarkets will therefore have to be taken seriously.

Knowledge is money. Since the separation of near market and basic research and the drive to obtain direct funding from the growers for the advice they receive, an atmosphere of secrecy has developed. This may limit the types of simulation model to which the project has access but more importantly may make it harder for systems to be adequately tested in the field prior to any eventual future commercial development.

Time is an increasingly precious commodity for all growers. The larger ones are contracting out more and more of their time-consuming field walking tasks to consultants. No grower will willingly take on a system which requires the investment of more time unless it offers substantial benefits.

Conclusion

It would seem that the first target for the prediction/forecasting models should be the advisor and the co-operative groups. The systems should be designed to be usable by the farm and harvest managers in the co-operative groups and by any advisor/consultant. As each of these groups has their own systems in place a good deal of flexibility in tailoring input and output fields may be required. Key areas for decision support systems are scheduling, harvest prediction, pest/disease forecasting, nutrition and planting distances. Any models in this area which do not require unusual meteorological data as input would be a good starting point.

References
DTI. 1990. *A Guide to Usability. Usability Now!* (Open University . Milton Keynes)
Greenwood, D.J., Draycott, A. and Rahn, C.R. 1994, 'WELLN', Horticultural Development Council, Petersfield.
Parker, C.G. 1994, GRIME: Results of the first phase user requirements survey, Project Report no. 2. May 1994.

Methodology

Current trends in Human Error Analysis Technique Development

Barry Kirwan

Industrial Ergonomics Group
School of Manufacturing & Mechanical Engineering
University of Birmingham
B15 2TT

This paper summarises the apparent trends in human error analysis, based on a qualitative review of thirty-five Human Error Identification (HEI) techniques. It concentrates on observable trends in the nature and approach of the techniques either recently developed, or undergoing development, and their implications for ergonomics. Three trends are noted: an increase in cognitive simulation methods; a focus on error of commission approaches; and attempts to model operating crew interactions. The general utility of HEI techniques for ergonomics is discussed.

1. Background

Ergonomics has always been concerned with error reduction, but generally appears to shy away from formal Human Error Identification (HEI) approaches. Such approaches examine tasks and identify possible errors that can occur, often as a function of poor interface design, training and procedures, and other ergonomics factors. Human error is usually formally assessed via Human Reliability Assessment (HRA), with HEI occurring first, and then any identified errors being fed forward into a risk assessment, and their probabilities or likelihoods are quantified by one of a number of Human Reliability Quantification (HRQ) methods. If particular identified errors are predicted to occur sufficiently frequently to have a dominating influence on unacceptable risk, then error reduction analysis (ERA) is carried out, either based on the identified causes of the error, or based on defending against the error's impact within the human-machine system configuration. ERA can (and frequently does) call upon the discipline of ergonomics to determine the best way to reduce the negative human impact on the system. Whilst HRQ and (particularly) ergonomics are relatively well-developed, however, HEI is not. A recent two-part review (Kirwan 1992a & 1992b) examined twelve techniques available for HEI, but none of these fully satisfied HEI needs in risk assessment applications. This paper reports on a study of 35 HEI techniques, and shows the major trends in this rapidly changing field, and briefly discusses the potential utility of HEI for ergonomists. The techniques are listed below in Table 1.

Table 1 HEI Techniques reviewed (Kirwan, 1995)

THERP*	Technique for Human Error Rate Prediction (*Swain & Guttmann, 1983*)
HAZOP*	HAZard and Operability Study technique (*Kletz, 1974*)
SHERPA*	Systematic Human Error Reduction and Prediction Approach (*Embrey, 1986*)
SRK*	Skill, Rule and Knowledge-based behaviour model (*Rasmussen et al, 1981*)
GEMS*	Generic Error Modelling System (*Reason, 1987; 1990*)
PHECA*	Potential Human Error Causes Analysis (Whalley, 1988)
Murphy Diagrams*	(*Pew et al, 1981*)
CADA*	Critical Action and Decision Approach (*Gall, 1990*)
HRMS*	Human Reliability Management System (*Kirwan, 1990*)
IMAS*	Influence Modelling and Assessment System (*Embrey, 1986*)
CMA*	Confusion Matrix Analysis (*Potash et al, 1981*)
CES*	Cognitive Environment Simulation (*Woods et al, 1990*)
INTENT	[not an acronym] (*Gertman, 1991*)
SNEAK	[not an acronym] (*Hahn and deVries, 1991*)
EOCA	Error of Commission Analysis (*Kirwan et al, 1995*)
PREDICT	PRocedure to Review and Evaluate Dependency In Complex Technologies (*Williams and Munley, 1992*)
PHEA	Predictive Human Error Analysis technique (*Embrey, 1993*)
TEACHER/SIERRA	Technique for Evaluating and Assessing the Contribution of Human Error to Risk [using the] Systems Induced Error Approach (*Embrey, 1993*)
SCHEMA	Systematic Critical Human Error Management Approach (*Livingston,Wright and Embrey, 1992*)
CREAM	Cognitive Reliability and Error Analysis Method (*Hollnagel &Embrey,1994*)
COMET	COMmission Event Trees (*Blackman, 1991*)
TAFEI	Task Analysis For Error Identification (*Baber and Stanton, 1991*)
COGENT	COGnitive EveNt Tree (*Gertman, 1993*)
COSIMO	COgnitive SImulation MOdel (*Cacciabue et al, 1992*)
TALENT	Task Analysis-Linked EvaluatioN Technique (*Ryan, 1988*)
PRMA**	Procedure Response Matrix Approach (*Parry, 1994*)
DYLAM	DYnamic Logical Analysing Methodology (*Amendola et al, 1985*)
INTEROPS	INTEgrated Reactor OPerator System (*Schryver, 1992*)
HEMECA	Human Error Mode Effect & Criticality Analysis (*Whittingham &Reed, 1989*)
TOPPE	Team Operations Performance and Procedure Evaluation (*Beith et al, 1991*)
ADSA**	Accident Dynamic Sequence Analysis (*Hsueh et al, 1994*)
CREWSIM	CREW SIMulation (*Dang and Siu, 1994*)
CAMEO/TAT	Cognitive Action Modelling of Erring Operator/Task Analysis Tool (*Fujita et al, 1994*)
CREWPRO**	CREW PROblem solving simulation (*Shen et al, 1994*)
SRS-HRA	Savannah River Site HRA (*Vail et al, 1994*)

*All the above references are in Kirwan (1995). References at the end of this paper are given for those techniques discussed in more detail. Those with a single asterisk are described in Kirwan (1992a/b). Acronyms marked with an '**' are this author's acronyms for the techniques, since the authors did not supply one in the original reference reviewed.*

2. Trends

2.1 Cognitive simulation

The first notable trend is the development of a range of computer simulation-based models: whether these are based on an information-processing theoretic approach, or a symbolic processing framework. Such simulation approaches invariably originate from the cognitive science domain rather than Human Factors. These simulations model the operator's thought processes, and offer potentially powerful ways of determining how human operators will respond in emergency scenarios, typically in complex environments such as nuclear power plants. However, the models are rarely finalised to the point where they can be easily implemented for HRA risk assessment purposes, and are highly resource-costly to develop, and in some cases even to apply. These are the most challenging HEI approaches, since they are attempting to mimic and predict human thought processes. In the case of at least one of these (CES), initial validation attempts appeared promising, though CES worked faster than the real expert operator.

ADSA is one of the latest simulation approaches (at the prototype stage) designed to identify a range of diagnosis and decision-making error modes such as fallacy, the taking of procedural short-cuts, and delayed response. What is particularly interesting about this approach is that the Performance Shaping Factors (PSF) in the model are linked to particular Psychological Error Mechanisms (PEMs) (e.g. the PSF *time pressure* leading to the PEM of taking a *short-cut*). This represents an advance as the simulation approaches become (apparently) more able to generate realistic cognitive External Error Modes (EEMs) that have been observed to occur in real events and incidents.

INTEROPS is another cognitive simulation model, this time using the simulation system SAINT, plus a knowledge base with a hypothesis-generating and testing system. It can apparently model a number of cognitive error mechanisms (e.g. confirmation bias, cognitive tunnel vision). Stress is modelled as a function of time spent in diagnosis, and ignoring evidence (tunnel vision) increases as a function of time. This represents a departure from other simulation approaches, and INTEROPS is based on an information processing approach rather than a symbolic processing theoretic approach (as COSIMO and CES are).

COSIMO is a symbolic processing theoretic approach, and (like CES) uses a blackboard Artificial Intelligence architecture, but works more based on 'frames' or 'scripts' than the pure rule-base underpinning CES. This means that COSIMO falls more in line with Reason's (1990) 'schema'-based modelling of cognitive activity and error. COSIMO is high on resources, and is not easy to integrate with risk assessment.

2.2 Errors of Commission

Error of commission analytic techniques represent another significant trend, with a range of techniques and approaches being suggested for a range of industries. An error of commission (EOC) is an unrequired act, i.e. one that was never intended (e.g. lifting up the undercarriage on a plane while the plane is still on the runway). Such errors are very frequently due to a subtle interplay between poor ergonomics design aspects, and can sometimes prove disastrous. Some techniques for identifying such errors are table-top simulation approaches, others relying on databases of ergonomics guidance to help identify the potential for slips (unintended acts) leading to errors of commission.

SNEAK relies on an Ergonomics database, and is philosophically interesting in that it is attempting to derive EOCs at a very detailed level, and is considering essentially previously unconsidered connections between events, actions, and system states. It is also potentially a very useful tool, since ergonomics design deficiencies are used to determine where and when sneak paths will occur. It is however a highly resource-intensive technique, and cannot be applied until the operational stage of the system.

2.3 Crew Interactions

The third noticeable trend is the attempt to model operating crew interactions, mainly within simulation methods, in terms of modelling communication errors and team members' confidence in their colleagues abilities and judgements. This modelling level represents the most psychologically ambitious so far, and the most realistic in terms of actual crew co-ordination tasks in a nuclear power emergency.

CREWSIM is currently the only simulation model that models (albeit crudely) team interactions in a dynamic event. Three members of the crew are simulated. CREWSIM models attention resources control, such that diagnostic activities will be suspended while other activities must be attended to. Again, as with ADSA, this represents an increase in modelling the dynamic aspects of the evolution of a scenario. CREWSIM also allows the consideration of different goal priorities within the operator, and in particular focuses on transitions from the basic generic emergency procedure in a certain reactor system design to other procedures (e.g. to Steam Generator Tube Rupture). Communication and confidence in other crew members are also intended to be modellable in an extension to CREWSIM called CREWPRO. These represent ambitious but significant enhancements of the external validity or realism of modelling (though of course such modelling requires validation

TOPPE is not primarily a HEI tool, but identifies procedural errors and team performance problems when carrying out Emergency Operating Procedures evaluations and walk-throughs. The approach is interesting since it offers a way of assessing potential team performance problems during training and procedural evaluation sessions. Such errors could be used to develop a crew error and communication error taxonomy, which would be useful for a range of industries, including aviation, naval, process control and military applications.

2.4 HEI and Ergonomics

As a more general and important trend, a number of techniques are clearly attempting to be more model-driven in their approach, rather than being purely taxonomic in nature, as was the case less than a decade ago. The two dominant models utilised at present are information processing and symbolic processing theory. This is coupled with a trend towards examining error potential in more depth than ever before, whether in a model-driven way or via more powerful task analytic or simulation approaches. These two general trends represent a shift in HRA towards more psychologically and ergonomically valid approaches. Ergonomics insights and recommendations are therefore becoming more realisable with such approaches, and these approaches themselves are using more ergonomics information in their databases and methods for isolating error forms. HEI is currently drawing significantly from the three fields of cognitive science, psychology, and ergonomics.

3. Summary on HEI methods: state of the art

There is still no single technique available at present which would be optimal on all the qualitative criteria cited in Kirwan (1992a & b, and 1995): there is therefore no clear 'best' technique currently available. Many of the techniques are still not highly structured, suggesting that the majority of the techniques are still seen as aiding the assessor rather than being fully prescriptive and model-driven, and implying that HEI is still an art rather than a well-definable science.

In terms of theoretical validity, there is a marked influence of the Skill, Rule and Knowledge model of Rasmussen and co-workers (1981) on many of the techniques (SRK, GEMS, PHECA, MURPHY, SHERPA, CADA, HRMS, and to a limited extent CES). This seems to have been the dominant model in HEI, although it may be that there is a trend developing more towards newer models like Generic Error Modelling System (Reason, 1990) in terms of defining, in particular, the cognitive error forms that techniques should be able to address. Techniques which are model-based do tend to also consider PSF and PEMs as well as EEMs, whereas non-model-based approaches (simple error taxonomies) only consider EEMs. For the simulation models, there are two main camps: information processing theory and symbolic processing theory. It is interesting to note however, that in terms of the errors they can model, there is considerable overlap (i.e. both model types appear able to model the dominant cognitive error forms).

The five techniques evaluated that appear currently available for use in HRA and risk assessment in the UK are THERP, HAZOP, SHERPA, CMA, and HRMS. This means that none of the recent developments referred to in this paper are really ready for practical HRA applications in risk assessments. It should be noted at this point that a trend of concern is for techniques to appear at the prototype stage and then to disappear within one or two years, never actually realising application. Thus, although the trends noted above are of interest and show welcome shifts towards more theoretical models and approaches in the difficult area of HEI, such shifts are ultimately of no consequence if the models never amount to more than academic curiosities. Therefore, whilst some techniques have been around for a decade or so, about half of the techniques are only one or two years old, and are still at the prototype stage. This current flood of HEI techniques onto the market does significantly reflect the concern for the improvement of HEI in HRA and risk assessment, and also the desire amongst developers for more theoretically plausible techniques, theory in this case coming from the domains of Psychology, Cognitive Science, and Ergonomics. However, even given the current plethora of techniques under development, it is not clear that only those will survive that are adequate according to an ergonomics model or theory. The determination of which approaches reach full development will be based ultimately on practicability criteria rather than academic ones, i.e. the usefulness and 'added value' such techniques lend to the risk assessment process, and of course their perceived cost-effectiveness, or sometimes simply their cost. This significantly constrains those involved in developing such approaches. Thus, whilst HEI is currently swinging more towards ergonomics than ever before, only time will tell if this shift will be a permanent one, or merely the swing of the pendulum.

Finally, although ergonomics clearly has something to offer HEI, the question might be raised as to what the reciprocal benefits are. Firstly, HEI techniques do not belong exclusively to HRA. A small number of the techniques are used by ergonomists qualitatively to evaluate interface design, workload, etc.. Secondly, the synergistic benefit from ergonomics and HEI is the ability to identify high priority interface, procedural, and training areas where ergonomics adequacy must be maximised in order to maximise safety. A recent example of this approach is shown in Brown et al (1994), wherein risk assessment has been used to identify key equipment items (valves, circuit breakers, etc..) on plant, where errors (especially EOCs) would have significant impacts on risk. Once these have been identified, ergonomists maximise the ergonomics design of these items. HEI should therefore be considered a tool in the Ergonomist's toolkit, and a means of prioritising ergonomics interventions.

References

Beith, B.H., Vail, R.E., and Drake, M. (1991) The team operations performance and procedure evaluation (TOPPE) technique as a method for evaluating emergency operating procedures. In *Human Factors Society Annual Conference Proceedings*, pp. 405-411.

Brown, W.S., Higgins, J.C., and O'Hara, J.M. (1994) Local control stations: Human engineering issues and insights. NUREG/CR-6146. Washington DC: USNRC.

Cacciabue, P.C., Decortis, F., Drozdowicz, B., Masson, M., and Nordvik, J.P. (1992) COSIMO: a cognitive simulation model of human decision making and behaviour in accident management of complex plants. IEEE Transactions on Systems, Man, and Cybernetics, **22,** No.5, 1058-1074.

Dang, V.N. and Siu, N.O. (1994) Simulating operator cognition for risk analysis: current models and CREWSIM. Paper in *PSAM-II Proceedings*, San Diego, California, March 20-25, pp. 066-7 - 066-13.

Gertman, D.I. (1991) INTENT: a method for calculating HEP estimates for decision-based errors. *Proceedings of the 35th Annual Human Factors Meeting*, San Francisco, September 2-6, pp. 1090 - 1094.

Hahn, H.A. and deVries. J.A. (1991) Identification of human errors of commission using Sneak Analysis. *Proceedings of the Human Factors Society 35th Annual Meeting*, San Francisco, September 2-6, pp. 1080 - 1084.

Hsueh, K-S., Soth, L. and Mosleh, A. (1994) A simulation study of errors of commission in nuclear power accidents. In *PSAM-II Proceedings*, San Diego, California, March 20-25, pp. 066-1 - 066-6.

Kirwan, B. (1992a) Human error identification in HRA. Part 1: Overview of approaches. Applied Ergonomics, **23**(5), 299-318.

Kirwan, B. (1992b) Human error identification in HRA. Part 2: Detailed comparison of techniques. Applied Ergonomics, **23**(6),371-381

Kirwan, B (1995) Review of Human Error Identification techniques for use in Nuclear Power and Reprocessing HRA/PSA: Volume II. IMC GNSR Project HF/GNSR/22. Industrial Ergonomics Group, University of Birmingham, March.

Rasmussen J., Pedersen, O.M., Carnino, A., Griffon, M., Mancini, C., and Gagnolet, P. (1981) - Classification system for reporting events involving human malfunction. Riso-M-2240, DK-4000, Riso National Laboratories, Denmark.

Reason, J.T. (1990) *Human Error.* Cambridge University Press, Cambridge, UK.

Schryver, J.C. (1992) Operator model-based design and evaluation of advanced systems: computational models. *Proceedings of the IEEE Fourth Conference on Human Factors and Power Plants*, pp. 121-127.

Shen, S-H., Smidts, C., and Mosleh, A. (1994) Elements of a model for operator problem solving and decision making in abnormal conditions. In *PSAM-II Proceedings*, San Diego, California, March 20-25, pp. 060-7 - 060-12.

Woods, D.D., Pople, H.E., and Roth, E.M. (1990) The cognitive environment simulation as a tool for modelling human performance and reliability, USNRC, Nureg/CR-5213, Washington D.C.

KITCHEN TO OPERATING TABLE - ERGONOMICS IN SELECTING OPERATING TABLE SYSTEMS

RJ Graves,* JM Gardner, M Anderson, ST Doherty, A Seaton,* J Ross,* and R Porter**

*Department of Environmental & Occupational Medicine, University of Aberdeen
**Department of Orthopaedics, University of Aberdeen

This paper describes the development of a methodology to help non-ergonomists to select Operating Table Systems (OTS). A review was undertaken of the ergonomics of operating theatres, of OTS used in some Scottish hospitals, of types of surgical procedures, of reported accidents, and of the use of OTS. Further activities included a postal survey of users and main OTS types, detailed ergonomic evaluations of frequently used surgical procedures, and a review of OTS selection procedures. The results were used to develop a handbook to assist users to select OTS. This consists of guidance on technical and surgical requirements, on simple ergonomic issues and on demonstrations and trials of short listed systems. Aide memoirs and worksheets are used during the selection process. During the development of the guidance, consultation was undertaken with users and a seminar held with users and manufacturers.

Introduction

The study was commissioned by the CSA (Common Services Agency), of the Scottish Home Office, which was interested in the extent of problems and injuries that were related to operating table system design, and in particular the selection processes involved in the purchase of such systems. Operating table systems (OTS) are a large capital expense and they are expected to have a long useful life. It is important that the selected operating table system carries out the required function without detriment to the patient and theatre staff. Therefore selecting operating table system with optimum characteristics is of paramount importance.

Ergonomic research in operating theatres has been limited to the transfer of instruments, the design of surgical instruments, evaluation of surgical chairs, heart rates of surgeons, and the layout of anaesthetic machines.

Not surprisingly, the main users of operating tables are; surgeons, anaesthetists, nurses, Operating Department Assistants (ODA's), and orderlies, as well as patients. To meet the human and medical requirements of surgery, the design of tables and theatres has become more functionally complex. This means one table can be used to perform a

117

variety of different surgical procedures, and with its ancillary equipment can be regarded as an Operating Table System (OTS). Fundamental elements which influence the design of the operating table are the medical implications associated with the positioning of the patient for specific surgical procedures. However it is not clear whether due consideration has been taken of the ergonomic implications of the tasks associated with surgical procedures.

The study had as its principal objective the development of an ergonomic methodology for the non-specialist evaluation of operating table systems (Graves et al, 1994a). This paper is in two parts. The first part deals with the approach taken, the results and conclusions from the initial project activities outlined above, while the second part outlines the thinking underlying the development of the structure and content of the guidance handbook, based upon the findings from the studies undertaken in the first part, and the results of a consultation during the development of the guidance.

Approach and Findings

Two major areas of study were identified. First, user requirements in relation to Operating Table design, and second, the decision making process associated with selecting Operating Table Systems (OTS). A number of project activities were undertaken to obtain information in relation to these major areas of study. These involved; carrying out a review of the literature on the ergonomics of operating theatres and ergonomic techniques for detailed evaluations of surgical procedures, reviewing table systems used in a sample of Scottish hospitals, reviewing the frequency of different types of surgical procedures, examining reported accidents related to the use of OTS, carrying out a postal survey of users from a sample of hospitals covering main OTS types to identify user requirements and problems, conducting detailed ergonomic evaluations of the most frequently used surgical procedures, and reviewing the OTS selection procedures used in Scottish hospitals.

For the review of operating procedures a detailed computer printout was obtained from Grampian Information Services listing all surgical operating procedures carried out in the Aberdeen hospitals from January 1st 1990 to December 31st 1990. These procedures were categorised according to the surgical speciality. For the year 1990, a total of 35,002 recorded operating procedures took place.

The five specialities performing the most procedures - general surgery, gynaecology, ENT, orthopaedics and urology, were analysed in greater detail. From the records of the procedures the site and nature of all the operations performed and hence the patient position likely to have been used for each procedure were determined in consultation with those with previous experience and familiarity with anaesthetic work, the surgical procedures performed and the patient positions required. From this data it was possible to determine that the commonest positions used by surgeons for placing patients for surgery were: supine flat, lithotomy, supine head up (reverse Trendelenburg), supine head down (Trendelenburg), Lloyd-Davies, lateral and prone.

The accident statistics for a Health Board's hospitals for the years 1985 to 1991 were reviewed and accidents reported from theatres accounted for between 3 and 4 percent of the total reported to the Health Board. The commonest accidents in theatre related to operating table usage were back injuries due to patient transfer, cuts, bumps and grazes when moving table, attachments falling off tables injuring fingers and toes, and, fingers getting crushed between tables and other equipment.

The aim of another study was to evaluate the user requirements of operating table systems and the effects such systems have on the health of their users. As part of this, detailed task analyses were carried out to determine the main user tasks. These were found to be grouped into pre, per and post-operative tasks, and for anaesthetists and ODA's pre-operative tasks can be further subdivided into pre, per and post-induction tasks. The way in which these tasks are performed varies from user to user to some degree, but they vary mainly due to the differing requirements of the various operating positions used, the most commonly used being supine flat, supine head up (reverse Trendelenburg), supine head down (Trendelenburg), lithotomy, Lloyd Davies, prone and lateral, as discussed earlier.

Following the task analyses, a questionnaire for each user group was drafted to identify the problems encountered performing the main tasks, in the commonly used patient positions, using the operating tables currently in service. A task oriented questionnaire was developed for each user group to determine problems, perceived health effects, near-miss incidents and opinions on the systems. Anthropometric details were asked for, as well as some other details.

Originally it was planned to select a random sample of users at one Scottish hospital as the study group. After consultation with the hospital's theatre management, the draft questionnaire was piloted to a sample of users and several problems were highlighted. From this it was concluded that it would have been extremely difficult to administer this questionnaire to a sufficient number of users, and the views of the users would have been those of a generally old system, soon to be replaced. As a result, the questionnaire was shortened and modified so that it could be self-administered. It was piloted again at the hospital and further minor adjustments made. With the help of the CSA, it was decided to expand the study to include three other Scottish hospitals using modern operating table systems.

513 questionnaires were issued and 397 were completed and returned, giving a response rate of 77%. The results showed the percentage of reported back problems. The highest percentages overall were those of the anaesthetists and ODA's, followed by the orderlies. However care needs to be taken in interpretation because of the sample sizes of the latter two categories. It is interesting to note that back problems were a high proportion of all the health complaints.

Detailed ergonomic evaluations of the most frequently used surgical procedures identified earlier were carried out to identify the frequency and source of awkward postures which may lead to discomfort and/or long term disorders. This involved taking video in theatre of a sample of the most common surgical procedures identified earlier. Video recordings of a selected number of surgical procedures (17 operations), in the first Scottish Hospital, were carried out in order to identify any ergonomic problems. The postures were then analysed out of theatre by examining the video material, and where necessary by observing and estimating angles of body parts from freeze framing. Posture targets (Corlett et al 1979) and the Upper Extremity Posture and Work Element Recording Sheet (Armstrong, Joseph and Goldstein 1982) were used to record working postures. A worksheet was specially developed for this project and it was used to analyse trunk and head postures and the OWAS method of posture analysis and criteria, obtained from a literature search, was used to assess the amount of risk in the postures and activities.

The ergonomic evaluations highlighted awkward postures which were associated with ergonomic issues in workspace layout which could be addressed by considering issues such as having improved seating design with backrests, footrests, easily and quickly adjustable height feature, armrests, etc., having easy access to operation site and easy reach to instrument trolleys, having adequate foot access and leg clearance when sitting at the operating table system, and improving the position of the surgical microscope and other ancillary equipment. This may reduce the poor postures that result from inadequate workspace layout, improving the design of the surgical instruments, in particular the design of the laparoscopic instruments. Aspects such as the forces required, the number and frequency of extreme hand and arm movements could be much improved by ergonomic redesign.

The decision making processes of the CSA and the Commissioning Teams of the four hospitals were analysed to establish the procedures used in the selection of operating table systems. Information from this part of the project was obtained by carrying out structured interviews. The information was obtained through interviews with CSA, theatre managers and input from the Theatre Users Committee and theatre staff. A number of users of operating table system are not consulted prior to purchase for their requirements, and a high proportion would like to be. This means that it would be advisable to find a means of widening and structuring the consultative process.

On the basis of the findings it was concluded finally there was a general need to have a structured approach to Operating Table System selection which covered issues such as ergonomic features and a guidance handbook should be developed which contained help on the issues to be considered in the selection process and some simple consideration of ergonomic principles.

The Development of the Guidance

The information from these project activities was then used to develop a guidance handbook (see Graves et al, 1994b) to assist users in asking questions identified as being of importance while considering which operating table system to select. The results of the study showed that procedures used in the selection of operating table systems in the four Scottish hospitals and the procedures used by the CSA varied in the number of stages, the amount of detail obtained and the level of user involvement. The questionnaire results and the detailed ergonomic case studies highlighted a number of problems with operating table systems which reinforced the need for some form of standardised procedure to aid in the selection of operating table systems.

The findings suggested initially that a multistage, structured procedure was needed for helping users in the selection of operating table systems. With the benefit of informal discussions, a seminar presentation to hospital staff, manufacturers and CSA staff, a case study in and assistance from a Consultant Surgeon at a location which carried out a wide range of surgical procedures, a three stage process was finally decided upon, which is incorporated in the guidance hand book to service these needs (Graves et al, op cit.).

This guidance consists of three stages. The first stage covers the technical and surgical requirements of OTS. The second deals with simple ergonomic issues and the third with on-site demonstrations and trials of the short listed systems. The following describes the thinking behind the main stages of the procedure and it is anticipated that input from the CSA could be at any of the stages.

Depending upon the situation, some of the detailed activities contained within the different stages may not be felt to be necessary, especially if time constraints are tight.

Also some of the more detailed analysis may only apply when a new theatre complex is to be built or the total refurbishment of an existing theatre complex, is required. Each of the stages will be considered now in more detail.

The first stage involves general discussions on what is needed in terms of operating table systems. It is likely that these general discussions would involve theatre managers, theatre staff (representing surgeons, anaesthetists, nurses, ODA's and orderlies), hospital engineers and CSA. All hospitals which participated in the study had a Theatre Users Committee which was made up of a selection of theatre staff and theatre managers.

Allocation of tasks to theatre users committee members is an important part of the first stage. This stage involves collecting information for the next stages of the selection process. Information which can be useful in this first stage includes: knowledge of the surgical procedures because these will dictate the amount and degree of OTS adjustment, and the need for specialised attachments, manufacturers literature, the relevant British Standards, information from theatre staff, information from other hospitals and theatre managers.

The CSA can help with gathering the above information. As the ergonomic studies of surgical procedures showed there were awkward postures, it was felt that even at this early stage information should be available to help select those tables which could be adjusted to suit a range of users. The range of surgical procedures identified as typical from the study was available. Simple ergonomic assessments were carried out using users' size data to determine the amount of adjustability necessary to cover approximately 90 percent of a British male and female population and provided as a reference for both sitting and standing patient access.

It was found from the study that the hospitals studied had no formal procedure for documenting the basic requirements of Operating Table Systems. As a consequence a number of aide memoirs and worksheets were developed to provide help to the users in recording their observations. In addition, a number of worksheets were developed to help remind the user to consider different types of information which should be recorded at this stage in the selection process. The aim of this stage of the process is to document the basic requirements of OTS, thus setting the scene for the rest of the stages.

The next stage involved helping to check on some simple ergonomic aspects of OTS design. This included the impact of sitting and/or standing on table adjustability and factors affecting access to the surgical site. It was felt that this should be provided in a way which would need limited ergonomics knowledge and expertise, with the option for more detailed ergonomics assistance if necessary. To this end, more detailed information on the issues involved were placed in appendices for reference if needed. This simple check on ergonomic features provides information on adjustability and access for the tables short listed from the first stage. This should help in selecting the final short list for on-site demonstrations and assessment.

The final stage involves the demonstration of a selection of operating table systems within the theatre environment and/or formal on-site assessment trials. The short list of operating table systems should be demonstrated within a theatre setting and cover the types of application identified earlier. This should involve a manufacturer's representative demonstrating the operating table systems to theatre staff, Theatre Users Committee members and theatre managers. This could be followed by more formal trials if necessary.

In a similar way to the previous stages, worksheets are provided to help the Theatre Users Committee structure their questions and record the answers. Included in this stage are questions on training and maintenance requirements. Measurements of position of controls, displays, height range, mattress thickness, parameters of the base and parameters of the attachments should be obtained at this stage. An operating table system checklist was designed to be used here. The outcome of the operating table demonstrations is the selection of the most suitable systems to be used for a trial period.

The input from the CSA depended upon the: hospital, situation, operating table systems that are required, time allocated and the overall cost. It is proposed that input from the CSA should be at any stage of the decision making process. It is important that the CSA is kept up-to-date about progress throughout the selection process and they should be informed of any difficulties encountered. The CSA input can provide guidance when needed and feedback from hospitals which have used this selection process.

During the development of the prototype guidance, consultation was undertaken by discussions with users, by the use of a seminar outlining the approach held with users and manufacturers and its informal use in two examples of selecting OTS. The seminar took place towards the end of the project hosted by the CSA to provide preliminary results from the postal questionnaire and an initial outline of the approach to the proposed guidance handbook. The participants included representatives from the manufacturers, surgical, anaesthetic, nursing, and CSA staff, from throughout Scotland. The objectives of the seminar were to provide an awareness of ergonomics principles in design, outline the results of the study on selected operating systems and an ergonomics approach to improving operating table design, provide a forum for consultation and development of the prototype ergonomics approach to operating table evaluation, and to encourage participation in improving operating table design.

The key issues which were identified to the audience in optimising the Operating Table System selection decision making process were having a set procedure for collecting and analysing information obtained from theatre staff and other sources, taking account of the task requirements, ensuring appropriate involvement of CSA, ensuring systematic feedback from theatre staff, documenting requirements and results from operating table demonstrations and user trials. The outcome of the seminar showed general support for a structured approach to the selection of OTS, and taking account of ergonomic principles in OTS design. Informal trials of the guidance indicated that it appeared to contain the information necessary for users to ask key questions while selecting OTS. The CSA are now in the process of publishing the guidance.

References

Graves, R.J. Anderson, M. Gardner, J. Doherty, S. Seaton, A. Ross, J. Porter. R. 1994a Development of an ergonomic methodology for the evaluation of operating table systems. Common Services Agency, Grant Ref No. EEV/C148. Final Report. Department of Environmental and Occupational Medicine, University of Aberdeen: Aberdeen.

Graves, R.J. Gardner, J. Anderson, M. S. Seaton, A. Ross, J. Porter. R. 1994b Ergonomic guidance for selecting operating table systems. Common Services Agency, Grant Ref No. EEV/C148. Department of Environmental and Occupational Medicine, University of Aberdeen: Aberdeen.

UNEXPECTED HUMAN SYSTEM INTERACTIONS AND THE MANAGEMENT OF SAFETY

Peter R. Michael & Nigel P. de Bray

Systems Design Dept. FPC 510
British Aerospace Defence Ltd (Dynamics Division)
Filton, Bristol BS12 7QW

This paper outlines a methodology which extends ergonomic analysis to explore complex and idiosyncratic human actions whether involved as designers, managers, operators or users. A combination of soft systems theory and personal construct theory is proposed, which may be adapted to analyze an engineered system and the parent organisation in which it resides for vulnerability to any "unexpected" human-system interactions which may impinge upon safety.

Introduction

It is an unfortunate truth that human fallibility plays a significant role in the majority of accidents. This may be as a result of poor design, inadequate maintenance, or incorrect operation. The contribution of ergonomics towards the design of safe systems has become particularly significant in recent years. Aspects of human performance are now available in validated models of the human operator which were once only described in qualitative terms or based upon empirical data. Thus the "standard" behaviour of humans, whether individually or as a team, is now well understood.

However the idiosyncratic or non-rational human processes which are difficult to quantify are both the strength and the weakness of the human. Indeed most would argue that this is the essence of being "human"! Thus the human, whether in the role of designer, user, operator, maintainer, manager or bystander must be considered in the total safety audit.

Even for accidents apparently caused by one person there are usually many contributory factors linked to training, selection and social interactions which led up to the event. A group of people may interact in many unexpected ways. Complex situational failures are often not amenable to current analysis techniques, hence a methodology is needed to fill this gap. Thus the focus of mainstream Ergonomics is moving from "hard" human factors engineering to include human aspects of the whole working environment and the management organisation. The Manprint initiative is an example of this new emphasis and the methodology outlined here should be regarded as a supportive technique which gives the ergonomist different perspective. The method proposed in this paper assumes some familiarity with "soft systems" and repertory grid methods. This methodology is widely applicable whether examining an engineering plant or a leisure complex.

Conceptual Models of Problem Situation

The "soft-systems" methodology of Checkland (1981) is used to analyze the total problem situation considering the complexities of an engineered system residing within its organisational environment. This is used to build up a **conceptual model** of potentially critical interactions. The conceptual model captures the essence each issue and the interactions between issues (viz: Staff Selection & Training, Ownership, Motivation, Team Operations, Change Process, Personnel etc.). Here the ergonomist working as a consultant has to utilise the information held by the "local experts" viz: the managers, supervisors and users whilst not succumbing to their prejudices. Often they will either not perceive or admit to the possibility of "unexpected" human behaviour causing hazards.

Checkland's methodology is a flexible iterative process which involves building up *conceptual models* of the current problem situation, and a *root definition of the relevant systems* involved which might lead to a feasible and desirable outcome.

The idea of a conceptual model is sometimes difficult to grasp but it is often useful to visualise it as a multi-layer process flow chart combining all relevant issues. However some of the processes may be informational or financial and include intangible concepts such as "influence" and "confidence". The advantage of a "soft systems" approach is that does not constrain the outcome of the analysis and allows divergent thinking.

Some top level interactions are illustrated as a conceptual model in Figure 1. This provides a starting point to evolve a more detailed picture of the complexity involving issues such as staff training and motivation etc. The problem of ownership and change is illustrated. This model picks up the successive root definitions in an iterative process of refinement and clarification as the situation becomes more clearly understood. This iterative process is repeated until a satisfactory root definition of the problem set is achieved along with a good conceptual model. All the important issues can be captured and expressed in the conceptual model for use later in the analysis.

Developing the Actors

The following procedure, based on knowledge elicitation methods is now used. The ergonomist should now use his "local experts" to devise a group of about twelve hypothetical "actors" based on different people who *might* be working as operators, managers or users in the organisation. The ergonomist must use his elicitation skills to prompt each local expert in a neutral manner whilst relating back to the issues highlighted in the conceptual model phase. Any stereotypes or prejudices held by these local experts may be crucial to the problem and must not be challenged. Political correctness is out!

These "actors" may be purely imaginary or based on people who currently or previously held the posts, but will usually have composite characteristics of a number of individuals. The "actors" are considered in combinations of three at a time. Two of the "actors" are compared and contrasted with the third and the differences between them identified. These differences may be any human characteristic, e.g. accuracy, flexibility, strength, intelligence, friendliness, motivation and skill level. The characteristics need not have formal names, but must be identifiable as differences between individuals which are also of some relevance to the conceptual model developed earlier.

The characteristics of the "actors" can then be analyzed by formal repertory grid methods (Fransella & Bannister 1977) and refined to ensure that most of the significant areas have been covered. Sometimes this can be done more quickly using less formal techniques and there should be no attempt to establish a detailed personality profile of each "actor". *This builds up a set of characteristics or personal constructs which the local experts see as being potentially relevant to the problem set within the organisation.*

The Failure Repertoire

The Failure Repertoire is developed by putting each "actor" in a situation and working through a variety of potentially hazardous experiences. The local expert decides how the "actor" or combinations of "actors" might respond, that is to say :-

Character(s) x Situation = Potential Incident

Much of this ground will have been covered already in obtaining the personal constructs.   More importantly for the current theme we should consider any non-normative attributes that might cause an incident:- hooliganism, influence of drugs, illness, infirmity, or simply disorientation at a first encounter. Anyone crossing the English channel takes a while to acclimatise to the differences!   [An Italian lorry driver stopped in the fast lane of a motorway in order to ask for directions. News Report 8 Dec '94]. During the pilot studies the use of "actors" was found to facilitate the extraction of the failure repertoire in a way that was neither difficult nor confrontational.

There is a view that the unexpected aspects of human behaviour can be largely controlled by correct training and operating procedures. However Figure 2 attempts to capture the scope of the problem by summarising results of a pilot study showing how different aspects of human behaviour contribute to errors and violations of operating procedures. Here the viewpoint of the "actor" was shown to be significant.

(Character x Viewpoint) x Situation = Incident

Viewpoints such as dedication, alienation, competition and ownership should have emerged during the conceptual model stage in the analysis. However new viewpoints can be introduced here and if necessary the earlier analysis re-iterated to check e.g.:-

Distraction: *"This supervisor is very worried about her child"*
Arrogance: *"This operator knows it all (and ignores instructions)"*

Several differing viewpoints may emerge from the local experts during these sessions which are not always reconcilable. This may lead to more than one conceptual model each of which should be regarded as equally valid.

Testing The Failure Repertoire In The Organisational Context

Once the Failure Repertoire is established then the way in which the organisation copes with these individual differences can be examined. This is achieved by returning to the conceptual model and analysing each relevant issue for each failure repertoire diagnosed.

Here the simplest approach is to divide the conceptual model into subsets dealing with each issue: Staff Training; Ownership, Motivation, Team Operations, Change Process, Personnel etc.   Then to build up a Failure Interaction Matrix of the failure repertoire against the conceptual model issues. Each cell in the Failure Interaction Matrix is examined and ones where a short fall or critical event might occur can be highlighted. Then for each highlighted cell an action response must be generated to define how the procedures or the management system should be modified to reduce the risk potential. If necessary each critical cell could be assigned a priority rating to give some indication of progress in improving the organisation. This part of the process is of course crucial and is only described in outline here.

Concluding Remarks

This paper has shown how a combination of techniques well established in psychology and systems analysis, may be adapted to examine human linked critical events in complex situations thereby extending the tools available to the ergonomist or safety expert. Whilst this methodology has only been tested in pilot studies to date, it is both flexible and simple. The advantage is it helps to promote divergent thinking to analyze both the engineered system and importantly the organisation in which it resides. In theory there is no human generated event that it could not uncover and examine provided that the interviewing skills of the consultant ergonomist and the imagination of the local experts are sufficient.

This methodology however is not rigorous nor should it ever be because that would limit its capability to examine new situations. What is needed for the future is a validated set of "actors" which can provide a more substantive basis for the analysis. This would shorten the time and improve the quality of analysis considerably.

References

Checkland P. 1981 *Systems Thinking and Systems Practice*
J Wiley & Sons; Chichester
Fransella F.& Bannister D. 1977 *A Manual For Repertory Grid Techniques*
Academic Press; London IBSN 0-12-265450-1

Note: The views represented in this paper are those of the authors and not necessarily those of British Aerospace PLC.

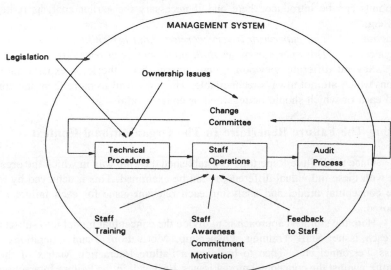

Figure 1 Conceptual Model of a Safety Management System

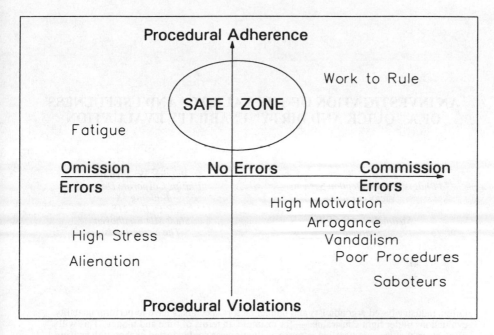

Figure 2 Procedures & Accidents   (Hypothetical)

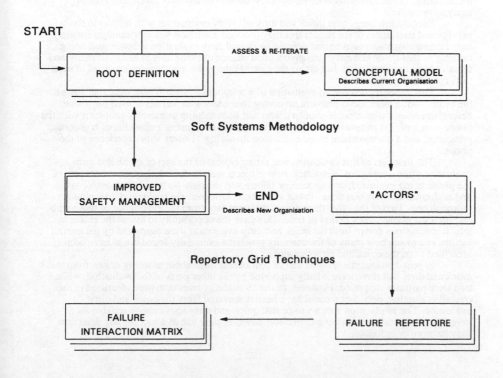

Figure 3 Summary Of Methodology

AN INVESTIGATION OF THE VALIDITY AND USEFULNESS OF A "QUICK AND DIRTY" USABILITY EVALUATION

Kevin C. Kerr
Philips Communication Systems
Victoria Place
Airdrie
Strathclyde ML6 9BL
Scotland

Patrick W. Jordan
Philips Corporate Design
Building SX
P.O. Box 218
5600 MD Eindhoven
The Netherlands

Often, industry based ergonomists and human factors specialists have to perform usability evaluations under tight constraints — for example, in terms of time and money. This will often mean that it is not possible to make evaluations as sophisticated or thorough as would be wished. A common approach in such circumstances is to perform an informal, or "quick and dirty", evaluation. This means making compromises in evaluation design in order to make savings in time and money. Typically, compromises might include: number of subjects participating, criteria for subject involvement, measures of usability taken, and type of analysis performed.

Speculation hangs over quick and dirty usability evaluations with respect to the validity and usefulness of the results that they produce. Common wisdom amongst industry based ergonomists seems to be that doing a quick and dirty evaluation is better than doing nothing — that quick and dirty tests give a good chance of being able to assess usability and identify bugs in an interface. But is this the case? Perhaps the inaccuracies involved in such tests can actually mislead.

This paper reports on the evaluation of a telephone system. There were four parts to the evaluation: a think aloud protocol involving four users who had never used the phone before, interviews with 'trouble-shooters' who had been helping users solve problems with the phone over a period of several months, an 'expert' evaluation using a checklist of ergonomic principles, and a questionnaire based evaluation involving 31 users with experience of the phone.

The first part of this evaluation was rather typical of the sort of quick and dirty evaluation often carried out in industry. Few subjects were involved, they were not users of the phone being evaluated, task success or failure was the only performance measure taken, and no formal analysis was done. Taken as a whole, however, the evaluation was quite a thorough one. Two of the other three stages gathered data from the experience of those who had used the phone over a period of time. This gave the opportunity to assess the extent to which conclusions drawn from the quick and dirty evaluation were supported by the overall results, and to see how many of the usability problems eventually found could have been identified from the quick and dirty test alone.

A post-hoc analysis showed that of four conclusions about usability drawn from the quick and dirty test three were wholly supported by the other parts of the evaluation, while two were partially supported. However, of the 19 usability bugs that were identified in the overall evaluation, only 5 or 6 could have been discovered from the quick and dirty evaluation. The implication appears to be that quick and dirty tests may be reliable for investigating specific aspects of an interface, however their use as a tool for bug detection may be more questionable.

Introduction

The importance of empirical usability evaluation is increasingly being recognised by industry. This is reflected in the number of human factors specialists now employed in the commercial sector, the inclusion of usability criteria in product specifications, user involvement in the product creation process and the promotion of ergonomic design as a selling point in product marketing. However, industry based ergonomists are still asked to carry out usability evaluations under (often sharp) constraints, in particular in terms of the amount of time available to carry out an evaluation and report back to commissioners.

Very often, these constraints will mean that compromises have to be made in the design and running of evaluations. These compromises will mean that some of the criteria that would be strictly adhered to in an 'scientifically' rigorous evaluation will be relaxed. For example, the subjects participating in the experiment may not be representative of the end users. Instead of recruiting a carefully selected sample, the evaluator often has to resort to asking those closest at hand (e.g., colleagues) to participate.

Other factors which might be compromised include the measures of usability taken, the rigour of the experimental controls and balances and the surroundings in which the evaluations are conducted. Informal evaluations will often rely on somewhat 'blunt' measures of performance, such as simply recording task success or failure, rather than measures such as time on task or error rates. Often, the emphasis may be on collecting qualitative data, such as subjects' comments or responses to open ended questions. Rigorous experimental designs will often include task order balancing between subjects. This is often lost in informal designs where it is common for all participants to be presented with tasks in the same order at the possible risk of introducing unchecked learning effects. Similarly, inappropriate surroundings may effect the ecological validity of what is being done. Informal tests are often carried out in usability laboratories, where conditions may be very different from the 'real life' scenarios in which the product will be used. Jordan and Thomas (1994), for example, point out that a car stereo which seems usable when being used in a laboratory may be far less so when trying to control a car at the same time.

However, despite these sorts of limitations, there seems to be almost a consensus amongst industry based ergonomists that it is better to perform a quick and dirty usability evaluation than to perform no evaluation at all. This attitude can be reinforced by those commissioning usability evaluations. Commissioners may simply want results as quickly as possible and may not necessarily fully appreciate the issues that effect the validity and usefulness of the results. But the question remains, how useful are quick and dirty evaluations — are they useful in assessing usability and highlighting possible usability bugs or is there a danger that they can produce misleading results leading to misguided usability assessments and design decisions? '

This paper reports an investigation of the validity and usefulness of a quick and dirty evaluation of a business telephone. Two main research questions were addressed:

• Did the outcomes of the quick and dirty evaluation lead to correct inferences about the usability of the telephone for particular tasks?
• Did the outcomes of the quick and dirty evaluation reveal a significant proportion of the usability bugs associated with the telephone?

The approach taken was to compare the outcomes of an informal usability evaluation of the telephone with the outcomes of a more rigorous evaluation of the same product.

Method

The evaluation of a business telephone consisted of four parts.

1. Think aloud protocol. In a laboratory based study, four subjects, who had never used the phone before, were set a series of tasks. Success or failure on these tasks was recorded as were subjects verbalisations during evaluation session. Each subject was set the tasks in the same order, thus there was no attempt to balance for possible learning effects. A number of symbols were used as labels on the telephone — users were asked to complete a questionnaire to indicate what they thought each of these symbols stood for.

The tasks set were: Phone an internal extension, phone an external number, and a series of programming tasks(e.g., program a telephone number under a function key).

The symbols included on the questionnaire were for the following functions: to hold a call, microphone mute, store, message waiting, incoming call, transmit, radial last number, recall, follow me and data call in progress.

2. Interviews with 'trouble-shooters'. The telephone had already been in use within an office environment. As the telephone was fairly complex, support was available for users in the form of trouble-shooters who would help or advise people when they had problems using the phone. Questionnaires were issued to the trouble-shooters asking about the problems that users had had and the complaints that users had made about the telephone. Trouble-shooters were also asked to give their opinions as to how the design of the telephone could be improved from a user's viewpoint.

3. 'Expert' appraisal. The phone was analysed in terms of its adherence to, or violation of, the principles of design for usability. Examples of some of the design criteria against which the phone was checked can be found in Ravden and Johnson (1989). This part of the evaluation was, then, non empirical.

4. Survey of users. Thirty-one users with experience of the telephone took part in a survey. Firstly, they were asked to fill in a feature checklist to indicate which features they used at all and which of these they used frequently. For a description of feature checklists as a usability evaluation technique see Edgerton (in press). They then completed a usability questionnaire which was largely based on the System Usability Scale (Brooke, in press). The aim of this was to get an overall picture of the usability of the telephone. A series of further questionnaires then asked about factors connected with learnability, feedback, and functionality. The users were then given a sheet of paper containing pictures of the symbols used on the phone and were asked to identify what each of them stood for. Finally, the users were asked if they thought that there were any functions missing from the phone and were asked for any general comments or criticisms of the phone which they may have had.

When considered in isolation, the first part of this study — the think aloud protocols — is fairly similar to the sort of quick and dirty evaluation that is often carried out in a commercial environment. The main measure of usability was task success or failure supplemented with qualitative data from subjects' verbalisations, the users had never previously used the telephone, no attempt to control for learning effects was made in the task ordering, and the evaluation was conducted under laboratory conditions.

The other stages of the study, when taken in combination are more representative of the more rigorous type of evaluation that human factors specialists might ideally like to carry out. In particular, stages 2 and 4 involved gathering data from users who had used the telephone in a work context (although in stage 2 this data came via the trouble-shooters). If stages 2 and 4 are taken to be representative of a comparatively rigorous empirical evaluation and stage 1 is taken to represent a typical quick and dirty evaluation, a comparison of the outcomes of stage 1 with the outcomes of stages 2 and 4 might be a means of comparing an informal evaluation with a more rigorous empirical one. As stage 3 was not empirical it will not be considered further for the purposes of this paper.

Results And Conclusions From Quick And Dirty Evaluation (Stage 1)

The telephone 's symbols, used on the questionnaire, were labels for the following functions: hold, mute, store, message waiting, incoming call, transmit, recall, follow me, data call in progress. For seven of these none of the subjects were able to correctly identify what the symbols stood for, one of the symbols was identified correctly by one subject only and the other by three of the four subjects. Based on this quick and dirty test, then, it might be predicted that, in general, there would be a problem understanding what the symbols stood for.

No subjects had problems placing a call either internally or externally, thus the second prediction from the quick and dirty study would be that there would be no problem with these basic functions. However, the final task — programming a number under a function key — caused considerable problems. Each of the subjects either gave up or needed assistance from the experimenter. Thus, the prediction would be that programming tasks would cause users problems.

The main issue to arise from subjects verbalisations during the quick and dirty evaluation concerned the user manual. All four subjects had problems finding the information they needed to complete the tasks set. The associated prediction, then, would be that the user manual was 'unfriendly'.

Comparison With Rigorous Evaluation (Stages 2 And 4)

Having used the outcomes of the quick and dirty stage of the evaluation to make a number of predictions, data from stages 2 and 4 can be used to see if the predictions are in line with what would have been predicted from a more rigorous evaluation. This section reviews the apparent validity of the predictions made on the basis of stage 1 of the evaluation based on the outcomes of stages 2 and 4.

Prediction 1: Symbols used were not meaningful. This prediction received partial support from stages 2 and 4. From the interviews with the trouble-shooters (stage 2) it emerged that users often asked about the meaning of icons. However, of the 31 users who filled in the questionnaire about the symbols (stage 4) between 22 and 30 users correctly identified the meaning of each symbol. Thus it would appear that problems with the symbols may only occur at the early stages of interaction.

Prediction 2: Making standard internal or external calls would cause no problems. This was largely supported by the outcomes of stages 2 and 4 of the evaluation. The trouble-shooters didn't mention any problems relating to these basic functions. Similarly there were no complaints or comments about this from the users, except that it should, perhaps, be made easier to distinguish between an internal and an external call.

Prediction 3: Programming tasks would prove difficult. This was supported by the outcomes of stage 2 of the evaluation, as three of the most common queries reported by the trouble-shooters related to programming tasks. Also, two of the four tasks that the trouble-shooters said users had trouble understanding related to programming and of the five design changes that the trouble-shooters suggested two related to programming. However, there was nothing in stage 4 of the evaluation to support this prediction.

Prediction 4: Manual is not user friendly. This prediction was supported by the outcomes of both stage 2 and stage 4 of the evaluation. Trouble-shooters noted that there had been difficulty in using the manual. Responses to the user questionnaires indicated that the manual was rated rather negatively and amongst users' comments and criticisms were comments indicating that the manual should be made more user friendly.

The predictions made on the basis of the quick and dirty evaluation are summarised in table 1, along with indications of whether or not they were supported by the outcomes stages 2 and 4 of the evaluation. Three of the predictions made were supported by the outcomes of both stages 2 and 4, the other two were supported by one or the other.

PREDICTION FROM STAGE 1	SUPPORT FROM STAGE 2	SUPPORT FROM STAGE 4
1. Symbols not meaningful.	Yes	No
2. Internal and external calls no problem.	Yes	Yes
3. Programming tasks difficult.	Yes	No
4. Manual not user friendly	Yes	Yes

Table 1. Predictions made from stage 1 of the evaluation. Were they supported by the outcomes of stages 2 and 4?

Usability Bugs

The second research question to be addressed in this paper relates to the proportion of usability bugs that were revealed by the quick and dirty usability evaluation. In all, nineteen usability bugs were identified from stages 2 and 4 of the evaluation, five of which could have been picked up from the outcomes of the quick and dirty evaluation and one of which could possibly have been picked up, depending on how the analyst interpreted the data.

In stage 2, trouble-shooters comments about users' problems, and the users comments were used as a means of identifying bugs. In stage 4, bugs were identified from users' responses to open ended questions asking for comments or criticism that they might have with respect to the phone. It might also be argued that data gathered from the feature checklists could give an indication of possible bugs — perhaps functions that were rarely or never used were avoided due to usability problems. However, this is not necessarily the reason why they were avoided — it might be that the users simply had little or no need for these functions — thus the feature checklists were not considered for this part of the analysis.

Table 2 lists the bugs that were found from stages 2 and 4 of the evaluation and indicates whether or not they could also have been identified from the quick and dirty evaluation. Bugs discovered in the quick and dirty analysis could be identified from three potential sources — incorrect responses on the symbols questionnaire, problems in task completion or failure to complete tasks, and verbalisations during task completion. Where bugs could have been identified from the quick and dirty evaluation, the source of identification is listed.

USABILITY BUG (DETECTED IN STAGES 2 AND 4)	DETECTED IN STAGE 1?	SOURCE
Programming	Yes	Tasks
Symbols	Yes	Symbol questionnaire
Call pick up	No	—
Holding a call	Yes	Symbol questionnaire
Manual	Yes	Verbalisations
Feel of buttons	No	—
Speaker volume	No	—
Connection cords Ageing	No	—
Feel of handset	No	—
LCD display	No	—
Use of colour	No	—
Operator function	No	—
Transfer	Yes	Symbol questionnaire
Feedback on screen	?	Tasks / Verbalisations
Inconsistencies	No	—
Screen size	No	—
LED visibility	No	—
Key layout	No	—
Distinguishing internal/external calls	No	—

Table 2. Usability bugs. Could the bugs identified in stages 2 and 4 of the evaluation have been identified from the outcomes of stage 1?

The reason for the doubt over whether poor feedback on the screen could have been identified is that,while it was clear that users didn't always know what the results of their actions were, it was not necessarily clear that this had come about through poor on-screen feedback. It could have been, for example, that the telephone was designed in such a way that the functions of the keys were not explicit and that this was causing problems. It might equally have been that the problems arose due to other types of feedback that were inadequate. Certainly, a post hoc look at the outcomes of the quick and dirty study indicate that inadequate screen based feedback would be a possible explanation for these problems, but on the evidence of the quick and dirty study alone this would not be the only reasonable explanation.

Other than the issue of feedback and the issue of the users indicating that the manual was difficult to use, the other bugs discovered from stage 1 related directly to the tasks set (as in the difficulties with programming) or to the symbols that were asked about on the questionnaire. It would, presumably, be predictable that users might have trouble with holding a call as the symbol for this was one that subjects had difficulty in identifying. Indeed, a prediction from stage 1 was that the symbols as a whole would prove difficult to identify, thus it might also have been predictable that users would have problems with the transfer function as the button required to activate this is also marked with a symbol (although this symbol was not one of those asked about in stage 1).

Although a large majority of the bugs listed could probably not have been identified from stage 1 of this evaluation, it may still, in principle, be possible to detect most of these sorts of bugs from a quick and dirty evaluation. Perhaps if more subjects had been included or if a wider range of tasks had been set, then more bugs would have been discovered. So, the low bug detection rate may not be an indictment of quick and dirty evaluations in general, but simply a reflection of the design of this particular evaluation. Issues associated with developments over a period of time — such as the ageing of the telephone cord — are unlikely to be picked up over a short session with novice users. However, if the users chosen for the quick and dirty evaluation had had previous experience of use with the system, then it seems possible that these sorts of bugs could be picked up.

It should be noted, however, that the remedies suggested above for improving the bug detection rate of informal evaluations, each require a greater degree of rigour to be applied to aspects of the evaluation design — setting a more appropriate range of tasks and using a larger sample of more representative users. These suggestions, though, effectively amount to making the evaluation less quick and dirty. The implication may be, then, that rates of bug detection are likely to improve as the design of the evaluation moves along the continuum away from informality and towards rigour. Presumably, there will be a trade-off between the costs involved in rigorous evaluation and the degree and value of bug detection. Perhaps a future research goal might be to look at the parameters operating on this trade off, so that usability practitioners have a basis on which to set the degree of rigour and formality to be designed into an evaluation.

Conclusions

The predictions made about usability on the basis of the quick and dirty evaluation were largely supported by the outcomes of the other parts of the evaluation. This indicates that the quick and dirty test was effective as a means of addressing specific usability issues. However, this quick and dirty evaluation was far less effective at picking up bugs, revealing only about a quarter of those found from the more rigorous parts of the study.

Clearly, it would be rash to try and generalise on the basis of this one-off study. However, as a first pass at this issue, the results suggest that quick and dirty evaluation is likely to be effective when the central usability issues can be clearly identified and a set of tasks defined to address these issues. However, as a general scan for bugs quick and dirty evaluations may be less effective.

References

Brooke, J., in press. SUS — a quick and dirty usability scale. In P.W. Jordan, D.B. Thomas, B.A. Weerdmeester and I.L. McClelland (eds.), Usability Evaluation in Industry. London: Taylor and Francis.

Edgerton, E.A., in press. Feature checklists: a cost effective method for 'in the field' usability evaluation. In P.W. Jordan, D.B. Thomas, B.A. Weerdmeester and I.L. McClelland (eds.), Usability Evaluation in Industry. London: Taylor and Francis.

Jordan, P.W., 1994. Ecological validity in laboratory based usability evaluations. In Proceedings of the Human Factors and Ergonomics Society Conference 1994. California: Human Factors and Ergonomics Society.

Ravden, S.J and Johnson, G.I., 1989. Evaluating Usability of Human-Computer Interfaces: a Practical Method. Chichester: Ellis-Horwood.

TOWARDS LOW-COST VISUAL SEARCH STUDIES

Derek Scott

Interactive Systems Centre,
University of Ulster at Magee,
Derry City,
N. Ireland   BT48 7JL

By way of a reaction against the tendency towards ever-increasingly expensive equipment in a constantly decreasing financial situation, this paper presents six pilot studies concerning aspects of visual search by using readily available and "free" materials. A gender-differences paradigm was employed. Tests included a word search, degraded (mosaic) faces, facial recognition from eyes-only displays, two studies of commercial advertisement conspicuity, and identifying differences between two similar prints.

Introduction

As ergonomists working either within the academic community or within industrial establishments are likely to be aware, financial constraints abound. At the same time, competitive and state-of-the-art research calls for increasingly sophisticated and expensive equipment. Within teaching establishments such as experimental psychology laboratory courses, there is a general trend towards spending a substantial portion of an equipment budget on networked computer suites and the associated software and technical support demands. Visits to many university teaching laboratories clearly show that for much of the time PCs and Macs are standing idle, saved by being given an occasional "outing" by a particularly diligent student practicing her word-processing skills. Although it would be a regressive step to advocate a total return to the days of paper-and-pencil experimentation, it may well be time to pause for thought and reappraise what can be achieved with readily available materials. As we all know, it is more important to ask the right questions, and to be methodologically sound, rather than to have the "best" equipment.

This paper presents six pilot studies, offered more by way of advocating a reaction against this tide of expenditure, than for the importance of testing hypotheses. A gender differences paradigm was adopted. The tests employed included a word search, facial recognition of degraded (by the mosaic effect) images, facial recognition from eyes-only displays, two studies concerning commercial advertisement conspicuity, and identify differences between two similar prints.

Study one: Word search

This study employed a word grid taken from a competition within a popular Sunday newspaper magazine. The task essentially was to locate and circle eight words within an array which could be read in any direction, including diagonally. Conveniently, four of these words would generally be regarded as male oriented words (e.g. toolset, camcorder) and four as largely female oriented (e.g. fridge, perfume). It was hypothesised that due the feature recognition "pop-out" effect (Treisman, 1985), that males would more readily perceive male-oriented words than female-oriented words and, conversely, that search times for females would be quicker for female-oriented words than for male-oriented words.

Twelve subjects (6 males, 6 females) took part. The list of words to be located was balanced between males and females so as to vary the order of presentation; i.e. a male-oriented word was not always presented first. The Experimenter recorded times required to locate each word, and as such the order of location was also recorded.

Analysing for order located and also for cumulative times for male- versus female-oriented words produced no particular differences between the performance of male and female subjects.

Study two: Facial recognition of degraded images

A previous study (Scott, 1994) examined how degraded facial images (such as those perceived from low bandwidth video-communications systems) might be inadequately perceived by the viewer. The present study was designed to test whether males versus females could recognise familiar (male versus female) faces when degraded by the "mosaic" procedure.

Three males (mean age = 28, sd = 5.8) and three females ($\bar{X}$ = 36, sd = 5.3) participated.

Table 1: Results from visual search performance (mean times in seconds with standard deviations in parentheses) of male and female subjects recognising male and female degraded facial images.

	Male Subjects	Female Subjects
Male Faces	18.917 (2.538)	6.625 (3.498)
Female Faces	12.983 (8.698)	9.542 (5.818)
Combined	14.271 (5.777)	9.583 (4.294)

The small N involved did not warrant inferential statistical analysis, but the (very limited) data suggests that these males were superior at identifying distorted facial images of people generally, in contrast with the female participants. The male subjects were almost three times better at recognising the degraded male faces than were the females.

Study three: Eyes recognition

Although not properly a visual search study, this pilot is included here as part of the battery of gender differences studies. Once again, however, this study explored the use of readily obtainable materials; namely part of a weekly competition in a Sunday newspaper magazine.

The standard colour prints of the eyes (only) of well known figures were pasted onto postcards. The eyes of seven males (e.g. Jean-Claude van Damme and Tom O'Connor) and seven females (e.g. Jerry Hall and Sarah Ferguson) were involved. Subjects (3 male, 3 female; mean age 31 (sd = 7)) had simply to correctly identify the person whose eyes were shown..

Table 2: Mean recognition times (seconds) for the combined average performances (discounting those over 30 seconds) and error rates (including performances cut-off at 30s), with standard deviations in parentheses.

	Gender of eyes shown			
	Male		Female	
	Times	Errors	Times	Errors
Male Subjects	10.298 (12.860)	5.333 (2.082)	11.375 (6.187)	4.667 (1.527)
Female Subjects	3.250 (0.053)	5.000 (1.000)	3.389 (1.398)	5.000 (1.000)

Whilst female subjects appear to have excelled their male counterparts in speed of recognition of the eyes, irrespective of gender, there are clearly no differences concerning whether the stimulus materials were male or female. It seems likely that some males/females might be more likely to recognise the opposite gender where subject and object are of similar ages, and also where there is a degree of sexual attractiveness. For instance, and concerning displays used in this study, a young male might be more likely to recognise the eyes of Linda Lusadi, a young female those of Patrick Swayze, and more mature males and females the features of Terry Wogan or Elizabeth Taylor. It should also be added that the floor effect concerning the high number of errors recorded was regarded as surprising, particularly as there was no degradation involved in terms of size or colour from the original prints.

Study four: Books and videos

As with the aforementioned study, book-club, and more recently, video-club, enrollment advertisements must consider their target audience and advertisement strategists go to great lengths to capture the potential market. It is worth considering, particularly as the pages of the magazines which typically house these advertisements tend to be skimmed over, whether particular record, book or video covers stand out within visual attention.

Materials consisted of full-colour page advertisement from magazines. There were six pages relating to book clubs and two pages relating to video clubs. From each page, subjects were required to locate one book/video with a feminine image package (e.g. "Delia: Summer Collection" or "Sister Act") and one with a male visual presentation (e.g. "SAS Omnibus" or "Terminator 2").

There was little difference between means (and standard deviations) for males looking for male- versus female-oriented covers (14.4s (sd = 10.9) vs 12.0s (sd = 10.7), respectively), or between means for females looking at male- versus female-oriented covers (11.7s (sd = 11.0) vs 8.6s (sd = 8.3), respectively).

Study five: Consumer (catalogue) goods search

The consumer population is often bombarded with what may be regarded, on the one hand, as junk mail or page wasting magazine space yet, on the other hand, as an opportunity for unemployed house-bound people to constructively earn money. One form of attracting agents is that of catalogue sales advertisements where, as part of the visual advertisement, the reader is asked to choose a gift from an array of goods. Although both male-oriented and female-oriented goods are on display to attract the consumer, the question remains---particularly as it might be expected that the typical responder be a female-- whether there is a gender bias in these displays. In other words, are some of the items displayed more conspicuous to men and do other, more female-oriented items, stand out more with female consumers?

In this study, subjects (3 male, 4 female) were presented with sixteen arrays of consumer "gifts" which they were required to search through for specified items. Half of these were male-oriented items (e.g. "7 piece gardening set" and " household toolkit" and half female-oriented items (e.g. "3 piece casserole set" and a "Ladyshave"). Recognition times were recorded by stopwatch.

Search time means for each male and female subjects on search times for male and female objects were subjected to a two-way ANOVA, which showed no main effects or interaction (see Table 3).

From this very provisional study, it can not therefore be concluded study that in this sort of commercial array which is designed for its visual appeal that what may be regarded as gender-specific domestic items do particularly pop-out within visual attention.

Table 3: Overall mean of the mean times (seconds) from each block of
five subjects (standard deviations in parentheses).

		Object gender orientation	
		Male	Female
Subject gender	M	5.974 (2.036)	6.148 (2.290)
	F	5.186 (3.064)	4.993 (1.907)

Study 6: Eye movement monitoring

Eye movement monitoring studies of a two dimensional format (see Scott, 1993; for review) necessitate not only equipment costing many thousands of pounds, but also unless equipment costing tens of thousands of pounds is available , requires an invasive procedure such as the scleral search coil technique, whereby a local anaesthetic needs to be applied to the subject's eye to reduce the severe discomforfort and the attendant risk of scratching the cornea whilst fitting and removing the search coil.

An approach was made to allow subjects to self-monitor their eye movements. Equipment consisted of ten different pairs of photographs differing between the pairs in that one of each pair had been modified by darkroom techniques to have at least four minor differences to its partner. Once more, these were taken from trivial competitions in a magazine. Three of these pairs consisted of feminine type scenes, three containing only males, three of couples, and one of a man (Julian Clarey) holding a dog! Three female and two male subjects were instructed (along with a demonstration) to follow their eye movements whilst searching for dissimilarities with a pen, and not to allow the pen to leave the paper. At the time of writing these qualitative data from 50 trials await analyses, but it is expected to quantify this for gender differences by way of transition matrix analysis as the present author has been involved in relation to using the scleral search coil technique (Ponsoda et al., 1994). Essentially, this procedure provides ratios of where, after a saccadic eye movement in one direction and a fixation, the direction of the subsequent saccade is likely to be.

References
Scott, D. 1993, Visual search in modern human-computer interfaces. *Behaviour and Information Technology*, **12**, 174-189.

Scott, D. (1994) Video telecommunications and non-verbal cues. Submitted to *Displays*.

Ponsoda,V., Scott, D. & Findlay, J.M. (1994). A probability vector and transition matrix analysis of visual search on VDU-based tasks. Accepted for publication in *Acta Psychologica*.

Treisman, A. (1985) Features and objects in visual processing. *Scientific American*, **246**, 106-115.

APPLYING ARISTOTLE'S THEORY OF POETICS TO DESIGN

Steven Clarke   **Patrick W Jordan**   **Gilbert Cockton**

Department of Computing   *Philips Corporate Design*   *Department of Computing*
Science   *Philips International B.V.*   *Science*
University of Glasgow   *Building SX*   *University of Glasgow*
Glasgow   *Postbus 218*   *Glasgow*
G12 8QQ   *5600 MD Eindhoven*   *G12 8QQ*
                               The Netherlands

In the book, 'Computers As Theatre', Laurel (1991) considers the potential
contribution that Aristotle's theory of Poetics can make to software design. This
paper reports a study in which Laurel's ideas were used as the basis for the design
of an information retrieval system. There were two main research questions: did
the principles outlined form a practical basis for making design decisions and did
software designed according to these principles bring measurable benefits to the
users? In order to address these questions a designer's experiences with applying
the principles throughout the design of an information retrieval system were
recorded and the usability of the finished system was compared to that of a
'traditionally' designed system via an empirical evaluation. The outcomes of this
evaluation suggested a number of benefits in designing according to these
principles — in particular in terms of making an interface more engaging and
enjoyable to use. However, the designer felt that some of the principles outlined
were difficult to translate into concrete design solutions.

Introduction

In the book 'Computers as Theatre', Laurel places a large emphasis on the
representation of action. Laurel highlights the representation of action as the major element
which theatre and human computer interaction have in common. Given that this
representation of action is one of the fundamental elements of both theatrical principles and,
in Laurel's view, the design of human computer interfaces, Laurel suggests that benefits may
accrue from applying theatrical principles to the design of an interface. Her view is that the
designer should view the computer as a medium through which interaction and collaboration
between human and computer can take place. Both human and computer exist together in this
'virtual world' where all that is represented is all that is important. The representation should
suggest actions and assumptions made for both human and computer.

Unlike 'traditional' HCI, which focuses on representing objects and the methods for
interacting with those objects, representing action requires designers of human computer
interfaces to adopt a different style of reasoning. Human computer interfaces could represent
action in the same way a play does. Plays can engage their audience to such an extent that the
audiences become so engrossed in the play they believe it is actually happening. Anything
which happens in the play which seems unbelievable may be immediately recognised as such
by the audience. This level of engagement can be to the playwright's advantage, since the
audience may believe anything which seems remotely believable, such as plot twists. It also
enables the audience to make educated guesses as to what will happen next. User interface
designers have apparently not yet been able to produce such an engaging feeling in the users
of their interfaces, but since both user interfaces and plays are concerned with representing
action, could human computer interfaces become just as engaging as plays?

Laurel suggests that an approach to achieving this might be to use Aristotle's notion of the structure of plays and apply it to human–computer interaction. Knowledge about an object's structure may make it easier both to evaluate that object in terms of what it is supposed to do and to design the object.

The Structure of Human–Computer Interactions

In his theory of poetics, Aristotle defined six elements which together constitute a play. According to him, these elements must be carefully designed so that they combine in a pleasing, engaging and cathartic manner. Laurel has taken these six elements and placed them in the context of designing human computer activity. A brief description of these elements, as well as examples of their possible application in human–computer interface design, follows.

Enactment: This refers to the senses being employed in the representation. Sensory representations should be chosen carefully so that they are consistent with the whole representation. Human–computer interface designers should choose carefully what sensory phenomena they employ to represent the interface. Using sound in an interface for use by users who are hard of hearing or use of colour in an interface used by colour blind people would be extreme examples of poor choices for the element of enactment.

Pattern: All sensory phenomena employed by the representation should be structured into some kind of pattern. Patterns are pleasurable to perceive whether or not they are used as semiotic devices. Human–computer interface designers should structure the representational phenomena appropriately, or, in other words, they should strive to produce a consistent interface.

Language: A computer interface employs some kind of language in order to facilitate some task. The languages used need not be based on words, but instead can be based on signs and symbols, non-verbal sounds or animation sequences. In human–computer interaction, individual elements of the enactment (graphical signs, symbols, words) can be used to communicate between the computer and it's users.

Thought: Thought is important in that it shapes what an agent communicates. Computer based agents do not have to think; at most they just have to appear as if they are thinking. If agents are successfully portrayed as thinking then other agents, both human and computer, may be able to infer possible actions the agent may take.

Character and Agency: Computer based agents are a representation of bundles of code which when executed perform some operation on behalf of the user. An agent's character and traits determines what it thinks about and therefore in turn, what it may do. Human agents and computer based agents should cooperate to achieve certain goals, normally determined by the human agent.

The Whole Action: The object of a dramatic representation, according to Aristotle, is action. The action is much more important than the characters. Characters are present in order to portray the action. Each character's individual actions help the whole action progress from beginning, through middle, to end. Any single action which does not have a meaningful role with respect to the whole action is a gratuitous action and should therefore not be included. Designers should aim not simply to represent the objects of an application, such as documents and folders, but more importantly, the whole action which relates these objects such as writing a book.

Any representation of action which is adhering to Aristotle's theories should contain these six elements. Each element has ramifications for the feel and look of the representation. These elements alone however are unlikely to produce a satisfying representation. In order to satisfy their audience/users, the playwright/designer must also be aware of the different techniques available for *orchestrating action*.

Orchestrating Action

Laurel discusses how, like a play, the potential for action is almost limitless before the computer has booted up or the play has started. We may know what the play is "about" through the programme. Likewise we may know what we can do with our computer since we know what software we can run on it. But in both cases we are almost never sure how the action will progress and how it will end.

In a play, the potential for action decreases constantly as actors appear on stage, as the dialogue progresses. One character's actions will normally constrain another character's actions and so on until the final climax when normally the protagonist has only one final course of action open to him or her. The same can be said for computer applications. When a user opens a word processor for example, the potential for action is decreased since the user can now only perform word processing actions. But because most computers now have some form of multi tasking, whereby users can use different applications at the same time, there are multiple potential's for action. Each opened application opens new potential for action but that potential will always decrease as the user continues with his or her work. One could imagine analysing the action from a higher conceptual level than individual applications. If the work the user is doing in all the open applications is related then there becomes only one potential for action which is decreasing as the user uses all the applications.

It is important to be aware of this progression from almost limitless potential for action to zero potential since bad design decisions can hinder the user in progressing to the completed, "zero potential" state for action.

Constraints

One of the most fundamental aspects of creating a "virtual world" and helping the users make the progression from limitless to zero potential for action is the appropriate use of constraints. Constraints can be anything from gentle suggestions to alarming dialogue boxes They can be used to further the action, since if the representation is good enough users will understand the constraints and not attempt actions which are unsupported. Therefore it is better to place constraints in terms of the virtual world.

Evaluating Laurel's approach

A study was undertaken in the context of an undergraduate honours project to evaluate Laurel's approach as described above. The aims of the study were to determine if the application of Aristotle's principles formed a practical basis for making design decisions and if software designed according to these principles brought benefits to the users which might not otherwise have been the case. There were, therefore, two parts to the study. The first part involved building a system according to these principles and recording the designer's experiences in trying to apply the principles. This would help to indicate whether this sort of approach could prove practical. The second part of the study was a user evaluation of the resulting system. The aim of this evaluation was to see if this approach to design would bring measurable benefits to the user.

Applying the principles to a design.

A small information retrieval system was designed and built guided by the approach advocated by Laurel. The principles influenced the system as well as the manner in which it was designed in a number of ways. This section describes the system built and particular design decisions which came about through application of Aristotle's principles.

The Film and Theatre Studies Department, University of Glasgow, had identified the need for a computer based system to support students when preparing essays, presentations etc. At the time, the only available resource for this purpose was a Library. This was felt to be unsatisfactory since the library could only afford to keep one or two copies of each of the relevant source materials yet there are consistently, on average, two or more students who need access to these materials. Furthermore, once a student has access to the material, much time is wasted as the student searches for the relevant section or page within the book or magazine. Due to these difficulties, it was proposed that a computer based system be implemented that would better support students in this process. It was decided that the first author of this paper would design and implement a small but usable prototype system to address this need and that the system would be designed according to the approach suggested by Laurel. The system would present students with a variety of material concerning the playwright Edward Bond and his play 'Saved'. 'Saved' gained notoriety in the 1960s for its depiction of violent youths and was subsequently banned from public performance only to be critically acclaimed a few years later when the ban was removed. The system was to present material on the playwright, the play and the reaction to it.

The first stage in the design was to determine what action should be represented by the system. The library supported accessing documents but as described above this may not have been supported appropriately for this context of use. More importantly however, students may not be interested in how they access documents, they may be more interested in the information presented in the documents and how that information might be of use. Therefore, the computer based system would have to be designed with an emphasis on how best to *experience* the information, rather than how to access it. Laurel uses an information retrieval system as an example to illustrate definition of the 'whole action' . Through reading this example the designer concluded that experiencing, rather than accessing information, should also be the whole action for this system.

This suggested to the designer that information should be presented in its original form since this would preserve all the information present in the particular medium and give the student a richer experience of the information. Instead of presenting transcripts of videos, the actual video should be shown. It was also decided that computer based agents should be employed which guide students, not only to appropriate documents, but also to sections within those documents, thereby removing much of the burden of accessing documents from the student. The agents suggest documents or parts of documents to the user by examining the document the user is currently looking at and determining whether there are other related documents which might be of interest. It was decided to implement three agents. The agents were interested in the playwright and the play, the production of the play, and the reaction to the play respectively. Therefore, the playwright agent would only suggest further documents if the current document was discussing something to do with the playwright and if there were other related documents. It was felt that through using three instead of one agent, users should be better able to determine what sort of document the agent will suggest before it actually suggests it. The user could then save time viewing an agent's suggestions by simply ignoring suggestions from any particular agent that the user does not share a common interest with. These agents, together with the human agent, would form the characters which would help the action progress to its completion or catharsis.

It was decided that the whole system would be designed in accordance with a library metaphor, which would allow students to browse at their will but also to take full advantage of the 'librarians' (the computer based agents) who would guide the students to relevant material. In keeping with the library metaphor a note taking facility was provided that would allow students to copy any text from any document but would prevent them from cutting that text. All notes were stored for later use, rather than storing the most recently taken note which is the normal behaviour of most cut and paste style text editors, since the designer felt that the former behaviour better matched the way students take notes in a library. Figure 1 shows a sample screen of the system.

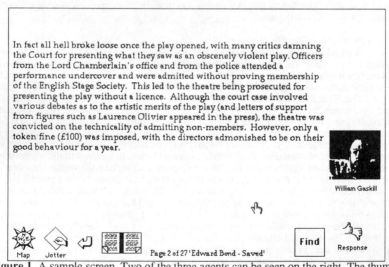

Figure 1. A sample screen. Two of the three agents can be seen on the right. The thumbs up/thumbs down symbol represents the agent concerned with reaction to the play, whilst William Gaskell is the agent concerned with the production of the play.

Trying to apply Aristotle's principles influenced the design process itself. Due to the emphasis on the whole action, it was necessary to try to gain an understanding of how the students approached their work and prepared for essays. Without such an understanding it would have been difficult to produce a system that fully met the students' needs. An iterative approach to design was taken, showing the system to students and staff and watching them as they used the system, recording critical incidents and comments made. This approach facilitated the designer in learning about students' work by watching them, and the students, in turn, were able to contribute to the design process by discussing the various possibilities for the next iteration of the system.

Evaluating benefits to the users.

Once the system was complete it was evaluated against a similar system which had been designed and built without using Laurel's approach by another designer. There were some similarities between the two systems, such as the method for taking notes, however, the differences between the two were enough to make comparison meaningful.

One of the difficulties in designing the evaluation was to find a method which would capture students' subjective experience of using the interface. Traditional empirical techniques generally test interfaces for phenomena such as task performance, error rates etc., whereas here it was important to capture subjective measures of the user. The most appropriate technique available seemed to be the *think-aloud protocol*. In this method, users are asked to explain what they are thinking about while using the interface and so a lot of information can be collected about what the user knows, what they expect to happen etc. It was important to see how they described using both systems, to see if one seemed more engaging than the other. This would be apparent in the way they referred to the system. If they explicitly referred to the computer throughout their interaction with the system then the interaction could not be described as very engaging. If they referred to themselves a lot instead of the computer however, e.g. "I am going to see what the William Gaskill agent has to say about this document" then the interaction might be described as quite engaging.

Study

Eight subjects were asked to perform a number of information retrieval based tasks on both the 'new' system (i.e. the one designed in accordance with Laurel's approach) and the 'traditional' system. and to describe what they were thinking and how they felt while doing so. Vivzi (1992) reports that when using a think-aloud protocol, 80% of usability problems are detected with four or five subjects. Furthermore, additional subjects are less likely to reveal new information and the most severe usability problems are likely to be detected by the first few subjects. Since the intention was to investigate more than only usability problems it was felt that eight subjects would be more appropriate. After using each system, subjects completed a usability questionnaire which was used to calculate a usability score for each system (the higher the better). Subjects then answered some questions on their experiences of using the system.

The eight subjects were all second year computing science students. The subjects were divided into groups of four. The first group used the new system first and the second group used the traditional system first.

Results

The questionnaires returned some interesting results. Average usability scores for each system were the same (36 out of 50, n = 8). After filling out the questionnaire, every subject, when asked, said that they preferred using the new stack, even though four of those subjects had given the traditional stack better marks in the questionnaire. The results from the questionnaire are presented in figure 2.

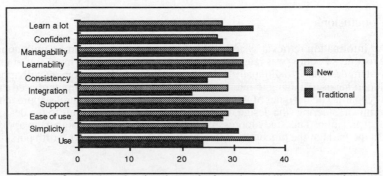

Figure 2: Bar chart showing the results from the usability questionnaires.

On the whole, reaction to the agents in the new system was positive but some subjects reported that they often felt that the agents took too much control away from them. It was generally agreed amongst subjects that agents would be more useful if they could be used when the user requested rather than only when the agent made itself available. Analysis of notes taken during the think alouds and questions asked of the subjects after the evaluation showed that they felt that the new system was much more engaging than the traditional system, all subjects rating it higher in terms of engagement than the traditional system.

As well as evaluating benefits to the users, the designer's experiences with using Laurel's approach were also recorded.

Discussion

The results indicate that users preferred the new interface to the traditional one, and that they felt more engaged whilst using it. From their comments, however, it was clear that users were still very conscious that they were interacting with a computer, and were, thus, not so engaged as they might have been if watching a film or a play. A probable reason for this is that users' interactions were still via keyboard and mouse. These may have been constant reminders that they were using a computer, making it difficult for them to immerse themselves in the 'virtual world' being represented.

Subjects' reactions to the new stack were positive. All said that they preferred it to the traditionally designed system and found it more fun to use. However, this preference was not reflected in the ratings given on the usability questionnaire, where there was no overall difference in the ratings given to the systems. When the responses to each of the questions were broken down individually, it appeared that the traditional system was perceived as being less complex and more learnable than the new design. However, subjects indicated that they would be more inclined to use the new system and felt that it was better integrated and more consistent.

The approach outlined by Laurel, then, helped to guide the designer towards building a more engaging interface, that was fun for the users. However, when reporting on his experiences of designing the interface in accordance with this approach, the designer noted that he felt that the principles as outlined were only enough to 'steer' him in the right direction, rather than forming concrete and unambiguous design guidelines. There appears, then, to be a lot of scope for interpretation within the approach as it now stands. The designer also noted that it took him a considerable time to formulate his approach to the design in terms of how to use the theory of poetics in order to help him to make design decisions.

Perhaps the most constructive way to consider Laurel's approach, then, would be to treat it as a means of steering the direction of a design and of highlighting the benefits of engagement and agency, rather than as a means of supplying designers with a concrete basis for design decisions.

Conclusions

An information retrieval system, designed with guidance from Laurel's approach, based on Aristotle's Poetics, was rated by users as more engaging and more fun to use than a similar system designed 'traditionally'. However, the level of engagement still appeared to be limited due to the users having to interact through the traditional computer based media of mouse and keyboard. The designer of the new system felt that whilst Laurel's approach was useful in terms of 'steering' the design, there was still much that was left to the designer's own interpretation. This study represents a 'first pass' (for this designer) in applying this approach On the basis of the outcomes of this study, the approach looks worth pursuing further.

References
Laurel, B. 1991, *Computers as Theatre*, 1st edn (Addison Wesley).
Vivzi. R.A. 1992, Refining the test phase of usability evaluation: How many subjects is enough?, *Human Factors*, **34** (4).

... BUT HOW MUCH EXTRA WOULD YOU PAY FOR IT? AN INFORMAL TECHNIQUE FOR SETTING PRIORITIES IN REQUIREMENTS CAPTURE

Patrick W. Jordan
Philips Corporate Design
Philips International B.V.
Building SX
Postbus 218
5600 MD Eindhoven
The Netherlands

D. Bruce Thomas
Philips Corporate Design
Philips International B.V.
Building SX
Postbus 218
5600 MD Eindhoven
The Netherlands

A semi-structured interview was used as part of a 'quick and dirty' study aimed at requirements capture for a product incorporating a printer. Three issues were of particular interest with respect to the printer: print quality, ease of cartridge loading, and 'eco-friendliness'. The aim was to evaluate the comparative level of importance that users attached to each of these issues.

Six users were asked to put a monetary value on the importance of each of the issues. They were told that the product in its basic form would cost around Dfl 1500 and asked how much extra they would be prepared to pay for good performance with respect to each of the three issues under consideration. Results indicated that print quality was most important (mean perceived value (mpv) = Dfl 358, sd = Dfl 168), followed by eco-friendliness (mpv = Dfl 300, sd = Dfl 137) with ease of loading last (mpv = Dfl 125, sd = Dfl 121). As might be expected with such a small sample size, these differences did not quite prove statistically significant when analysed with a non-parametric Friedman test ($p = 0.14$).

Using this 'costing' technique gives respondents a chance to anchor their responses in a fairly concrete context — that of making a purchase decision. This may give advantages over open-ended questions where users are asked for somewhat 'abstract' responses or rating scale questionnaires where the basis for the choice of response category may not always be clear.

Introduction

In the development of a product incorporating a printer, a decision needed to be made concerning the printer mechanism to be used. The product management identified their ideal solution as being to use a compact ink jet mechanism. However, at the time, a suitable mechanism was not available.

The alternatives to be considered were a thermal transfer mechanism and a large ink jet mechanism. The thermal transfer mechanism was cheaper and smaller, but concern was expressed concerning the reliability of the technology, the potentially poorer print quality and the possibility of a greater negative impact on the environment. Of these two, the large ink jet mechanism was considered to be technically the better solution, but it had the major disadvantage of being extremely bulky.

Five issues were identified as being critical to the decision as to the choice between the large ink jet mechanism and the thermal transfer mechanism:

- size
- print quality
- cost
- acceptance to manufacturers building products incorporating the product in development
- acceptance to end users.

In view of the importance of end user acceptance, ergonomists were invited to participate in the decision making process. Three issues were identified which it was considered might be of potential concern to users, and therefore influential to the success of the product on the market:
- print quality
- environmental friendliness
- ease of loading the ink cartridge or thermal transfer roll.

The intention was that having established the comparative importance of these issues to the users, the product management would be able to predict the technical solution which would be most acceptable, based on their knowledge of the properties of the alternatives under consideration. Ergonomists may not be involved very often in decision making concerning technology other than for input and output devices. This, then, was a fairly novel situation for which standard measures were not readily available.

Experience with questionnaires used in usability testing has suggested that simply to ask users for preferences might not provide sufficiently discriminatory results. When numerical scales are used to try to get an understanding of the relative importance of various issues, there may be a tendency for respondents to mark around the mid-point or just to the positive side of the mid-point. This tends to lead to bunching of the data, producing insufficiently discriminatory results.

When open ended questions are asked about users' values there is a danger that respondents be inclined to give answers which they feel will be 'socially acceptable', rather than those that really represent their underlying feelings or which reflect their actual behaviour. For example, whilst many people might claim that environmental factors are very important to them, this will not necessarily be reflected in a choice for more environmentally friendly products.

This paper describes a technique developed for use in requirements capture to get an idea of users' priorities — putting a cost value to the importance of an property. Users were asked to put a financial value to the importance of each of the three issues of concern identified. This was done through asking how much extra participants would pay for products exhibiting a high level of performance with respect to the issues mentioned. The premise was not that users would necessarily be expected to pay precisely what they said they would when confronted with a real purchase choice, rather it was hoped that the values put on issues would at least give an idea of their comparative importance.

Method and results

A small informal study was designed to investigate issues of potential importance to the users of a product containing a printer (neither the product itself nor the type of product can be mentioned for reasons of confidentiality). Six users participated, all of whom had experience of the type of product in question. Semi-structured interviewing was the basis of the study. Participants were interviewed individually.

After an open ended question relating to factors affecting purchase choice, users were asked about each of the three issues that were the subject of this investigation.

Print quality
Participants were shown print-outs of varying degrees of quality: a low quality print-out (fax), moderate quality (ink jet) and high quality (laser). They were then asked what would be acceptable in the context of the type of product under

investigation. Five of the six felt that fax quality would not be enough. However, all thought that ink jet quality would be sufficient. When asked how important print quality was in the context of this product all said that it mattered.

Ease of use
In the context of this product, ease of use was strongly associated with ease of loading and unloading the cartridge in the printer. To get an idea of this type of task users tried removing and replacing the cartridge of two types of ink jet printer — internal cartridge and external cartridge. All participants were able to complete the task successfully. Most participants said that the importance to them of easy cartridge changing would be dependent on how often the cartridge would have to be changed.

Ecological issues
When asked about the importance of ecological issues, all respondents said that they though that they were important in the context of this product. Responses ranged from quite important to very important. Participants identified the following issues as being central when considering eco-friendliness:

- Materials from which product manufactured.
- Type of ink used (chemicals used in it etc.).
- Power consumption of product.
- Paper type.
- Cartridge type.

Putting a value to the issues
As a rough test of the comparative importance of the three main issues that were looked at, participants were asked to put a monetary value on each of them. They were asked how much extra they would pay for good performance with respect to each of these issues. Participants were told that the product in its basic form would cost around about 1500 Dutch Guilders (Dfl). They were then asked how much extra they would be willing to pay for, respectively, good print quality, easy cartridge change and eco-friendliness. A clear pattern emerged. Print quality was the most valued issue, followed by eco-friendliness, followed by ease of cartridge changing. Table 1 summarises the responses, giving the mean perceived added value (mpv) attached to each issue as well as the number of respondents giving the issues 1st, 2nd and 3rd place ratings. As might be expected with such a small sample size, these differences did not quite prove statistically significant when analysed with a non-parametric Friedman test ($p = 0.14$). One participant put the same added value on ease of cartridge change and eco-friendliness. These were ranked joint second — thus the higher number of second place rankings than third place.

Table 1. Mean perceived added value and rankings per issue.

ISSUE	MPV (Dfl)	1ST RANKINGS	2ND RANKINGS	3RD RANKINGS
Print quality	358 (sd = 168)	4	1	1
Ease of cartridge change	125 (sd = 121)	0	2	4
Eco-friedliness	300 (sd = 137)	2	4	0

Discussion

This study shares many of the characteristics of the sorts of 'quick and dirty' user requirements investigation that ergonomists are often asked to carry out in a commercial environment. There was a very short time available (2 days) from design of the study to analysis of the data. This ruled out the possibility of any complex experimental design and inevitably restricted the possibilities for through controls and balances on data gathering. The commissioners of the investigation had to make a quick decision as to a suitable mechanism for the printer and had specified that an investigation based on the three criteria mentioned would form the basis of their decision. The intention was to discover what users' priorities were and then to decide which technology was most suitable on the basis of the commissioners' knowledge about the match between the properties of the technologies and the users' priorities.

The question for the investigation was straightforward — finding the comparative importance of three issues. The costing technique was quick to administer and by using price as a trade-off against other properties gave a benchmark for the importance of the issues looked at. The technique yields quantitative data. For some commissioners this may carry more weight than qualitative data. In particular it may appear more easy to interpret in a situation where a decision involving trade-offs was going to have to be made. Qualitative data such as, for example, that yielded from an open ended questionnaire, is perhaps more open to interpretation than numeric data provided directly by the subjects. In this study the responses to the open ended questions seemed to confirm that all three of the issues looked at were rated as being important, however the comparative importance of each is would be rather open to interpretation on the basis of responses to these questions alone.

Perhaps the major advantage of asking users how much extra they would pay for something is that it gives the possibility of anchoring responses in a context that is real to users (and to the commissioners) — making a purchase choice. Answering 'philosophical' open ended questions about how important something is, or being asked to rate something on a numerical scale, may not give users' such a meaningful reference point for making their judgements. After all, people are likely to have clearer ideas about the significance of spending an additional Dfl 100 on a product then they are about the difference between, say "important" and "very important" on a rating scale.

Limitations on interpretation

Although asking users about how much they would pay for something might be useful as an indicator of the comparative importance of various properties, it would almost certainly be inappropriate to interpret the figures themselves as really representing what people would pay. In order to have confidence in such figures it would be necessary to have evidence from real purchase choice situations. Whilst people may talk about spending a particular quantity of money on something, they may hesitate to actually spend it. In any case, issues of how much people would actually be prepared to pay for something seem to be in the domain of market research rather than ergonomics. It is important, then, that commissioners are not encouraged to take the results at face value, but simply as a guide to comparative importance.

Conclusion

As a means of establishing the comparative importance of various issues, respondents to a semi-structured interview were asked how much extra they would pay for a product if it exhibited certain properties. In this case, the issues raised were in the context of a product containing a printer. The investigation looked at the relative importance of three properties — print quality, ease of cartridge change and eco-friendliness. Results indicated that print quality was the most important issue for users, followed by eco-friendliness, with ease of cartridge change last.

Asking how much they would pay for something gives respondents the chance to anchor their responses in a concrete context — making a purchase. This may give advantages over open ended questions where somewhat 'philosophical' judgements may be given and over rating scales where the criteria for choosing a point on the scale may not be obvious.

Field Measurements of Seated Vibration Dosage

Dryver R. Huston[1], Michael R. Werner[1], James P. Tranowski[2], and Gerry Weisman[2]

[1]Mechanical Engineering Department,
[2]Vermont Rehabilitation Engineering Center,
University of Vermont,
Burlington, VT 05405

This paper describes the results of an effort to assess the vibration environment of vehicle operators. It has been hypothesized that whole-body vibration is a contributory cause in the low number of low-back pain claims and disabilities of vehicle operators. However, there are very few definitive studies that can make a direct link between the vibration environment and the observed back injuries. Some of the difficulties in performing such a study has been that there is no agreed upon vibration dosage measure. The focus of this study is to develop an inexpensive yet robust field vibration acquisition, data analysis and dosage recording system. The prototype data acquisition system is a portable PC with integrated circuit accelerometers as the transducers. The vibration data are acquired and analyzed in pseudo-real-time. The data are then converted to an equivalent dosage by frequency domain conversion and comparison with ISO 2631/1 standard. The result of prototype testing with an electric wheelchair will be reported.

Introduction:

Numerous studies have determined that there may be some correlation between whole body vibration exposure and low back pain Bongers, Hulshof, Dijkstra and Boshuizen 1990 and Boshuizen, Bongers and Hulshof (1990). The majority of these studies, however, stopped short of establishing a causal relationship Hulshof, Veldhuijzen van Zanten (1987). Difficulty in making such a determination may be due to the nature of these studies. Epidemiological studies frequently include little or no vibration data. When this data is included, it is generally not individual dosages but rather broad approximations of the actual dose Hulshof et al. (1987). Representative samples are taken from typical vehicles driven under normal circumstances. Resulting values are then multiplied by the number of years of driving experience the driver has to obtain a cumulative dose. The result is that the cited vibration dose is an approximation of the actual dose a rider receives Johanning, Wilder, Landrigan and Pope (1991). In order to establish a level of causality, individual long term dose must be obtained for both an appropriate control group and a group of test subjects. These dosages must then be considered when making observations regarding disease or degenerative changes of the spine.

The expense of tracking whole body vibration exposure histories on a large number of individuals is prohibitive. Units specifically designed to perform this task are currently available. The majority of these units consist of sound level monitoring equipment modified to record vibration exposure. Unfortunately these packages usually require a specially trained operator with Per unit costs that generally exceed 10,000 US. dollars. This makes most of them unsuitable for long term monitoring of a large number of subjects.

A number of vibration dose recording schemes have been published over the past two decades. Most of these are based on empirical levels of physical tolerance and/or comfort. ISO 2631/1 (ISO/TC 108. ISO 2631/1-1985(E)) is one of these and has gained fairly broad acceptance internationally. Because of this the ISO 2631/1 reporting strategy has been chosen as a model for reporting dosage levels. Frequency domain conversion and inclusion into 1/3 octave bands in accordance with ISO 2631/1 will facilitate convenient comparison between standard exposure limits and vibration dose.

The use of a laptop personal computer incorporating a PCMCIA type data acquisition card and inexpensive micro-machined integrated circuit accelerometers allows for a significantly less expensive, robust and more versatile system. Current technology will facilitate pseudo-real-time data analysis and reporting all in a user friendly graphical environment.

Equipment:

A Toshiba® 1950cs 50 MHz 486DX computer was chosen as a computer platform for this system. Cost, battery life, speed, and size were principal factors in making this decision. Power for the entire system is provided by the vehicle cigarette lighter via a 12VDC to 115VAC 50 watt standard adapter. The standard wall power adapter unit provided with the laptop computer is then plugged into the 115VAC supply. An adapter was built to hijack 18VDC from the Toshiba ® to supply power to the accelerometer amplifiers. Three IC Sensor® micro-machined integrated circuit accelerometers were chosen primarily because of their performance characteristics. The nature of the construction of piezoresistive accelerometers provides high over-range protection making them very robust. They provide DC response, are relatively inexpensive and in surface mount configuration they are very small. This was obviously a concern as the operator had to sit on the triaxial transducer pack. The three accelerometers were arranged on a mounting block in a standard X,Y, Z configuration. Simple temperature compensated quad OP-amp amplifier circuits for these accelerometers had to be made in house according to circuit information provided by the manufacturer. Initial models did not prove to be as robust as was thought

Results:

Representative samples of raw data in the time and frequency domain are shown below. Conversion to 1/3 octave bins and comparison with ISO 2631/1 standards was completed using a post processing routine during the initial system testing. Typical time histories and frequency spectra are shown in figures 2 and 3.

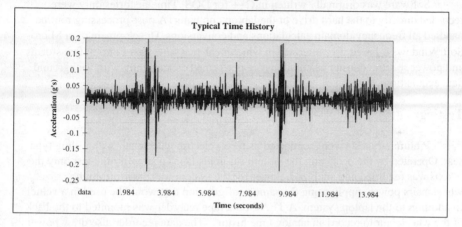

Figure 1. Typical Time History

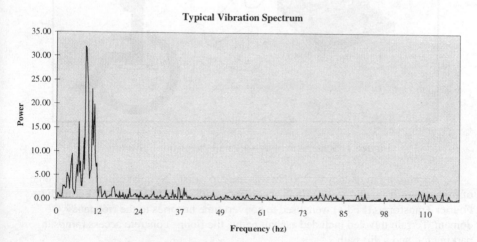

Figure 2. Typical Vibration Spectrum

necessary, so a custom printed circuit board was designed and will be built that will subsequently be fitted with the proper components for each individual accelerometer. Signals from the amplifiers were then passed to a Computer Boards Inc.® DAS-08 PCMCIA type A/D data acquisition card. All connecting cabling is shielded and the laptop and amplifier were housed in an aluminum attaché case mounted behind the drivers seat.

Software was originally written in C++ for DOS. Time histories only were recorded directly to the hard drive of the laptop computer. A post-processing routine handled all frequency domain calculations and conversions. Development of a Micro-Soft Windows® based user interface in which near-real-time Fast Fourier Transforms and power spectral density calculations are performed concurrently with background data acquisition.

Testing:

Preliminary tests were conducted using an electric wheelchair with a sling type seat. Operated by the occupant, the system sat in his lap. To simplify the test only the 'z' axis was recorded and analyzed. Power for the system was drawn from the wheelchairs power supply using the same configuration that would be used in a vehicle. In addition to the laptop system, A TEAC data tape recorder was mounted to the back of the wheelchair to record an analog time history. The data recorder also drew power from the wheelchair. A schematic of the system is shown below.

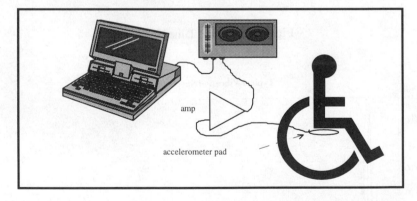

Figure 1. Schematic of Wheelchair Acquisition System

A sampling frequency of 250 Hz was used to satisfy Nyquist criteria for a range of frequencies from of 1 to 120 Hz in accordance with ISO 2631/1. A 1024 point Fast Fourier Transforms (FFT's) were used to convert time histories to the frequency domain. Terrain traveled included asphalt paths, tile floors, concrete access ramps, a parking lot, and a dirt path.

Conclusions:

Due primarily to the drastic reduction in cost and increase in capabilities of the portable computer, a PC based, near-real-time vibration dosimeter is now practically and fiscally feasible. Micromachined IC based accelerometers are accurate, inexpensive, small and rugged. Associated amplifier circuitry is simple and inexpensive to build. Data acquisition cards are readily available from a number of vendors at comparatively low cost in the PCMCIA format. These cards have all the capabilities of their larger ISA and VESA counterparts. Data storage costs for these platforms have fallen below $1 per megabyte allowing for longer continuous operating times.
Microsoft Windows® based software is currently being developed that will perform all of the conversion and comparison data analysis onboard in real time. Upon completion this prototype system will be implemented in the campus shuttle busses. Results will be reported.

References

Bongers, P.M.;Hulshof, Carel T.J.; Dijkstra, L.; Boshuizen, H.C. 1990, *Back Disorders in Crane Operators Exposed to Whole-Body Vibration*, Arch Occup Environ Health, **60**, 129-137.

Boshuizen, Hendreik C.; Bongers, Paulien M.; Hulshof, Carel T.J. 1990, *Self Reported Back Pain in Tractor Drivers Exposed to Whole-Body Vibration,* Int Arch Occup Environ Health, **62**, 109-115.

Hulshof, Carek; Veldhuijzen van Zanten, Brino 1987, *Whole-Body Vibration and Low-Back Pain: A Review Of Epidemiological Studies,* Int Arch Occup Environ Health, **59**, 205-220.

Johanning, Eckardt; Wilder, David G.; Landrigan, Phillip J.; Pope Malcolm H. 1991, *Whole Body Vibration Exposure in Subway Cars and Review of Adverse Health Effects,* Journal of Occupational Medicine, **33** (5), 605-613.

APPLICATION OF THE EXPERT SYSTEM IN ERGONOMIC DIAGNOSIS OF WORKSTATIONS

Ewa Górska

Institute of Organization for Production Systems
Warsaw University of Technology
ul. Narbutta 85, 02-524 Warsaw
Poland

The evaluation of workstations and working conditions is equally important both for the personnel employed and for the manufacturing process itself. A complete evaluation carried out according to relevant standards and regulations is a time and money consuming task. This is why it cannot be widely used in everyday practice. For this purpose, a practical method was developed and verified in different factories.

With the knowledge accumulated in it and the information possessed by the evaluator, the system makes it possible to evaluate various sets of ergonomic factors, workstations and to determine the priorities in the field of modernization.

Introduction

A complete evaluation of workstations is the basic condition indispensable to increase the effectivness of production and to make strategic decisions in management.

The overall condition of an enterprise as well as all the shortcomings resulting from the manufacturing process organisation and from the activities performed by those who are in charge of the said process can be best seen at a workstation.

An analysis of the results obtained from the evaluation of workstations makes the current management easier and, consequently, it is also easier to control the overall factory manufacturing system. Therefore, on the one hand, it is necessary to formulate the procedure in accordance with the factors used for the evaluation of a workstation are selected and, on the other hand, to define precisely the tool which would lead to a fast and effective diagnosis of the state of an enterprise.

There is a need to develop a simple diagnostic tool which would be useful for the evaluation of workstations and in which a properly programmed computer would take over the main task.

Analysis of the methods used so far indicates that there is lack of a method which would include a variant selection of factors - it can be found neither in the publications available on the subject nor in practical research. Too little attention has been paid so far to methods for selecting factors that are used as the input material for the research and evaluation of workstations. The same concerns scientifically justified criteria which should be applied in the aforesaid selection.

These problems are found difficult in the case of typical, conventional algorithmic approaches and in this connection they are traditionally solved by means of "hand" methods by human-experts.

The Expert Systems have been used in the organization and management for a long time now but they still have not found their right place at the level of workstations.

Due to its features, the considered problem of selecting factors used for the evaluation of workstations can be solved applying the methodology of Artificial Intelligence. Till now this problem has normally been solved by experts basing on their knowledge and intuition combined with experience. Symbolic inference based on heuristic knowledge was used to solve the problem. At the same time it should be noted that the task is not trite at all and that the problem is of great significance.

On the other hand, regarding the fact that the system can be applied in various enterprises and that its application can lead to envisaged profits (both in terms of money and economic ones), it has been found out that the development of an Expert System for the problem formulated in the paper is fully justified.

Description of the method

The System for Evaluation of Workstations, like other Expert Systems, is based on the assumption that special equipment, complicated documentation and experienced staff are not necessary to carry out the evaluation.

The system takes into consideration the major factors that are decisive as regards its efficiency, operability and applicability to particular organisational units, departments and enterprises. To ensure respective flexibility of the evaluation system, an algorithm has been accepted requesting the initial conditions to be selected first, followed by the proper evaluation.

At the first stage (selection of the evaluation factors being representative for given manufacturing conditions and aims formulated by an enterprise that wants to achieve them):
- the problem has to be defined, it means the aim of evaluation should be determined,
- evaluation criteria set up,
- set of data and selection procedures has to be determined,
- a set of significant factors is selected.

At the second stage (development of a method for the evaluation of workstations basing on the ascertained states of the selected factors and defining directions and actions aimed at improving its efficiency) the selected factors are evaluated at the following steps:
- registration of data (facts),
- evaluation of data,
- interpretation of the evaluation results and diagnosis of the level of ergonomics of workstations.

The factors relevant to ergonomics were selected from many evaluation criteria given in check lists, questionnaires, forms, evaluation charts, tests, and experts' opinions.

Incorrect selection of the number of factors to be considered in analysis may result in excessive error of the evaluation (when the number is too small) and in increased amount of work, time an costs (when the number is too big). This can make the process of taking decisions on modernization more difficult and decrease their rightness.

For these reasons, it is necessary to limit the set of factors, however, ensuring the requested accuracy of the evaluation. The process of selecting factors is the fundamental activity while collecting necessary information for appropriate evaluation of ergonomics of a workstation and for further design activity.

The second stage of building up the system of evaluation was the development of the method of evaluation of workstations. The stage covers a number of activities aimed at the development of a computerized evaluation system.

For application of a computer to registration, evaluation and analysis of information, it is necessary:
- to set up questions relevant to individual factors,
- to evaluate selected individual factors,
- to identify the quality of a work station under study from the view point of ergonomics, using a numerical indicator.

Each individual factor in the set has an additional procedure defining requirements, numerical values and information that makes its evaluation easier. The procedures are of an open type and they can be complemented and brought up to date, if necessary. The set of factors, relevant questions and the procedure compose the basic investigation tool in an enterprise.

The idea of an Expert System

While developing an Expert System, it is of prime importance to decide what knowledge should be stored in the Knowledge Base as the most essential for the diagnosis and what data should be included in the data base of the Expert System. It is also important how they should be formalized and represented.

Data Base includes all the factors that reflect all the important components that allow each workstation fulfil its functions.

The set includes the most essential factors which can be used for the diagnosis of workstations.

The final form of the set results from systematization and standardization of views and approaches represented by authors of various methods of evaluating workstations and addressed either to a definite problem or to a selected enterprise. Owing to the above, the information included in the Data Base is universal and can be applied to evaluation of any workstation or any enterprise.

By way of example, the set may include the following factors and corresponding questions:

- height of the working place
    Does the height of the working place ensure comfortable work both in the sitting and standing position interchangeably (e.g. on a high stool)?

- the space at a workstation
    Does the free space at a workstation allow the worker to move freely?
    and the like.

The system Knowledge Base includes the expertise provided by specialists who professionally deal with the workstation organization. The Base contains a set of criteria that characterise the object under investigation and the principles in accordance with which the criteria are selected.

The extraction of knowledge was performed by conducting numerous interviews with experts, questionnaires, observations of the problem solution process, segregation of the acquired knowledge and coding it in a way that is easily comprehensible for a computer.

The experts included:

- scientists from Warsaw University of Technology, the Central Institute of Work Protection, the Institute of Industrial Design, the Institute of Engineering Industry who professionally deal with the problems related to organization of workstations,

- employees of factories: production engineers, designers, engineers-organisers, physiologists, economists, foremen of particular departments, employees representing the work safety service.

Experts were presented with a set of factors used for the evaluation of workstations in the form of a questionnaire.
The experts were asked questions and shared their opinions as concerns the information indispensable to determine the importance of factors subjected to evaluation and the statement of their own grounds used for the hierarchical arrangement of the factors.

Rules in the base of knowledge are presented with their degree of probability, which is then considered in inference.
An examplary rule looks as follows:

RULE 27

IF	level of mechanization and automation	is	automatic
THEN	height of the working place		Probability = 5/10
and	tool ergonomics		Probability = 1/10
and	aesthetics of workstation		Probability = 7/10
and	visibility of indicators and control devices		Probability = 9/10
and	legibility of indicators and control devices		Probability = 9/10
and	lighting		Probability = 7/10
and	noise		Probability = 9/10
and	mechanical vibrations		Probability = 9/10
and	radiation		Probability = 1/10
and	thermic comfort		Probability = 7/10
and	air pollution		Probability = 7/10
and	chemical substances		Probability = 1/10

To make the system fully deserve the name of an Expert System (i.e. the one which behaves like an expert as regards the conclusion drawn as well as the minuteness of details and adequacy of the questions being asked) it should have so-called detailed knowledge coded in its Knowledge Base. This means, among other things, that questions addressed to the user will corresspond with the level of knowledge possessed by him.

For example, the system will not be trying to get an answer to a direct question. Should work efficiency be increased in the enterprise?

It will obtain the conclusion by itself as a result of an answer to a number of questions, including such questions as:

> Do you have problems with the selling of your products?
> Do you have problems with competitors?
> Do you have problems with the staff?
> Do your problems result from low wages?
> Are you satisfied with the profits obtained by your enterprise?

The corresponding rule has the following form:

RULE 9
IF   you have problems with the sale of your products   yes
and   you have problems with competitors   yes
THEN   increase in productivity   Probability = 10/10
and   improve quality of the manufactured products   Probability = 10/10

Implementation of the Expert System

Regarding the requirements formulated with respect to an Expert System, a decision was made to use the skeleton Expert System EXSYS, offered by EXSYS Inc., as an implementation tool.
It is a system of rules where particular rules have the following form:

$$\text{IF - condition}$$
$$\text{THEN - conclusion}$$
$$\text{ELSE - conclusion}$$

The purpose of the system created by means of the EXSYS system is to find the best solution of the problem basing on the data (concerning an enterprise) that have been introduced by the user.
If there are several solutions possible, the system organizes them in accordance to the degree of probability.
During a consultation session the system asks the user questions and expects to get answers chosen from the list of possible answers.
In such a system as mentioned above the role of the user is limited to :
- providing a computer consultation on the characteristics of an enterprise,
- providing answers to the questions asked by the system,
- accomplishment of rationalizations suggested by computer.

Conclusions

The main purpose of the research is automation of the diagnosis process of the condition of an enterprise in terms of ergonomics. The diagnosis is two-stage and it includes:
- selection of factors from a set of factors characterizing the ergonomic state and adequate for given production conditions,
- ergonomic evaluation of an enterprise using the selected factors.

The paper presents the idea of a computer system created within the formula of an expert system, which is supposed to replace a man-expert in the process of diagnosis.

Naturally, the system can be developed into a complete diagnostic system by building up the superstructure of the system. The basic purpose of the superstructure will be the construction of a system which would function in the following way:
- the system will ask the user questions thus obtaining information on the market conditions under which the enterprise operates and its financial standing,
- it will provide diagnosis and make further decisions concerning modernization steps,
- it will provide instructions and design assumptions of directed improvements.

References

Balagurusamy, E., Howe, J. 1990, Expert Systems for Management and Engineering, Ellis Horwood.

Famili, A., Nau D.S., Kim S.H. 1992, Artifical Intelligence Applications in Manufacturing, The MIT Press.

Górska, E. 1993, Practical System for Evaluation of the Organizational Level of a Work Station From the View Point of Ergonomics and Work's Arduousness. In W.S. Marras, W. Korwowski, J.L. Smith, L. Pacholski: *The Ergonomics of Manual Work*, (Taylor & Francis, London-Washington DC.) 535-538.

Górska, E. 1993, Elastic System for Evaluation of the Organizational Level of a Work Station, Doctoral dissertation, Poznań-Warszawa.

Górska, E. 1994, Ergonomics Investigations in Market Economy Conditions. In *Second International Conference on Science and Technology "Current Problems of Fundamental"*. Hayes-Roth, F., Waterman D.S., Lenat D.B. eds, 1983, Building Expert Systems, Teknowledge Series in Knowledge Engineering. Addison-Wesley.

Meyer, W. 1990, Expert Systems in Factory Management. Knowledge Based CIM. Ellis Horwood Limited: Chichester.

Kusiak, A. 1990, Intelligent Manufacturiung Systems. Prentice Hall International Series in Industrial and Systems Engineering.

Waterman, D.A. 1986, A Guide to Expert Systems. The Teknowledge Series in Knowledge Engineering, (Addison-Wesley).

SOFT MODELLING OF THE ERGONOMICITY OF THE MULTIAGENT MANUFACTURING SYSTEMS

Leszek Pacholski & Małgorzata Wejman

Technical University of Poznań
Institute of Management Engineering
Strzelecka 11, 60-965 Poznań, Poland

At the Technical University of Poznań has been worked out and put into practice the methodical conception of ergonomicity level evaluation of the multiagent manufacturing systems, based on subjective psychosomatic load symptoms research (PLS method). Explorated instrument used for elaborate a formal model is theory of fuzzy sets.

Introduction

The treatment of the subject of investigations and ergonomic methodology in scientific categories requires the creation of a theory formed in a formalized way, including descriptions and functional dependences as well as logical rules.

The usefulness of classical methods of formal modelling of technical systems became rooted in the mentality of scientists to much a degree that these methods also began to be transformed into the field of knowledge, comprising within range the human factor apart from the technical one. Such transformation of "hard" methods to investigate phenomena, processes and systems determined by the cooperation of the human factor with the technical one, are criticized more and more frequently as well by theoreticians as by practicians engaged in fields of knowledge whose subject is man.

Here, inadequacy of the character of "technical" modelling classical methods, for the "humanizing" character of investigated problems, is the basic critical argument. The variables, entering the "humanizing" model, can be not always expressed in the form of numerical parameters. Therefore, in the stage of the model conceptualization the variables, the values which cannot be expressed in the form of numbers, are rejected or exchanged artificially for the measurable one.

The consequences of such formal manipulations, performed on input variables, can be often very essential for the final shape of investigated phenomena, processes and systems determined by the cooperation of the human factor with the technical one.

The above mentioned difficulties, connected with the satisfaction of demands for measurability of variables, entering the model which characterizes the methodological and objective systems of modern ergonomics, created the necessity to search for suitable formal solutions on the ground of the fuzzy sets theory.

Method descriprion

The method beeing presented has been worked out at the Technical University of Poznań. It enables to ergonomic evaluate the multiagent manufacturing systems in the complex way taking into account so-called softness of formal modelling. The method is based on making a complex ergonomic diagnosis of the multiagent manufacturing systems at definite time intervals with the aid of an ergonomic check-list, designed by the authors of this paper.

The evaluation of the degree of conformity of the man-machine diagnosing unit to the ergonomic requirements is determined by the method of psychosomatic load symptoms (PLS method).

PLS method is based on subjective psychosomatic load symptoms research. The ergonomic check-list makes the set of ergonomical decisive criteria. Of course, method PLS is subjective to the full.

The check-list has been worked out so that consisted symptoms could correlate successively with particular problems from 1 - 4 areas:

- ergonomicity of technology and exploitation,
- ergonomicity of work space,
- ergonomicity of work profit by information and steering,
- ergonomicity of physical environment.

Ergonomicity of technology and exploitation lie in protecting health and assure safety employees, and chip in to their good feeling and make easier realize goal of their work. Ergonomicity of work space depend on harmonize posture of body with its power and motion of body. Ergonomicity profit by information and steering - kind of informations should be agreement with worker's perceptions possibilities, steering elements should be choose and lay down agreement with possibilities of human hands. Physical environment of work should be keep in this manner to its factors haven't pernicious influence for people; it regard factors like:

* noise,
* microclimate,
* mechanical vibration,
* air pollution,
* static electricity,
* electromagnicity radiation,
* ultrasounds, infrasounds.

A lengt of PLS check-list allowed for - for example - physical environment factors presentation, for noise and microclimate in a work place.

For noise:
    Are you feeling the following symptoms during the work:
- headaches,
- feeling tiredness, nervousness,
- impossibility of concentration,
- impossibility of understanding,
- generally feeling of discomfort, causeed, in your opinion, by noise at your work position.

For microclimate:
    Are you feeling the following symptoms during the work:
- hotness, sweating,
- coldness,
- stuffiness,
- discomfort coused by draughts,
- rheumatic pains of limbs,
- are you complaining of frequent colds,
- are you feeling that the level of moisture is too high.

The difficulties, connected with the fulfilment of the condition of precision as well as that of measurability of all the ergonomic requirements, brought about the necessity of treating the above mentioned set of indexes as a class of elements with fuzzy limits, i.e. in which there is no sharp boundary between the elements belonging and those not belonging to this class.
The above formulation is an informal definition of the fuzzy set.

Therefore, if $X = \{x\}$ will be a set of system indexes, then the fuzzy set A in X is characterized by the membership function x definited on X and assuming the values in the $[0,1]$ interval, i.e.:

$$x : X \rightarrow [0,1]$$

The A fuzzy set of diagostics variables is represented as a set of ordered pairs:

$$A = \{[\, f(x)\,,x\,] \quad ; \quad x \in X \}$$

The ergonomic variable, membershiped to the premised linguistic alphabet, describing the so-called ergonomicity of the diagnosing man-machine unit is the result of above mentioned evaluation. The ergonomic diagnostic variables have been interpreted as measures of membership, in accordance with the fuzzy sets theory.
The "soft" character of the standard, reflecting a definite ergonomic requirement, is the fundamental methodical premise of such treatment of the ergonomic variables.
The linguistic alphabet used in practice (very high, high, medium, low and very low level of unergonomicity) was determined in the artificial space representation definited of the natural numbers interval from 1 to 9.

As the result of the workers evaluation aggregation, it is therefore possible to determine the formal parameter of ergonomicity of a definitee man-machine units.

The formal parameter of ergonomicity assumes the numerical value represented as the fuzzy set A characterized by the membership function A(x) and maximilized the aggregate Q of the form:

$$Q\ [E(1)] = \underset{x \in X}{Max}\ [A(x) \wedge B_i^{E(i)}(x)\},$$

where B_i - the fuzzy set representing the linguistic evaluation of the level E(1) by "i" worker.

Conclusions

Practical applications of the ergonomicity level evaluation method based on the fuzzy sets theory, was carried out and put into practice at several Polish productive plants. PLS method allowed for procurement a teal image of an ergonomicity of work, beeing felt by workers, with a very high probability. Synthetical ergonomicity indexes Q imaged the real ergonomicity level on every workstands and in whole multiagent manufacturing system area in high probability too. In the opposite to the description in a classical categories, the nature of a formalization in fuzzy sets category isn't synonymous not average, but very individual and adapted for reliality.

The result of researches and computations was detail procurement about ergonomicity level in manufacturing systems. It allowed for a right direction and hierarchy of correction functions.

References

Pacholski, L. 1986, An application of fuzzy methods in the complex ergonomic diagnostics in industrial production systems, in W.Karwowski and A. Mital (ed.), *Application of Fuzzy Sets Theory in Human Factors,* (Elsevier, Amsterdam) 211-225.

Pacholski, L., Górska, M. (Wejman, M.), 1991, Formal modelling of the ergonomicity level evaluation in industrial production systems, in Y. Queinec and F. Daniellou (ed), *Designing for Everyone,* (Taylor & Francis, London) 1282-1284.

Wejman, M., 1993, The method of psychosomatic load symptoms (PLS) application in ergonomic assessment of seamstresses workstands, in W. S. Marras, W. Karwowski, J. L. Smith, L. Pacholski (ed), *The Ergonomics of Manual Work,* (Taylor & Francis, London - Washington DC) 407-409.

ERGONOMIC CHECK-LIST - THE POSSIBILITY
OF RATIONAL REDUCTION THE NUMBER OF ITEMS

Edwin Tytyk

High Shool of Education
Institute of Technology
65-077 Zielona Góra
ul. Wojska Polskiego 69
Poland

One of the main methodological difficulties that arise when using ergonomic check-lists is their great extensiveness.The hypothesis has been formed that it is possible to reduce rationally the number of criteria (items) without a visible fall of the gnostic valute of such a set. The analogy may be here the hierarchization of estimation factors suggested by an Italian economist V.Perato, known as a "20-80 Principle". As a result of investigation, from the preliminary list of all criteria, 44% found themselves in zone "A" (the most important ones, securing 80% of estimation relevancy) and the next 23% of its quantity in zone "B". For design decisions, the ergonomic list of criteria should be reduced.

Check-lists in the estimation and design processes

The basic tool of ergonomic system researches are different kinds of check-lists. The most important of them is the Ergonomic System Analysis Check-lists (ESAC) presented at the 2-nd Congress of the International Ergonomic Association in Dortmund, 1964. Up till now, the list has been a questionary on which a lot of improved versions of check-lists used both in diagnostic researches and ergonomic design are based.

For the needs of design a lot of specialized lists comprising the so-called "target - aiming questions" have been worked out in different countries. There is quite a wide range of literature on this topic.

Estimation criteria included in check-lists used for the estimation differ formally and logically from decisive criteria comprised in the lists of target aiming questions. This difference depends on the difference which separates the diagnostic process from the design process.

In the diagnostic process there is a finite (most often real) object of analysis and estimation, in the design process, on the other hand, the object is "in statu nascendi" as an unfinished and incomplete mental product. Therefore in design criteria the most important is their informative and implicative aspect; in diagnostic criteria, however, it is the imperative aspect.

The complexity of diagnostic researches lies in treating the studied object in the categories of the system with account taken of all its complexity and connections with the environment. The smallest object of ergonomic complex researches is the workstation, understood as an elementary Man-Machine-System. The object of complex ergonomic design is understood in the same way. Moreover, the attention should be paid to the fact that approach to ergonomic problems requires that account is taken of characteristic dualism of this system based on the differentiation between:

Operative Man - Machine - System
and
Maintenance Man - Machine - System.

The second of the systems is becoming more important in the contemporary industry because of fast automation and robotization of many technological processes of machining and assembly.

The hitherto existing directions of improving check-lists

The basic advantage of ESAC - its complex character is compensated by great extensiveness of the list hindering its practical application. Attempts of operationalization and formalization of check-lists have lasted up till now comprising both actions on the sets of "hard" criteria and on the sets of "fuzzy" criteria. National versions of ESAC readjusted to the specificity and the level of industry development have been made in some countries, among others in Poland. The earliest way of reducing the number of items was object specialization of the check-lists and referring them to a strictly determined set of analysed objects. Another way was substantial specialization, i.e. narrowing the thematic area of items to the definite sector of ergonomic problems, i.e. purposeful return to the notion of check-lists (with laid down basis for building ESAC). In that case, of course, specjalized check-lists lost their complex value.

Among the ergonomic check-lists that are well - known and used in Poland, worthy of notice, is so called Ergonomic Problem List prepared by Profesor Leszek Pacholski. It enables a sequential selection of items from the list on the basis of the answers ealier received and, what is very important, the answers do not have a descriptive character. Thanks to the proper formulation of items, there is a strictly determined repertoire of answers: "Yes", "No", "Does not refer". It enables numerical coding of answers and thanks to that, a computer can be used for the analysis of the system. As a result, one can considerably shorten the time of the system study and the analysis of the research results in comparison with researches carried out by means of the ESAC.

The new method of hierarchization and selection of check-lists items

Criteria making up a definite check-list are heterogeneous considering their essence, the ability of quantification and relevancy. Undoubtedly, there is a group of the most important criteria deciding on the final estimation, and a group of criteria of little importance having no great influence on the final estimation. The above mentioned opinion may be treated as the first methodological assumption. It can be supported by rules of statistics and moreover it is also intuitively proper. The second assumption is that the above mentioned remark refers both to the sets of diagnostic and design criteria.

There are logical premises for formulating a hypothesis that a small group of criteria having the highest ranges of importance secures a very high probability of getting an accurate estimation (diagnostic criteria) or an accurate choice (design criteria). Thanks to that, it would be possible to reduce rationally the number of criteria of a definite check-list without a visible fall of gnostic value of such a set.

The analogy may be here the hierarchization of estimation factors suggested by an Italian economist Vilfredo Pareto known as a "20-80 principle". It says that 20% of elements of given set are responsible for 80% of intensity of the studied feature (e.g. 20% of details create 80% of expense).

On the basis of these premises, emprical piloting studies, the aim of which was to state the purposefulness of further resarches in that field, were carried out. The researches were conducted at nine workstations realizing the technologies of casting, hot forging and metal machining. The preliminary list of 64 criteria was made on the basic of ESAC and its derivatives. Ranking of importance of criteria was carried out with the help of the modified Delphic method (with disregard of the postponed self-correction of respondents' answers. The researches were carried out at work of production workers taking into account their opinions, and also estimations of the foreman, the worker from the industrial safety section and the production engineer. On the basis of the obtained results, a diagram of a curve of Pareto's type with three zones was made (Fig.1).

Zone "A" comprised eight most important criteria securing 80% of estimation relevancy, zone "B" - the next four criteria. Criteria belonging to zones "A" and "B" put together secured 90% of estimation relevancy. Six criteria were classified to zone "C", other criteria proved to be entirely insignificant.

The above mentioned results of piloting research dealt with diagnostic criteria. However, there are no basis to think that an analogical hypothesis cannot be formulated for decisive criteria used in design processes.

A preliminary list of decisive criteria for Ergonomic Aided Design of machines and workstations for industry has been worked out. The list includes 21 items comprising the basic ergonomic problems in technology of machine building. The obtained results are shown in Fig.2.

The diagram shows that hypothesis on a possibility of reducing the size of the decision criteria set used in ergonomic design process, were empirical not confirmed. It means, design criteria, especially the ones of a general character, cannot be reduced, because at different stages and steps of a design process each time other criteria are important, relating to exact determined tasks. The decision criteria have equivalent ranks of importance, that also may confirm a correct preselection of them.

This situation does not facilitate of using of CAD procedures in ergonomic design, because it is necessary to apply a great number of criteria in design process, which moreover are heterogeneous in respect of their matter and ability to quantification. The CAD procedures, with ergonomic optimization blocks into their structures, can be easily created nowadays, above all as a help for solving relevant and simple and detailed design problems (e.g. work area or control device layout).

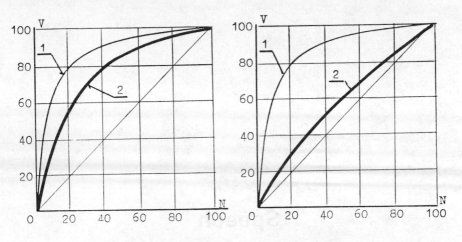

V - cumulated % of relevancy values of criteria
N - cumulated % of the number of criteria
1 - theoretical Pareto's distribution
2 - emprical Pareto's distribution

Figure 1. Pareto's distribution   Figure 2. Pareto's distribution
for diagnostic criteria   for design criteria

References:

Pacholski, L., 1977, Methodology of Ergonomic Diagnosing in the Enterprises of Furniture Industry. Issue: Rozprawy 81, Technical University of Poznań Publishers (in Polish).
Tytyk, E., 1991, The Methodology of the Ergonomic Design in Machine Building. Issue: Rozprawy 252, Technical University of Poznań Publishers (in Polish).

Speech

USER'S PERCEPTION IN AUTOMATIC TELEPHONE ANSWERING

M. Zajicek, K. Brownsey

The Speech Project,
School of Computing and Mathematical Sciences,
Oxford Brookes University, Oxford OX3 OBP, UK,
Tel: 0865 483683, Fax: 0865 483666,
Email: MZAJICEK@uk.ac.brookes

This paper addresses the problem of user's perception when engaging in a computer generated dialogue for automatic telephone answering. We show that callers will apply different contexts and perceptions to questions used in the dialogues. They will also have different perceptions of the underlying structure of the organisation they are contacting. The paper shows the means by which we have incorporated these differences into our dialogue building mechanism, as dialogues are created.

Introduction

We are looking at the problem of handing calls to large organisations, where the calls may be made by people with only a vague idea of the organisations structure, even though they may see their own goals, for the telephone interaction, as fairly clear. This is done by dynamically creating dialogue responses, depending on the caller's utterances where the system's speech recognition capabilities are limited.

The caller is only able to use a small vocabulary (of the order of 10 words) because the speech recognition components must cater for a wide range of different voices. The interaction takes the form of the caller answering questions which are dynamically constructed by the system in order to guide them to a useful end point corresponding to part of the organisation that can meet the caller's goals.

The paper starts a brief background to computerised telephone answering, and then discuss the way in which a system, based on an associative network and using fuzzy sets can be used to develop dialogues dynamically.

A major problem when creating spoken dialogues is that the English language is particularly context based when compared to other European languages. This means that the meaning of sentences can differ widely depending on the user's perception and the context of the question. The main aim of this paper is to show how we have incorporated user variability of perception into out dialogue builder.

Goal question matrices are used to deal with user variability of perception. Here callers with definite goals in phoning a large organisation are able to provide a quantitative measure of the their perception of the meaning of questions used in the dynamically created dialogue. We show the construction of a group of matrices for a set of goals and questions.

The nature of computerised telephone answering dialogues

Background

At present, commercially available standard menu type dialogues provide a choice of options for the caller. These are predefined and deal only with calls in which the set of user goals are, in the main, predictable. So they can be anticipated within a fixed dialogue structure. Systems using fixed dialogue structures are capable of handling only a small proportion of incoming of calls to large organisations, where the caller goals are likely to be vague.

Previously we have worked on a more negotiated dialogue structure which functions on a 'Yes', 'No', 'Don't Know' response handling a degree of caller uncertainty about end destination, see Zajicek and Brownsey (1993) and Brownsey, Zajicek and Hewitt (1994).

We have investigated the conceptualisation employed by callers in dealing with computerised calls, see Zajicek and Brownsey (1994). However the routing of callers by this method is still carried out within static dialogue structure. Essentially the logic employed is still embedded within the dialogue.

Current work

In our current research we are looking at using dynamic dialogue construction to locate a suitable destination for a call in a large organisation. The dialogue is developed from knowledge, held in fact structures, derived from the basic information attached to an organisation. These are held within an associative network.

The purpose of these structures is to provide the dialogue mechanism with knowledge to dynamically generate new and appropriate questions, guided by the history of the dialogue.

It is beyond the scope of this paper to describe the mechanism of the associative network. A more detailed description is given in Zajicek and Brownsey (1995). Suffice to say that the facts for a large hospital for example will concern membership of ward teams or research groups, clinic times, transport to hospital details, etc. etc. Relationships between entities, such as doctors and diseases, will also be stored. The associative network emulates the structure of facts and relationships as they would be seen by a very knowledgeable telephone receptionist.

An example of the type of ill defined call to a large organisation would be somebody phoning a large teaching hospital to find the time of the next fun run for Epilepsy research.

The telephone answering system represents the unlikely combination of a receptionist with complete knowledge of all information relevant to the organisation, and a caller who can only answer 'Yes', 'No', and 'Don't Know' to her questions.

In making a call to an organisation the user's goal is to speak to a person holding information relevant to their enquiry, leave a message, or hear a pre-recorded message. The system uses the underlying structure of facts about the organisation to construct questions to guide the user to a useful end point. During this process the system establishes likely hypotheses about the user's goal. It also presents the user with facts that are intended to augment their relevant knowledge of the organisation, and their perception of it.

The construction of question sentences

Thus we aim to build dialogues that question callers to establish their goals in making a call and so infer an end point that would be helpful to them. The dialogue emulates the questioning of a telephone receptionist as they try to establish where a call should be directed based on their knowledge of the organisation and the caller's responses.

The human receptionist involved in this kind of dialogue will ask general questions first. When these cannot be answered more information about the organisation will be offered. As she accumulates information about the caller's goal in making the call, questions can become more specific but will still be open as the direction in which the dialogue is proceeding might not be the right one. Our dynamic question generator seeks to reproduce this form of analysis of user response.

Computerised question sentence building

The information held in the associative net, together with the dialogue history so far, are used to dynamically create questions.

The system focuses on several nodes in the associative network at any stage of the dialogue. These will be nodes that it considers relevant to the line of questioning, and it will take these several nodes to construct a question sentence. Often the questions may be 'obvious' - to the human observer. So, if we have a node representing 'classes', as in 'evening classes', and one representing 'organiser', a reasonable question might be 'do you wish to speak to an organiser of classes ?'. But if we had a node representing 'organisation' and one representing 'finances' perhaps it may not be clear whether the appropriate question was 'do you want information about organisation of finances ?' or 'do you want information about finance of organisations ?'. Nodes are weighted using the set of responses. This weighting is used to select and structure the elements for a question sentence.

Perception and use of basket words

The type of question sentences to be constructed should be such that from some, or all, of the possible answers, the system can make inferences, albeit of a fuzzy kind, concerning the user's goal.

We use 'basket' terms, in questions, for groups of goals. For example a caller calling a hospital trying to find out the time of the fun run for their favourite charity might be asked 'Is it a medical matter?'. The term 'medical is a 'basket' word which includes some references to medical staff and many other things.

Callers will have different perceptions of whether their call is medical. Some people might think that medical implies illness and only 'ill' people, others might think it applies to everything in a hospital. Neither is wrong, but both are dependent on the caller's perception of the term medical. Hence we have a degree of membership of an entity in the set of 'medical things'. So basket words correspond to fuzzy sets.

The system's model of the organisation underlies the interaction between user and system. This model is a network with endpoints. The system must build a model of the user goals and map it onto an endpoint in the organisation. Seen from the system's model, the same *stated* user goal - e.g. speaking to the person organising maternity classes - may be variable between users.

There may be two distinct specific goals when two different users state "I wish to speak to the person organising maternity classes". One may mean the person who is in charge of the content of the classes, the class plan and so on. The other may mean the administrator, responsible for the paperwork, handling of finances etc. Of course it may be that both roles meet in the same person but if not, there may be a difference between the actual, as opposed to stated, goals. Hence the desired endpoints of the system's model may also be different. So variability between actual goals, associated with the same stated goal, must be taken into account.

Humans are accustomed to dealing with such variability because, compared with machines, humans are good at 'uncertain' or 'fuzzy' reasoning.

There is a whole swathe of formalisms devised for machines to deal with informalities such as uncertainty or fuzziness, including classical probability theory and fuzzy logic for example. Because the uncertainty here can be modelled using fuzzy sets and fuzzy measure we have taken an approach based on fuzzy measure theory see Klir and Folger (1988).

The goal and question matrix set

The aim of the goal and question matrix set is to obtain a measure of the possible variability, as discussed above, of the user goals. To do this we take a set of questions, involving basket terms, and a set of possible *stated* user goals. From the system point of view, the stated user goals will involve one or more of the basket word sets.

We model the user as seeing that their goals also involve membership of these top level sets, via the basket terms. Of course, we do not anticipate that any one user's perceived pattern of membership matches the system pattern or that of any other user necessarily. We can use the response from a variety of users to get a measure of the membership of these top levels sets.

We carried out an experiment on 12 student users, from a variety of disciplines, using the example of calls to a large hospital. Each subject participated in 6 trials. In each trial the subject was given a goal - e.g. to find out who organises maternity classes - and then was asked to reply 'yes', 'no' or 'don't know' to six questions, each involving a basket term - e.g. medical matter. The order of goals was varied between subjects, and the order of questions varied within the goals for each subject.

The goals were as follows.

A	- to find out who organises maternity classes
B	- to find out the date and time of the Fun Run for Children in Need
C	- to find out how to get to the hospital
D	- to find out which ward a patient would be on
E	- to find out visiting hours
F	- to contact the Kidney Disease Research Group

The questions asked were as follows

1	- are you calling about a medical matter ?
2	- are you calling about a personal matter ?
3	- do you have an administrative query ?
4	- is your call about financial matters ?
5	- do you need access information ?
6	- do you want transport information ?

Goal

Question		A	B	C	D	E	F
	1	10	0	4	8	2	9
	2	3	1	5	11	8	1
	3	7	7	2	2	4	7
	4	2	3	0	0	0	0
	5	1	0	5	4	8	1
	6	0	1	11	1	0	0

Figure 1. - Matrix 1 - 'Yes' responses

Goal

Question		A	B	C	D	E	F
	1	1	12	7	3	8	1
	2	7	10	3	1	1	10
	3	3	2	9	7	4	3
	4	9	5	12	12	12	12
	5	10	12	2	6	3	9
	6	11	10	0	11	11	12

Figure 2. - Matrix 2 - 'No' responses

Goal

Question		A	B	C	D	E	F
	1	1	0	1	1	2	2
	2	2	1	4	0	3	1
	3	2	3	1	3	4	2
	4	1	4	0	0	0	0
	5	1	0	5	2	1	2
	6	1	1	1	0	1	0

Figure 3. - Matrix 3 - 'Don't Know' Responses

Goal

Question		A	B	C	D	E	F
	1	Y	N	N	Y	N	Y
	2	N	N	N	N	Y	N
	3	Y	Y	N	Y	Y	Y
	4	N	N	N	N	N	Y
	5	N	N	N	Y	Y	N
	6	N	N	Y	N	N	N

Figure 4. - Matrix 4
Relevance of questions to goals as interpreted from the model

The results are given in three matrices, Figure 1, Figure 2 and Figure 3, one for each of 'Yes', 'No' and 'Don't Know'. Note that the figures for each cell position across the three matrices add to 12, any response other than 'Yes' or 'No' being taken as a 'Don't Know'.

Figure 4 shows how the model of the organisational data would score, using a straight Yes/No, and interpreting relevance through the links in the network representation. We can use the complete set of matrices to assess responses to other questions, where the system uses the same relevance interpretation. This provides the basis for the use of fuzzy set theory to derive weightings. These are essentially fuzzy measures, which can be combined, using, for example, Dempster's rule of combination see Gordon and Shortliffe.

Conclusions

We have shown how a major usability problem in human-computer speech driven dialogues has been addressed by researchers. The problem is that of caller's perception of the meaning of computer generated sentences, and has been tackled by employing empirical user based data generated by the Goal Question Matrix Set.

We can generalise this approach to other applications. For example, querying a data base by telephone is a similar problem. The authors are currently investigating the somewhat different area where context sensitive computer dialogue is used, is in hypertext applications, where human perception of relevant text can vary.

References

Brownsey, K., Zajicek, M. and Hewitt, J. 1994, A structure for user oriented dialogues in computer aided telephony, Interacting with Computers, Vol. **6.4**

Gordon, J. and Shortliffe, E. 1984, The Dempster-Shafer theory of evidence, *Uncertain Reasoning (Ed. Shafer, G. and Pearl, J.)* (Morgan-Kaufmann)

Klir, G. and Folger, T 1988, *Fuzzy sets, uncertainty and information,* (Prentice-hall International)

Zajicek, M. and Brownsey, K. 1993, Methods for traversing a pre-recorded speech message network to optimise dialogue in telephone answering systems, *Proceedings Eurospeech '93.*Vol. **2**, 1351-1354

Zajicek, M. and Brownsey, K. 1994, Conceptual modelling in negotiative dialogues for computer aided telephony, *Proceedings. Seventh European Conference on Cognitive Ergonomics,*(Gesellschaft fur Mathematic und Datenverarbeitung MBH), 47-56

Zajicek, M. and Brownsey, K. 1995, An approach to dynamic question construction in computerised telephone answering systems, *Proceedings of HFT'95*

SPEECH RECOGNITION AND KEYBOARD INPUT FOR CONTROL AND STRESS REPORTING IN ATC SIMULATION

H. David
Eurocontrol Experimental Centre
91222 Bretigny-sur-Orge, France

S. Pledger
Loughborough University of Technology
Loughborough Leicestershire
England

Eight Air Traffic Controllers carried out eleven exercises on a TRACON/Pro autonomous real-time simulator, employing the Subjective Workload Assessment Technique (SWAT) and Instantaneous Self Assessment (ISA) methods, reported by voice or keyboard, while controlling the simulated traffic by voice (VERBEX Commander VAT31 recognition system) or keyboard. The mean first-time recognition rate for spoken ATC orders was 79%, rising to 94% before the controller reverted to keyboard input. Subjective opinions, as measured by NASA-TLX (Task Load Index) and by questionnaires, showed that voice recognition was generally preferred, and performance records showed that, except in one case, it was more efficient than keyboard input. SWAT was found to be more intrusive than ISA. Mean ISA ratings were highly correlated with post-exercise NASA/TLX ratings.

Introduction

Attempts have been made to carry out systematic studies of the psychological and/or physiological effects of carrying out Air Traffic Control on the controller: - David (1985a), David (1985b) and Vanwonterghem and Rabit (1989).are examples. Unfortunately, these attempts all formed part of full-scale Real-Time simulations. This makes it very difficult to maintain adequate control over the circumstances of the experiment, so that balanced experimental designs, and, above all, a sufficient variety of suitable participants have always been difficult to obtain. Vanwonterghem and Verboven (1988) used of a simulation game in place of the full-scale Real-Time Simulator, but participants in that study expressed a feeling that it was not representative of 'real' Air traffic Control.

It has therefore been necessary to find a 'model' for ATC simulation, in a more readily available form than the .complex and expensive Real-Time Simulator, but having a greater degree of realism than the computer games then available. In particular, a primary difference between 'games' and real-life ATC is that control is exercised in reality through R/T links to aircraft and telephone links to adjacent sectors or centres.

In this study, TRACON/Pro was used to evaluate the potential deterioration in performance associated with the introduction of SWAT or ISA measures of strain, using voice or keyboard control of the simulator, and voice or keyboard reporting of the SWAT or ISA. The major system components are described in the relevant EEC report, currently in preparation and in Pledger (1994).

Experimental Hypotheses

- The TRACON/Pro simulator may be used as an experimental test-bed.
- On-line tasks are intrusive.
- SWAT is more intrusive than ISA.
- Speech control is more efficient than keyboard control
- Speech control is preferable to keyboard control
- Intrusiveness is greater when the same mode is used for control and reporting.

Experimental Design

Eight controllers completed eleven measured exercises, with a different traffic sample (matched for size and difficulty) in each exercise. The first five measured exercises used one control mode (speech or key) and the second five the other. The final exercise replicated the first, although a different traffic sample was used.

The first measured exercise in each block of five and the final exercise involved no on-line measurement. The remaining four employed each on-line measurement method, using each reporting method.

The orders of presentation of measurement types and reporting methods were balanced within the set of measured exercises for each controller, and the orders of presentation of control modes and exercises were balanced between controllers to minimise learning effects.

Experimental Procedure

The available space for experimental trials was an office in a temporary building This space was air-conditioned, but lacked effective sound insulation. (In practice, the noise from adjacent offices and low-flying aircraft did not affect either the controllers or the voice recognition system.) Lighting was controlled by a slatted blind over the external window, and by blocking the internal window with cardboard.

Each controller was given a pre-briefing which emphasized that it was the system, not the controller, that was being tested. They were then trained in the use of TRACON/Pro, and the voice recognition system was trained for the controller's voice (53 words, 99 phrases - about one hour). Controllers then carried out training exercises, with samples increasing from 10 to 22 aircraft until they failed three times to complete a trial without separation errors, or reached the maximum traffic level., using the first control mode. They then carried out the first block of five measured trials, repeated the training with the second input mode, carried out the second block of five trials, and finally carried out the last trial with the initial input mode.

Each trial took approximately 45 minutes, including run-up and run-down times, when aircraft were entering or leaving the simulated area. Controllers carried out eleven measured and up to twenty training exercises, over a period of three to five consecutive days.

Care was taken to minimise the intrusiveness of the experimenter, including the use of a second monitor to avoid the need for the experimenter to look over the controller's shoulder, and careful siting of the experimenter and recording equipment out of the main line of vision of the controller. Controllers were allowed knowledge of the results of each run if they asked for it. Controllers completed a final questionnaire, in the presence of the experimenter, and of the equipment.

Analysis

A total of 31 parameters were derived for each run, and were analysed by analysis of variance,supplemented by correlation and regression analysis of subjective assessment components and scores.

The primary measure of potential intrusiveness is the decrement between the controller's theoretical maximum score and his actual score. If on-line measures are intrusive, then we should see a decrement in performance when they are being taken, and, if they are perceived as being intrusive, a corresponding effect on the NASA-TLX scores. There were, in fact, significantly more hand-over errors when on-line measurements were being taken, although the total number of errors is not large in itself.

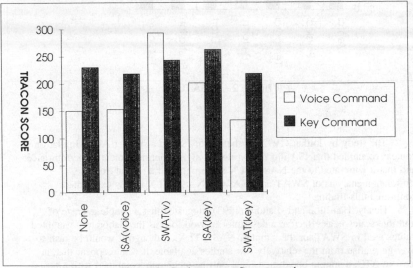

Figure 1 - Performance Decrement

Figure 1 shows the overall mean performance decrement for each of the ten experimental conditions. The performance decrement, the total number of errors, and the number of hand-over errors were all particularly high when the SWAT test results were reported by voice. or when ISA was reported by pressing a key.

Figure 2 summarizes the analyses of the NASA-TLX scores and their 'test/no test' components, showing a significant overall increase in the NASA-TLX score and four of its components. Mental Demand, Temporal Demand and Effort show just significant (< 5%) increments, while Frustration and the overall TLX score show highly significant differences (< 0.1 %).

Although there was a general subjective impression that SWAT was generally more intrusive than ISA (seven out of eight controllers would prefer ISA to SWAT if they could choose),there was no significant difference in NASA/TLX ratings. (One controller commented : "SWAT will always be difficult to use in a Real-Time ATC system where absolute priority must be given to the task in hand i.e. where it could be used to produce meaningful results where a controller is under pressure, its use by the controller is impossible." Another reported that he abandoned the attempt to make three separate ratings when under stress, supporting the conclusion that SWAT required too much mental effort in periods of high workload.)

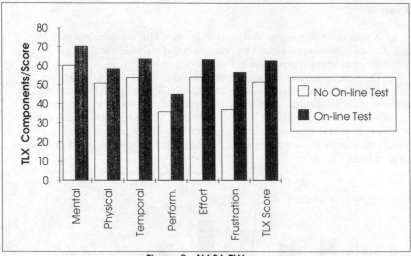

Figure 2 - NASA TLX

The study by Jordan (1992) of the use of ISA in a simpler ATC simulation context concluded that ISA did not produce a significant decrement in performance, and that it correlated to the NASA-TLX score obtained after the exercise. Correlation analysis of SWAT and ISA with NASA-TLX in this experiment confirmed this finding.

Hendy, Hamilton and Landry (1993) suggested that a simple average of multiple scales is as effective a descriptor of workload as the elaborately weighted values used in SWAT, or the simpler NASA-TLX. Although it would be rash to over-generalise from the relatively few studies available, it would appear that, at least in ATC, a simple one-dimensional on-line technique, combined with a post-exercise multi-dimensional measure, where each scale is analysed as well as the derived index is probably the most effective choice of techniques.

Although controllers preferred voice control to keying in general, both for reporting on-line measures and for controlling, the only statistically significant effect of reporting mode was an interaction of voice reporting and voice control in the number of missed values. This is probably an artifact of the system, since the means of recording were not identical.. There was no statistical evidence that voice control was more efficient than key control.

The use of SWAT reporting by voice in association with voice command resulted in a particularly high decrement in performance. (This decrement was common to all controllers who showed decrements - it was not due to poor performance by one or two individuals.) If this case is excluded, voice control appears to be generally more efficient.

Controllers generally considered that voice control was preferable to keyboard control, stating that keyboard control required them to withdraw their attention from the screen, (See Nijhuis 1993). The performance of the voice recognition system was considerably better than that observed in the previous EEC study (Prosser and Schmidt 1988), even though the controllers were carrying out Air Traffic Control, rather than (as in that study) simply speaking pre-determined phrases. (The mean recognition rate for complete ATC instructions was 79% at the first attempt, rising to 94% before the controller reverted to keying.) Controllers, however, considered it insufficient for real life control, except for non-urgent activities (such as controlling display settings.) They were more willing to accept it for real-time simulation, particularly for non-critical activities. Several controllers remarked that the system had difficulty recognising certain specific aircraft call-signs. This produced the appearance that one aircraft was failing to respond to orders - which, one controller commented verbally, was in fact more realistic than intended. Concern was also expressed that, on some occasions, the wrong aircraft responded to an order - e.g. TWA149 responding in place of TWA159. In reality, controllers usually notice an unexpected voice in this circumstance - where only one voice has been used to provide the speech generation samples, the difference may not be noticed.

The overall results of this experiment do not provide a clear answer to the question whether the use of the same mode for control and stress reporting is more or less disruptive.. Examination of Figure 1 suggests that the nature of the test is a significant intervening factor. For ISA the effects of command mode and test mode appear to be nearly linear - there appears to be virtually no interaction.. For SWAT there appears to be a marked deterioration of performance when using voice reporting, and a substantial decrement of performance when the same mode is used for command and on-line test reporting.

As controllers remarked, SWAT required more actual 'thinking about' than ISA, although there is no clear reason why the allocation of mental resources should be affected by using a common input modality for both tasks.

Discussion

This experiment has shown that it is possible for one undergraduate student to run serious experiments using the TRACON/Pro to present Air Traffic Controllers with a realistic ATC task, requiring no help to carry out the experiment. (Help was required with the setting up of the software, and is duly acknowledged.)

Although extensive training was undertaken before this experiment, and there was no measurable increment in performance during the experiment, some controllers felt that they needed more training.

Although controllers were not generally physically affected by this task, some controllers found it tiring. Since they carried out as much as six hours of active control during one day, this is hardly surprising, although it confirms that they were fully occupied. Five out of eight controllers reported some soreness of the eyes. This fact, although not in itself serious, may have implications in view of the changing nature of ATC work.

In 24 of the 88 measured exercises the controller made no errors. Although this is a tribute to the efficiency and dedication of the participating controllers, it is a considerable inconvenience for experimental purposes. Controllers were working hard during the measured exercises. It would be possible to increase the number of errors in future exercises by increasing the traffic load, but this risks forcing controllers into the unnatural working modes associated with 'losing the picture'. It would be preferable in future to carry out longer exercises with the same traffic density

Conclusions

- The TRACON/Pro simulator may be used as an experimental test-bed.
- On-line measurements are intrusive, and SWAT is more intrusive than ISA.
- Speech control is more efficient than, and is preferred to, keyboard control, but the speech recognition system is not yet sufficiently reliable for use in real life ATC or Real-Time ATC simulation .
- the SWAT test was particularly disruptive when spoken responses were required in association with spoken control.

Acknowledgements

The Authors wish to acknowledge the technical assistance provided by Mm. H. Hering and G. Coatleven, and the administrative aid of Mme. J. Roelofsen. They gratefully acknowledge the cooperation and constructive attitudes of the controllers taking part in the study. They wish to thank the Director-General of EUROCONTROL for permission to publish this paper.

References

David Hugh, 1985a, Measurement of Controllers' Mental State in a Real-Time Simulation Environment, EEC Report No 183

David Hugh, 1985b, Measurement of Air Traffic Controllers' Eye Movements in Real-Time Simulation, EEC Report No 187

Hendy, Keith C,Hamilton Kevin M, Landry Lois N., 1993, Measuring Subjective Workload; When is one scale better than many?, **34**(5), *Human Factors*, Dec 1993 , pp 579-601

Jordan C S, 1992, Experimental study of the effect of an instantaneous self assessment workload recorder on task performance, DRA TM (CADS) 92011 (HMSO London)

Nijhuis H B, 1993, Workload in Air Traffic Control Communication, in *Contemporary Ergonomics 1993 'Ergonomics and Energy', Ergonomics Society Annual Conference Proceedings,Edinburgh, Scotland*, April 1993 (Taylor and Francis London), ISBN 0-7484-0070-2 , pp 284-289

Pledger Simon, 1994, Effects of Self-Assessment on Performing Air Traffic Control, Thesis for the Diploma in Professional Studies, Loughborough University of Technology

Prosser, M and Schmit P, 1988,Integration d'un Terminal d'Entree Vocale dans l'environnement de Simulation du CEE, EEC Report No. 204

Vanwonterghem Kamiel and Rabit Monique, 1989, Activite Electrique Cerebrale et Charge Mentale chez les Controleurs du Trafic Aerien, EEC Report No. 228

Vanwonterghem Kamiel and Verboven Jacqueline, 1988, Brain Activity in a Simulated ATC Task, EEC Report No. 219

OPTIMISATION OF FEEDBACK POSITION IN THE ENTRY OF DIGIT STRINGS BY VOICE

Kate S. Hone and Chris Baber

Industrial Ergonomics Group
School of Manufacturing and Mechanical Engineering
University of Birmingham
Edgbaston
Birmingham
B15 2TT

A simulation was run to determine the optimum feedback position and error correction method for the task of entering a six digit string into a speech recognition device. MicroSAINT was used to model six different strategies for the digit entry task over a range of recognition accuracies. These strategies included four different feedback positions, and two different error correction strategies. It was found that when recognition accuracy was high all six strategies performed well, with entry of all six digits combined with a repetition method of error correction performing best at the highest recognition accuracy. As recognition accuracy was reduced other methods became optimum. Therefore, it is possible to tailor the feedback position and correction method employed to the recognition accuracy of the system used.

Introduction

There are very many applications which involve the entry of numeric data in the form of digit strings, and while keyboards still provide the major input medium, the use of speech input has frequently been proposed as a viable alternative, particularly in applications where there could be an advantage from hands and/or eyes free interaction. As in all speech input applications, however, the problem of dealing with recognition errors must be addressed, usually through the provision of feedback and a means of error correction. Unfortunately there seems to be little consensus over how the error correction should be achieved, or over the point at which it should occur within the digit string. In the current paper, therefore, previous approaches to this issue are reviewed in an attempt to determine why differences of opinion exist between researchers. Following this the advantages of a simulation approach to the problem are discussed and a simulation study into the optimum feedback position for digit strings is described.

Much of the literature on feedback and error correction strategies in automatic speech recognition (ASR) has considered the problem of digit entry. The approaches taken have ranged from pure guesswork, through experimental trials and subjective assessments, to mathematical modelling. Often researchers have compared so called 'terminal' feedback strategies to 'concurrent' feedback, with a majority coming out in favour of 'terminal' strategies. There are various reasons for this apparent preference. Witten (1982) for instance, favours terminal feedback on the basis that the flow of communication is not disrupted; Waterworth (1989) prefers it because of user preference and performance times; Frankish and Noyes (1990) and Hapeshi, Hudson and Jones (1988) prefer terminal feedback because concurrent feedback seems to cause short term memory disruption (at least when feedback is auditory). However, not all researchers agree that terminal feedback is best. Schurick et al (1985) for instance, conclude that concurrent feedback is preferable to terminal feedback on the basis of both transaction time and subjective data. We would argue that these differences between studies can be largely accounted for by differences between the lengths of the digit strings entered and the recognition accuracies obtained. Thus these studies largely fail to account adequately for the complex interdependencies which exist between such factors.

String length is obviously an important determinant of input efficiency when terminal feedback is used. Terminal feedback implies feedback which occurs after a complete string of digits has been entered, but in practice this could mean feedback after as few as two digits or after very many digits have been input. The longer the input string entered before feedback is given, the more efficient the interaction will be, at least before the time taken for error correction is taken into account. On the other hand the more digits are input at once, the higher the chance of a recognition error occurring, as this probability is equal to $(1-P^N)$, where P is the probability of a single digit being input correctly and N is the number of digits in the string. As the initial recognition accuracy becomes lower, so the probability of a correct string being entered decreases rapidly. Thus the time taken to correct the string becomes increasingly important in determining the optimum feedback position as recognition accuracy falls. Ainsworth (1988) recognised this and used mathematical modelling to determine the optimum string length before feedback occurred, based on the recognition accuracy and processing time of the recogniser used. He considered the situation where correction following a misrecognition involves repetition of the whole string. He used the overhead ratio of the system[1] and the probability of a recognition error occurring within a string to develop a mathematical formula which predicts the string length which will produce the lowest mean input time per digit. He then tested his formula against results achieved with a real recognition device and concluded that the system functioned approximately as predicted by the theory. With the particular overhead ratio value of the system that he used, Ainsworth (1988) found that the actual optimum string lengths at 95%, 90%, 80% and 60% recognition accuracies were 4,2,2 and 1 respectively, against predicted values of 4, 2, 1 and 1. Thus the only deviation from the expected values was at 80% and in fact at this accuracy there was actually no significant difference between the mean digit entry times for 1 and 2 digit strings so the results still support the efficacy of Ainsworth's modelling approach.

Noyes and Frankish (1994) also investigated the effect of recognition accuracy on optimum feedback position. They ran an experiment where subjects entered six digits with either terminal or concurrent feedback. Thus feedback and error correction could occur either after six digits or after one digit. At the value of recognition accuracy that they obtained (which was approximately 96%) they found no difference between the efficiency of these two strategies. Extrapolating mathematically from their results, however, they predicted that at higher recognition accuracies, terminal feedback (i.e. six digits entered

---

[1] The overhead ratio refers to the ratio between the time to say digits and get spoken feedback, and the time to switch between recognition and synthesis modes.

together) would be most efficient, whereas as recognition accuracy was reduced concurrent feedback would become most efficient. Thus the pattern of results obtained is identical to Ainsworth's (1988) in that longer strings are most efficient when recognition accuracy is high, while concurrent feedback becomes optimum as recognition is reduced.

The work of both Ainsworth (1988) and Noyes and Frankish (1994) shows that mathematical modelling can be used to predict the optimum feedback position for digit entry (if parameters such as the processing time and accuracy of the speech recogniser are known), however, both consider only situations where the input string (of whatever length) is repeated following a misrecognition. Other correction strategies are available, however, for instance a set of entered digits can be selectively edited to correct any misrecognition errors. Some experimental work has looked at editing approaches, but only from the point of view of subjective satisfaction. An example is work by Schurick, Beverly, Williges and Maynard (1985) who found that subjects preferred to use a correction command which deleted the last term in a buffer store, rather than one which deleted the whole store. This was apparently because they believed that the latter strategy would waste too much time. However, as subjects were free to choose either strategy throughout the experiment, there is no relative transaction time data to support this claim.

Obviously to determine the optimum feedback position and correction strategy for the entry of digit strings by voice it would be necessary to compare different string lengths, different recognition accuracies and different correction strategies. No work to date has considered all of these issues. While both Ainsworth (1988) and Noyes and Frankish (1994) have used mathematical formulae to predict the optimum position for feedback and correction over a range of recognition accuracies, they only consider the correction method of string repetition and no alternatives such as editing methods. This is not surprising given the difficulty of deriving a mathematical formula to predict performance times with an editing strategy. This is difficult because the actual strategy used will depend on where an error occurs within a string, and different strategies will be used if different numbers of errors occur within a string. It is however, relatively easy to simulate both straightforward repetition strategies and more complex editing strategies using the task network modelling method 'MicroSAINT'. Thus in the current work MicroSAINT is used to predict optimum feedback position <u>and</u> error correction method for digit entry over a large range of recognition accuracies. Previous work (Hone and Baber, 1993) has shown that MicroSAINT produces results comparable to work with human subjects, and comparison of the results obtained to those of Noyes and Frankish also confirms the usefulness of the approach.

Method

MicroSAINT models were run of six different methods of inputting six digits by voice. All models assumed that feedback was auditory and the timings used were based on the real experimental data of Noyes and Frankish (1994). The methods modelled were:
1) concurrent feedback, i.e. feedback after each digit [CONCURRENT]
2) feedback after 2 digits with correction by string repetition [2 DIGIT]
3) feedback after 2 digits with correction by editing [2 DIGIT-EDIT]
4) feedback after 3 digits with correction by string repetition [3 DIGIT]
5) feedback after 3 digits with correction by editing [3 DIGIT-EDIT]
6) terminal feedback [TERMINAL]

With the non editing strategies it was assumed that the user entered a string of digits (of length 1,2,3 or 6 digits) and then received feedback on what had been recognised. If a misrecognition was detected users are assumed to say "try again", and then repeat the whole digit string (of whatever length). With the editing strategies it was assumed that the user could move to the start of a string by saying "try again" and could move backwards and forwards within the string using the commands "back" and "forward". A string was assumed to be accepted on the command "enter", and the whole digit input accepted with the "OK" command.

A major assumption in the derivation of the models was that command words (such as "try again", "ok", "back", etc) would not be confused with data words (the digits 0-9). This assumption is considered valid as Noyes and Frankish (1994) found a negligible number of such confusions in their work. Furthermore, analysis of the vocabulary with the HENR system (Moore, 1977) which can predict possible confusions over a range of accuracies, suggested that misrecognition of data words for command words and vice versa would be highly unlikely even a the lower recognition accuracies modelled.

Each model was run 256 times[2] at recognition accuracies of 99%, 95%, 90%, 85%, 80% and 75%.

Results

The predicted mean transaction times for each of the six methods of inputting six digits are shown in graph 1.

Graph 1: Mean transaction times

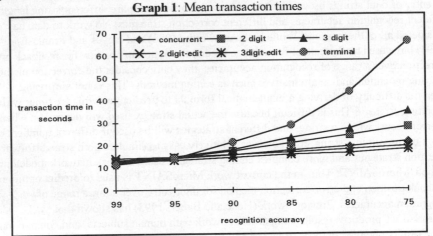

The maximum predicted transaction times for the six methods are plotted on graph 2.

Graph 2: Maximum transaction times

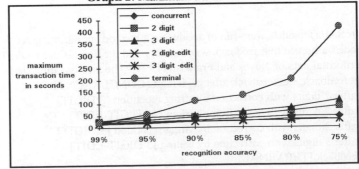

---

[2] This value was chosen as this was the number of trials per method used by Frankish and Noyes (1994)

Discussion

It is clear from looking at graph 1 that when recognition accuracy is high there is little to choose between methods. Considering first only the relative performance of terminal and concurrent feedback, it is clear that while at the highest recognition accuracy of 99% terminal feedback is the slightly faster strategy, once recognition accuracy falls below 95% concurrent becomes the more efficient. This mirrors the results obtained by Noyes and Frankish (1994), and in fact the MicroSAINT predictions plotted on graph 1 for terminal and concurrent feedback accord very well with the graph of mean times predicted by Noyes and Frankish (1994) using mathematical modelling.

Considering now the strategies involving correction by string repetition, the fastest at 99% accuracy is terminal feedback, the next fastest is feedback after 3 digits, then feedback after 2 digits and the slowest is concurrent feedback. However this situation is exactly reversed at the lower accuracies. The mean transaction times for these four methods intersect between 94.5% and 98% recognition accuracy, however it is always the case that either concurrent or terminal feedback is most efficient (terminal above 94.5%, concurrent below). Thus the results would suggest that there would be no advantage from giving feedback after 2 or 3 digit strings with correction by repetition. However, when 2 and 3 digit strings with correction by editing are modelled these are shown to be more efficient than concurrent feedback over the whole range of recognition accuracies and more efficient than terminal feedback below about 98% recognition accuracy. Of the editing strategies, feedback after 3 digits was the more efficient. The results suggest that for good performance over a large range of accuracies feedback after 3 digit-edit strategy is most efficient. However, this strategy may entail the user loosing track of where he/she is within an interaction. In the current work we have assumed auditory feedback and it is hard to imagine that in practice a user would find it easy to navigate back and forth to perform corrections within a buffer store that they have only heard presented.

The maximum transaction time data generated shows the same pattern as the mean time data. However the intersections between methods occur at much higher recognition accuracies than for mean times. Also the mean times rise much more rapidly so that at 75% accuracy with the terminal feedback condition there is a possible maximum of about seven minutes for the entry of six digits, a value which would clearly be unacceptable to any user. This highlights the importance of subjective opinion. Obviously using a modelling technique to predict transaction times as was done here does not allow the collection of subjective data as such. However, the work reported here does carry the implicit assumption that users of automatic speech systems will prefer to use feedback and correction strategies which allow them to input data at the most efficient rate given the accuracy and processing capacity of the recogniser that they are using. There is in fact some good evidence that humans optimise transaction time in their interactions with each other, which suggests that they would welcome the use of optimum strategies for speech input. An example of such evidence is the work of Clark and Schaefer (1987) who studied directory enquiry conversations. They noted that when a conversation has been characterised by frequent misrecognitions, speakers are more likely to reduce the number of digits per chunk of telephone number which they communicate. Less anecdotal is evidence provided by Ainsworth (1988), he applied the same mathematical theory that he derived for optimising digit string length for speech systems to human-human communication of digits. Interestingly the overhead ratio which exist between two people predicts the use of a grouping of three to six digits at 99% recognition, which accords well with human practice. However, this string length could equally be due to a speaker reacting to the short term memory capacity of their listener. We plan future work to investigate this issue further.

Conclusion

The results of the current experiment suggest that MicroSAINT models can be used to predict the optimum feedback position and correction method for the input of digits by speech. Once the processing time of the speech recognition system is known, graphs can be produced which would allow designers to choose the best strategy based on the expected recognition accuracy of that recogniser. Alternatively, if a range of recognition accuracies is expected, a strategy which produces consistent times across the range may be chosen. The current work suggests that a strategy based on editing inputs will produce consistently quick transaction times over a large range of recognition accuracies. However, such methods will require more user training, and in the case of auditory feedback systems may lead to users being unable to keep track of where they are within an interaction.

References

Ainsworth, W.A. (1988) Optimisation of string length for spoken digit input with error correction. *International Journal of Man-Machine Studies*, **28**. 573-581.

Clark, H.H. and Schaefer, E.F. (1987) Collaborating on contributions to conversations. *Language and Cognitive Processes*, **2**(1). 19-41.

Hapeshi, K., Hudson, S. and Jones, D.M. (1988) Data feedback and STM. *Contemporary Ergonomics 1988*, Ed. E.D. Megaw. Taylor and Francis, London. 105-110.

Hone, K.S. and Baber, C. (1993) Using task networks to model error correction dialogues for automatic speech recognition.*Contemporary Ergonomics, 1993*, Ed. E. Lovesey.

Noyes, J.M. and Frankish, C.R. (1994) Errors and error correction in automatic speech recognition systems. *Ergonomics*, **37**(11). 1943-1957.

Moore, R.K. (1977) Evaluating speech recognisers. *IEEE Trans ASSP* **25**. 178-183.

Schurick, T.M., Beverly, H., Williges, B.H. and Maynard, J.F. (1985) User feedback requirements with automatic speech recognition. *Ergonomics*, **28**. 1543-1555.

Waterworth, J.A. (1989) Interactive strategies for conversational computer systems. In *The Structure of Multimodal Dialogue*, Ed. M.M. Taylor, F. Neel and D.G. Bouwhuis. Elsevier Science Publishers B.V. (North-Holland).

Witten, I.H. (1982) *Principles of Computer Speech*. Academic Press, New York.

Backs

AN ASSESSMENT OF MUSCULOSKELETAL STRAIN EXPERIENCED BY NURSES IN THE WORKPLACE

Luz Hernandez,[*] **Ash Genaidy,**[**] **Sue Davis**[**]
Ali Alhemoud[**] **and Lin Guo**[**]

[*]*Corpoven, Purto La Cruz, Anzoategui, Venezuela*
[**]*University Of Cincinnati, Cincinnati, Ohio, U.S.A.*

This study was conducted to determine whether the activities performed by fourteen registered nurses subject the different body joints to varied levels of musculoskeletal strain. While the present study confirmed the presence of significant lower back problems among nurses, the findings strongly suggest that nurses experienced significant musculoskeletal strain in other body parts/joints.

Introduction

It is well established that nurses experience high incident rates of work-related injuries and illnesses. In the United States, the Bureau of Labor Statistics (1982-1990) reported that:

- Incident rates for nurses are about twice as much the national average;
- Nursing ranks in the top three professions in terms of the highest work-related injuries and illnesses (see table 1).

Also, many researchers emphasized that nurses are exposed to a high risk of occupational lower back disorders due to patient handling (e.g., Jensen, 1987; Garret et al., 1992).

Based on observation of work practices, it was hypothesized that the job requirements of nursing affect not only the lower back but also other body joints such as the neck and shoulders. Thus, this study was conducted to:

- Determine the degree of musculoskeletal strain experienced by nurses for different body parts/joints (i.e., fingers, wrist/hand, elbow, shoulder, neck, upper back, lower back, hip/thigh, knee, ankle/foot and toes).
- Compare the level of musculoskeletal strain among different body parts/joints.
- Determine whether or not there is a progressive increase in the level of musculoskeletal strain with time.

Table 1. Occupational injury/illness rates for nursing
in the United States (Bureau of Labor Statistics, 1982-1990)

Year	IR-1*	IR-1**	Rank	IR-2*	IR-2**	IR-3*	IR-3**
1982	7.7	10.1	2	3.5	5.7	58.7	95.1
1983	7.6	11.0	3	3.4	6.0	58.5	98.2
1984	8.0	11.6	3	3.7	6.5	63.4	121.4
1986	7.9	13.5	1	3.6	7.7	65.8	129.0
1987	8.3	14.2	2	3.8	8.0	69.9	158.9
1988	8.6	15.0	2	4.0	8.7	76.1	180.6
1989	8.6	15.5	2	4.0	8.8	78.0	181.6
1990	8.8	15.6	2	4.1	9.1	84.0	190.3

Note: IR-1 -- injury/illness rates for total number of cases; IR-2 -- injury/
illness rates for lost work day cases; IR-3 -- injury/illness rates for lost work
days; * -- private sector; ** -- nursing and personal care facilities

Methods and procedures

Subjects

Fourteen nurses volunteered to participate in this study. On the average, the study
population served 9 years as registered nurses. The average physical characteristics were:
- Age: 42.4 years ($\pm$ 10.2);
- Height: 168 cm ($\pm$ 9);
- Weight: 69.9 kg ($\pm$ 19.5).

The subjects rated their health as 8.1 ($\pm$ 1.9) where: 0 is poor health and 10 is excellent
health.

Assessment of musculoskeletal strain

Musculoskeletal strain was assessed using the Nordic musculoskeletal history
survey (Kuorinka et al., 1987) and the "0-10" Borg's scale of physical exertion (Borg,
1982). The history survey dealt with previous musculoskeletal aches, pains and injuries in
various body parts/joints. The Borg's scale was adapted to assess the degree of discomfort
perceived on-the-job in the right and left sides of eleven body parts/joints (i.e., fingers,
wrist/hand, elbow, shoulder, neck, upper back, lower back, hip/thigh, knee, ankle/feet and
toes). The details of the Borg's scale are presented in table 1.

The discomfort data was collected over six days in two weeks, at the beginning,
middle and end of the shift. The history survey was filled out on another day.

Statistical Analysis

The data collected on musculoskeletal strain was subjected to both descriptive and
inferential statistics.

Table 2. Modified Borg's scale for
assessing musculoskeletal discomfort.

0	no discomfort	
0.5	very, very weak	(just noticeable)
1	very weak	
2	weak	(light)
3	moderate	
4	somewhat strong	
5	strong	(heavy)
6		
7	very strong	
8		
9		
10	very, very strong	(almost maximum)

Results

Musculoskeletal history survey

The analysis of musculoskeletal history survey showed the following findings based on the frequency of musculoskeletal strain (i.e., population percentage reporting aches/pains/injuries during the course of life or the last twelve months prior to the start of the study):

(1) High frequency of aches/pains/injuries were reported by nurses for different body parts/joints during the course of their life (lower back: 93%; neck: 64%; shoulder: 64% and wrist/hand: 50%). In general, the spine (i.e., neck, upper back, lower back) scored the highest frequencies, followed by the upper extremities (fingers, wrist/hand, elbow, shoulder), then the lower extremities (hip/thigh, knee, ankle/feet, toes).

(2) Of those percentages of aches/pains/injuries reported by nurses during the course of life, significant proportions were attributed to the last twelve months prior to the start of the study (lower back: 92%; neck: 100%; shoulder: 88%; and wrist/hand: 50%). All other body parts showed that more than 50% of the study participants had aches/pains/injuries in the last 12 months. In general, the spine scored the highest frequencies, followed by the lower extremities, then the upper extremities.

(3) With respect to the location of aches/pains/injuries, most of the subjects reported both the right and left sides of the spine (neck: 56%; upper back: 75% ; lower back: 84%), knees (80%) and ankle (100%) during the last 12 months. Lower values were documented for other body parts/joints (shoulder: 33%; elbow: 25%; wrist/hand: 14%; fingers: 17%; hip/thigh: 25%; toes: 0%)

The analysis of musculoskeletal history survey showed the following findings based on the severity of musculoskeletal strain (i.e., percentage of nurses reporting aches/pains/injuries over a period of days):

(1) On the average, over 30% of the nurses reporting aches/pains/injuries in the last 12 months had musculoskeletal strain for over 30 days (fingers: 40%; wrist/hand: 67%; shoulder: 37%; neck: 33%; upper back: 25%; lower back: 41%; ankle/foot: 33%).

(2) The only body part that caused interruption in the nurses' normal activities for more than 30 days was the lower back (8%).

(3) Of those percentages which reported aches/pains/injuries to both sides of the body in the last seven days, the following results were obtained: shoulder -- 12%; lower back -- 33%; hip/thighs -- 25%; knees -- 25%; ankle/feet -- 17%; toes -- 100%.

Musculoskeletal discomfort data

The musculoskeletal discomfort data indicated that:

(1) The level of discomfort progressively increased throughout the day for all body parts/joints.

(2) The discomfort data were different for the eleven body parts/joints studied.

(3) 91% of the subjects reported discomfort levels in the 0-2 range with 7% in the 3-5 range and 2% in the 6-10 range.

(4) The discomfort data in the 0-2 range was uniformly distributed across all body parts/joints (4-5%).

(5) In the range of 3-5, the lower back had the highest percentage (9.5%) followed by neck (7.5%), shoulder (6.5%), knees (6.5%).

(6) In the range of 6-10, the ankle/foot resulted in the highest percentage (10%) followed by toes (9%), neck (8%) and lower back (5.5%).

The analysis of variance results (see table 2) indicated that the main effects of body part/joint and time were significant at the 5% level significant demonstrating that: (1) the discomfort perceived by nurses on-the-job increased significantly during the course of time and (2) nurses experienced varied levels of discomfort among different body parts/joints. The interaction of body part/joint and time was not found significant at the 5% level suggesting that discomfort increased over time the same way for all body parts. The main effect of day and its interaction with body part/joint and time was not significant at the 5% level. This indicates that the discomfort data collected are consistent from day to day with the same trend throughout the day for each body part/joint. It should be pointed out that the differences between various body parts/joints in terms of perceived discomfort are considered medium. The time effect was small while the day effect was negligible (Keppel, 1991).

Relationship between musculoskeletal history survey and discomfort data

Regression analysis showed that the strongest correlation ($r = 0.67$) values were found between:

(1) The population percentage reporting discomfort in the 0-2 range (right side) and the population percentage reporting aches, pains or injuries in the last 12 months;

(2) The population percentage reporting discomfort in the 3-5 range (right side) and the population percentage reporting aches, pains or injuries in the last 12 months.

Table 3. Analysis of variance for musculoskeletal discomfort data.

Source	DF	Sum of Squares	Mean Square	F-Value	Pr > F	ω^2(%)*
Model	210	2136.44	10.13	5.48	0.0001	
Error	5017	9276.20	1.85			
Total	5227	11412.64				
Subjects	13	1317.39				
Body Part	21	496.58	23.65	12.93	0.0001	8.20
Time	2	142.52	71.26	38.97	0.0001	2.60
Day	2	4.34	2.17	1.19	0.3050	0.01
Body Part*Time	42	70.99	1.69	0.92	0.6111	0.00
Body Part*Day	42	46.75	1.11	0.61	0.9784	0.00
Time*Day	4	3.45	0.86	0.47	0.7567	0.00
Time*Day*Body Part	84	45.56	0.54	0.29	1.0000	0.00

* Relative magnitude of treatment effect

Discussion

While the present study supports previous reports of the presence of significant lower back problems among nurses, the findings strongly suggest that nurses experienced significant musculoskeletal strain in other body parts/joints. In general, the lower back, neck, shoulder and wrist/hand recorded the highest scores in terms of frequency and severity of musculoskeletal strain. Other body parts scored high values in terms of severity but not frequency. Collectively, these results suggest that corrective measures should not concentrate on relieving the load on the lumbar spine and ignore the load on other joints of the body such as the neck and the upper extremity joints.

The musculoskeletal history and discomfort surveys demonstrated the importance of the spine as a source of strain. Overall, the highest discomfort scores were reported for the spine specially the lower back. The results of both the musculoskeletal history and discomfort surveys did not point out whether the lower or upper extremities are exposed to a higher degree of strain. Thus, further investigation is warranted.

The results of musculoskeletal discomfort survey showed that 91% of the nurses reported scores in the 0-2 range, 7% in the 3-5 range, and 2% in the 6-10 range. The reporting of low scores may be attributed to several factors. For example, nurses perform a variety of activities. Few activities are considered heavy such as patient lifting. The majority of other activities requires medium to light effort like walking and record keeping. Therefore, the discomfort responses obtained at discrete intervals are the average of all these efforts. Moreover, the low scores may emphasize the cumulative pathogenesis nature of musculoskeletal problems meaning that the individual application of stress seems to be harmless in itself over a very short duration but effects accumulate over time. Thus, people tolerate prolonged exposure to musculoskeletal discomfort. These results are also supported by the modest correlation found between musculoskeletal history and discomfort surveys.

It should be pointed out that the results of this study are preliminary in nature. In order to settle the issue of whether nurses experience significant musculoskeletal strain in multiple body joints, the sample size of nurses surveyed in this pilot study should be increased. According to the discomfort data collected, the statistical power of our preliminary investigation indicated that our ability to detect treatment effects was 35%. The sample size should be increased to at least 40 or 60 nurses if our ability should improve to 80% or 90%.

References

Borg, G.A.V. 1982, *Psychophysical bases of perceived exertion*, Medicine and Science in Sports and Exercise, **14 (5)**, 377-381.

Bureau of Labor Statistics 1982-1990, *Occupational Injuries and Illnesses in the United States*, (U.S. Government Printing Office).

Garret, B., Singiser, D., Banks, S. 1992, *Back injuries among nursing personnel -- The relationship of personal characteristics, risk factors and nursing practices*, AAOHN Journal, **40**, 510-516.

Jensen, R. 1987, *Disabling back injuries among nursing personnel: research needs and justification*, Research in Nursing and Health, **10**, 29-38.

Keppel, G. 1991, *Design and analysis: a researcher's handbook*, (Prentice Hall, Englewood Cliffs, New Jersey).

Kuorinka, I., Jonsson, B., Kilbom, A., Vinterberg, H., Biering-Sorensen, F., Andersson, G. and Jorgensen, K. 1987, *Standardized nordic questionnaires for the analysis of musculoskeletal symptoms*, Applied Ergonomics, **18 (3)**, 223-237.

Development and Testing of a Back Workload Assessment System

Gerald Weisman, Jim Tranowski, Dan Gottesman, and Todd MacKenzie

Vermont Rehabilitation Engineering Center
University of Vermont
Burlington, Vermont, 05401
USA

Reducing the risk of low back pain and injuries in the workplace and accommodating low-back impaired workers requires detailed information about biomechanical stresses. A hardware and software system, the Workload Assessment System (WAS) has been developed providing an objective, quantifiable description of the biomechanical stresses on the body as one performs actual work tasks. The WAS consists of three independent transducers; a triaxial electrogoniometer measuring the posture of the lumbar spine, an electromyography system measuring the activity (RMS) of the erector spinae muscles and a foot load measurement system measuring the load under each foot. Unique methods for analyzing and presenting the data of the 7 channels, known as the RUBIX CUBE and the STAT-TABLET, have been developed.

Introduction

Reducing the risk of low back pain and injuries in the workplace and accommodating low-back impaired workers requires detailed information about biomechanical stresses on the job. Describing these stresses is the first step toward reducing them. A biomechanical profile of job conditions allows for the identification of job task demands that pose health or safety risks to workers. The availability of such biomechanical data for given occupations, or specific workplaces, would also enable physicians to make more informed recommendations about whether or not an injured worker is ready to return to work. Additionally, developing practical methodologies to quantify physical job stresses would allow for much needed epidemiological studies. A hardware and software system, the Workload Assessment System (WAS) has been developed providing an objective, quantifiable description of the biomechanical stresses on the body as one performs actual work tasks.

Methods

Measuring 3D Trunk Motion

This project has developed a hardware and software system (the electrogoniometer) that is capable of measuring and analyzing trunk motion in the three cardinal planes.[1] The goniometer is an electromechanical device that is harnessed to a subject via two plates. One is situated on the mid-scapular region and centered on the thoracic spine and the other centered on the sacrum. The plates are joined by a mechanical linkage which allow continuous movement between the sacral and mid-scapular plates, thus allowing the wearer to move freely. The sacral plate houses three rotational potentiometers, the axes of which are aligned with three physiologic planes: sagital (flexion-extension), coronal (lateral bending), and transverse (axial rotation). (Figure 1)

An algorithm has been developed to enable the presentation of goniometer data in time bins, making it possible to depict the quality of time spent in various postures. The algorithm forms a table of postures called the TCUBE (values of flexion, lateral bending and rotation). The rate at which the data-logger samples the goniometer (and the other hardware) varies. However, most of the studies conducted used a sampling rate of 10 Hz. The TCUBE data can be mapped and analyzed according to the researcher's or clinician's wishes. The resolution of the TCUBE bins is 5 degrees; however, once in the PC, the raw, binned (5-degree) data can be mapped to different size bins (e.g., 10-degree bins) forming a PCUBE. Up to nine time bins can be created with any value of time in order to appreciate the static or dynamic quality of the postures held. Data are then analyzed according the time bins selected and a RUBIX CUBE is formed. The RUBIX CUBE essentially defines the time spent in all postures (three dimensions) as well as the quality of the time spent in those postures, e.g., static or dynamic. The RUBIX CUBE represents five variables, three spatial dimensions, total time spent in postures and the distribution of time in static versus dynamic postures. In order to present these data on paper (two dimensions), the data must be either collapsed or sliced. For example, if flexion is shown versus rotation, all the lateral bending postures can be summed (collapsing) or flexion versus rotation can be presented for a given value of lateral bending (slicing) Figure 2 is a representation of output from the software package showing two physical dimensions and six time bins. Similar RUBIX CUBE algorithms have been developed for the presentation of EMG and limb load data.

While the algorithm described above allows for the presentation of goniometer and other data in a graphics format, it does not allow for easy statistical analysis of the data. Another algorithm, the STAT-TABLET, has been developed to set limits for adopted postures and to enable statistical analysis to be carried out. Temporally, postures are classified as being either dynamic or static. Dynamic postures are here defined as those held for five seconds or less, whereas static postures are those maintained for more than five seconds. Spatial analysis of the STAT-TABLET involves classifying postures into four specific categories: **Neutral**-Postures within five degrees of the upright posture; **Moderate**-Postures in one or two dimensions that are between 5 and 15 degrees from the upright posture; **3D Moderate**-Postures that combine all three dimensions that are between 5 and 15 degrees from the upright posture; **Extreme**-Postures that are more than 15 degrees in any direction from the upright posture.

The software for the STAT-TABLET allow the values, and thus the definitions, of the

---

1. Now commercially available as the BackTracker™ from Isotechnologies, Hillsborough, NC, USA

classes to be changed. The values chosen above have been selected on the basis of ergonomic guidelines. Information on postures in each cell include the total time maintained, in seconds (percentage), the mean time maintained, as well as the minimum and maximum times.

EMG Measurements

The system for collecting and analyzing EMG data has been designed to minimize or eliminate some of the problems often faced by researchers and clinicians who use EMG recordings. The Fasstech (Burlington, MA, USA) active electrodes, along with amplification and RMS hardware, make it possible to obtain a signal clearly associated with postural changes and loading of the spine. Preamplifiers are attached directly to the electrodes reducing the problems of interference from environmental signals. The hardware also has built in analog filters which help to eliminate interference that would otherwise degrade the data. (Figure 3)

The Polycorder data-logger (Omnidata, Salt Lake City, Utah, USA) is used to sample and collect data. RMS EMG data were sampled at 10 Hz, allowing for data collection for up to five and one-half hours. Resting EMG and submaximal exertion samples are collected at the beginning, middle, and end of each data-collection session to provide a basis for comparing EMG values across time and across subjects or workers.

EMG data analysis involves determining the length of time specific exertions are held. The magnitude of the EMG signal is determined using a sliding time window. The duration of the window is specified by the user, and can be as short as a few data points (less than 1/2 second) or as long as 60 seconds. The smoothed function for magnitude is processed further into a chronological file of amplitude data for both left and right sides. Magnitudes are assigned to one of eight magnitude bins. As in the presentation of goniometer data, up to nine time bins can be defined to assess the length of time exertions are held. It is this binned data that is used to create the graphical image described below.

A graphical presentation method similar to that described above for the goniometer data has been developed to present the data for one or more data collection sessions in a single graphic. The X and Y axis represent EMG magnitude on the right and left sides, respectively. Each of these are divided into a number of bins. Within each cell is a four-bar graph, with each bar representing a specific time period. The bar height represents the number of samples or the total amount of time in the category. The graphic clearly illustrates whether the activities were primarily symmetrical, whether there was a lot of time spent in high level exertions, and whether the exertions were sustained or quick.

Limb Load Measurements

A Limb Load Monitor (LLM), developed by the Rehab Engineering Center at Moss Rehabilitation Hospital in Philadelphia to provide feedback in gait training, is used to measure gross load on the body. In gait training, a specific level of force is set and any force generated above the preset level sets off an audio feedback signal. The system consists of a capacitive pad worn inside the shoe and a small electronic package that interprets the signal from the pad and can be correlated with force on the pad. (Figure 4) Analysis of the LLM data is similar to that used in analyzing EMG data. Data are presented graphically in such a way as to illustrate the magnitude and time durations of the loads on the feet during work activities. Data are normalized to the subject's body weight. Laboratory tests of the system found the average absolute errors observed were 9.4 pounds for men and 7.1 pounds for women.

Discussion/Summary

One of the primary purposes of performing a job analysis is to define the physical demands required of the worker to perform specific tasks. The US Department of Labor defines 28 different physical demand factors. These factors include such things as standing, walking, sitting, lifting, carrying, pushing, pulling and balancing. The frequency of each factor is considered. The scale used includes; not present, occasional (under 20 percent of the time), frequent (between 20 percent and 80 percent of the time) and constant (over 80 percent of the time). These frequencies must be estimated by the observer. These estimations are imprecise and are subject to observer variability. Additionally, jobs are rated as to their overall strenuousness. The five strenuousness rating are used to quantify the physical demands. These ratings include, sedentary work, light work, medium work, heavy work and very heavy work.

The WAS system makes it possible to measure not only postures but also forces and loads associated with these postures. The electrogoniometer, which measures three-dimensional trunk motion, represents a significant improvement over previous methods of assessing workplace postures. The goniometer is easy to use and allows workers to carry on their normal activities, without being tethered to monitoring equipment or hampered in their actions. Data can be collected for long periods of time without direct observation. The analysis is automated and efficient. This information can be used to modify or redesign jobs, either the specific tasks or the work environment, minimizing awkward, uncomfortable, or potentially injurious postures.

The EMG techniques used in the WAS are unique in their ability to provide information about the forces necessary to perform specific jobs and tasks over long periods of time. Previously EMG had been used in job analysis to assess specific elements of tasks for short time periods. The WAS system collects and analyzes EMG data over long periods of time, indicating and describing the exertions required to perform the job tasks.

Using existing technology, the system also incorporates a method of assessing the gross loads on the body during job task performance. The Limb Load Monitor and its data collection and analysis software have proven reliable for measuring loads in the workplace. In addition to providing information about loads experienced while lifting or carrying, the Limb Load Monitor also provides information about workers performing tasks while walking or sitting.

The WAS, taken as a whole, provides a method for analyzing many of the physical demands of a job. It provides in an objective, quantifiable way, a means of describing these physical demands.

Acknowledgements

This work was funded by a grant from the National Institute of Disability and Rehabilitation Research, US Department of Education (H133E30014-95).

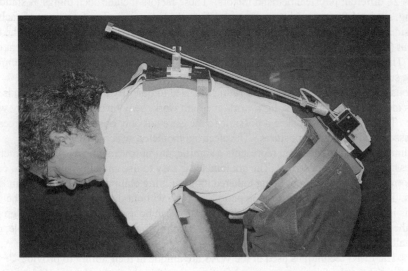

Figure 1. Electrogoniometer

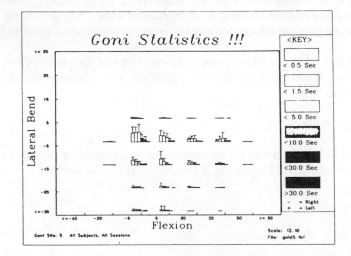

Figure 2. RUBIX CUBE with posture data

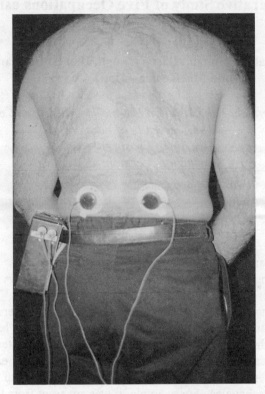

Figure 3. RMS EMG collection system

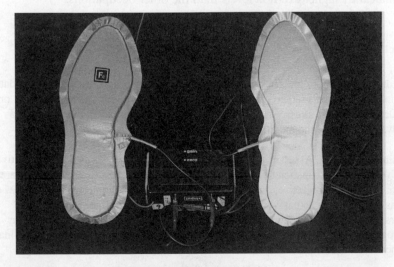

Figure 4. Limb Load Monitor

A Comparative Study of Five Occupations using the Workload Assessment System

Gerald Weisman, Larry Haugh, Dan Gottesman, Todd MacKenzie, Jim Tranowski

The Vermont Rehabilitation Engineering Research Center
The University of Vermont
One South Prospect Street
Burlington, Vermont 05041
USA

A comparative study of five different occupations was conducted using a Workload Assessment System (WAS) previously developed to provide an inventory of biomechanical stresses. Occupations included construction workers, custodial workers, x-ray technicians, warehouse pickers and warehouse stockers. At least 5 workers in each of the occupational settings were instrumented with the WAS and data was collected at 10HZ. Significant differences among the different worksites were apparent from all three of the measurement techniques. For example, it was apparent from the data that the X-ray technicians tended to be more sedentary than the other workers. In contrast, the stock clerks perform tasks requiring significantly higher exertion levels than the other occupations.

Introduction

Reducing the risk of low back pain and injuries in the workplace and accommodating low back impaired workers requires detailed information about biomechanical stresses on the job. Describing these stresses is the first step toward reducing them thus leading to effective worksite design and modifications.

Using a Workload Assessment System (WAS) previously described a comparative study of five different occupations was conducted. The WAS is a hardware and software system that provides an objective, quantifiable description of the biomechanical stresses on the body as one performs actual work tasks. The WAS consists of three independent transducers; a triaxial electrogoniometer to measure the posture of the lumbar spine, an electromyography system to measure erector spinae muscle activity (RMS) and a foot load measurement system to measure the load under each foot.

Methods

Subjects

A total of 29 subjects were recruited from five different worksites representing different experiences of low back injuries. These worksites included:

Site 1	Custodial (cleaning) workers	7 subjects
Site 2	Construction workers	6 subjects
Site 3	X-Ray technicians	6 subjects
Site 4	Warehouse workers (Item pickers)	5 subjects
Site 5	Warehouse workers (Stock clerks)	5 subjects

Data Collection

Data were collected using the WAS for two hours during the morning and again for two hours during the afternoon of the same day. All data was collected at 10 HZ. Tasks performed during data collection time were representative of the subjects' typical duties.

Data Analysis

The dependent variables of the study include; 1) posture, including range of motion and time history, 2) EMG of the erector spinae muscles and 3) limb load. Independent variables include occupation, incidence of low back injuries, job tasks, gender and age. This paper reports on analysis to illustrate differences in biomechanical stresses among the five different occupations. Therefore, using "occupation" or "site" as the independent variable, analyses were conducted to determine differences in posture, EMG and limb load.

Each dependent variable was analyzed separately. A number of different hypotheses were developed based on the form of the RUBIX CUBE for EMG and limb load and the form of the STAT-TABLET for posture. Differences for the five different sites were tested using a one-way ANOVA analysis.

Results

EMG

The RUBIX CUBE representation of the data depicts the magnitude of the muscle exertions for both the right and left erector spinae muscle groups. The magnitude scale represents the value of the contraction normalized to a standard contraction where 100 equals the value of the standard contraction. It is expected that the heaviest jobs, i.e. those that require significant heavy lifting and/or carrying would have a lot of data in those bins above 100. The height of the bars on the plot represent the relative time spent in that particular contraction.

Symmetrical contractions of the erector spinae representing movements made in the sagittal plane are represented by data bins along the diagonal originating at the [0-25, 0-25] bin. The further away from the diagonal the more asymmetrical the contractions and thus the movements.

Time bins are used to depict the length of time of specific contractions. Six time bins are used to depict contraction times of less than .5 seconds up to greater than 30 seconds. Sustained contractions are represented by data in the "longer" time bins while highly repetitive tasks are indicated by data in the "shorter" time bins.

The differences in the EMG RUBIX CUBE plots for the five different sites were tested using a one-way ANOVA analysis. Figures 1 and 2 show RUBIX CUBE plots for data collected at the X-ray and stock clerk sites respectively. For each site, data from all subjects and all sessions are included. Differences among different worksites, which represent differing requirements and working conditions, are apparent from the data plots. As shown in the figures, X-ray technicians were more sedentary than other workers. They experienced significantly longer times at the lowest contraction levels while concomitantly experiencing the least amount of time at the higher contraction levels. In contrast, stock clerks perform tasks that require significantly higher exertion levels than the other occupations.

Significant differences were shown between the occupations in the symmetry of the tasks. For those tasks requiring the longest contraction times, the X-ray technicians' tasks were symmetrical. Custodial workers experienced symmetrical contractions, but for exertions of shorter duration.

Limb Load

The RUBIX CUBE representation of the Limb Load data depicts the magnitude of the gross load under the right and left feet. The magnitude scale represents the value of the load normalized to body weight where 100 equals body weight. It is expected that the heaviest jobs, i.e. those that require significant heavy lifting and/or carrying would have a lot of data in those bins above 100. The height of the bars on the plot represent the relative time spent at that particular load level.

Symmetrical loading under the feet, i.e. equal loading under each foot, are represented by data in bins along the diagonal originating at the [0-25, 0-25] bin. The further away from the diagonal the more asymmetrical the loading.

The differences in the Limb Load RUBIX CUBE plots for the five different sites were tested using a one-way ANOVA analysis. Figures 3 & 4 show data collected from two of the sites. In each case, data from all subjects and all sessions at the site are included. Limb Load analyses showed that there were significant differences among the different worksites. The warehouse pickers clearly had the "heaviest" tasks, i.e. those requiring the greatest amount of lifting and carrying. The symmetry of the various tasks also differs among the occupations. Stock clerks experienced a significantly greater amount of short duration symmetrical loading while X-ray technicians experienced long duration symmetrical loadings. These findings are consistent with those of the EMG recordings during mostly sedentary tasks. Specificactivities,such as walking, can be identified using these data. Warehouse pickers clearly demonstrate a high amount of asymmetrical, short-duration loadings, which are representative of walking. In contrast, construction workers ranked lowest in the amount of time spent in these conditions.

Posture

The RUBIX CUBE representation of the posture from data collected using the electrogoniometer depicts the amount of time spent in postures in 3 dimensional space and considers the dynamic/static nature of the posture. The time domain is represented by six different time bins which represent the amount of time a particular posture was held. The magnitude scale represents the value of the range of motion in a particular cardinal plane, i.e. flexion-extension, lateral bending or rotation. The [-5,5] bin represents the "neutral" posture. Results are depicted for the flexion-extension and lateral bending ranges of motion. The rotation range of motion has been "collapsed," i.e. is summed in the data depicted. It is expected that those jobs requiring the greatest range of motion will have RUBIX CUBE plots with a lot of data in the extreme bins. The height of the bars on the plot represent the relative time spent in particular postures.

Asymmetrical postures, i.e. those outside the sagittal plane, are depicted as data in bins on either side of the lateral bend neutral position. The overall symmetry of the job can be ascertained by observing the relative amount of time spent in postures on either side of the neutral position. STAT-TABLETS were generated for each of the sites, combining the data from all session of all subjects.

Analysis, using a one-way ANOVA, showed that there were significant differences among the sites in terms of posture. The findings are consistent with those of the EMG and Limb Load data. Postures assumed by X-ray technicians illustrate the sedentary nature of much of the job. Construction workers and stock clerks spent more time in extreme postures, while custodial workers spent the least.

Discussion

The need to describe and quantify biomechanical stresses in the workplace has been recognized by NIOSH (1981, 1986). A biomechanical profile of job conditions allows for the identification of job task demands that pose health or safety risks to workers and increase the potential for beneficial workplace modifications. Chaffin et al. (1977) suggest that the physiological and biomechanical analysis of different modes of materials handling, and the training of workers in the correct methods of handling loads, could largely obviate the problems of low back pain in industry. The availability of such biomechanical data for given occupations, or specific worksites, would also enable physicians and other health care workers to make more informed recommendations about whether or not an injured worker is ready to return to work. Additionally, developing practical methodologies to quantify physical job stresses would allow for much needed epidemiological studies.

The comparative study of different occupational settings documents differences in the physical demands for different jobs and validates the WAS system's ability to discriminate between jobs. The EMG data illustrate the sedentary nature of the X-ray technicians' job, showing that most muscle activity of these workers involved minimal contractions over long periods of time. While X-ray technicians may be at risk for back injury from lifting patients, they do not lift as often as many other kinds of workers. The Limb Load data provide similar information. Jobs that are predominantly sedentary, such as those in X-ray technology, are characterized by the relatively large amounts of time during which the body experiences only minimal loads. On the other hand, warehouse workers sustain far greater loads, particularly during lifting and carrying tasks. In both EMG and the Limb Load data, differences in symmetry of tasks and loads are also evident.

The posture data also illustrate differences in the kinds and durations of postures required of the workers in the different jobs. The results of the posture data are consistent with that found in the EMG data: jobs requiring the most awkward postures are the same jobs that require the greatest exertions.

Acknowledgements

The work was funded by a grant from the National Institute of Disability and Rehabilitation Research, US Department of Education (H133E30014-95).

References

Chaffin DB, Herrin GD, Keyserling WM & Garg A. 1977, A method for evaluating the biomechanical stresses resulting from manual materials handling jobs, Amer Ind Hyg Assoc J, 38:662-675

National Institute for Occupational Safety and Health. 1981, A work practices guide for manual lifting, Tech. Report No. 81-122, U.S. Dept. of Health and Human Services (NIOSH), Cincinnati, Ohio

National Institute for Occupational Safety and Health. 1986, Proposed national strategies for the prevention of leading work-related diseases and injuries, U.S. Dept. of Health and Human Services (NIOSH), Cincinnati, Ohio

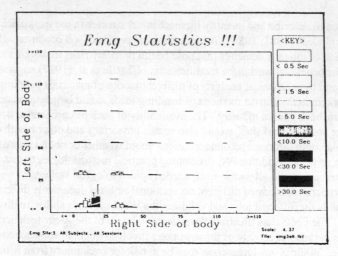

Figure 1. X-ray technicians

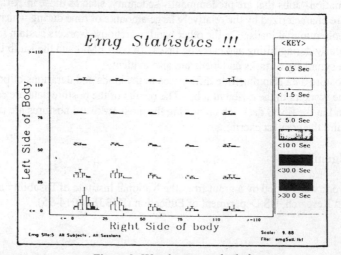

Figure 2. Warehouse stock clerks

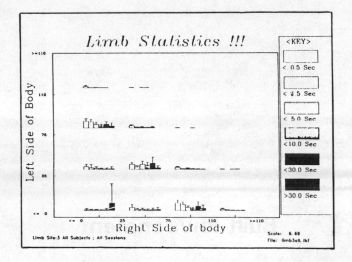

Figure 4. X-ray technicians

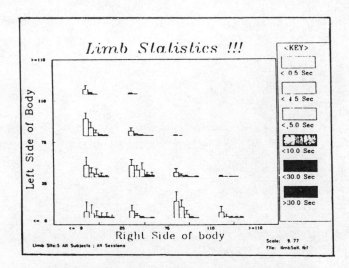

Figure 5. Warehouse stock clerks

Built Environment

Sick Building Syndrome and Occupational Stress

Joanne O. Crawford   **Leslie H. Hawkins**

Industrial Ergonomics Group   *Robens Institute of Occupational*
School of Manufacturing and   *Health and Safety*
Mechanical Engineering   *University of Surrey*
University of Birmingham   *Guildford*
Edgbaston   *Surrey*
Birmingham   *GU2 5YW*
B15 2TT

Sick Building Syndrome (SBS) has been associated with numerous factors
including for example, ventilation and indoor air pollution and the sources
are clearly multifactorial. This study examines the symptoms and causes of
SBS in emergency control room staff. The population were assessed using a
health and comfort questionnaire, environmental monitoring and an
occupational stress questionnaire. The results found that there was a high
prevalence of SBS complaints, staff attributed their complaints to a dry
working environment. Since new technology and new control rooms had
been introduced, there were reported to be an increase in some of the SBS
symptoms. The staff were experiencing a stress effect attributed to
organisational and managerial sources. One conclusion found in the study
was that the interaction between SBS and stress needs further examination.

Introduction

Sick Building Syndrome (SBS) has been defined as "a syndrome of non-specific
malaise the onset of which is associated with occupancy of certain modern buildings"
(WHO 1983). However, not all the staff will have the same, type number or severity of
symptoms". The symptoms include dry symptoms such as stuffy nose, dry throat, and dry
skin; allergic symptoms such as blocked, runny or itchy nose, watering or itchy eyes.
Asthma symptoms including tightness of the chest and wheezing and symptoms such as
headache and lethargy. (Robertson et al 1985, Leeman 1988). The symptoms suffered
at work have one factor in common in that they are strongly associated with the occupant's
building and are relieved when individuals leave that particular building. It is apparent
however that symptoms such as those listed will occur in other type of work and home
environments apart from office buildings.

Sick Building Syndrome has been associated with numerous factors including for
example, ventilation and indoor air pollution. Research to date has not provided a
satisfactory answer to the source of SBS, although it is clearly multifactorial.

The role of occupational stress and SBS has been examined by a number of
authors. Colligan (1981) used the theories of Selye to suggest that indoor pollutants or

dry stuffy air caused an increase in arousal and anxiety leading to a stress reaction. Carlton-Foss (1984) examined the role of the thermal environment and stress. Again using Selye's theories, he suggested that certain types of personalities are more likely to complain of discomfort and illness at work and to blame environmental factors. When occupants were asked to evaluate the thermal comfort, the evaluations correlated more with the personality measurements than with environmental measurements. Karasek (et al 1981) suggested a Jobs-Demand Control model to explain occupational stress and SBS. Where the source of stress is the lack of control over the environment and that lack of control and dissatisfaction over the environment was linked with symptoms. Similar results were found by Wilson and Hedge 1987. Pennebaker and Watson (1988) theorised that SBS symptoms are related to a state of negative affectivity. This theory was suggested as self-reported stress is highly correlated with physical symptoms and negative mood states. Hedge (et al 1992) postulated that the susceptibility of a person to develop SBS symptoms is related to the stress that they experience and whether it is job stress or the fear of or reaction to exposure to indoor pollutants is unclear.

This study examines symptoms and causes of SBS and Occupational Stress in emergency control room staff. Before the study took place, there was a perception among control room staff that they were suffering a number of ill-health complaints similar to SBS symptoms and also suffered occupational stress. The job of the control room staff is to deal with emergency telephone calls, mobilise manpower and equipment and contact other relevant services. Work equipment includes telephones, radios and VDUs. The job is characterised by periods of activity and inactivity and a shiftwork pattern. New computer technology had been introduced in the control rooms over the 5 years prior to the study and in many this has happened simultaneously with the building of a new control room. What was not apparent was whether the ill-health complaints were related the new control rooms or the demands of the new technology.

The aims of the study were to:-
1. Determine specific health problems and complaints.
2. Determine what staff attributed their health complaints to.
3. Investigate the relationship between new technology and health complaints.
4. Measurement of environmental parameters including temperature, relative humidity, air-flow, sound levels and illumination levels.
5. Obtain objective evidence such as sickness absence or retirement date.
6. Investigate occupational stress and the link between job stress and health complaints.

Method

A health and comfort questionnaire which had previously been piloted was distributed to all control staff of an emergency service across the UK a month before the control room visits took place. The number of questionnaires distributed was 1797. Six control rooms were visited consisting of two rural control rooms (A and B), two metropolitan control rooms (C and D) and two county control rooms (E and F) during the months of December and January. All of the control rooms visited apart from Control Room A had air-conditioning and had been built in the last 5 years before the study. From the questionnaire data a Sick Building Syndrome Score (SBSS) was calculated. The SBSS is calculated by adding the number of responses of often or sometimes for each of the symptoms, dividing this by the total number of responses to the symptom and adding all the scores for the relevant symptoms. Symptoms included in the calculation were headache, dizziness, eyestrain, sore dry eyes, nasal congestion, lethargy, tension, allergy and skin rashes.

An environmental survey was carried out in each of the control rooms visited and included measurements of temperature and relative humidity (Squirrel monitor Grant Instruments), air-flow (air-flow meter model TA 3000T), noise levels (Cirrus Sound Level Meter) and illumination levels (Hagner Lux Meter). Environmental parameters were monitored over a 24 hour period. Sickness absence data was collected from 5 of the control rooms visited and the percentage of total days lost calculated from 4 of the data sets. Due to the quality of the data it was only possible to assess the percentage of total days lost from these four. Retirement data was obtained from a central source of all the control room staff. The retirement data did not give enough information for analysis. The Occupational Stress Indicator (OSI), a group assessment, was administered to staff in the six control rooms visited.

Results

The number of questionnaires returned on a national level were 1114 giving a response rate of 62%. The response rate of the individual control rooms visited varied between 64% and 81%.

The environmental survey found that temperature in each of the control rooms was normally above the recommended level and fluctuated , relative humidity in 4 out of 6 control rooms was below recommended levels, air-flow was lower than recommended values in 3 of the control rooms visited (BSI 1990). The main complaints about air quality from the questionnaire were that it was unpleasant, stuffy, dry and the temperature fluctuated. Daytime illumination levels were found to range between 12 lux and 350 lux in the control rooms. The questionnaire data found that artificial lighting was in use 24 hours per day and there were a number of reports of glare and reflection problems. Sound levels were found to be between 55 and 70 dBA and a number of questionnaire respondents found noise irritating and distracting on day shifts or both shifts.

The prevalence of symptoms reported often or sometimes in each of the control rooms and at a national level indicate that the most common symptoms are headaches, eyestrain, sore dry eyes, nasal congestion, lethargy and tension with more than 70% of respondents complaining of these types of symptoms sometimes or often. There were no patterns found in symptom when different control rooms were compared. The Sick Building Sickness Score was calculated for each of the control rooms and at a national level and the results are shown in Table 1.

Table 1. Building Sickness Scores

Control Room	Building Sickness Score
A	2.75
B	2.00
C	1.73
D	1.77
E	1.90
F	1.22
National	1.64

Questionnaire results found that occupants of the control rooms mostly attributed their health complaints to a dry stuffy working atmosphere. Respondents were asked whether symptoms had changed since the introduction of new technology. The most common responses were an increase in the frequency of SBS symptoms (18.5% at a national level) and an increase in headaches and eyestrain (31.6% at a national level). Since working in the new control room environment respondents reported an increase in the frequency of SBS symptoms (33.2% at a national level), more tired and lethargic (18.4% at a national level) and a general increase in feelings of ill-health (14.8% at a national level).

The sickness absence data from the four control rooms was analysed. The percentage of total days lost varied between 1% and 8%. In two of the control rooms sickness absence did increase the year after new technology was introduced.

The OSI results from the six control rooms visited found that the sources of stress were management, relationships with other people and the organisational structures at work. The OSI also assesses the individual at work with regard to job satisfaction, type "A" behaviour and total control. The results are shown in Table 2.
The results found that all of the groups studied had high levels of job satisfaction, had a higher than expected level of type "A" behaviours and a lack of control at work. The methods of coping with stress at work included social support from colleagues, task strategies and job involvement. The effect of stress on the control room staff was found to be higher than expected levels of physical and mental ill-health. For example, headaches, indigestion, palpitations, inability to sleep and depression.

Table 2 Individual Stress Scores

Control room	Job Satisfaction 88.3 ± 3.7 (1)	Type "A" 45.9 ± 4.9 (1)	Total Control 33.9 ± 3.7 (1)(2)
A	110 ± 15.48	74.4 ± 9.32	62.45 ± 5.75
B	109.75 ± 15.48	69.69 ± 9.01	62.4 ± 5.75
C	91.87 ± 23.05	70.5 ± 11.61	62.3 ± 5.97
D	91.8 ± 12.26	68.8 ± 10.97	64.5 ± 7.05
E	92.27 ± 25.8	68.8 ± 6.68	64.1 ±7.64
F	100.5 ± 12.11	69.3 ± 8.3	61.6 ± 5.12

(1) = Validated norms

(2) = Note that a higher figure in total control indicates an increased feeling of loss of control

Discussion

The control rooms studied were chosen to represent 3 different levels of workload. It was not aimed to directly compare each of the control rooms but to try and assess if there were any common problems between groups, e.g., metropolitan or rural.

The collection of both questionnaire and environmental data can only be seen as representative of a point in time. As the surveys were carried out in January and December, it cannot be assumed that the same environmental problems would occur at other times of year.

The fact that environmental parameters in some of the control rooms were outside of recommended levels (BSI 1990) and the temperature fluctuated, is likely to have contributed to the occurrence of some SBS symptoms. Wyon (1973) and Jaakola et al (1990) suggested that fluctuating temperatures contributed to sleepiness and fatigue. Smith and Webb (1991) found that steam humidification reduced the prevalence of some of the dry SBS symptoms, e.g., dry eyes, dry throats. The staff in the control rooms mostly attributed the cause of symptoms to a dry working atmosphere. Environmental monitoring data showed that low relative humidity was a common occurrence and would support this perception.

The prevalence of symptoms across all control room staff was found to be high. However no consistent patterns of symptom reporting were obtained from the control rooms. This may be due to inherent building differences rather than workload levels. The SBSS calculated from each control room was found to be highest in the only non-air-conditioned environment. However there were no specific patterns found in SBSS.

The response to the questions of health since the introduction of new technology or since moving into a new control room did not provide any clear answer. The data obtained was historical and it is impossible to separate whether the reported increase in symptoms is due to the demands of new technology or the new environment which accompanied this.

Sickness absence data did not give any clear indications of any loss in time since the new control rooms had been built. Typically sickness absence data included injury and maternity leave statistics, but little information causes that might relate to SBS symptoms. Due to the types of symptoms within SBS, it may be better to assess the number of people leaving work early and the symptoms that cause them to leave rather than whole days absence. However this type of information is rarely kept.

The OSI found that there was a stress effect occurring in each of the control rooms visited. What was interesting from the results that the source of the stress was not the job itself - indeed there was a high level of job satisfaction - rather the organisational and managerial climate. The staff who took part in the stress survey were also found to score higher than validated norms for Type "A" personality. This may contribute to the overall stress score as people within this category generally show more stress prone behaviours and stress related problems. What is uncertain is whether this type of trait is necessary for the work being carried out, or why there was a high proportion of Type "A". From the OSI, the physical health effects found for stress were lethargy, palpitations, headaches, inability to sleep and indigestion. The last two could also be associated with shiftwork. There were also mental effects attributed to stress such as negative attitudes towards other job factors. Hedge et al (1992 and 1993) suggest that work stress influences SBS symptoms. If this is the case then many of the symptoms could be attributed to stress at work rather than environmental conditions, for example, headaches, tension, irritation, and nasal congestion. Hedge (et al 1988) found that self-reported work related illnesses are linked to dissatisfaction with environmental factors and job stress, not job satisfaction. This study appears to confirm this suggestion.

To explain the effects of stress and SBS Figure 1. shows a proposed model of SBS and occupational stress. The model indicates that not only do environmental factors, building characteristics and personal and job factors contribute to SBS, but they also contribute to stress levels at work. The effect of SBS and occupational stress on health has similar symptom outcomes. It is therefore difficult to separate the two and both SBS and stress affect each other, for example the symptoms affecting stress levels and stress affecting levels of symptom reporting..

Conclusion

The study of control room staff found the following:-
1. The types of health complaints occurring were SBS symptoms, with over 70% of occupants reporting having them often or sometimes.
2. The main attributions of the health complaints were air quality and a dry atmosphere.
3. Historical evidence suggested an increase in SBS symptoms including eye problems and headaches with the introduction of new technology.
4. Environmental parameters were not always within recommended levels (BSI 1990).
5. Sickness absence statistics did not show any patterns or increase in reporting time off work.
6. A model is suggested which predicts an interaction between stress and SBS.

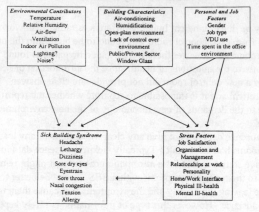

Figure 1. Proposed Model of Sick Building Syndrome and Occupational Stress

References

BSI, 1990, BS7179 *Ergonomics of design and use of visual display terminals (VDTs) in offices. Part 6. Code of Practice for the design of VDT work environments.* British Standards Institute, London.

Carlton-Foss, J.A., 1984, Comfort and discomfort in office environmental problems. Annals of the American Conference on Industrial Hygiene, 10, 93-112.

Colligan, M.J., 1981, The psychological effects of indoor air pollution. Bulletin of the New York Academy of Medicine, 57, 1014-1025.

Cooper, C., Sloan, S.J. and Williams, S., 1988. *Occupational Stress Management Guide.* NFER-Nelson, Oxford.

Hedge, A., 1988, Job stress, job satisfaction and work-related illness in offices. *Proceedings of the Human Factors Society - 32nd Annual Meeting*, 777-779.

Hedge, A., Erickson, W.A. and Rubin, G., 1992, Effects of personal and occupational factors in sick building syndrome reports in air-conditioned offices. *Stress and Well-being at Work*, pp 286-298, Ed Quick, J.C., Murphy, L.R. and Hurrell, J.J., American Psychological Association.

Hedge, A., Erickson, W.A. and Rubin, G., 1993, Why do gender, job stress, job satisfaction, perceived indoor air quality and VDT use influence reports of the sick building syndrome in offices. In *Work With Display Units '92*, Eds Luczak, H., Cakir, A. and Cakir, G., Elsevier Science, Amsterdam.

Jaakkola, J.J.K., Reinikainen, L.M., Heinonen, O.P., Majanen, A. and Seppanen, O., 1990, Indoor air quality requirements for healthy office buildings: Recommendations based on an epidemiological study. Paper presented at Conseil Internationale du Batiment Rotterdam.

Karasek, R.A., Baker, D.B., Marxer, F., et al, 1981, Job decision latitude, job demands and coronary heart disease. A prospective study of Swedish men. American Journal of Public Health, 71, 694-705.

Leeman, A., 1988, The sick building problem. *Seminar on Engineering Solutions to the Sick Building Syndrome*, Institute of Mechanical Engineers, June, 1-6.

Pennebaker, J.W., & Watson D., 1988, Self-reports and physiological measures in the workplace. *Occupational Stress: Issues and Developments in Research*, Ed Hurrell J.J., Murphy, L.R., Sauter, S.L. and Cooper C.L., pp 184-199. Taylor & Francis.

Robertson, A.S., Burge, P.S., Hedge, A., Sims, F.S., Gill, J., Finnegan, M., Pickering, C.A.C. and Dalton, G., 1985, Comparison of health problems related to work and environmental measurements in two office buildings with different ventilation systems. British Medical Journal, 291, August, 373-376.

WHO, 1983, *Indoor Air Pollutants: Exposure and Health Effects*, Euro Reports and Studies, 78, 23-26.

Wilson, S. and Hedge, A., 1987, *A Study of Building Sickness.* (Building Use Studies Ltd).

Wyon, D. P., Asgeirsdottir, T.H., Jensen, P. K. and Fanger P. O., 1973, The effects of ambient temperature swings on comfort, performance and behaviour. Archives des Sciences Physiologique, 27, 441-458.

SURFACE TEMPERATURES AND THE THERMAL SENSATION AND DISCOMFORT OF HANDRAILS

by

Lisa Halabi and Ken Parsons

Department of Human Sciences
Loughborough University of Technology
United Kingdom

An experiment was conducted to determine a data base of subjective data concerning the thermal sensations obtained when bare hands touch materials of different surface temperatures. Forty eight subjects were exposed to four materials, wood, aluminium, nylon and stainless steel in the form of handrails, sixteen in each of three temperature conditions, 15 C, 20 C or 25 C. Subjects were requested to feel the materials for a short period whilst rating their thermal sensation and discomfort. A mathematical model, used for predicting contact temperature and further developed by Parsons(1993), was also examined. The results were presented in the form of centiles to quantify individual differences and provide data that can be used for the selection of materials.

Introduction

Contact between bare skin and solid surfaces may cause thermal discomfort depending upon the temperature of the materials and the material types. Bare skin touching metal at low temperatures will cause a cold sensation and discomfort for example. The sensation and discomfort felt may be important in the design and construction of handrails or handles of vehicles or floor materials in a swimming pool or other areas where people walk on surfaces with bare feet. It would be helpful to designers of floors, handrails etc. if fundamental ergonomics data were available to allow the prediction of thermal sensation and discomfort caused by contact with a range of materials at a range of temperatures. This paper presents a study that makes a contribution to establishing such a database.

Method

Forty eight subjects were used in a laboratory experiment where four materials in the form of handrails were placed in a thermal chamber at three temperatures, 15 C, 20 C and 25 C. Sixteen subjects provided subjective responses for each temperature, after touching the materials by placing their right hand through a hole in the thermal chamber wall and grasping the handrail. For each of the three experimental sessions four subjects (two male, two female) touched each of the materials first, in an order determined by a Latin square. That is four groups of four subjects per temperature. Subjects were undergraduate and postgraduate students, predominantly young with a mean age of 23.7 years.

Subjects conducted the experiment in groups of four. They arrived at the laboratory and rested in a thermal neutral room for at least ten minutes before the start of the experiment. The handrails were placed in the thermal chamber several hours before the start of the experiment to ensure that they had equilibrated with the chamber temperature. When in a thermal neutral condition, subjects were called out of the thermal neutral room to the test area one at a time. Prior to this the subject's hand temperature was recorded using a thermistor placed in the hand. The subject had previously read the instruction sheet. He or she placed the right hand through the hole in the chamber (shielded by a foam rubber cover) and grasped the hand rail. Immediate impression was given of both thermal sensation and thermal discomfort by referring to scales placed directly in front of them. These are shown in Table 1.

Table 1. Scales of thermal sensation and discomfort used in the experiment

Sensation	Comfort
7 Hot	4 Very uncomfortable
6 Warm	3 Uncomfortable
5 Slightly Warm	2 Slightly Uncomfortable
4 Neutral	1 Not Uncomfortable
3 Slightly Cool	
2 Cool	
1 Cold	

After immediate impressions had been recorded the subject kept the hand on the handrail and gave further sensation and comfort ratings after

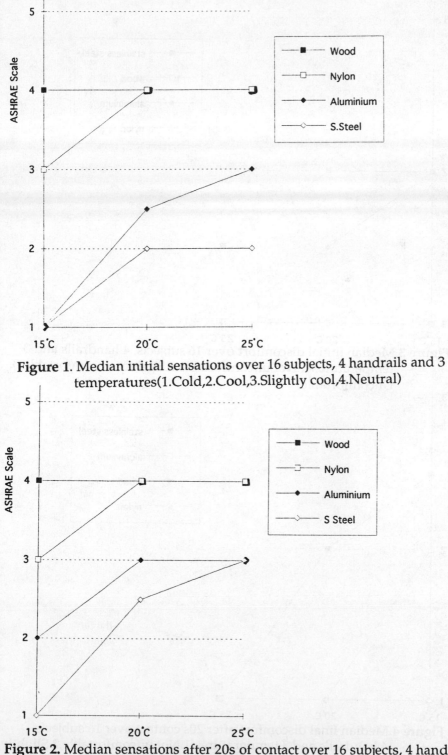

Figure 1. Median initial sensations over 16 subjects, 4 handrails and 3 temperatures(1.Cold,2.Cool,3.Slightly cool,4.Neutral)

Figure 2. Median sensations after 20s of contact over 16 subjects, 4 handrails and 3 temperatures.(1.Cold,2.Cool,3.Slightly cool,4.Neutral)

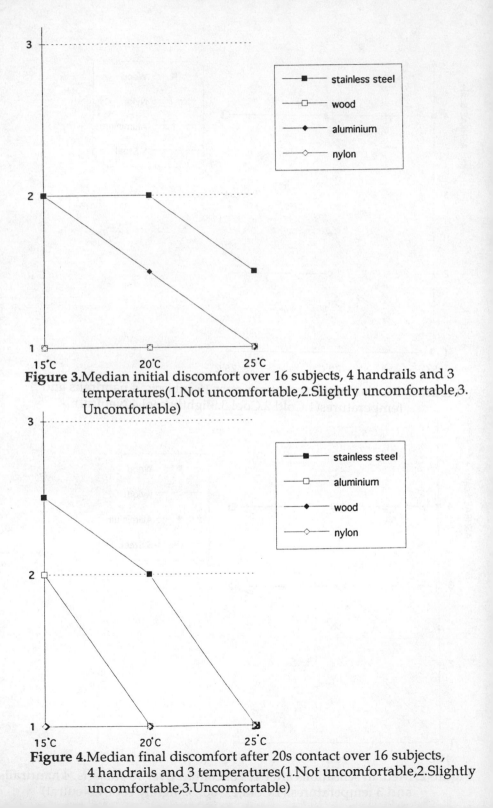

Figure 3. Median initial discomfort over 16 subjects, 4 handrails and 3 temperatures(1.Not uncomfortable,2.Slightly uncomfortable,3. Uncomfortable)

Figure 4. Median final discomfort after 20s contact over 16 subjects, 4 handrails and 3 temperatures(1.Not uncomfortable,2.Slightly uncomfortable,3.Uncomfortable)

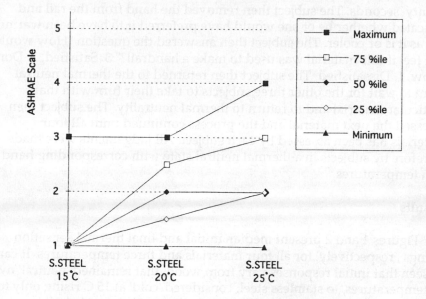

Figure 5. Range of sensation ratings obtained from 16 subjects on initial
contact with a stainless steel handrail.(1.Cold,2.Cool,3.Slightly
cool,4.Neutral)

Surface temperature Th	Contact temperature °C	Initial sensation vote 7 point scale
wood 25°C	31	4 neutral
wood 20°C	29	4 neutral
wood 15°C	28	4 neutral
s.steel 25°C	26	2 cool
aluminium 25°C	25	3 slightly cool
aluminium 20°C	21	2.5 cool/sl cool
s.steel 20°C	21	2 cool
s.steel 15°C	17	1 cold
aluminium 15°C	16	1 cold

Table 2. Comparison of calculated contact temperature with median initial
sensation ratings.

twenty seconds. The subject then removed the hand from the rail and indicated whether he or she would have preferred it to have been warmer, just as it is or cooler. The subject then answered the question "How would you feel if this material was used to make a handrail?" 3. Satisfied, 2. Don't Know, 1. Dissatisfied. The subject then returned to the thermal neutral room to wait for the other three subjects to take their turn with that particular material and to return to thermal neutrality. The subject then assessed the next material and the process continued until all four materials has been assessed by each subject. All judgements were made therefore by subjects in a thermal neutral state with corresponding hand skin temperatures.

Results

Figures 1 and 2 present median initial and final thermal sensation ratings , respectively, for all four materials and three temperatures. It can be seen that initial responses vary from wood, that remained 'neutral' over all temperatures, to stainless steel, considered 'cold' at 15 C rising only to 'cool' at 25 C. Sensations after 20 seconds (figure 2) showed similar results, however the two metal handrails showed a trend towards being less cold than initial responses. Figures 3 and 4 show median initial and final discomfort ratings respectively. It can be seen that wood and nylon were not uncomfortable. Steel was the most uncomfortable at 15 C. Figure 5 presents the range of ratings obtained over all subjects and demonstrates individual variation.

A mathematical model of heat transfer allows the prediction of the 'contact temperature' at the interface of the human skin and a solid surface. This model and its practical developments (leading to equivalent contact temperature, t_{ceq})are described in Parsons(1993). This temperature depends upon the thermal conductivity, specific heat and density of the materials. Calculations of contact temperature are compared with median initial sensation ratings in Table 2. It can be seen that contact temperature may provide the basis for a scale for determining sensations caused by contact with a wide range of temperatures and material types.

References

Parsons K C (1993) Human thermal environments, Taylor & Francis, London. ISBN 0-7484-0041-9.

Acknowledgements

The authors acknowledge the technical support of Mr Trevor Cole of the Human Thermal Environments Laboratory, Loughborough University.

ANALYSING ACCESS AND USE
IN THE BUILT ENVIRONMENT

John Mitchell

Principal Lecturer, Health Research Institute, Sheffield Hallam University,
Collegiate Crescent Campus, Sheffield S10 2BP
Phone 0114 2532404 Fax 0114 2532430

Architectural barriers prevent disabled consumers from using the
built environment fully and independently. Despite the familiarity of
disabled consumers with these barriers, they are rarely reported and
the building industry therefore has little basis for remedial action.
Consumer evaluations are currently undertaken by means of either
'access checklists' or 'user-trials'. The effectiveness of each of each
method is discussed and suggestions are made for developing
improved versions to allow consumers to evaluate the built
environment effectively and reliably.

Introduction

The Effects of a Disabling Environment

All systems place some demand on their users. If the level of demand is low then
virtually all users can operate the system fully and effectively. However, if the level of
demand is high then some would-be users will be unable to operate the system
independently. In such circumstances the system could be said to disable its user.

Dependence in important areas of life imposes heavy and continuing social and
financial penalties on those who are directly affected and also on their families, on care
and support services and on the Exchequer. This country has a relatively high
proportion of elderly and disabled people (Martin, 1988; Social Trends, 1988) and the
consequences of avoidable dependence are therefore likely to be considerable.

The quality of the built environment is of particular significance since many
people spend the majority of their lives within it. An environment which does not
match the needs and capacities of its users is likely to disable them and reduce their
autonomy. Those affected will find that they cannot undertake essential activities, for
example, they may be unable to use public transport, shops, schools, workplaces, or
even parts of their homes (Fenwick, 1977).

As a result, it would be prudent to ensure that the built environment is so well
matched to the needs and capacities of its users that avoidable dependence is minimised.
Once achieved, the quality of the future built environment and its components can be

protected by appropriate standards, regulations and guidance. The replacement of obsolete 'disabling' facilities by new 'user-friendly' equivalents would allow the level of enforced dependence to be reduced over time.

Consumer Evaluation of the Built Environment

Unfortunately, access surveys by disabled people have shown that new buildings and facilities still fail to provide them with unrestricted access and use (Maclean, 1991 a,b,c, and 1993; Mitchell, 1994, 1995; Murray, 1993).

However, though these barriers were familiar to disabled people, the planners of the buildings concerned were unaware of their existence until the survey findings had been published.

Such disparities of perception suggest that planners may not be receiving effective feedback from their consumers. Few consumer evaluations of buildings and facilities have so far been published and few of these have received publicity outside their immediate locality.

While many areas of consumption are regularly tested and reported by the Consumers' Association, evaluations of the built environment are rarely included (Access for Disabled Consumers, in press). In the absence of comprehensive, detailed reporting of barriers in the built environment it would be hardly surprising if the building industry was unaware of their nature, extent and consequences.

Without routine evaluation of buildings, consumers have few ways of alerting the building industry to any problems they face. As a result, even those which could be readily resolved, remain unreported and unaddressed.

The paucity of consumer feedback is also likely to make it more difficult for the building industry to establish priorities for remedial action in terms of the problems which currently exist, the feasibility and cost of different solutions and the likely impact on the marketplace of these solutions.

Such priorities would be easier to establish if consumers could readily evaluate and report on the quality of the built environments which they use. The value of these finding would, however, depend on the quality and suitability of the methods used.

Critique of Existing Methods

To provide effective and reliable feedback of consumer problems in the built environment it is suggested that the methods used should:
- be capable of detecting any barrier
- provide factual, replicable data
- accurately reflect patterns of building use.

Buildings are currently either evaluated by means of either 'access checklists' (eg Royal Association for Disability and Rehabilitation, 1993) or 'user-trials' (eg Maclean, 1991). Neither methodology succeeds in fully meeting these criteria.

Access checklists are lists of known barriers which are noted and reported during surveys. Their standardised format has several advantages. Surveys can be undertaken rapidly, they can be undertaken by able-bodied as well as disabled people since no particular insights are required and the findings can be rapidly analysed and presented. The information from checklist surveys is often used to compile 'access guides' to the accessibility and usability of facilities in particular cities (eg Derby City Council, 1993).

However, checklists have certain important limitations. Since they can only detect familiar barriers they are mainly effective in identifying physical barriers and less effective at identifying sensory or cognitive barriers. Furthermore, though versions have been produced for a wide range of different public buildings and facilities such as theatres, libraries and canteens, none appear to have been produced which would enable consumers to evaluate the quality of private homes.

'User-trials' do not use standardised checklists and therefore offer greater flexibility. Instead of relying on visual checks to identify barriers, testers attempt to use the facilities fully and independently. If a tester succeeds in the attempt, this implies that no insurpassable barrier is present. However, if a task cannot be completed, this is taken as revealing a barrier to either access or use.

Depending on the capacities of the survey team, user-trials can be used to evaluate any type of building or facility and to detect any type of barrier regardless of its familiarity or its nature. Though the findings of individual testers appear to be reasonably self-consistent, it is difficult to standardise findings or to compare the findings of different groups. Indeed, if the same building was to be surveyed by teams with dissimilar patterns of functional capacity, the findings would inevitably reflect these differences.

Accordingly, there is a need for greater standardisation of user-trial surveys in two particular respects.

Firstly, since survey findings are directly affected by the composition of a survey team, it is important to describe the range of capacities within its membership (eg Maclean, 1991). This will allow those using the data to assess how well the composition of a particular team accords with the patterns of functional limitation within the population as a whole (Mitchell, 1995).

Secondly, there is scope for clearer definition of the processes by which consumers use buildings or facilities. These processes fall naturally into a sequence of stages which begin with the approach to the building, extend to mobility within the building and include the use of facilities within the building.

Implicit versions of this sequence can be found in publications by those providing design guidance (eg British Standards Institution 5810; Department of the Environment, 1985, 1992: Hull City Council, 1993; Palfreyman, 1993) as well as explicit versions in the reports of disabled consumers (Maclean, 1991 a, b, c and 1993, Mitchell, 1994; Murray, 1993).

There is also scope for linkage between the two methods of survey. User trials can be used to identify the nature and scope of existing barriers in new and older buildings and facilities. Once these barriers have been comprehensively described it will then be possible to rework access checklists so that they can be used to detect all barriers in a much wider range of environments than is possible at present.

References

British Standards Institution, BS 5810, *Access for the Disabled to Buildings*
Consumers' Association, in press, *Access for Disabled Consumers*, ('Which?', London)
Department of the Environment, 1985, 1992, Building Regulations (Part M)
Derby City Council, 1993, *The Derby Access Guide for Disabled People*, (Derby City Council, Derby)
Fenwick, D, 1977, *Wheelchairs and their Users*, (OPCS, London)
Hull City Council, 1993, *Access for the Disabled: A Standard Guide*, Hull
Maclean, L.
1991 a, *The Waltheof Centre,* (PAVIC Publications, Sheffield)
1991 b, *The Sheffield Arena,* (PAVIC Publications, Sheffield)
1991 c, *The Don Valley Stadium*, (PAVIC Publications, Sheffield)
1993, *The Hillsborough Leisure Centre*, (PAVIC Publications, Sheffield)

Martin, J. 1988, *The prevalence of disability among adults*, (OPCS, London)

Mitchell, J. 1994, *Sheffield's Supertram*, (PAVIC Publications, Sheffield)

Mitchell, J. 1995, *Do planning regulations ensure that new buildings can be used by all?*, Contemporary Ergonomics

Murray, S. (ed.), 1993, *Accessible Issues relating to Metrolink*, (Metrolink Consultative Committee, Manchester)

Palfreyman, T. 1993, *Designing for Accessibility*, (Centre on Environment for the Handicapped, London)

Royal Association for Disability and Rehabilitation, 1993, *Access Guide-Making Instructions*, (RADAR, London)

Social Trends, 1988, Table 6, (OPCS, HMSO, London) **52** Summer 20

"DO PLANNING REGULATIONS ENSURE THAT NEW BUILDINGS CAN BE USED BY ALL?"

John Mitchell

Principal Lecturer, Health Research Institute, Sheffield Hallam University,
Collegiate Crescent Campus, Sheffield S10 2BP
Phone 0114 2532404 Fax 0114 2532430

People with limited mobility, hearing, sight and use of their hands form more than fourteen per cent of the population in the United Kingdom. Though planning regulations and guidance are intended to ensure that they have full access and use of the built environment, evidence from recent access surveys has called their effectiveness into question. Examples are provided of the nature and effects of various barriers in new buildings and facilities. It is suggested that the effectiveness of planning regulations could be increased if consumer evaluation of the built environment was undertaken more thoroughly and regularly.

Introduction

By definition, public facilities are intended for the use of the public as a whole. However, the importance of a good match between facilities and their users has not always been recognised. Indeed, until recent times public facilities such as railway stations, town halls, libraries and theatres, were constructed on the assumption that all members of the public are fully able-bodied and unencumbered.

Though this assumption may be correct for most members of the public it does not apply to elderly or disabled people who are impeded by such features as stairs or heavy doors. Indeed, some members of the public, for example wheelchair users, are completely excluded by such barriers.

The recognition that poorly designed architectural features can discriminate against disabled people led to a general reconsideration of design practice in this country and overseas. This included the concept of 'Barrier-free' design and the adoption of Part M of the Building Regulations and improved design standards. In Sweden, the Building Ordinance (1966) was enacted which requires that new buildings are fully accessible and usable by all members of the public. 'Access check-lists' have been developed to enable users to test the accessibility of local facilities.

The construction of new buildings and facilities provides opportunities for improving the match with users and their needs. It had been assumed that architectural barriers would gradually disappear as new, accessible facilities replaced older, less satisfactory versions. Unfortunately, surveys of new buildings and systems undertaken by disabled people reveal that this confidence is misplaced (Maclean 1991 a, b, c and 1993; Mitchell, 1994; Murray, 1993).

Subsequently, similar findings have been obtained in surveys undertaken on behalf of
the Consumers' Association (Consumers' Association, Access for Disabled Consumers,
in press),

Not only do newly constructed buildings and facilities contain many familiar
physical barriers, they also contain other features which can only be used by those who
can see, communicate verbally, hear, use their hands, manoeuvre in restricted spaces
and read written English.

The accessibility and usability of these buildings were tested by teams of disabled
people, who attempted to use them fully and independently, on exactly the same basis as
any other member of the public. The results were recorded and used to identify any
feature which could not be fully used by all members of the team.

The buildings surveyed included recreational and sporting facilities and the first
section of Sheffield's Supertram system. All buildings were completed in or after 1991
and complied fully with design guidance and building regulations.

This paper will review the extent to which planning procedures can ensure that
newly constructed facilities are sufficiently well matched to the capacities of the public
that full access and use can be guaranteed.

Functional Capacities of the General Public

Barriers place demands on their users which can only be surmounted by those
with sufficient physical, sensory or cognitive capacities. In the built environment, their
effect is to 'filter out' anybody who cannot cope with their demands. For example, if
directional information is provided exclusively in written English, only those who can
see, read or understand English will be able to use the facility without assistance from
others.

A large proportion of the general public have been found to have limited physical
and sensory function. Tables 1 and 2 set out the relative numbers of people across the
age and functional spectra.

Table 1. Percentages of the UK Population who are outside working age

Pre-school	(6.4%)
Under 16	20.5%
Over 65	15.7%
Over 75	(6.5%)
Total	36.2%

Source: OPCS, 1988, Social Trends, Table 6, Page 20, No 52, Summer, HMSO,
London

Table 2. Functional Limitations within the UK Population

Mobility	10%
Hearing	6%
Sight	5%
Limited use of hands	4%
Wheelchair users	<1%
All disabilities	14%

Note: Figures do not include people with dyslexia or cognitive limitations

Sources: Martin J et al, 1988, The Prevalence of Disability among Adults, OPCS
Fenwick D, 1977, Wheelchairs and their Users, OPCS

These figures suggest a wide range of 'non-standard' functional capacities in the population. Low functional capacity cannot be regarded as a rare and unpredictable occurrence which falls outside the scope of routine planning. It forms part of the normal distribution of capacities within the population and, as such, is as 'normal' as high functional capacity.

For most individuals, high functional capacity is not a steady state throughout life. Peak capacity is only reached after a number of years and its transience is illustrated by its decline when activity levels are reduced, when short term illness intervenes and as a result of ageing, injury and sickness.

Despite widespread public perceptions, wheelchair users make up a relatively small part of the numbers of disabled people. Though people with limited mobility are the most prevalent group amounting to ten people in every hundred, the numbers of those with limited hearing, sight and manual dexterity have far-reaching implications for the planning and delivery of products and services.

Functional Capacities within the Survey Teams

The survey teams used to evaluate facilities in Sheffield (eg Maclean, 1991) varied in number from 5 to 15. Four core members took part in all surveys. Their capacities correspond to the five most prevalent functional limitations and also include dyslexia as shown in Table 3 below:

Table 3. Functional Capacities of Core Survey Team

	Crutches	Hearing	Sight	Hands	Wheelchair	Dyslexia
A	X					
B	X			X		X
C		X		X	X	
D		X	X			

Members A and B use crutches and cannot walk more than one hundred metres without stopping to rest. Hearing loss for members C and D is not extreme but both have difficulty in following a conversation if other people are speaking at the same time. Member D is blind and uses a guide dog for mobility. Member B cannot read. Members B and C have limited reach and have difficulty in sorting change. Member C uses a manually propelled wheelchair.

When larger survey teams were available, the effect was to confirm the findings of the core group, unless the new members had specific functional limitations which were not already represented. For example, further information and insights were provided by the participation of members who use powered wheelchairs, or who have very restricted reach or hearing.

Survey Methods - Process Analysis

Access checklists were not used during the surveys since these can only detect barriers that have already been identified (Mitchell, 1995).

Instead, it was decided to identify and analyse the process by which consumers use the built environment. Accordingly, each team member attempted to use the facilities fully and independently on exactly the same basis as any other member of the public.

The following sequence was used by the teams during their surveys:

Getting to the building/system
        Members approached the building by either public or transport, selected the most appropriate alighting or parking point, locating the entrance and reaching the main entrance.

Getting into and around the building/system
        Members entered the building and found their way around it. Internal doors, lifts, information and signage, reception desks and kiosks and, emergency procedures were considered at this stage.

Using the main facilities
        The main facilities within a building/system, such as the spectator and sporting facilities in the sports venues and the ticket machines and trams in the Supertram system, were found and used.

Using the 'Backup' facilities
        Essential but subsidiary facilities such as buffets, telephones and toilets, were found and used.
        Following each survey, members' feedback, comments and discussion were recorded and analysed.

Findings

        The surveys revealed that many standard designs and layouts in the built environment are unsuitable for use by some members of the public with disabilities.
        With two important exceptions, steps had not been widely used in the design of the sporting venues or the Supertram system. Many barriers of other types were however found in the design of building components such as doorsets, lifts, signage and kiosks. Further problems were found in the layout of facilities such as changing rooms, hygiene facilities, buffets and reception areas.
        Though no steps were found at the entrances to the sporting venues or to Supertram platforms, they nevertheless provided the only means of reaching the tiered seating in the sports venues and the main passenger seating areas on Supertram. As a result, access is denied to all who cannot negotiate steps or manoeuvre in restricted spaces.
        People with different patterns of disability found that they were impeded by quite different features of buildings. For example:

Doorsets
        People with limited mobility and manual dexterity found difficulty in using heavy or cumbersome doors. There was sometimes too little manoeuvring space or wheelchair users were impeded by protruding door handles.
        Problems for blind people were caused by the lack of a consistent strategy for positioning push plates or pull-handles. It was also not clear whether to attempt to use the right or left hand part of a doorset.

Lifts
        All the lifts in the sporting venues were difficult for the team to use. Blind users had no directional information to enable them to find the lift, the call button or the control panel within the lift. The absence of audible signals prevented them from selecting the floor they required and from leaving when it had been reached. The

absence of usable information inside lifts is likely to cause problems for blind and for deaf visitors during an emergency.

Wheelchair users found that the location of call buttons alongside lift doors in a recess made it difficult or impossible for them to call the lift. They also found that the placement of controls in a single corner of a lift and at a standard height often made their use difficult or impossible. The installation of lifts with internal dimensions of only 1,500mm prevents them from turning round inside the lift and compels them to decide whether to reverse into or out of the lift.

Signage

The absence of good-quality signage in both the sports venues and the Supertram system impeded all members of the teams. The absence of tactile signage and 'guidance paths' however completely prevented blind people from finding any of the buildings, the facilities within them or the entrances to Supertram platforms. However, an excellent quality guidance path has been included on Supertram platforms which includes tactile indicators which allow passengers, including blind people, to position themselves at the points at which the tram doors will open.

Kiosk and information desks

Noise levels at these points made it difficult for many members of the teams to gain the information they required. This was particularly difficult for people with limited hearing and there were no technical means of overcoming these problems.

Communication difficulties also affected wheelchair users who were generally much lower than the staff at desks. In addition, the absence of 'cutaways' at desks and kiosks prevented them from getting close to the desk and made it difficult for them to reach when paying and taking change.

Escape and emergency systems

Emergency information was not available throughout the sports venues. Since it was usually presented in either visual or auditory form it would have been difficult for blind or deaf people to acquire information during an emergency.

Primary and Secondary Barriers and their Elimination through Consumer Feedback

These shortcomings act as barriers to members of the public because they require functional capacities which many people do not possess. They can be regarded as 'Primary' barriers if they prevent members of the public from getting into a building or as 'Secondary' barriers if they prevent them from using any of the facilities it contains.

Their continuing existence in new buildings and facilities has serious implications both for consumers and for the building industry since they demonstrate that the use of the best available guidance does not guarantee that even newly constructed parts of the built environment are fully accessible and usable.

It is difficult to explain the continuation of these problems since both the Department of the Environment and the British Standards Institution actively seek inputs from disabled people and their organisations whenever they develop or update their guidance (DoE 1990, 1992; BS 0). An alternative explanation (Mitchell, 1995) is that more effective methods are required for consumer evaluation of the built environment.

It is appropriate that disabled people should take the initiative in developing methods for evaluating the built environment. Though they are particularly affected by and aware of primary and secondary barriers, the effects of such barriers limit the functional effectiveness of many other members of the public. A barrier-free built environment is ultimately to the benefit of all.

References

British Standards Institution, BS 0, *A Standard for Standards*

Consumers' Association, in press, *Access for Disabled Consumers*, ('Which?', London)

Department of the Environment, 1990, 1992, Building Regulations (Part M)

Fenwick, D, 1977, *Wheelchairs and their Users*, (OPCS, London)

Maclean, L.

1991 a, *The Waltheof Centre,* (PAVIC Publications, Sheffield)

1991 b, *The Sheffield Arena,* (PAVIC Publications, Sheffield)

1991 c, *The Don Valley Stadium*, (PAVIC Publications, Sheffield)

1993, *The Hillsborough Leisure Centre*, (PAVIC Publications, Sheffield)

Martin, J. 1988, *The prevalence of disability among adults*, (OPCS, London)

Mitchell, J. 1994, *Sheffield's Supertram*, (PAVIC Publications, Sheffield)

Mitchell, J. 1995, *Analysing Access and Use in the Built Environment*, Contemporary Ergonomics

Murray, S. (ed.), 1993, *Accessible Issues relating to Metrolink*, (Metrolink Consultative Committee, Manchester)

Social Trends, 1988, Table 6, (OPCS, HMSO, London) **52** Summer 20

Swedish Government, 1966, *The Building Ordinance.* In M. Beckman, 1977, *Building for Everyone*, Stockholm

Risk Assessment

IS RISK ASSESSMENT A NECESSARY DECISION-MAKING TOOL FOR ALL ORGANISATIONS?

Deborah Walker and Sue Cox

*Centre for Hazard and Risk Management,
Loughborough University of Technology,
Loughborough, Leicestershire LE11 3TU*

Risk assessment is an essential feature of recent health and safety regulation in the European Community. It is a proactive process which should ensure that the employer's response to risk control is proportional to the magnitude of the risk. Through a range of training courses and research activities, the Centre for Hazard and Risk Management has gathered data on the implementation and management of risk assessment in a wide range of UK organisations. This paper considers the role of risk assessment in health and safety management. It will seek to support, through practical experience, the hypothesis that, although risk assessment is a vital element in the safety management process, its application varies depending on a variety of factors including size of organisation, industrial sector, economic influence, perceived risks and management attitudes.

Introduction

Quantified risk assessment was originally developed as an engineering technique used in the nuclear and aerospace industries. Its roots can be traced to ideas put forward by Pugsley (1942) and later developed by Freudenthal (1947). The techniques of quantified risk assessment have become increasingly important over the years and are used today in many high risk industries. More recently, the practice of risk assessment has developed to include qualitative methods which are particularly useful in low risk environments. Such methodologies have also been used in major hazard environments to support general hazard and risk assessment.

The introduction of the Management of Health and Safety at Work Regulations (MHSWR) (1992) brought with them the requirement for employers to conduct risk assessments of all their activities. The Centre for Hazard and Risk Management (CHaRM) at Loughborough University of Technology developed a protocol for qualitative risk assessment which enabled organisations to comply with this legislation. This protocol, published in the Risk Assessment Toolkit, Cox (1992), has been presented to over 250 organisations at public access and in-company courses.

Participants were subsequently contacted to provide follow-up information on their use of these generic techniques. The toolkit has also been used as the basis for the development of risk assessment software, which has been produced in collaboration with a software house. The product design has been based on extensive consultation with a number of organisations who have bought into the Toolkit concept and who have developed individual risk assessment systems based on its protocol. Further developments on the Toolkit protocol are planned on the basis of a follow-up survey (ibid).

In parallel with these system design activities, CHaRM is carrying out complementary research on health and safety in small to medium sized enterprises (SMEs). This research includes an investigation of methods of health and safety management, the application of risk assessment techniques and the development of self-audit systems.

This paper will first discuss the role of risk assessment in health and safety management. It will then review the application of the Risk Assessment Toolkit, Cox (1992), in a variety of organisations and industrial enterprises in the United Kingdom. Some preliminary evidence on risk assessment in SMEs also is presented. The authors will argue that the technique of risk assessment is not only a necessary but also an essential health and safety decision-making tool for all organisations.

The Role of Risk Assessment in Health and Safety Management

The concept of health and safety management is encapsulated in the publication 'Successful Health and Safety Management', HS(G)65, Health and Safety Executive (HSE) (1991). The advice in this document has generally found favour with organisational management and is now being used to support the development of health and safety management systems. HS(G)65 describes six key elements in the process of health and safety management. They include:

a) **Policy**, which sets out the organisation's general approach, intentions and objectives towards health and safety issues;

b) **Organising**, which is the process of designing and establishing the structures, responsibilities and relationships which shape the total work environment;

c) **Planning**, the organisational process which is used to determine the methods by which specific objectives should be set and how resources are allocated;

d) **Implementing**, which focuses on the practical management actions and the necessary employee behaviours required to achieve success;

e) **Performance measurement**, which is the process by which information is gathered which reflects progress towards health and safety goals; and

f) **Audit and performance review**, which includes the review of necessary information and the processes of reflection. These are the final steps in the health and safety management cycle.

Risk assessment may be considered to be the focus of planning. The outputs of a risk assessment provide a description of the health and safety hazards, an assessment of the risks to all those who may be affected by such hazards and the necessary control actions. In practice this information should enable managers to take appropriate actions to eliminate any unnecessary hazards and to reduce those risks which cannot be removed if the risk assessment process is fully integrated into the health and safety management cycle.

The need for organisations to carry out risk assessment and develop health and safety management systems has further been given impetus by the introduction of the new regulations requiring 'management' approaches, for example the MHSWR (1992) and the EC Framework Directive 89/391/EEC, EC (1989). However, more than a year after the deadline set for its implementation (end of 1992), five European Union member states still have not incorporated the 1989 Community framework directive into their natural laws, Vogel (1994). There is also some evidence from the authors' research that, even in the U.K., compliance is patchy.

Current knowledge about the number and types of organisations actually adopting a health and safety management approach and carrying out risk assessments is limited. A number of projects, including the Lead Authority project, Foster (1993), and the European Foundation project, Vogel (1994), are addressing this shortfall. A preliminary analysis of data gathered from activities carried out by CHaRM (both anecdotal and questionnaire) highlight a number of key issues in relation to health and safety management and the risk assessment process.

These data were gained from the following sources:
a)   one day seminars and workshops on the Risk Assessment Toolkit;
b)   in-house risk assessment system development workshops and software; and
c)   questionnaire surveys.
These are described below.

Risk Assessment Toolkit Workshops

CHaRM has run a series of workshops (in-company and public) at regular intervals since June 1992. These workshops have provided the basis for organisational risk assessment systems to enable compliance with the MHSWR (1992). A breakdown of organisations taking up this training opportunity is shown in Table 1.

Table 1. Analysis of organisations participating in risk assessment training

Business Sector (by SIC code)		Company Size		
		Large	Medium	Small
0	Agriculture, forestry and fishing	0	0	0
1	Energy and water supply industries	7	0	0
2	Extraction of minerals and ores other than fuels; manufacture of metals, mineral products and chemicals.	21	8	0
3	Metal goods, engineering and vehicle industries.	16	8	0
4	Other manufacturing industries.	14	20	0
5	Construction	3	3	0
6	Distribution, hotels and catering; repairs	0	3	0
7	Transport and communication	15	0	0
8	Banking, finance, insurance, business services and leasing	2	2	0
9	Other services (total)	33	27	4
	- health care	3	6	0
	- education	0	3	2
	- local government	11	6	0
	- central government	16	4	0
	- emergency services	3	0	0
	- miscellaneous	0	8	2

The organisations have been broken down by business sector and company size. The definitions for company size are based on those used by the European Union, Boyle and McGrath (1994), and the HSE (1994a). Accordingly, small to medium sized enterprises are those who employ less than 500, have a net turnover of less than ECU38m and less than one third of the ownership is ascribed to a parent organisation or financial institution. Small firms are defined as those with 50 or less employees.

These data show that uptake of CHaRM training is spread across most of the industrial and service sectors with the exception of SIC 0. The majority of organisations were large or medium-sized. The absence of small-sized organisations is not surprising as this type of training is relatively expensive. Follow-up phone calls and questionnaires confirmed that many companies attending training have subsequently gone on to implement risk assessment in varying degrees through a variety of methods. This feedback confirmed the importance of risk assessment in the development of health and safety systems. It also highlighted the benefits for future policy and decision making.

Risk Assessment Software/Systems Development

CHaRM has recently developed RMS, risk management software in collaboration with Warwick IC Systems (1995). The software was developed in consultation with a number of organisations who have used the Risk Assessment Toolkit as the basis for their own risk assessments. In accordance with the trade name (RMS), this software not only facilitates the recording of risk assessments but also assists in managing the control strategies that are the result of a risk assessment. This is illustrated by the main features of the software which are described in Table 2.

Table 2. Features of RMS

Feature	Description
Risk Assessment	This covers identification of hazards followed by estimation and evaluation of risk
Actions/Controls	Actions and control strategies can be recorded to help ensure that on-going requirements can be effectively managed.
Personnel Records	This incorporates employment history, training and document issue to provide effective control of employee competence and exposure to defined work environments.
Accidents	Details of an incident can be recorded to enable effective and structured analysis.
Claims/litigations and notices	This allows for the reporting and analysis of claims, litigations and enforcement notices.

Analysis of user needs has led to the development of several proprietary risk assessment systems including software such as RMS. These developments reflect a recognition that risk assessment is a vital element within the Risk Management Process, Cox and Tait (1991).

This complex process of risk management can be difficult to manage effectively using simple paper systems. RMS has been designed to encompass both risk assessment and risk management elements to accommodate complex problem solving.

SME Survey

The SME Research Group within CHaRM has been collecting data via semi-structured interview and questionnaire from a small sample of SMEs (300) in the Charnwood district of Leicestershire; this is an initial pilot study of a larger survey. The study has looked at safety awareness, methods of health and safety management, risk assessment techniques and self-audit systems in SMEs. These data have been supplemented by a 'think tank' workshop, when a group of 'experts' discussed health and safety management in SMEs, Vassie and Cox (1994). The responses highlighted the fact that the majority of SMEs do not manage health and safety and therefore do not carry out risk assessments. Where SMEs do carry out risk assessment there are a variety of factors influencing their activities. Some of these factors are summarised in Table 3.

Table 3. Factors influencing the activities of SMEs

Factor	Comments
Trade/Professional Associations	Such organisations may have produced industry specific guidance and methods of risk assessment.
Enforcement Visit	Enforcement officers can provide advice or require evidence of risk assessment.
Management Awareness	Positive management attitude is often present when those involved have worked with larger organisations who have a good health and safety record or where managers have participated in some health and safety training.
Size of Organisation	Size of organisation is important since at some stage a health and safety specialist will be employed.
Type of Ownership	Organisations that are subsidiaries of larger companies are often sent information and instruction from a head office function.
Insurance Companies	Insurance companies may require a risk assessment before insuring an organisation. This is often linked to use of external consultants.
Degree of Risk	Organisations operating in high risk areas are more likely to have carried out risk assessments.
Intuitive Risk Assessment	Risk assessments can be carried out 'intuitively' in that the hazards have been identified and the risks controlled, but not documented.

Discussion

The data from the various activities described in this paper highlight a number of trends and influences. First, risk assessment is increasingly regarded as a vital and integral part of health and safety management and is therefore part of the decision making process within an organisation. This is also reflected in current legislative trends, HSE advice and guidance, expressed industry needs and the subsequent development of RMS.

This paper provides some evidence which suggests that not all organisations are currently performing risk assessments. This is reinforced by data from other studies currently being carried out in the European Community, Vogel (1994).

Second, where risk assessment does occur there are a number of factors which influence this process. These include:

a)   size of organisation - larger organisations are more likely to carry out risk assessment;
b)   industrial sector - this may be linked to risks and perceived risks;
c)   management awareness and commitment - this sets the health and safety culture;
d)   economic influences - particularly acute with SMEs; and
e)   HSE and enforcement officer visits.

Application of risk assessment within SMEs is particularly difficult to assess and the positive influences (see Table 3) presented in this paper are associated with limited success. There are many barriers to implementation that must be overcome. One factor that may be developed is the use of partnerships or networks between SMEs. Organisations such as trade associations, professional bodies, Chamber of Commerce and Business Clubs could also play a vital role in facilitating this process. However, the current situation in health and safety management is extremely fluid, HSE (1994b). Much effort is being extended to raise awareness through educative companies and advisory centres throughout Europe. As large and medium sized organisations demonstrate successes and cascade their enthusiasm for risk assessment the authors are confident that its uptake will penetrate all market sectors.

References
Boyle, J. and McGrath, D., 1994, How to get research funding in Europe, *Scientific Computing 2*, July, 33-37.
Cox, S.J. and Tait, N.R.S., 1991, *Reliability, Safety and Risk Management - An integrated approach* (Butterworth Heinemann)
Cox, S., 1992, The Risk Assessment Toolkit (Loughborough University).
EC, 1989, Council Directive of 12th June on the introduction of measures to encourage improvements in the safety and health of workers at work (89/391/EEC).
Foster, A., 1993, Local authority Enforcement: Lead Authorities, *Health and Safety Bulletin 210*, June, 9-11.
Freudenthal, A.M., 1947, The safety of structures *Trans.Am.Soc.Civil Engrs.*, 125-127.
Health and Safety Executive, 1991, Successful health and safety management HS(G)65.
Health and Safety Executive, 1994a, Review of regulation working paper No. 1 small firms and the self-employed.
Health and Safety Executive, 1994b, Review of health and safety regulation HSC12.
HMSO, 1992, Management of Health and Safety at Work Regulations and Approved Code of Practice.
Pugsley, A.G., 1942, A philosophy of aeroplane strength factors, Reports and Memoranda No. 1906 (Aeronautical Research Council, London).
Vassie, L. and Cox, S., 1994, Progress report on health and safety: voluntary schemes in a European context, CHaRM report to the European Union 1995.
Vogel, L., 1994, Prevention at the workplace: an initial review of how the 1989 Community framework directive is being implemented (European Trade Union Technical Bureau for Health and Safety, Brussels).
Walker, D. and Cox, S., 1994, Feedback on risk assessment toolkit training, CHaRM internal report 1994/01.
Warwick IC Systems and CHaRM, 1995, RMS Risk Management software (Warwick IC Systems, Ripley, Derbyshire, Tel. 01773-512656) launched 25 January 1995.

MACRO-ERGONOMIC RISK ASSESSMENT IN THE NUCLEAR REMEDIATION INDUSTRY

Lin Guo,* Ash Genaidy,* Doran Christensen** And Kevin Huntington**

*University of Cincinnati, Cincinnati, Ohio, U.S.A.
**FERMCO, U.S. Department of Energy, Cincinnati, Ohio, U.S.A.*

The objective of this investigation was to develop a practical tool to assess ergonomic stresses in the workplace. The tool was used to evaluate jobs in a nuclear remediation facility.

Introduction

In the past few years, awareness about the impact of ergonomic stresses upon the workforce has been heightened in many countries around the world. Legislation has passed which requires ergonomic control programs. In the United States, the Occupational Health and Safety Administration is in the process of promulgating an Ergonomic Standard which will affect every business in America. Worksite analysis is a key component of this standard. Thus, a practical tool to assess ergonomic stresses is warranted and is the subject of this investigation. A macroscopic- (macro-) ergonomic risk assessment tool was developed and used to evaluate jobs in a nuclear remediation facility.

Development of macro-ergonomic risk assessment tool

Structure

Based on a review of the risk assessment methods in the published literature, it was evident that there was a lack of criteria for developing practical ergonomic risk assessment tools. Thus, the following criteria were adopted for this purpose:

- Scope -- The approach should be comprehensive and incorporate the majority of risk factors found in the workplace.
- Simplicity -- The analyst should know how to use this tool after a brief training session even though he/she may not have a profound knowledge of ergonomics.
- Practicality -- This criterion implies that the assessment tool should be cost-effective. From an industrial standpoint, cost is governed by four factors: scheduling and time

required; personnel involved; materials/equipment needed; cost. To remain cost-effective, the tool should allow the user to perform the job analysis with minimal resource utilization.

- Usefulness -- The information collected should be useful in many ways to minimize ergonomic stresses. For example, management can prioritize jobs in terms of the control action required. Medical service departments can use the assessment data for fit-for-duty and injury/illness evaluation. Supervisors can make better decisions about job assignments specially for light-duty tasks. Human resource departments can incorporate the information in the design of pre-placement tests and on-the-job training. The various company functions can make a sound decision about the provision of reasonable accommodations for compliance with the Americans with Disability Act (ADA).
- Reliability -- The methodology should be founded on the basis of good scientific evidence. Also, it should be well defined to ensure consistency of data collected.
- Job-specificity -- The tool should specifically reflect job requirements in terms of tasks and elements within tasks.

The risk assessment tool assumes that any job requires one or more of the following task categories: manual material handling, hand work, postural loading, sensory requirements, environmental factors, usage of personal protective equipment. The six task categories consist of a total of 64 task elements (e.g., lifting, carrying).

Each job (e.g., carpenter), task (e.g., manual material handling) and task element (e.g., lifting) is evaluated on the basis of Relative Stress Index (RSI). This is a "0-10" scale where "0" denotes hazardous/unsafe/unproductive and "10" means safe/non-hazardous/productive. Trigger points are developed to evaluate jobs, tasks and task elements:

- Red zone -- Immediate action is required (RSI's of "0-2.5").
- Yellow zone -- A temporizing approach is required (RSI's of "2.6-7.5") including further evaluation.
- Green zone -- No action is required (RSI's of "7.6" or greater) because the job/task/task element is safe/non-hazardous/productive.

RSI is based on a quantitative evaluation of job requirements which take into account multiple variables such as load, repetition, task duration, travel distance, etc. A series of mathematical equations for computing RSI values was derived based on the assumption that job variables had both main and interactive effects.

Data gathering

The procedures used to collect data are discussed in the following section:

- Preliminary information gathering -- Preliminary information essential for accurate data collection includes list of jobs and job descriptions, job safety analyses, injury data and other relevant information.
- Job definition -- The list of jobs provided by the company can be helpful but not necessarily sufficient because they may be defined differently than that which is required for ergonomic evaluation. According to the Department of Labor (1982), a job is a single position or a group of positions, the major work activities and objectives of which are similar in terms of worker actions, methodologies, materials, products, and/or worker characteristics. The list of jobs and other preliminary information provided by a company should be carefully reviewed and redefined if necessary.

MMH/HAND WORK/SENSORY FACTORS	DUR	REP	WT	TD	HD
1. Lifting/Lowering	F	R	5-11	F-W	0.2-0.3
2. Carrying	O	R	5-11	0-1.5	0-0.1
3. Pushing	F	NR	12-23		
4. Pulling	NP				
5. Handling	F	R	3-5		
6. Fingering	S	NR	0-1		
7. Standing	F	R			
8. Sitting	O	NR			
9. Reclining	NP				
10. Walking	O	NR			
11. Climbing Stairs/Ramps	S	NR			
12. Climbing Ladders/Scaffolds	NP				
13. Balancing	NP				
14. Kneeling	NP				
15. Crouching	NP				
16. Crawling	NP				
17. Stooping	O	NR			
18. Trunk/Head Twisting	O	NR			
19. Forearm Twisting	NR				
20. Reaching Above Shoulder	S	NR			
21. Reaching Below Shoulder	F	NR			
22. Bending Head/Forearm/Hand	F	R			
23. Near Vision (< 0.5m)	F				
24. Midrange Vision (0.6m - 6m)	F				
25. Far Vision (> 6m)	S				
26. Depth Perception	S				
27. Basic Color Vision	NP				
28. Intermediate Color Vision	NP				
29. Color Shade Vision	NP				
30. Comprehend/Articulate Speech -- 2.5m	NP				
31. Comprehend/Articulate Speech -- 5m	NP				
32. Comprehend/Articulate Speech -- high ambient noise	NP				
33. Feeling/Touching	NP				
34. Tasting/Smelling	NP				

DUR=Duration
REP=Repetition
WT=Weight/force (kg)
TD=Travel distance (m)
HD=Horizontal distance(m)

ENVIRONMENTAL/PPE FACTORS	DUR
35. Outdoor	O
36. Indoor	F
37. Extreme Cold	NP
38. Extreme Heat	NP
39. Wet	NP
40. Humid	NP
41. Noise	R
42. Vibration	NP
43. Respiratory/Pulmonary Irritants/Sensitizers	NP
44. Contact/Skin Irritants/Sensitizers	NP
45. Radiant Energy	NP
46. Electrical Energy/Shock	NP
47. Work at Heights	NP
48. Work Below Ground	NP
49. Work in Confined Spaces	NP
50. Work with Immediately Dangerous/Lethal materials	NP
51. Work closely with Others Physically/Cooperatively	C
52. Work alone with Accessible Aid	NP
53. Work with poorly Accessible Aid	C
54. Protracted or Irregular Hours of Work	NP
55. Operation of Heavy/Hazardous Vehicles/Equipment	NP
56. Other Physical Hazard (specify)	NP
57. Other Chemical Hazard (specify)	NP
58. Other Biological Hazard (specify)	NP
59. Other Radiological Hazard (specify)	NP
60. PPE -- Respirator	NP
61. PPE -- Body Suit	NP
62. PPE -- Hearing Protectors	NP
63. PPE -- Vision Protection	NP
64. PPE -- With Close Skin Contact	NP

Job: Mailroom clerk
Supervisor: Bob Smith (05/27/94)
Worker: Jack Williams (06/10/94)
Interviewer: Ash Genaidy

Figure 1. An example of a completed macro-ergonomic risk assessment tool

- Supervisor interviews -- Management and supervisory interviews are essential for accurate data collection. The supervisor should be informed of study purpose and general assessment procedures (figure 1) prior to data collection. If the interviewee supervises more than one job, an interview sheet should be used for each job. Also, the following points should be taken into account during the interview process: assume the worst-case scenario for different levels of a job variable (e.g., load); correctly define tasks for each job; and take as many notes as possible to ensure capture of all relevant information.
- Worker interviews -- Interview should be conducted with knowledgeable workers who perform the jobs to validate the data collected from supervisors.
- Supervisor and employee reviews -- Reviews of the data collected should be conducted to further validate and assess its accuracy.
- Findings preparation/presentation -- The findings should be presented in a variety of ways including comparison of jobs, tasks and task elements.

Testing of macro-ergonomic risk assessment tool

To validate the macroscopic risk assessment tool, a pilot study was conducted in a nuclear remediation facility over a three-month period. Prior to the start, a list of 30 jobs were submitted by the company for investigation. The number of jobs was later reduced to 26 after redefinition.

Interviews were conducted with supervisors and workers. RSI values were then computed for each job, task and task element. Figures 2 and 3 present the comparisons of the jobs and manual material handling. Job comparisons showed that 24 out of 26 jobs fell into the yellow zone. Painter had the lowest RSI (3.8) and traffic administration had the highest RSI (8.6). Of the six task categories defined in this study, the majority of jobs fell within the yellow zone. A number of job tasks fell into the red zone meaning immediate action should be taken to remediate the hazards.

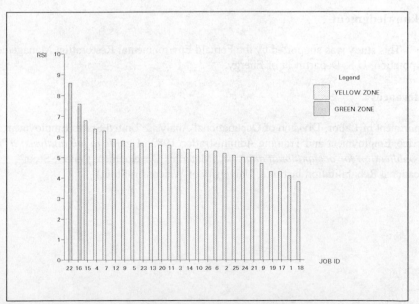

Figure 2. Comparison of all jobs

        A carpenter job is shown as an example in figure 4. Hand work scored the lowest RSI value. Also, the sensory and personal protective equipment tasks were in the red zone. A closer examination indicated that the carpenter job required a high degree of fingering, near vision, color vision, feeling, touching, vision protection, skin protection, etc. Thus a control action is required for this job.

        After the RSI's were computed, a complete report was developed which contained job/task/task element comparisons, job description and the detailed information provided in the checklist.

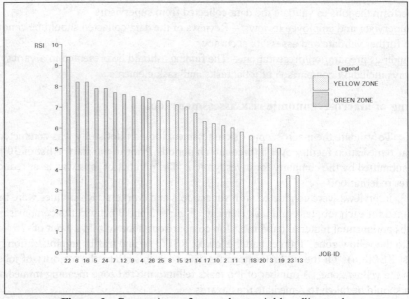

Figure 3. Comparison of manual material handling tasks

Acknowledgment

        This study was supported by the Fernald Environmental Restoration Management Corporation, U.S. Department of Energy.

References

Department of Labor, Division of Occupational Analysis, United States Employment Service, Employment and Training Administration 1982, *A guide to job analysis: a "how-to" publication for occupational analysis*, (Materials Development Center, Stout Vocational Rehabilitation Institute, University of Wisconsin-Stout).

Job: Carpenter

Overall	MMH	Handwork	Posture	Sensory	Environ.	PPE
4.1	5.8	2.0	5.3	1.8	4.7	1.6

FACTOR	RSI	FACTOR	RSI
1. Lifting/Lowering	6.7	33. Feeling/Touching	0.0
2. Carrying	5.2	34. Tasting/Smelling	4.0
3. Pushing	5.5	35. Outdoor	2.0
4. Pulling	5.5	36. Indoor	2.0
5. Handling	3.1	37. Extreme Cold	10.0
6. Fingering	1.0	38. Extreme Heat	2.0
7. Standing	5.3	39. Wet	4.0
8. Sitting	10.0	40. Humid	4.0
9. Reclining	10.0	41. Noise	2.0
10. Walking	0.0	42. Vibration	4.0
11. Climbing Stairs/Ramps	4.3	43. Respiratory/Pulmonary Irritants/Sensitizers	2.0
12. Climbing Ladders/Scaffolds	4.3	44. Contact/Skin Irritants/Sensitizers	4.0
13. Balancing	4.3	45. Radiant Energy	6.0
14. Kneeling	5.3	46. Electrical Energy/Shock	2.0
15. Crouching	5.3	47. Work at Heights	2.0
16. Crawling	6.3	48. Work Below Ground	4.0
17. Stooping	5.3	49. Work in Confined Spaces	6.0
18. Trunk/Head Twisting	1.3	50. Work with Immediately Dangerous/Lethal materials	10.0
19. Forearm Twisting	7.3	51. Work closely with Others Physically/Cooperatively	10.0
20. Reaching Above Shoulder	5.3	52. Work alone with Accessible Aid	6.0
21. Reaching Below Shoulder	4.3	53. Work with Poorly Accessible Aid	4.0
22. Bending Head/Forearm/Hand	5.3	54. Protracted or Irregular Hours of Work	2.0
23. Near Vision (< 0.5m)	0.0	55. Operation of heavy/hazardous vehicles/equipment	10.0
24. Midrange Vision (0.6m - 6m)	2.0	56. Other Physical Hazard (specify)	10.0
25. Far Vision (> 6m)	2.0	57. Other Chemical Hazard (specify)	2.0
26. Depth Perception	0.0	58. Other Biological Hazard (specify)	6.0
27. Basic Color Vision	2.0	59. Other Radiological Hazard (specify)	2.0
28. Intermediate Color Vision	2.0	60. PPE -- Respirator	2.0
29. Color Shade Vision	4.0	61. PPE -- Body Suit	2.0
30. Comprehend/Articulate Speech -- 2.5m	2.0	62. PPE -- Hearing Protectors	4.0
31. Comprehend/Articulate Speech -- 5m	2.0	63. PPE -- Vision Protection	0.0
32. Comprehend/Articulate Speech -- high ambient noise	2.0	64. PPE -- With Close Skin Contact	0.0

Figure 4. Macro-ergonomic risk assessment for carpenter

Control Rooms

VisiWall: Designing an interactive multi-screen display for an electrical power utility control room

D. Darvill, J. Brace, L. Ryan & J. Gnocato

*MPR Teltech Ltd.
8999 Nelson Way,
Burnaby, B.C. Canada, V5A 4B5*

This paper is a report of the human factors activities that led to the mock-up and specification of a user interface design for the a large multi-screen display in an electrical power utility control room. The *VisiWall* system is a combination of hardware and software that provides an environment in which multiple users can view information on a large multi-screen display. Users can also seamlessly interact with and move objects between the large screen display and their local console displays. In this paper, the control room environment and three activities related to the design of a user-centred and usable interface are described, i.e., a task and user needs analysis, the development of a user interface mock-up, and a usability evaluation of this *VisiWall* mock-up.

Acknowledgement: The *VisiWall* Project is sponsored by PRECARN Associates and involves a collaboration of MPR Teltech Ltd., M3i Systems Inc., and B.C. Hydro.

Introduction

Managing large complex supervisory control environments, such as an electric power utility generation and distribution system, is becoming an increasingly difficult task. Operations staff must quickly and effectively respond to system changes. To do so, operators rely on some form of visual overview of the system to maintain an awareness of the state of the system elements. To provide this essential overview, there is a growing interest in the installation of large, multi-screen, projection or CRT displays in control rooms. These displays can provide an interactive, flexible, large-scale overview of the vast amounts of information commonly available about the system under control (Axelson, 1994; Tani, Horita, Yamaashi, Tanikoshi and Futakawa, 1994).

Traditionally, many organisations have used electro-mechanical mosaic wall displays to provide the required overview of their system or network. However, these mosaic wall displays have a number of drawbacks:

1. They require a large space to display sufficient detail of the operation; often taking up an entire wall in a control room.

2. The layout of the network or process is quite inflexible; network elements often must be taped, painted or mounted in fixed locations.

3. It is seldom easy for the operators to interact with the wall board; especially when they may require a ladder to reach some locations.

4. The amount and type of information that can be displayed is limited because of the electro-mechanical nature of the display.

5. Updating or rearranging information about the network itself can be difficult because of the fixed physical requirements of the wall board layout.

6. The wall board displays are primarily read-only information sources about limited pre-selected aspects of the network or process being monitored.

Properly used, large arrays of multi-screen projection or CRT displays can be installed in control rooms to mitigate most of the drawbacks of traditional mosaic wall board displays.

A goal of the *VisiWall* project was to develop a prototype visual wall display for use in an electrical power utility control room. The *VisiWall* system would provide the network operators with a flexible, seamless, interactive workspace that could be used to monitor network status as well as other purposes (Gnocato, 1994).

The project also provided an opportunity to explore the possible benefits that such an interactive environment would provide compared to the more traditional environment of an electro-mechanical mosaic wall board and multiple supervisory control system CRT displays.

Control Room Study

The Southern Interior Control Centre, where this study was conducted, is a relatively new B.C. Hydro installation with ten experienced operators and two apprentices. It includes a large mosaic wall board (approximately 11 m long and 4 m high) that provides the current status of key network elements at a glance. This overview information and additional details are also available on three supervisory control system CRT displays placed on a console in front of the operator.

The *VisiWall* prototype will be installed in the existing control room environment where it will operate in parallel with the existing displays. The prototype will consist of three edge matched large screen projection displays laid out horizontally (approximately 4 m long and 1 m high) and a local CRT display placed on the operator's console. X-Wall software by M3i Systems allows the large screen displays and the CRT display to be utilised as a single contiguous workspace (Gnocato, 1994).

The goal of the control room study (CRS) was to learn as much as possible in a relatively short time about the background, training, tasks, needs and environment of the operators of the current supervisory control system (SCS). The operators were involved early in the design and development process in order to elicit their participation and knowledge and to improve the chances of their acceptance of the *VisiWall* prototype. The information gathered from the operators was used to design a mock-up of a potential user interface for the *VisiWall* prototype. Feedback from the operators about the task analysis activities and about their evaluation of the user interface mock-up was provided to the prototype developers as part of the design specification for the *VisiWall* prototype user interface.

Prior to any user interface development, basic information was gathered about the operators' tasks in their current control room setting. Using questionnaires, interviews, direct observations, training materials and procedural manuals, the operators' current procedures, tasks, information flow and problems were identified.

A general questionnaire was prepared based on initial familiarisation interviews with a sample of the operators and on findings from five control room studies conducted for another project to develop intelligent graphic interfaces (IGI Project) for real-time supervisory control room settings. These IGI control room studies identified a number of common problems within such settings: e.g., the difficulty of presenting the appropriate information to support problem solving and reduce workload, the navigation of large complex problem spaces, the recognition and management of alarms, the need to monitor the big picture while attempting to solve a specific problem, the effects of training and experience, and the need for communications among operators on the same shift, between shifts and with other staff (Dickinson, Jones, Dubs and Ryan, 1993).

The questionnaire and subsequent semi-structured interviews were used to acquire information about task difficulty, information flow, user needs and user preferences from all twelve operators. Observations of the operators were used to verify information and gather additional details of their tasks. Table 1 provides a list of issues that were raised by the operators and proposed solutions that were incorporated into the design of an interactive mock-up of a potential user interface for the *VisiWall* prototype (see Figures 1 and 2).

Table 1. Major issues raised by operators during interviews and solutions demonstrated with the user interface mock-up.

Operator Interviews (Issues raised)	User Interface Mock-up (Potential solutions)
1. Operators generally used an overview display when monitoring the network, but often abandoned this detailed overview when working on a specific problem. They then had to rely on the mosaic wall display.	- For monitoring the network, the *VisiWall* display shows different, but appropriate, levels of detail for each network node at all times, allowing all three SCS displays to be used for specific problem solving.
2. Difficulty moving mouse or trackball among the three SCS CRT displays.	- The mock-up cursor moved seamlessly between the large and local *VisiWall* displays.
3. Operators had a problem seeing all safety tags at zoomed out levels of the SCS schematic displays. There also were no tags shown on the mosaic wall display.	- A means was provided to place tags on items within the large mock-up display elements and to view tag information at both levels of detail on the local *VisiWall* display.
4. There is no direct connection between the information on the mosaic wall and the three SCS CRT displays. Operators must remember information from the mosaic display when locating appropriate displays for the SCS CRTs.	- The *VisiWall* mock-up allowed an operator to manage the views shown on the SCS displays by selecting a node label from the large screen display. By dragging the label through a display filter at the bottom of the large display and dropping it onto the representation of a particular SCS display on the *VisiWall* local display, the appropriate information would be displayed on the SCS CRT display.

5. Some operators would prefer that the mosaic display only show lights for those elements that are off-normal (i.e., was a dark board configuration).	- The mock-up was designed to show closed, normal element colours close to the background colour, but opened, off-normal elements would be a darker shade than either the normal elements or background. This ShadowBoard© concept highlights only elements that should be of interest to the operator.
6. Operators often require notes to themselves and must pass information about the system to others both during their shift and at shift changes.	- A method was provided to create and attach notes, called PostCards©, to network elements displayed on the large *VisiWall* display or to message areas on both large and local *VisiWall* displays.
7. Information on the mosaic wall could be out of date if system had failed and no updates were being processed.	- A System Alive Indicator was added to the large *VisiWall* display to indicate that the system was operational.

VisiWall User Interface Mock-up

The *VisiWall* user interface mock-up was based on relevant tasks and user information collected in the first part of the study (see Table 1) and on the technical capabilities of the proposed *VisiWall* hardware and software. MacroMedia Director running on a Macintosh computer was used as a rapid prototyping environment for the user interface mock-up (Brace, Darvill and Gnocato, 1994). This environment allowed the use of multiple monitors as one continuous display space. One monitor was configured to simulate the large multi-screen *VisiWall* display (see Figure 1); a second simulated the local *VisiWall* display on the operator's console and also simulated the three existing SCS displays (see Figure 2). See Table 1 for a brief description of the features included in the user interface mock-up (see figures) and their relation to the issues raised by the operators during the interviews.

The user interface mock-up was taken to the control centre, shown to eight potential users of the *VisiWall* prototype system, and their comments were recorded. The operators were also asked to perform some simple tasks with the mock-up and comments about these tasks were also recorded.

Information from these usability sessions was provided to the *VisiWall* prototype system developers (Brace, Darvill and Gnocato, 1994). In general, the operators were pleased with the functionality, look and feel of the user interface mock-up. Some examples of their comments follow:

1. Operators expressed concerns that the *VisiWall* prototype be linked to their existing SCS; e.g., they did not want to have to manage yet another separate database.

2. When multiple tags for a particular device are displayed on the large display the highest priority tag colour should be displayed. All tags should be the appropriate colour.

3. Timers on PostCards© would be useful.

4. Operators suggested that the PostCards© might be used to display Procedures or Help notes; this would be especially useful for apprentices.

5. In addition to the display filters, a pop-up menu was shown as a method to select the information to be displayed on the SCS CRT displays. No definite preferences were found for one selection method or the other.

6. The *VisiWall* system should automatically show additional details of any node that was normal and closed but then enters an off-normal condition.

7. The display of overview information on a normal closed node was found to be useful.

The modified mock-up, with user feedback, became the initial user interface specification for the design of the *VisiWall* prototype that will be installed in the control room. Further usability evaluations will be conducted using this prototype.

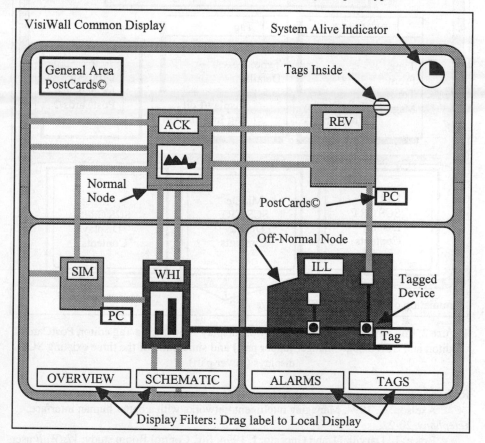

Figure 1. The *VisiWall* mock-up of the large screen common display showing System Alive Indicator, Shadow Board© effect, Normal and Off-Normal elements, Tags, PostCards©, and Display Filters (see text).

Summary

Using questionnaires, interviews, direct observations, training materials and procedural manuals, the goal of the control room study, i.e., to learn as much as possible in a relatively short time about the users, tasks and environment of the current supervisory control system, was achieved. The operators were interested in and participated in these activities with enthusiasm. Feedback from the operators about their tasks and about the user interface mock-up has been helpful to the prototype developers as they design the *VisiWall* prototype user interface.

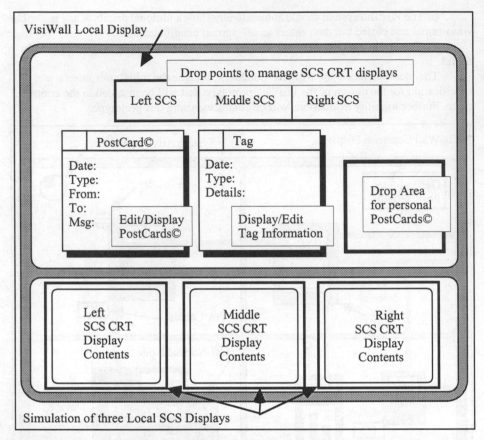

Figure 2. The mock-up of the local *VisiWall* display showing the Tag editor, PostCards©
editor, and SCS display manger (upper part) and simulation of the three existing SCS
displays (lower part).

References

Axelson, C. 1994, Managing intelligent networks with a better human interface.
Telephony, 20-22.

Brace, J., Darvill, D. and Gnocato, J. 1994, SIC Control Room study: *VisiWall* user
interface mock-up evaluation. *VisiWall Technical Report V94-004* (MPR Teltech,
Burnaby).

Dickinson, J., Jones, M., Dubs, S. and Ryan, L. 1993, IGI case study integration
report. *Intelligent Graphic Interface Project Technical Report IGI93-0005* (MPR Teltech,
Burnaby).

Gnocato, J. 1994, *VisiWall* specifications. *VisiWall Project Technical Report V94-003* (MPR Teltech, Burnaby).

Tani, M., Horita, M., Yamaashi, K., Tanikoshi, K. and Futakawa, M. 1994,
Courtyard: Integrating shared overview on a large screen and per-user detail on individual
screens. *Proceedings of CHI 94 Conference*, (ACM, New York), 44-50.

See no evil, hear no evil, speak no evil: verbal protocol analysis, eye movement analysis, and nuclear power plant diagnosis

Barry Kirwan*, Magnhild Kaarstad, Gunnar Hauland, and Knut Follesoe

*Industrial Ergonomics Group
School of Manufacturing &
Mechanical Engineering
University of Birmingham
B15 2TT

Halden Reactor Project
Institutt for Energiteknikk
Postboks 173, N-1751 Halden
Norway

Although many measures have been implemented to avoid recurrence of the Three Mile Island accident and similar potential accidents caused by diagnostic failure, it is still difficult to predict so-called cognitive errors such as misdiagnosis. Furthermore, empirical accounts of how licensed operators actually diagnose in realistic challenging nuclear power plant scenarios are few and far between. This makes it difficult both to predict risks from certain hazardous Nuclear Power Plant scenarios, and to design operator interfaces better to support the operator in such situations. The research referred to in this paper is part of an ongoing three year programme to investigate diagnosis and its failure mechanisms in realistic scenarios with suitably qualified operators. This paper focuses on the methodological issues facing anyone who wishes to investigate the dynamics of diagnosis.

Background

The ability to diagnose in abnormal events and emergencies is one of the primary functions of a nuclear power operator, yet our collective knowledge of how diagnosis actually works in practice, and therefore of how to support the operator most effectively, is scant indeed. A three-year project is therefore being undertaken at the OECD Halden Research Reactor Project in Norway, to investigate how better to support the operator(s) in such situations, and to evaluate what types of errors actually occur in realistic emergency scenarios. Such errors may amount to inefficiencies, such as inefficient information gathering activities which delay diagnosis, or may lead to more serious consequences such that the scenario is not fully diagnosed, or is misdiagnosed, perhaps leading then to the wrong actions being implemented (as happened at Three Mile Island). The first phase of the research, reported in this paper, is concerned with developing methodologies which enable us to study diagnosis in dynamically-evolving and complex scenarios, using real operators and a sophisticated real-time simulator facility. Such methodologies should enable us to follow and understand the operator's diagnosis in real time, and to determine errors and inefficiencies, and also to highlight aspects of the interface which are either supporting diagnostic behaviour, or contributing to the risk of diagnostic errors.

Three pilot studies have taken place in the Halden Man-Machine Laboratory (HAMMLAB) , a nuclear power plant simulator facility in Norway, which records all events in real time, whether those events originate from the system, the human operator, or due to the interaction of the two. Eleven real reactor operators have been used in these pilot studies, and a number of scenarios have been used, each one typically lasting 15 - 20 minutes (as this phase of the project is only focusing on the initial phase of diagnosis). These scenarios have varied in difficulty, but some have proven significantly challenging for experienced operators.

As part of this research, the primary question inevitably arises as to how diagnosis can be analysed during scenarios, since it is by its very nature a mental and hence generally unobservable phenomenon. Two main approaches have therefore been investigated, in two pilot studies, to develop a means of tracking diagnosis in real time, and to enable analysis not only of *how* diagnosis progressed in a particular way, but also *why*. The two data collection methods are **verbal protocol analysis** (VPA), and **eye movement tracking analysis** (EMTA), with a third analytic method referring to how such data can be meaningfully and accurately interpreted and described as diagnostic phenomena. The title of the paper therefore refers to the basic methodological conundra familiar to researchers who have attempted to use these methods: can you believe what a person is telling you ?; if a person is looking at an object, how do you know if that person is simultaneously seeing and thinking about that object ?; and when you interpret their diagnostic strategies, how do you know your interpretation is a true reflection of reality, as opposed to the diagnostic phenomena appearing merely how you wish them to appear (e.g. according to some contemporary theoretical model) ?. These questions are not easily answered, and the interim answer adopted in this study is that of a *context-driven* analytic approach. The three aspects of the methodology developed, namely verbal protocol analysis, eye movement tracking analysis, and interpretation of diagnostic strategies, trajectories and errors/inefficiencies, are outlined below.

Verbal Protocol Analysis

The purpose of this research is to gain insight into the operators' cognitive activities while performing complex tasks. One way of getting this information is to ask people to think aloud while they are doing the task. However, as we have no way of observing operators' mental behaviour directly, it is difficult to test whether there is a direct correlation between what people think and what they say. Verbal protocol analysis (VPA) is currently one of the richest means of investigating the nature of human cognition. VPA methodology consists of inferring the operator's cognitive processes from their own verbal reports, and can take three basic forms: verbal protocols can be given during task performance (*concurrent: VPC*); after task performance has been completed (*retrospective: VPR);* or during an interview period at fixed times during the task period (*interruptive: VPI*).

With VPC, the information being verbalised is directly accessible from short term memory, and is less susceptible than VPR to the potential biasing effects of longer term memory, wherein the subject may rationalise previous behaviour with the benefit of hindsight, and may also of course forget items that were actually strongly influencing performance at the time. VPI is a compromise between the two, since only several minutes elapse between each interruption. However in the context of diagnosis, VPI may give personnel time to reflect on the events they are seeing, which they might not have in the real situation. VPC is therefore expected to be the most valid approach, i.e. producing the most accurate account of what the operator was actually thinking, but itself has four main drawbacks: firstly people usually think faster than they speak; secondly some thought processes are difficult to verbalise (visual imagery and mathematical thought processes); thirdly people will not tend to report what is obvious to them; and fourthly, during the periods of high cognitive demands, the subject may actually go silent (stop verbalising), due to parallel processing limitations during excessive cognitive workload situations.

The first and second pilot studies found that, for a 15 minute scenario, a mixture of VPC and VPI was optimum, with most information being based on VPC, with VPI giving clues as to the overall diagnostic trajectory. An interactive debrief was used to fill in any gaps or inconsistencies perceived by the experimenters, or simply for corroboration purposes. Moreover, the two main methods delivered qualitatively different types of information. VPC was very useful in tracking the ongoing search processes when the operator was looking for symptoms and causes, etc.. Complementary to this, however, was that the VPI method enabled the operators to state their overall strategy, planning, and hypotheses, and this information was often crucial in enabling the experimenters to piece together the diagnosis, and to determine whether errors were occurring. No significant effects of 'reflection' on diagnostic performance were observed due to VPI.

Primary emphasis is therefore placed on VPC, with back-up and integrative information from VPI, and corroboration from debriefs. Even with this amount of information, however, particularly during high workload periods, information was being lost which held clues as to why errors actually occurred. If errors of omission occurred, almost by definition, the operator would not be able to tell us about it, and 'reconstruction' of the diagnostic trajectory (how they diagnosed a fault set) by the experimenters often left several competing hypotheses as to the causes of the errors. What was needed was principally to know what the subject actually saw, and, often, to know whether they did or did not see particular signals during the simulations. This led to experimentation with eye movement tracking (EMT) equipment.

Eye movement tracking

Eye tracker equipment allows the researcher to investigate where the operator is fixating at all times during the experiment. In particular, the eye tracker registers the position of the eye ball relative to the head, but by calibration procedures both before the experiment and during it, it is possible to determine what objects are being viewed, within a reasonably small fixation area (such as a valve on a VDU process mimic diagram). There are, however, two major problems with such data: firstly the amount of information recorded is vast, often with several saccades (movements) per second; and secondly (and more fundamentally) there is the difficulty of trying to decide if the subject was thinking about what his eyes were focused upon. However, as a general minimum, the eye tracker can identify what the subject appeared to be looking at, particularly where the subject also verbally refers to an object or a measurement, and also where the object or measurement focused upon makes sense, given the current diagnostic strategy, even if the subject is silent at this point. Importantly, the eye tracker tells you when something is not observed. Such information often provides clear clues as to the reasons for diagnostic failure.

Two analysis approaches were utilised with the eye tracker data. The first involved analysing the raw data second-by-second, to determine exactly what had been looked at, the number of fixations per object, and the sequence of registrations. This analytic approach was very highly resource intensive, with the analysis time to real time ratio being of the order of 100:1. An alternative method was therefore employed, which involved the analysis being driven by the limitations in the verbal protocol data: if the subject went silent for a period, or if his actions were not clear from what he said, the eye movement video-recordings were analysed for that section of the experimental trial. A typical example of incomplete verbalisation is where the operator says 'there is something wrong with that valve', but there are many valves on the screen. The eye movement recording, which also incorporates the verbal protocol soundtrack, makes clear the valve's identity in most cases. Therefore, to classify operators' diagnostic strategies and cognitive inefficiencies, eye tracking recording provides valuable supplementary evidence. In the pilot studies so far, the eye tracking data has led to firm clarification of 60% of incomplete or unclear verbalisations.

The analysis of eye movement data is currently more subjective than the experimenters would wish, and ways are being sought to render the analysis more rigorous and based more on the theory linking eye movements to cognition, but very little has so far appeared in the literature that has been found to be usable in the process control domain. Currently two or more analysts review the data and agree on an interpretation, or else agree that it is still uncertain as to what was going on. One further approach that may be tested is the use of auto-confrontation, in which the subject re-verbalises whilst watching his own eye movements on the video. However, the validity of this approach remains unproven, since it allows the subject the opportunity of 'filling in the gaps' in his original diagnosis, and has many of the problems associated with the validity of retrospective verbal protocols. At present, however, the experimenters are firmly convinced of the utility of the eye movement tracking technology for complementing the verbal protocol technique, and it is planned to use it in future experiments (some of which may necessitate two or more operators working together: if they both have eye trackers, this will lead to even more methodological issues).

Interpretation of cognitive activities

The project is primarily interested in three aspects of cognitive behaviour: diagnostic strategies, cognitive efficiencies, and quality of diagnosis. Types of fault-finding strategy have received relatively little attention in the literature where realistically complex scenarios are concerned (e.g. nuclear power plant diagnosis), a notable exception being the work of Rasmussen (1984), who differentiated between two major strategy types, namely topographic (top-down search checking the health of the system and then investigating any unhealthy sub-system), and symptomatic search (pattern-matching between observed symptoms and known fault patterns). Yoon and Hammer (1988) added hypothesis-driven search and data-driven search, the latter referring to a search strategy led purely by the alarm or other events as they occur, with the operator essentially waiting to see what appears in the data. This latter strategy often therefore leads into one of the other three strategies, once something meaningful has been detected or construed. In the three pilot studies to date, all four of these strategy types have been observed.

HAMMLAB records all interactions between the operator and the simulator interface, all the alarms and messages presented to the operator, and a log of the key process variables such as pressures and flows at all times during the simulation trial. Audio and video equipment also records operators' verbalisations, and EMT recording presents a detailed video medium showing where the operator looked during the scenario. Pre- and post-trial questionnaires (including NASA-TLX), as well as the VPI questionnaire, are also administered appropriately, and add another level of subjective information to the objective information recorded automatically. The verbal protocols are transcribed and translated from the video tapes to files and these files are merged with the corresponding event and variable logs from the simulator. This results in what is called a *merged protocol*, and onto this protocol is mapped the experimenters' interpretation of diagnostic strategies, errors, and inefficiencies.

However, the transformation of these raw data into pre-defined categories of diagnostic strategy etc. has proven to be challenging. The initial scoring system for strategies and errors, was based on a 'behaviouristic' analysis of the protocols, e.g. if an *'if.. then'* statement was made, then the subject was assumed to be in 'hypothesis-testing' mode, and if the subject was searching the primary loop, followed by the secondary loop, etc.., then it was assumed the subject was in a topographic search mode. Whilst reliability for such content analysis of the protocols was proving high, the experimenters grew uncomfortable with the results, since these results indicated that subjects were frequently changing between different strategies, and this had not been reported by other investigators in other studies. The experimenter became concerned that what appeared to be rapid and unsystematic changes in diagnostic strategies could in fact be described in terms of a more coherent overall strategy. Furthermore, this was not merely a preferential issue: what might appear to be an error given the first type of analysis might make more sense if the overall strategy was understood. What was missing from the analytic approach was the *intentionality* and *reasoning* (hermeneutics) of the operator's behaviour.

The analytic approach was therefore enhanced by translating the operator's problem solving behaviour into a narrative story, which followed the chronology of the operator's diagnosis, as a function of what he had detected and inferred/deduced from the signals received. The immediate difficulty with this approach is that this requires either the operators to elucidate significantly on everything they are doing and more importantly why they are doing it, or else it requires the experimenters to reach the level of at least novices in the domain being investigated. The latter approach was chosen, with the experimenters taking a crash course in nuclear power scenario handling. This approach greatly improved the narrative reconstruction process, and in instances in which the experimenters could not themselves infer why the operator had chosen a particular course of action, then the debrief was used to determine this. This meant that the narrative process had to proceed almost in real time, being written while the scenario was running. Despite this requiring significant training resources, and placing a cognitively demanding task on the experimenters as well as the subjects, the approach resulted in clearer pictures and reconstructions of the diagnosis and associated errors and inefficiencies, and the debriefs were used to validate the narratives. The EMT data were used to focus on particular performance areas for which the experimenters were initially unsure of what had happened, e.g. why the cause of an event had not been found. It was therefore felt that this 'context-driven' approach significantly enhanced the validity and insightfulness of the methodology.

Errors and Ergonomics Insights

Although only pilot studies have run at this stage, a number of errors and inefficiencies have already been detected. It should be stated that the scenarios are themselves complex and rare event scenarios, and that the operators are under some significant time pressure, in order to provoke diagnostic behaviour and error patterns, at this stage of the programme of research. Inefficiencies such as failure to generate sufficient hypotheses, incomplete searches of symptom sets, etc., have been recorded, as well as more serious error forms such as incomplete diagnosis (failing to determine all major causal events in a scenario), and misdiagnosis. The operators themselves confirmed the realism of the scenarios and the simulation, and were not surprised that errors were occurring.

With respect to design aspects supporting (or not supporting) diagnosis, certain aspects of the alarm interface appeared to be unsupportive of diagnosis of a particular scenario (Steam Generator Tube Rupture), due to the scrolling off of alarms before being visually accessed by the operators. One good example of the interface supporting the operator was one of the VDU formats concerned with the primary loop in the power plant, which proved a very effective support for topographic search patterns in particular, and it was recognised that a similar format picture for the secondary side of the reactor system would further enhance performance. The composite methodology adopted and developed during the pilot studies does therefore appear to be providing the types of insights desirable within the experimental programme.

Summary

In summary, the results showed firstly that both concurrent and interruptive VPs should be used, since they provided qualitatively different types of information in the tasks analysed. Eye movement tracking alone was effectively not meaningful as a source of information on diagnosis and diagnostic activities, but EMT together with VPA represented a powerful synergistic means of analysis, if the analysis was context-driven. After several analyses at different levels of diagnostic activity description, a narrative analytic approach was settled upon, which yielded the most meaningful and interpretable results and also gave indications of diagnostic strategies and errors. This analytic approach did however require a deeper level of technical understanding of the nuclear power plant than was first envisaged to be necessary. The next stage of the work will involve larger experiments and will begin to examine the effects of certain complexity variables on performance.

References

Rasmussen, J. (1984) Strategies of state identification and diagnosis in supervisory control tasks, and design of computer based support systems. In Rouse, W.B. (ed.), *Advances in Man-Machine System Research*, Volume 1, Greenwich, CT: JAI Press.
Yoon, W.C., and Hammer, J.M. (1988) Aiding the operator during novel fault diagnosis. IEEE Transactions on Systems, Man and Cybernetics, **18**, No. 1.

Training

TRAINING TRAINERS - DEVELOPING AN INSTRUCTOR'S COURSE ON MANUAL HANDLING FOR DRAYMEN

RJ Graves* And J Carrington-Porter**

*Human Factors & Work Design,
513A Lanark Road West, Balerno, Edinburgh
EH14 7DH

**Carlsberg Tetley - Alloa,
Alloa Brewery, Whins Road,
Alloa FK10 3RB

Training in how to lift correctly has been one means of controlling risk in manual handling. With the manual handling operations regulations such training needs to be extended to include recognition of risk factors associated with handling. In some parts of the brewing industry, trainee draymen undergo an induction course carried out by senior draymen during normal duties. This paper describes the development of a training pack to help Drayman Instructors teach during normal duties. It describes the training material and approach which "Tutors" needed to train Drayman instructors. Typical handling tasks were observed and existing written information was used to prepare the training material. The training pack contained all the material necessary for the Tutors and was evaluated on a two day course. The final version of the pack has been circulated within the industry and Tutors are training Drayman instructors.

Introduction

Training in how to lift correctly has been one means of controlling risk in manual handling. With the introduction of the Manual Handling Operations Regulations (MHOR -HSE, 1992), it has been recognised that such training needed to be extended. In particular workers should be able to recognise risk factors associated with handling.

In some parts of the brewing industry, trainee draymen undergo an induction course prior to further training while on deliveries. The latter training is carried out by senior draymen during their normal duties. This means that normal training aids are not available because this instruction is undertaken wherever the lorry may be at various times of the distribution cycle.

This paper describes the development of an approach to provide the Drayman instructor with a training pack which could be used during normal duties. This needed to cover general manual handling and ergonomic issues and provide them with a practical method of carrying this out during normal duties. In addition it describes the training material and approach for "Tutors" who were identified as needed to train the Drayman instructors.

Approach

A steering group was formed from within the client company to guide the development of the training material and ensure that it met the practical needs of the industry. The steering group consisted of the company's health and safety manger, the personnel/training manger, the local distribution manager, a consultant ergonomist, and the national health and safety manager. From discussions within the industry it was clear that the Drayman instructors pack needed to be comprehensive and self contained to be successful in providing support. Further, a requirement to train the people ("Tutors") who would train the instructors was identified.

The company does not have a dedicated training team. The training tends to occur at depots and it was recognised that Tutors needed to be brought up to a reasonable level of instructional ability. To this end, the concept behind the training package was twofold. First to produce a package which would contain enough information for a tutor to be trained in instructional technique with particular reference to manual handling for draymen. Second, to provide a self contained training pack which the Drayman instructor could use to train the drayman under instruction.

It was recognised that there may be little in the way of formal facilities for training. Consequently, the training pack was designed to be kept in a self contained plastic folder similar to the overall Tutors' pack. This would be kept by the Drayman instructor so that he could use the visual material and also any written material as a prompt when he took one to one instruction with the drayman under instruction.

The Development of the Training Pack

The company was developing its own guidance for distribution handling so that individual draymen would have a personal copy. This was an update and formalisation of existing good practice and training material. Its introduction contained information on the background to the importance of controlling manual handling risk in the brewing industry. Accident causes reported to the brewers' society over the period 1987-1991 were cited. These showed that there was about 1,000 per year and that over half of these were due to handling lifting or carrying. This emphasised the need for improved handling practice. Important principles from the MHOR were outlined and both employers and employees duties specified.

Other factors considered important by the Health and Safety Executive were covered. These included the working environment such as found in the customer's premises, the correct equipment to be used, the nature of the load, personal protection, and training.

Inevitably the content of the training pack mirrored the material in the draft industry guidance. However, as there were local practices which were more relevant to the local company, typical handling tasks carried out as part of their distribution were observed and existing written information was used to prepare the training material.

The final product of discussions and observations was a draft training pack that contained all the material necessary for the Tutors to train the Drayman instructors. In addition it contained the material to be used by the instructors when training the trainee. It was in three parts.

The first contained the introduction and guidelines for the use of the resource pack. The second contained the material for training the Tutors, and the third, the Drayman Instructors training pack material.

The general introduction described the breadth of handling covered by the training. These included the loading and unloading of vehicles used for transport, the stacking of containers in storage places whether this was a public house, hotel, retail or wholesale outlet of any sort, and the return of empties back to the depot.

It was emphasised that the pack was designed to provide the Tutor and Depot Instructors with the material required to train delivery crews in manual handling "best practice". Further, it acknowledged the company's reliance on the skills and commitment of individuals who know the job and have the respect of their co-workers. Although efforts were made to keep the pack simple to use, trainers were encouraged to introduce supplementary information or training aids which complemented their coaching style. The evaluation and recording of training activity was equally as important as the formal training sessions with trainers checking that delivery crews understood and retained information to benefit from the programme.

The company stated that the pack represented a considerable investment in resources and provided an opportunity to improve the working practice of delivery crews and minimise the risk of injury and physical fatigue. Finally it was accepted that although training was important there would be a gradual change in development of improvements in the way loads were manually handled in the future, with existing operations having to be assessed and improvements made where necessary.

The second section covered Tutors' course material; a timetable and lecture material. The timetable was designed to span two days because less would not have covered the material and it would have been difficult to release the Tutors for a longer training period.

The timetable included the following subjects: An Introduction to the Course; Revision on Ergonomics and Manual Handling Principles; Development and Practice of Instructional Skills; Principles of Instruction; Preparation and Presentation of Simple Instructional Tasks by Instructors (Company General Handling Procedures), an Overview of Company Principles of Safe Handling; and Revision and Practice in Instruction of Company Safe Handling Procedures.

The introduction to the Training Pack course outlined the development of the material and the context. The coverage of ergonomics and manual handling principles was felt to be essential in order to provide a sound basis upon which to introduce the risk factors and need for such training.

In the past, perhaps reflecting British industrial practice, this company has not tended to provide formal training on instructional skills at supervisor level and below. It was recognised early on in the development of the training approach that unless the Tutors and Instructor Draymen had satisfactory training skills, the introduction of this type of training would not be effective. In addition, unless the Tutor was able to demonstrate a capability to teach the handling skills expected of the Drayman Instructor, the latter would be less likely to be receptive. Finally, a need for formal assessment of both Tutors and Draymen Instructors was recognised.

Consequently a key component of the course was the development and practice of instructional skills for both the Tutor and Drayman Instructor coupled with training in the principles of instruction. For the Tutor this means having an understanding of the principles of instruction and how to train a Drayman Instructor on how to train a

Drayman trainee. For the Drayman Instructor this means being able to train the trainee on specific topics, for example how to carry out a traditional lift and lower of a case of bottles, or unload containers using best practice.

Therefore both needed to be able to demonstrate general training ability. This was accomplished by having sessions covering preparation and practice of simple instructional material starting with an "ice-breaker", the two minute talk beloved by those involved in developing instructional ability. Although the tutor needs to be aware of the principles of instruction in some depth, the focus for the Drayman Instructor was on being able to structure and carry out successful training in practical skills.

The military have tremendous experience in developing both instructional expertise and teaching practical skills. The cornerstone of the latter is the Lesson Plan, a structured aide memoire which forms the basis of any practical instructional period. This is a work sheet laid out with the following types of headings; Notes on preparation of subject matter, Aims, Lesson outline, and Confirmation of key points.

Section 3 contained the Instructors Training Pack. This consisted of introductory notes covering the information the Drayman Instructor needed to cover when teaching the trainee, and then specific guidance on handling of products, general handling activities, stillaging, and ullage removal. Stillaging refers to placing containers onto racks, and ullage removal involves moving containers which are likely to have liquid e.g. beer which was being returned. Both of these are perceived as high risk manual handling tasks. The former because full containers are uplifted onto these racks and the latter because of the unknown weight and response of the containers being returned.

As was noted earlier, there may be little in the way of formal facilities for training. The training pack was designed to be kept in a self contained plastic folder and kept by the Drayman Instructor so that he could use both visual and written material as a prompt when he took one to one instruction with the drayman under instruction. The visual material was obtained from video recordings of an experienced drayman instructor carrying out the task using best practice. The illustrations were "video grabbed" i.e. stills from the video recording were pasted into the training document. The use of this type of material allowed illustrations of postures and handling technique to be used which reflected local practice.

Figure 1 illustrates one of these taken from the Drayman Instructors training pack showing handling gas cylinders. The pages of the pack were laid out so that the illustrations with key steps was on the right hand side. On the left hand side was more detailed information to remind the instructor of further points to be taught to the trainee drayman. It was envisaged that, as the Drayman Instructor would be working one to one, he could use the pages as visual prompts to assist the trainee in understanding both the principles and the actions to be undertaken. All the handling practices were laid out in the training pack in a similar way.

Evaluating The Training Pack

The material and content was evaluated by running a two day course using potential Tutors as guinea pigs to act as Draymen Instructors. Members of the steering group took the role of Tutors and used the Training Pack material to train the Draymen Instructors. Nine potential Tutors took part. The objective of the first course was two fold. First to confirm that it was possible to use the contents of the Training Pack to

Figure 1 Extract from the Drayman Instructors training pack

3.2.2 PRODUCT HANDLING PROCEDURE   <u>Illustrations of handling</u>
gas cylinders

Step 1

When removing take hold of   Prevent damage
carrying handle, pull out from   or injury and
cage until you can hold   gas release
cylinder about three-quarters
way down with other hand.

Cylinders should not be
pulled out fully from the cage
or the bottom of the cylinder
allowed to drop onto the ground.

Step 2

Carry to vehicle with both   Balance load
hands to place into rack.   and reduce
Cylinder should be well   offset loading
balanced while carrying it.   on body

Step 3

Place one end into vehicle
rack while supporting it
with the other. Then push   Minimise
with both hands, and   body
move feet with the push.   twisting

instruct the participants and develop their instructional ability. Second to identify any improvements which could be made to the contents and approach.

The course and content was assessed formally by the participants using ratings of each of the components. An example of one of the questions is illustrated below.

Topic	Rating	LOW		MODERATE			HIGH	
The time spent practising skills learned		1	2	3	4	5	6	7

In addition to these types of questions, detailed debriefing sessions were held covering each course component. Generally the course and material were well received but improvements were suggested. These included more information in the material available to the Draymen Instructors. Changes were made to the pack and a second course was held, but this time the material was used as a basis to train Tutors

The second course took the majority of the trainee Tutors who had undertaken the first course through the material in the pack. The approach utilised exercises and role play under supervision to enable the Trainee Tutors to learn how to use the material to instruct the Draymen Instructors. So instead of being trainee Draymen Instructors, in turn each of the Trainee Tutors were required to act as a Tutor and prepare material to give to a colleague who acted as the Drayman Instructor.

Trainee Tutor's performance was assessed using formal assessment sheets. At the end of the course, a debriefing was held to assess the usefulness of the material from a Tutor's viewpoint.

The outcome from the discussions and courses was a final version of the Training Pack which was in three parts preceded by a Foreword. This Foreword highlighted the Training Pack's importance by stating that it was the result of a complete review of the training needs of the company's delivery crews. In addition it said, that this was part of a new training strategy, with regional Tutors being tasked with ensuring that depot based instructors maintained consistent standards of training effectiveness and as a consequence improve delivery crew manual handling practice.

It recognised the importance of developments in the field of ergonomics and the need for adherence to the handling methods to help minimise risk of injury or damage to property and form the basis from which new delivery technologies and systems could be developed. The inclusion of the Foreword was important because it was signed by the Director of Operations, so emphasising the acceptance of this approach at the highest level in the company.

The final version of the pack has been circulated within the industry throughout the United Kingdom, and Tutors are in the process of training the drayman instructors.

Reference
Health and Safety Executive 1992 Manual handling Guidance on regulations Manual handling operations regulations 1992. Health and Safety Executive.

TEACHING OLDER PEOPLE TO USE COMPUTERS: EVOLUTION AND EVALUATION OF A COURSE

David James[1], Faith Gibson[2], Gerry McCauley[3], Michelle Corby[1]
and
Karen Davidson[1].

[1]*Department of Psychology,*
[2]*Department of Applied Social Studies,*
University of Ulster,
Magee College,
LONDONDERRY BT49 OJG.
[3]*Causeway Institute of Further & Higher Education,*
BALLYMONEY,
County Antrim, N. Ireland.

Older people are partially isolated from society to the extent that they are ignorant of, and intimidated by, developments in information technology. A key to breaching this isolation may be to give older people sufficient reason for wanting to use computers. By offering the recording of personal life history as such a reason, a course and manual were developed through which we (and others) have successfully taught several cohorts of older people to use computers for word processing and other relatively simple functions. The course has primarily been evaluated in terms of a) objective and subjective measures of the skills imparted to participants; b) the effect of participation in the courses on attitudes to computers; d) the effect of course participation on general mental health; and e) critical evaluation of the courses and manuals by participants and tutors.

Introduction.

In a recent review, Sharit and Czaja (1994) highlighted the importance of issues concerning ageing and computer-based task performance to the development and practice of ergonomics in the work environment. Our own work with older computer users has stemmed from a still broader concern: computer-based technology is now so pervasive that to be computer-illiterate is, to a greater or lesser extent, to be forced to live at the margins of modern society.

An ironic aspect of the marginalisation induced by computer-illiteracy is that computer-based technology could provide uniquely powerful aids to keeping isolated individuals in touch with society. Given the relative ease-of-use of modern interfaces, it would take only a minimal degree of mastery of the technology to operate systems that could radically improve the quality of life for some of those, such as the disabled elderly, who currently feel most excluded from the lives of others. Furthermore, such is the flexibility of the potential applications of IT that there are few interests or talents which could not be enhanced by access to computer technology. Even outside the world of work,

262

therefore, there is good reason to help older people gain familiarity with IT. Why, then, has this sizeable group of potential computer user been relatively neglected?

Temple and Gavillet (1986) indicated a vicious circle of events that has led to the inaccessability of potentially-beneficial IT to many older people. Because the technology has been developed so recently, relatively few older people have had any formal training in IT, nor, in many cases, any direct contact with computers prior to being confronted by an application (such as an automated bank teller or a computerised library catalogue) which might well not have exuded user-friendliness. The confusion and hesitancy caused by encountering an unknown piece of technology can create anxiety and even animosity within the individual and these natural responses are interpreted by others as demonstrating a fundamental inability or unwillingness to use new technology. Thus, older people are to an extent "written off" by IT developers. Indeed, some computer companies that we have approached have emphasised that they wish their products to be associated in people's minds specifically with the younger generations.

Nevertheless, there are examples of IT developments intended to serve older people, but these are of little benefit if the intended users have an antipathy to the technology. We therefore considered that the best way of breaching the vicious circle might be to provide our elderly volunteers with a strong reason for wishing to use computers. We noted complementarities between the beneficial effects of reminiscence among older people, the enthusiasm that it engenders and the suitability of word processors for recording personal life history, where memories can be added to text as and when they are recalled without disrupting what has already been written. We therefore embarked on a pilot study in which we sought to teach a small group of elderly participants to use computers, initially as a means of recording their personal life histories. This pilot course has been developed to the point where it is now a regular part of the programmes offered by Further Education centres.

The development was an iterative process which involved the evaluation of several aspects of the programme. The evaluations were guided by four principal considerations. Firstly, it was reasonable to assume that the project's aim of providing optimal help, encouragement and conditions for older people to acquire basic computing skills could only be accomplished if the particular learning characteristics and requirements of this group were known. We therefore aimed to discover those aspects of the technology and the course which learners found most difficult or discouraging, and those which were easiest or most encouraging. Secondly, we were interested in possible incidental consequences of pursuing the course, such as the general mental health of the participants, the social ramifications of participating in the course, and possible changes in participants' attitudes to modern technology. Thirdly, the efficacy of this specific course as a means of introducing computing to older people had to be assessed. And finally, a major assumption at the start of the project was that older learners would require particularly salient motivation if they were to undertake a course in IT successfully (as Bourdelais, 1986, pointed out). We therefore tried to find out the aspects of the courses that were most important in encouraging participants to enrol and to continue attending. Assessment of the variables listed below, using the methods indicated, was attempted on the assumption that they were pertinent to the four factors outlined above. (The number of participants that we have been able to assess is not yet sufficient to allow a factor analysis of their responses, which might confirm or deny the importance and relative independence of the assumed four factors.   Such an analysis is planned for the future when numbers permit.)   The variables assessed were:

     1.   Learners' views on the ease-of-use of the hardware and software, (questionnaire survey and tutors' observation).   A comparison group of younger learners of word processing was also assessed;

     2.   the learners' views of the effectiveness and relative importance of the manual, software-based assistance and tutorial assistance as means of acquiring computing skills (questionnaire survey and observation by tutors);

     3.   the learners' assessment of the strengths and weaknesses of the course in all its aspects, (questionnaire survey);

     4.   the learners' views of the benefits (or otherwise) imparted by attending the course (questionnaire survey and observation by tutors);

     5.   the change in learners' attitudes to modern technology over the duration of the course (questionnaire survey);

     6.   the change in general mental health of the learners (and, in one instance, tutors) over the duration of the course (questionnaire survey and observation by tutors);

     7.   critical evaluation of the manual and software by comparison with accepted human-computer-interaction guidelines;

     8.   the learners' subjective estimates of the skill levels they have achieved as a result of taking part in the course (questionnaire survey and observation by tutors);

     9.   relatively objective measurement of skill levels achieved by learners at the conclusion of the course (observation of performance on a word-processing exercise).

Methods.

Participants
     Of the six participants in the pilot study, five were residents of an old people's home and one a neighbour or the proprietors.   In all subsequent courses, which have been run in six locations throughout Ireland and in Denmark, participants have been volunteers enroled through word of mouth or (mostly) through local publicity about the courses.   The vast majority have been totally computer-naïve.   Ages have ranged from 56 to 88, with a modal age of approximately 70 years.   Numbers have reached over 200 and on most courses women have slightly outnumbered men.   A group of 18-30 year old computer-naïve participants, all female, were used as a comparison group in the early stages.

Hardware, software and manual
     Initially, the courses used Microsoft Works on Apple Mac computers. Additionally, courses are now given using Microsoft Works for Windows, employing pc.s with 386 cpu.s or better.   The pilot study rapidly demonstrated that sharing machines was impracticable.   Within the above constraints, the equipment used has simply been whatever has been available at each location.   A disk with exercises to be carried out was issued to each participant.   A manual was produced as a result of experience teaching the pilot course and was subsequently modified in the light of further evaluations.   (A version has now been published; McAuley, McAuley, Gibson, James and Sturdy, 1994.)

Assessment materials and questionnaire administration
    In the investigation of possible changes in general mental health, the well-established GHQ was employed in either its 12- or 30-question form. All other questionnaires were based on the appropriate literature (e.g. the CAL literature, HCI literature, computer anxiety literature, etc.), were kept as brief as possible and were compiled in accordance with the principles and guidelines outlined by Sinclair (1975). (These questionnaires can be made available to interested parties.) The questionnaires were administered at the start and towards the end of courses, normally during the first and the penultimate sessions.

Teaching schedules and environments
    Courses have invariably been taught in two-hour sessions, with a break for tea and chat half way through. Sessions have either been once or twice weekly, the latter being the more popular. A full course has been consolidated at twenty sessions. The environment has always been as informal as possible, though the classroom layout of some locations could not be disguised.

Teaching methods
    Teaching methods have also been as informal as possible, with as much individual attention as the numbers of tutors and participants permitted. Writing personal histories can evoke strong emotions and time had to be available for these to be shared and talked about with tutors and fellow participants: an informal atmosphere was essential for these processes to take place.

Results and Discussion.

    Every participant who has been tested has shown an improvement in general mental health as measured by the GHQ over the duration of the course. Such improvements have sometimes been dramatic, but more often, since the starting point was usually satisfactory, very small. Nonetheless, the probability of all the tested participants (N = 41) showing a change in the same direction by chance is less than one in a billion. At an anecdotal level, the beneficial effects of the course was confirmed by the almost evangelical enthusiasm of participants and not-infrequent comments such as "it's transformed my life" and "it's the best thing that's happened".
    Formal assessment showed no significant differences between older and younger groups with respect to their opinions of the ease-of-use of the software and hardware, despite some members of the older groups being very frail and needing to acquire individual techniques to overcome physical problems. Most frequent difficulties reported were typographical errors. However, observation did suggest that a substantial proportion of the older groups took longer to acquire techniques: they tended to forget or discount the fact that they had asked the tutors or fellows for assistance (somewhat in line with the findings of Rabbitt and Abson, 1990, 1991).
    One hundred percent of respondents considered that the course was "just challenging enough". However, what each regarded as the strengths of the course varied from location to location. Thus, where much social interaction was possible, this was seen as important: where it was not, other features, such as "learning something new" took precedence. While the course manual (as developed at each stage) was reported as useful, tutorial assistance, co-operation with fellow participants or even referring to self-made notes which paraphrased the manual were all preferred to reference to the latter. Later versions of the manual therefore had room for the participants' own notes, a facility which was frequently

used. Some of the perceived benefits of the course have obvious relevance for the "marginalisation" concept: thus being able to talk to members of the younger generation about computers was an answer that occurred several times to an open question in the questionnaire, and in conversation. The commonest criticism of the course was that the sessions were too short and too infrequent. Twenty five percent of participants (mostly male) would have liked more technical information about computers. Criticisms of the manual, which were taken into account in later versions, included (i) a degree of inconsistency in the use of everyday or technical language; (ii) occasional assumptions of knowledge which the reader may not possess; (iii) some difficulty in finding answers to specific questions; and (iv) the sequence of instructions occasionally seemed inappropriate to the users' activity.

Since the majority of participants elected to take part in the course, it is not surprising that their attitude to modern technology was already reasonably positive. Changes in responses to the computer attitudes questionnaire over the duration of the course were not statistically significant, but were in a positive direction.

Finally, what level of competence has been acquired by participants? A striking feature has been the over-estimation, by the majority of participants, of their own level of competence. No-one has under-estimated her/his own ability. Nonetheless, all those surveyed have reached a level at which independent word-processing is possible, while a proportion (about 30%) have, through choice, progressed far beyond this. In this context, there is no reason to suppose that subjective estimation of one's ability is any less important than objective assessment.

Conclusions

However ingenious and potentially useful technological developments for older people may be, they are of little purpose if the intended user group shuns them. Unless we can breach the vicious circle, caused by lack of opportunity for computing experience and intimidation by novelty, in which many of the older generations are trapped, many developments in IT which could do so much to enhance the quality of life for this section of the population are doomed to disuse. However, when they are given good reason to do so, older computer-naïve individuals may not only acquire basic computing skills, but derive great pleasure and unexpected health benefits from doing so.

Encouraging older people to use IT is only one, albeit an important, part of the problem, however. There is a growing number of applications appropriate for this group, but if their potentials are to be fully realised, much thought also needs to be given to the provision of access to computers for those older people who wish to use them. While a few of our participants have bought their own machines and local branches of Age Concern have installed computer suites, these solutions are not available to all. In particular, the provision of networked machines for the growing numbers of older people who live on their own, is an area where the potential benefits are obvious.

A further question raised by this undertaking is what the most effective motivators would be for inducing those 40 - 65 year olds, whose computer illiteracy must diminish their employment prospects, to overcome their anxiety or embarrassment and undertake a course in computing. These questions are fundamentally ergonomic in nature, recognising that even the most potentially beneficial technological developments are useless unless their intended users are willing to employ them.

References
Bourdelais, F. 1986. Age is not a barrier to computing. Activities, Adaptation & Aging, **8**, 45-58.
McAuley, G., McAuley, J., Gibson, F., James, D.T.D. & Sturdy, D. 1994. *Essential Word Processing with Microsoft Works for Windows.* (NEC, Cambridge)
Rabbitt, P.M.A. and Abson, V. 1990. Lost and found: Some logical and methodological limitations of self-report questionnaires as tools to study cognitive ageing. British Journal of Psychology, **81**, 1-16.
Rabbitt, P.M.A. and Abson, V. 1991. Do older people know how good they are? British Journal of Psychology, **82**, 137-151.
Sharit, J. and Czaja, S.J. 1994. Ageing, computer-based task performance, and stress: issues and challenges. Ergonomics, **37**, 559-577.
Sinclair, M.A. 1975. Questionnaire design. Applied Ergonomics, **6**, 73-80.
Temple, L. and Gavillet , M. 1986. The development of computer confidence in seniors: an assessment of changes in computer anxiety and computer literacy. Activities, Adaptation & Aging, **8**, 63-76.

Acknowledgements
We are most grateful for the support and assistance we have received from the Year of Older People and Solidarity Between Generations of the European Community, the David Hobman Trust, Age Resource Awards, the McCrae Trust and the Faculty of Social and Health Sciences of the University of Ulster, North Antrim College of Further Education, Montague Nursing Home, North Eastern Education and Library Board, Northern Ireland Voluntary Trust Community Arts Awards, Workers' Education Association, Age Concern (N.I.), Age Concern (Coleraine) and Ulster Television (Adult Learners' Award).

Self Assessment Testing to Enhance Human Performance

Darwin P. Hunt
Human Performance Enhancement, Inc.
Las Cruces, New Mexico 88001 USA

Self Assessment Computer Analyzed Testing (SACAT)[1] which allows test takers to indicate their level of doubt or certainty about each answer is a spinoff from an extensive university basic research program on human learning, self assessment and performance. SACAT retains the advantages of multiple choice testing, remedies its disadvantages, and provides assessment, learning and instructional benefits. The benefits include (a) a multidimensional assessment and scoring of a person's "usable knowledge" on a topic, (b) the identification of topics about which individuals and groups are "misinformed", and (c) a measurement of the retainability of learned material. Misinformation not only is detrimental to safe and effective performance in real life situations, but also is counter productive as a foundation for more advanced training. Incorporating the concepts of usable knowledge, misinformation and retainability into testing methods should produce test scores which correspond more closely with later performance.

INTRODUCTION

In the current testing of knowledge misinformation is allowed to remain concealed. In the usual test of knowledge, it is determined whether a person can answer questions correctly. A wrong answer is taken simply to mean that the person is uninformed and has not yet learned the correct knowledge. However, the situation may be much worse than this. The person may be misinformed and strongly believe that the wrong answer which he or she selected is correct.

Because knowledge is such an important factor in determining a person's performance, the distinction between being uninformed and misinformed is important For example, if we are administering a licensing or certification test to a professional (say a aspiring key decision maker or physician), it is relevant to make the distinction between whether the person (a) has little confidence in the wrong answer which he or she gave and thus is not likely to use the knowledge in practicing the profession or (b has a strong belief that the selected wrong answer is correct and is highly likely to us the erroneous knowledge in making decisions.

The purpose of training is to prepare the trainees to perform later some task(or job more error-free, more effectively and with more satisfaction and understandin A large part of training is directed toward insuring that trainees acquire and retain th knowledge necessary and sufficient to perform the task(s) later. Misinformation not only leads to bad decisions and errors in performance, but also is counter productive as a foundation for more advanced learning.

The focus of this paper is on a method, called Self Assessment Computer Analyzed Testing (SACAT), to (1) provide a better measure of the acquisition and retainability of usable knowledge and (2) detect and identify misinformation with the general aim of improving the safety and effectiveness of human performance.

Performance and Knowledge. It should be recognized that knowledge is a concept (like gravity, learning and motivation), which is inferred from observing performanc

To determine whether a person possesses knowledge on, say, simple addition, we ask questions such as "2 + 2 = ?. Or we might pose the question as a response selection or multiple choice task, e.g., 2 + 2 = (a) 2, (b) 3, (c) 4 or (d) 5?

We then observe which alternative they select and infer that they do or do not possess knowledge about simple addition of single digit numbers. In anticipation of later discussion, it is noted that being correct is not a sufficient basis upon which to conclude that a person knows something. Furthermore, a person can possess considerable knowledge as a result of learning, but such knowledge remains a hidden potential until the person uses the knowledge to perform. Although knowledge is a necessary ingredient, other factors may limit its influence on performance.

Of course it is the intimate relationship between knowledge and performance which makes schools, training programs, universities, etc. important. Namely, knowledge which is acquired and retained as a result of education and training enables the person to perform at a higher level than would otherwise be the case.

There is a danger that the current emphasis on the direct measurement of task performance may be counter productive by diverting attention from, and neglecting, the measurement of knowledge. If a person's performance of some task is of low quality, the factor(s), such as inadequate knowledge, which limit the performance must be identified. Once identified, actions to remedy the deficiency can be taken.

The main points here are (1) knowledge is an essential factor in determining the quality of a person's performance and (2) how the relation between performance and knowledge is understood influences the conduct of training, testing and education. SACAT is aimed at providing a measurement of a person's knowledge which is more comprehensive and more useful than are other objective methods such as the traditional multiple choice test. It provides some special benefits for instruction, learning and performance as well as for assessment.

Self Assessment and Learning. The intimate relation between knowledge and self assessment has long been recognized. About 2500 years ago Confucius said, "When you know a thing, to recognize that you know it, and when you do not know a thing, to recognize that you do not know it. That is knowledge." However, there has been little transfer of this idea into the practice of testing and measurement of knowledge.

Over the past 15 years a number of studies dealing with self assessment have been conducted. The results (Hunt, 1982; Sams, 1989) indicate that learning is expedited by self assessment (SA) responding. Also, learning is significantly more rapid if the SA response is executed after, rather than before, the answer.

A signal detection analysis of the finding that learning is affected by the order in which the answer and the SA response are executed provides some insight into how to interpret the observation that SA responding enhances learning. In the signal detection analysis, one can conceptualize the subjects self assessment task as that of deciding whether a weak "signal of knowing" is present or absent. It is assumed that the SA responses of "Sure" and unsure provide reasonable estimates of the person's decision that the "signal of knowing" is present or absent. The accuracy of the SA responses is estimated by the calculation of the hit rate (HR) and false alarm rate (FAR) as the conditional probabilities of p(Sure | Correct) and p(Sure | Wrong).

This analysis found that subjects who give the answer first, followed by the SA response are better able to discriminate between knowing and not knowing the correct answer (d_r = 0.85 *vs* 0.52 for Answer-SA and SA-Answer, respectively).

Furthermore, this greater sensitivity is due entirely to a higher HR (0.61 *vs*

0.49) rather than the FAR (0.30 *vs* 0.30). The higher HR suggests that if a correct response is covertly selected, then its execution helps the learner to confirm its correctness. The finding that the FAR is not affected by the order of response execution suggests that the execution of a wrong answer has no affect on the accuracy with which it is identified by a person as a wrong answer.

MISINFORMATION AND USABLE KNOWLEDGE

It is commonly accepted that people behave in accordance with their knowledge; the more confident the belief then the more likely, more rapid and more reliable is the response. If a person is sure of the correctness of erroneous knowledge, then the performance of tasks which rely on this misinformation will likewise be in error. Therefore, from an educational and training point of view it is important to detect and identify misinformation.

Similarly, in the usual test of knowledge, if a person gives a correct answer, it is assumed that the person knows that specific material; and no distinction is made between whether the person is sure or unsure of his or her correct answers.

This line of thought leads to the concept of "usable" knowledge. Usable knowledge means here that a person is sufficiently sure of the correctness of the knowledge so that it will be used to make decisions, and to select and execute actions. Unusable knowledge is knowledge about which a person is not sure enough to use it as a basis for deciding or acting.

Figure 1 illustrates some relationships among correctness, sureness, and usability of knowledge. Usable knowledge may be either correct or incorrect. In either case the person would use the correct or incorrect knowledge as the basis for acting. If the knowledge is correct, then the person may be considered well informed. If the knowledge is incorrect, then the person is considered to be misinformed.

Criterion: Press <u>Key A</u> to extinguish an electrical fire.			
Person's knowledge is that <u>Key A</u> should be pressed to extinguish an electrical fire.		Person's knowledge is that <u>Key E</u> should be pressed to extinguish an electrical fire.	
Correct		Incorrect	
Sure	Unsure		Sure
Usable Knowledge	Unusable Knowledge		Usable Knowledge
Informed (to some level)		Uninformed	Misinformed

Figure 1. The relationship among the correctness of a person's knowledge, the sureness with which the person believes the knowledge to be correct, and the usability of the knowledge (adapted from Hunt & Furustig, 1989).

From this point of view, education and training should be directed toward helping the learner acquire and retain knowledge which is both correct and usable; being correct is not sufficient. The goals of education and training programs should not only specify the content areas which should be learned (which they now do), but also should indicate some minimum level of usability. It might also be helpful, in some content areas, to state a maximum level of misinformation which is acceptable.

Self Assessment Computer Analyzed Testing

The common multiple choice test, which is widely used in the U.S. to measure people's knowledge and is increasingly being employed in the UK, has many advantages which include objectivity, ease and economy of administering and scoring, the ability to measure knowledge in most content areas at most levels of knowledge and reliability.

However, the knowledge of a person has more dimensions than is represented by the percentage correct score on a multiple choice test. Incorporating the concepts of usable knowledge and misinformation into testing produces test scores which are more representative of the way in which knowledge contributes to a person's everyday decisions and performance in work, at home, and in play.

The observation that a person recognizes or recalls a correct response on a test does not justify a confident conclusion concerning whether the knowledge has been learned to a usable level (Figure 1). Similarly, if a person makes an incorrect response on a test, we do not know whether the person is uninformed or misinformed. SACAT provides remedies for both of these inadequacies.

An "Instructor's Summary" of the test results is provided for those who prefer not to inspect the detailed printouts. In addition to giving summary statements of how well the group of students performed on the test, the Instructor's Summary lists those specific test items about which students as a group are misinformed. On tests in introductory psychology (with 50-200 students per class), 5-10% or more of the test items have a high percentage of Sure-but-Wrong responses indicating misinformation. Others (Shuford, 1994) report misinformation is as high as 20%. A printout of the complete SACAT item analysis is also provided (Table 1).

It is concluded from human self assessment process theory (Hunt & Sams, 1988) that the quality of people's performance depends on both the knowledge they

Table 1. An analysis of the questions and test takers' answers and self assessments.

ANALYSIS OF QUESTIONS

TEST: History 201   INST: M.Jones   DATE: 1 Aug 1991

QUEST #	% CORR	% SA SCORE	SURE CORR	SURE WRONG	UNSURE CORR	UNSURE WRONG	NO SA RESP
1	17	49	3	17	13	67	0
2	73	74	7	0	67	27	0
3	63	64	10	7	53	30	0
4	23	72	0	3	23	73	0
5	97	84	57	0	40	3	0
6	50	60	20	10	30	40	0
7	33	54	13	13	20	53	0
8	83	75	15	0	67	17	0
9	57	67	7	0	50	43	0
10	63	72	10	0	53	37	0
11	23	66	0	3	23	73	0
12	50	65	7	3	43	43	3
13	13	59	0	7	13	77	3
14	60	68	10	0	47	40	3
15	43	57	3	7	40	47	3
16	37	73	10	0	23	63	3
17	100	75	30	0	67	0	3
18	67	73	10	0	57	30	3
19	77	79	20	0	53	23	3
20	77	80	30	0	43	23	3
21	80	71	23	7	50	13	7
22	47	75	7	3	37	50	3
23	27	57	7	10	17	63	3
24	47	76	10	0	33	53	3
25	60	76	13	0	47	37	3
26	57	76	23	3	33	37	3
27	23	75	0	3	23	70	3
28	10	44	0	23	10	63	3
29	13	84	0	3	13	77	7
30	7	81	0	3	3	87	7
.	.	.	.	.	.	.	.
.	.	.	.	.	.	.	.
44	7	89	0	0	3	90	7
45	23	59	0	3	20	73	3
46	30	75	0	0	30	63	7
47	10	67	0	0	10	83	7
48	67	76	7	0	57	33	3
49	67	73	10	0	53	33	3
50	20	78	3	0	13	77	7
51	30	71	0	0	30	67	3
52	53	61	0	0	53	40	7
53	47	79	10	0	33	50	7
54	50	69	0	3	47	43	7
55	33	68	0	0	33	60	7
56	33	72	0	3	33	57	7
57	17	69	0	7	17	67	10
58	43	71	3	3	40	47	7

possess and the confidence with which they possess it. Thus, in using SACAT, it seems proper to reward those test takers who are most accurate in assessing their own knowledge; developing the skill of self assessment is worth practicing and should be rewarded. To provide such a reward, and to provide an incentive for engaging in self assessment, the % Correct score can be increased, say, 3% for the accurate self assessors. A % Self Assessment (%SA) Score which is an index of the overall accuracy with which a person performs the self assessments has been developed.

The computer analysis of the answer sheets of SACAT provides a printout of the scores (Table 2) which can be displayed for knowledge of results to the test takers. The test takers who are more accurate in their self assessment and, thus, most deserving of the test bonus, are identified with an asterisk (*). A printout of the answers and self assessment responses of each test taker on each test item is also available.

Table 2. A listing of the test takers and their scores. An asterisk (*) indicates that the test taker was one of the more accurate in assessing the correctness of their own answers.

LISTING OF STUDENTS' SCORES
TEST: History 201   INST: M.Jones   DATE: 1 Aug 1991

ID NUMBER	% CORR	% SA SCORE	SURE CORR	SURE WRONG	UNSURE CORR	UNSURE WRONG	NOSA RESP
..89415	63	69	24	3	38	34	0
..51696	71	75	50	7	22	20	0
..58543	89	86	88	7	2	4	0
..58532*	61	69	14	4	46	36	0
..52535	71	75	41	7	30	21	0
..58580*	82	83	52	0	30	18	0
..52515*	38	61	25	25	13	38	0
..99938*	95	92	91	4	4	2	0
..52531	70	79	52	13	18	18	0
..99958	89	84	89	11	0	0	0
..58592	86	82	30	0	55	14	0
..52511	79	77	38	4	41	18	0
..12464*	54	71	21	2	32	45	0
..44574*	95	92	80	2	14	4	0
..58521	88	85	75	5	13	7	0
..58531	91	86	88	9	4	0	0
..52531	75	67	68	20	7	5	0
..52525	48	46	45	38	4	14	0
..52553	88	83	30	0	57	13	0
..52527	57	60	41	16	14	27	2
..26779*	95	91	91	4	4	2	0
..52539	55	61	46	18	9	27	0
..52529	79	79	45	4	34	18	0
..50074	68	73	38	5	30	27	0
..58529*	55	68	18	5	36	39	2
..58558*	82	85	66	4	16	14	0
..58580	70	65	50	14	20	16	0
Highest	95	92	91	38	57	45	2
Median	75	77	46	5	18	18	0
Lowest	38	46	14	0	0	0	0

OTHER FEATURES OF SACAT

Motivational effects. On post-test questionnaires about 40% of the students indicated that they study more to prepare for SACAT, e.g., "to be able to mark that I am sure of my answer," than they do for the usual multiple choice test. The possibility that an instructor can increase the time spent by students in studying the material simply by employing SACAT, is attractive.

Reduction of gender bias. Critics argue that traditional multiple choice tests are biased against various groups of people, such as females. To be gender biased means, here, that if a male and female know the same amount about the topics of the test, then one of them will obtain a lower (or higher) score on the test than the other due to a gender characteristic which is not relevant to knowledge. Our own research findings (Hassmen & Hunt, 1994) are (a) female students score lower than male students when the common multiple choice test is employed and (b) the difference in the percentage correct answers between male and female university students is reduced when SACAT is used (Table 3).

Improvement of Retention. The aim of most training is that, as a consequence of the learning which occurs, the trainee will later be able to perform some task or activity.

This requires the trainee to (1) acquire the necessary knowledge (and skills and attitudes) and (2) retain the learned material until a later time when it will be used to make decisions and to select and execute actions.

Recent unpublished preliminary research (Cabigon, 1993) indicates that the retention of responses which are correct at the end of learning is a monotonic increasing function of the level of sureness (Table 4). Only 25% of the correct responses about which the learner is "Not sure at all" are retained a week later, while approximately 90% of the correct responses about which the learner is "Extremely Sure" are retained.

This suggests that including self assessments, along with correctness, as part of the training criteria for determining whether a person is ready for graduation would improve the correlation between training assessments and later job performance.

Table 3. The mean number of correct answers on tests using the usual multiple choice test and using SACAT (n = 30 males, 30 fems. for each of the two tests).

Kind of Test	Gender of Person	Number Correct	Standard Deviation
MC	Female	23.9	0.54
	Male	29.2	0.67
SACAT	Female	27.8	0.53
	Male	29.7	0.55

Table 4. The percentage of the correct responses on the final learning trial which were retained a week later as a function of self assessment level on the final learning trial.

Level of sureness on the final learning trial	Percentage retained correctly a week later
"Not sure at all"	25%
"Very unsure"	75 %
"Somewhat sure"	75 %
"Very sure"	88 %
"Extremely sure"	91 %

REFERENCES

Cabigon, H.J.S. 1993, *Effects of self-assessment on retention of training.* Unpub. Master's thesis, New Mexico State University, Las Cruces, New Mexico.

Hassmen, P. & Hunt, D.P. 1994, Human self assessment in multiple choice testing. *Journal of Educational Measurement,* **31**, 149-160.

Hunt, D.P. and Furustig, H. 1989, Being informed, being misinformed and disinformation: A human learning and decision making approach. *Technical Report PM 56:238,* Karlstad: Institution 56 Manniska Maskin System.

Hunt, D.P. 1982, Effects of human self-assessment responding on learning. *Journal of Applied Psychology,* **67**, 75-82.

Hunt, D.P. and Sams, M. 1988, Human self assessment process theory. In D.Dornic and G.Ljunggren (Eds.) *Psychophysics in action.* Heidelberg, FRG: Springer-Verlag.

Hunt, D.P. 1993, Human self assessment: Theory and application to learning and testing. In D. Leclercq & J. Bruno (Eds.) *Item Banking: Interactive testing and self-assessment,* NATO Advanced Research Workshop, ASI Series F, Vol. 112, Berlin: Springer-Verlag.

Sams, M.R. 1989, *Effects of observational assessments and patterns of success-failure on self-confidence.* Unpublished Dissertation. N.M.S.U., Las Cruces, New Mexico.

Shuford, E. 1994, In Pursuit of the Fallacy: Resurrecting the Penalty. In D. Leclercq & J. Bruno (Eds.) *Item Banking: Interactive testing and self-assessment,* NATO Advanced Research Workshop, ASI Series F, Vol. 112, Berlin: Springer-Verlag.

Note: [1]Self Assessment Computer Analyzed Testing, SACAT and Multiple Choice Self Assessment Answer Sheets are registered trademarks and are copyrighted by HPE, Inc., 345 North Water Street, Las Cruces, NM 88001.

THE EFFECTIVENESS OF DISPLAY SCREEN USER TRAINING: IMPLICATIONS FOR A CORPORATE TRAINING PROGRAMME

Stephen Russell Hartley

Behavioural Advice Assistant,
Research and Intelligence Unit,
Cleveland County Council, Melrose House,
Middlesbrough, Cleveland TS1 2YW

The present study was conducted to examine whether a training package developed at Cleveland County Council would meet the requirements contained in Regulation 6(*1*) of the Health and Safety (Display Screen Equipment) Regulations 1992. Ninety-six display screen "users" were involved in the study which adopted a quasi-experimental separate-sample pretest-posttest control group design. Awareness and behaviour were examined one week prior to training, and one and four weeks after training had been delivered. The results of the study indicated that training had been effective in so far as changes in awareness and behaviour were evident one week after delegates had attended the training. However, other findings cast doubt on the 'adequacy' of the package. Implications for a corporate training programme are considered.

Introduction

The Health and Safety (Display Screen Equipment) Regulations 1992 came into force in the United Kingdom on the 1 January 1993, implementing Directive 90/270/EEC. This action formed part of a package of six pieces of legislation.

The objective of the DSE Regs. was to provide protection to employees who habitually use display screen equipment as a significant part of their normal work. For the purpose of the Regulations such individuals were defined as display screen "users". In protecting health and safety in the workplace emphasis was placed on the need to identify hazards and reduce the associated risk(s) to health and safety 'as far as is reasonably practicable.' It was noted that the likelihood of experiencing display screen-related health problems is the product of three factors. The first and probably the most important factor is the way in which the display screen is used. For example, a "user" who performs a copy typing task more or less continuously will be more likely to experience health problems than a "user" who works at the display screen for only a couple of hours each day. Second, individual differences in susceptibility to injury. For instance, two "users" might work at the display screen in a similar manner but differ considerably in the number of headaches and instances of visual fatigue which they report. Third, the suitability of the environment, furniture, equipment and software being used to perform the display screen work.

The Regulations adopted a number of complementary strategies one of which places particular emphasis on the role of the "user" in protecting themselves. Specifically, Regulation 6(*1*) places a legal requirement on the employer to '*ensure that he [the display screen "user"] is provided with adequate health and safety training in the use of any workstation upon which he may be required to work.*' Guidance provided by the Health and Safety Executive has indicated that this means that the employer should ensure that the "user" is necessarily competent to use the display screen in a safe and healthy way. Evidence based on observations at Cleveland County Council suggested that a majority of display screen "users" were not adequately competent. Specifically, they were not adequately aware of the risks to health associated with display screen work and the steps which could be taken to reduce them. This formed the basis of a decision to introduce a corporate training programme.

The argument that the effectiveness of training should be assessed was advanced on three grounds. First, Regulation 6(*1*) places a legal requirement on the employer to provide 'adequate' health and safety training. Second, it has been estimated that as many as 1500 County Council employees might be designated as "users". If each of these was to attend training this would entail a considerable investment. The utility of this could not be assumed. Finally, an examination of the effectiveness of the pilot training programme would identify ways in which it might either be possible or necessary to improve the programme. This should help maximise the benefits of the training programme.

By manipulating training attendance, the present study was designed to investigate whether, on the basis of the sample taken, training provided to display screen "users" was effective. It was hypothesized that if this was the case, changes in the competence of display screen "users" should be evident. In the context of this study competence was assessed directly by examining participants awareness of issues relating to use of the display screen, and indirectly by observing behaviour.

Method

Design

A total of 96 male and female employees of Cleveland County Council participated in the study. Each of these individuals had been provisionally identified as a display screen "user." Participants were randomly assigned (to quota) to one of six subgroups, Experimental conditions E_{b1}, E_1 and E_4, and Control conditions C_{b1}, C_1, and C_4 (see Figure 1). These differed in two respects: whether or not participants attended a training session, and when the awareness and behaviour of participants was assessed relative to the introduction of training.

A separate-sample experimental design was adopted on the grounds of practicality. Examination of the representation of participants in each conditions indicated that it was possible to accept the assumption of homogeneity of variance.

Procedure

Participants were asked to attend a 'review' session which took place away from their workplace, under laboratory style conditions.

The purpose and importance of the study was briefly explained to each participant. The participant was also informed that they should feel free to make any adjustments to the workstation and could make use of any equipment within the immediate vicinity.

Participants were asked to remove their jackets during the observation session where necessary[1].

---

[1]This had the aim of improving the accuracy with which posture could be recorded.

Figure 1. Table summarising the nature of each of the six conditions.

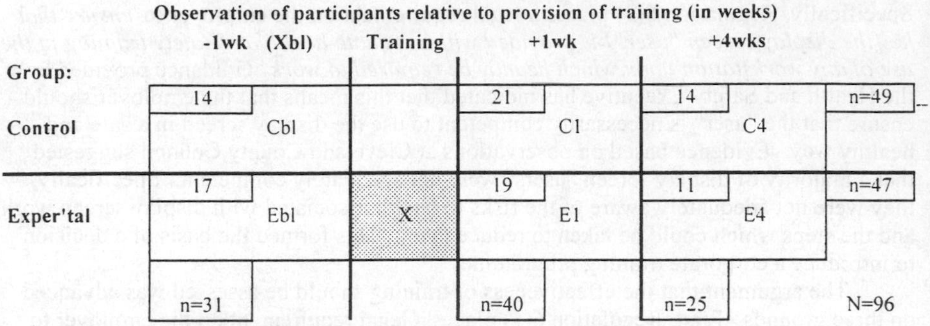

Observation of participants relative to provision of training (in weeks)

Group:	-1wk (Xbl)		Training	+1wk		+4wks	
Control	14	Cbl		21 C1	14 C4		n=49
Exper'tal	17	Ebl	X	19 E1	11 E4		n=47
	n=31		n=40		n=25		N=96

Seated at the workstation, the participant was presented with the Display Screen Equipment User Questionnaire[2] and a number of General Items. The two questionnaires were presented in turn using the display screen and required the participant to enter their response to each item using the keyboard. This part of the study lasted approximately 5 minutes and also provided opportunity for the participant to become familiar with the workstation[3]. The experimenter discretely recorded any changes which the participant made to the workstation and immediate working environment. Where the participant appeared to be having difficulty adjusting the chair or the display screen a brief description of the method of adjustment was provided.

In the second part of the study the participant performed a data entry task and a copy typing task[4]. The order in which the tasks were performed was determined at random. The posture of the participant performing each task was recorded at three pre-determined intervals using a camera (see Figure 2).

Finally, participants were asked to respond to the Display Screen Information Questionnaire. This had been designed to examine participants awareness of a number of issues relating to use of the display screen (eg. awareness of the risks to health associated with working at a display screen).

Each participant was debriefed upon completion of the study.

Results and discussion

The study suggests that training was effective in so far as differences in awareness and behaviour were evident one week after delegates had attended training. Specifically, delegates were more aware of their obligations under the Regulations, compared with

---

[2]The Display Screen Equipment User Questionnaire has been devised by Cleveland County Council Research and Intelligence Unit. In the present study the instrument was used to confirm the status of the participant as a display screen "user." Participants who did not meet this criterion were excluded from the analysis.

[3]A number of authors have expressed the view that five minutes is long enough for an individual to become familiar with a chair (ie. Grandjean, Hünting, Wotzka and Scharer, 1973; Schakel, Chidsey and Shipley, 1969).

[4]The copy typing and data entry tasks were selected for two reasons. First, these were regarded as tasks which might be typically performed at a workstation. Second, these tasks did not place excessive cognitive demands on the participant.

Figure 2. Example of photograph used to record posture.

"users" who had not attended training[5] (X_{bl} x=.355; E_1 x=1.053; t=3.71; p<.001). Furthermore, delegates were also significantly more aware of the health risks which are (X_{bl} x=3.355; E_1 x=6.579; t=7.39; p<.001) and are not currently associated with display screen work (X_{bl} x=.936; E_1 x=.316; t=-2.79; p=.008). In terms of behaviour, delegates made considerably more adjustments to the workstation compared with "users" who had not attended training (X_{bl} x=1.774; E_1 x=4.000; t=4.07; p<.001). Indeed, these differences were also observed four weeks after training attendance (X_{bl} x=1.774; E_4 x=3.364; t=2.46; p<.018). The number of adjustments made by participants in condition C_4 and group X_{bl} was not significantly different (X_{bl} x=1.774; C_4 x=2.143; t=.73; p=.468). This can be seen in Figure 3.

Despite this, doubt is cast on the adequacy of the pilot training programme for two main reasons. First, the number of adjustments which delegates made to the workstation was relatively small. For example, delegates who were observed one week after attending training made the largest mean number of adjustments (E_1 x=4.000; SD=2.186); the maximum number which could have been made was fourteen. The most important question is whether "users" would have behaved similarly if the study had been conducted in the workplace. Unfortunately, this study does not enable informed

---

[5]The data suggests that participants allocated to conditions E_{bl} and C_{bl} were not significantly different in terms of their awareness or behaviour. This implies that the two conditions could be combined to form a larger baseline group. This group is referred to as condition X_{bl} (n=31).

Figure 3. Adjustments made to the workstation by participants in the experimental and control conditions.

Observation of participants relative to provision of training (in weeks)

			-1wk	Training	+1wk Control	+1wk Exptal	+4wks Control	+4wks Exptal	
	Gp		Xbl		C1	E1	C4	E4	
		SD	1.67		1.97	2.19	1.29	2.29	SD
Gp	SD	mean	1.77		2.19	4.00	2.14	3.36	x
Xbl	1.67	1.77			NS	.001	NS	.018	
Training									
C1	1.97	2.19				.009	NS		
E1	2.19	4.00						NS	
C4	1.29	2.14						NS	
E4	2.29	3.36							

comment to be made. Second, whilst increased awareness was observed among delegates one week after they had attended training, these changes did not appear to have been maintained after four weeks. It was interesting to note the considerable discrepancy between this finding and the views of delegates recorded on a self-completion course evaluation questionnaire[6]. Item 8 of this posed the question 'do you think you need further training in this subject area or any related areas?' Of the 65 delegates who responded, 56 indicated that they did not consider that further training was necessary.

Having established that the pilot training programme did not adequately increase the competence of "users" who attended, there are a number of solutions which might be considered. These are based on the assumption that the information content and structure of the pilot training programme was appropriate. Evidence obtained from the self-report evaluation questionnaire suggests that this assumption is justified. First, the training session could be divided into two or more shorter sessions. This would reduce the amount of information presented to delegates at any one time and should improve retention. This arrangement would also benefit from the fact that information could be reinforced and summarised on a number of separate occasions. However, there are two primary difficulties with the intervention. First, the extent to which an individual's competence increases will be heavily dependent on the number of sessions which they attend. Experience suggests that average attendance will decline as the number of sessions delegates are required to attend increases. Of course this depends on whether attendance is made compulsory. Consequently increasing the number of sessions might actually have the net effect of reducing the overall effectiveness of the training programme. In addition, increasing the number of training sessions would increase the logistical problems associated with arranging those sessions in a large complex organisation such as Cleveland County Council.

Second, delegates could be provided with 'refresher training.' The difficulties of getting delegates to attend more than one training session have already been mentioned, but an alternative approach might be to make a computer-based instruction package available. The problem with this intervention is that it relies heavily on the motivation of

---

[6]It is Cleveland County Council policy that delegates should be asked to complete a standard course evaluation questionnaire after attending training.

the "user". It is quite likely that the package would be used by a relatively small number of delegates. This suggests that the intervention might be ineffective from an organisational perspective. Complementing a formal training programme with a computer-based instruction package might also have an added drawback. It is possible that supervisors would encourage "users" to use the computer-based package as an alternative to attending the formal training programme. This would not be a desirable situation. It is likely that the face-to-face situation would provide a more conducive forum in which to present up-to-date information and address the questions and concerns of delegates. It is true that a computer-based package might provide the answer to a question, but the depth and the way in which that answer is presented is likely to be limited compared with the options available to the trainer.

Finally, the awareness, attitudes and behaviour of "users" could be reinforced in the workplace by appropriate placement of posters or 'safety propaganda.' Experience suggests that this would be a useful, low-cost method of reinforcement, especially if the content was actively endorsed by managers and supervisors. An extensive review by Sell (1977) concluded that to be effective safety posters should:

- *Be specific to a particular task and situation;*

- *Back up a training programme;*

- *Give a positive instruction;*

- *Be placed close to where the desired action is to take place;*

- *Build on existing attitudes and knowledge;*

- *Emphasize nonsafety aspects.*

In the context of Cleveland County Council it is suggested that 'safety propaganda' would be the most effective additional intervention. In addition to posters the County Council is exploring the suitability of other media for reinforcing safe behaviour and good practice in the workplace. Coffee coasters and 'MS Windows wallpaper or screen savers' seem particularly appropriate due to their proximity to the place of work. These media are also frequently the focus of the display screen "user's" attention.

References

European Directive 90/270/EEC: Council Directive of 29 May 1990 on the minimum safety and health requirements for work with display screen equipment, *Official Journal of the European Communities,* **1(156)**, pp14-18

Grandjean, E., Hünting,.W., Wotzka, G. and Scharer, R. 1973, An ergonomic investigation of multipurpose chairs, *Human Factors,* **15(3)**, pp247-255.

Health and Safety Executive, 1992, *Display Screen Equipment Work: Guidance on Regulations,* (HMSO).

Sell, R.G. 1977, What does safety propaganda do for safety?, *Applied Ergonomics,* **84**, pp203-214.

Shakel. B., Chidsey, K.D., and Shipley, P. 1969, The assessment of chair comfort, *Ergonomics,* **12(2)**, pp269-306.

Drivers and Driving

AN ANALYSIS OF DRIVER REQUIREMENTS FOR MOTORWAY TRAFFIC INFORMATION

R. Graham, V.A. Mitchell and M. Ashby

HUSAT Research Institute
Loughborough University of Technology
The Elms, Elms Grove,
Loughborough, Leics., LE11 1RG

Current sources of road, weather and traffic information are often judged by drivers to be inaccurate and untimely. To ensure that emerging in-vehicle systems improve upon this situation, it is important to ascertain what information is required by drivers. A series of semi-structured interviews was conducted with frequent motorway users, in which they were asked about how they used currently-available information services and how these might be improved. Subjects were also presented with motorway scenarios and required to indicate the preferred content of traffic messages. The results have been used to assess how to tailor in-vehicle systems to meet the information requirements of drivers.

Introduction

Recent advances in network and short-range communications (SRC) have sparked the development of a series of advanced traffic information systems. In-vehicle systems such as TrafficMaster™ are already available in the UK and others are being developed, based on the SRC road-vehicle link which is likely to come into widespread use for automatic road tolling applications (Department of Transport, 1993). Traditional sources of traffic information, such as radio reports or variable road signs, are often judged by drivers to be inaccurate and untimely. To ensure that the emerging systems improve upon this situation, it is important to ascertain what information is needed by drivers. This requires an understanding of why drivers need traffic information and how its provision affects driving behaviours such as the choice of route. Towards this end, this paper presents an analysis of driver requirements for traffic-related information.

The work was carried out in the BRIMMI (Basic Research In Man Machine Interaction) project, part of the pan-European EUREKA PROMETHEUS programme, and was one of a series of on-going studies into optimising the man-machine interface of in-vehicle information systems. Such systems are able to provide the driver with information about road, weather and traffic 'events' (where an event is defined as "any deviation from the normal traffic equilibrium state"; RDS-ALERT, 1990). Previous research in the project has included road trials which investigated the assimilation and retention of information and produced recommendations for the optimal length and timing of messages (Graham & Mitchell, 1994). To complement these studies, the present study concentrated on the issue of message content and used an interview/ questionnaire approach.

Before attempting to optimise the content of in-vehicle messages to the needs and preferences of the driver, current practices for producing messages must be considered. A European standard for broadcast traffic messages has been developed within the DRIVE programme (project V1029) and named the ALERT-C protocol (RDS-ALERT, 1990). The protocol assumes that for standard traffic messages, five basic items of information must be provided. These are an event description (details of the problem and its severity), its location (the area or point location where the source of the problem is located), its extent (the areas or points also affected by the incident) its estimated duration (how long the problem is expected to last) and diversion advice (whether an alternative route is recommended). However, this project was concerned with providing a platform for coding messages, and did not consider which items of information were really required by drivers.

An issue related to optimising message content is the length of messages. Long, complex in-vehicle messages will increase eyes-off-the-road time and compromise driving safety. Furthermore, drivers may experience problems filtering long messages and retaining the necessary information until it is required. On the other hand, it is necessary to provide sufficient information about an event for drivers to make decisions regarding route efficiency and safety. Labiale (1990) found that only messages of 4 'information units' (for example "traffic jam on A40") could be recalled with 100% accuracy after a 30-second delay. This fell to 96% accuracy for messages of 7-9 units and 75% for 10-12 units. Similarly, Graham & Mitchell (1994) concluded that message screens containing 7 or more elements (an element being a distance, road number or speed limit, for example) were unacceptable in terms of safety, usefulness and driver acceptability.

Another consideration in the design of in-vehicle messages is what information is currently available to the driver, and when this is successfully used. Drivers' perceptions of problems or deficiencies with existing information providers can help optimise new systems. Furthermore, in-vehicle information must complement that available from traditional sources, being neither redundant nor contradictory. Streff and Wallace (1993) conducted a large-scale survey in which they found that more than 30% of respondents felt that radio traffic information arrived too late to use, and around a third found such messages irrelevant. Additionally, in unfamiliar areas, drivers did not know where to look for the sources of broadcast information. However, drivers did seem willing to act on the information provided when it was relevant, diverting from their chosen route for time savings of as little as 5-10 minutes. In another survey (Spyridakis et al, 1991), 76% of drivers said that traffic reports and messages (from any source) would sometimes or frequently affect their choice of route.

A final important factor in optimising the content of in-vehicle messages is the potential for customisation by individuals or certain driver groups. It is likely that the information required by drivers will vary according to user group; for example, a professional HGV driver will have different needs to a weekend leisure driver. Barfield et al (1991) argue that rather than treating commuters as a single, homogenous audience, they could be divided into four distinct sub-groups. Their survey found 21% of drivers to be 'route changers' (were willing to change routes on or before a major road), 23% were 'non-changers' (were unwilling to change departure time, route or mode of travel), 40% were 'route and time changers', and 16% were 'pre-trip changers' (were willing to make time, route or mode changes before leaving home). One solution would be to provide all information to all types of driver, but this is clearly unacceptable from an ergonomics point of view (Färber, 1993). Alternatively, it would be possible to cater for most of the user population in most situations with a reduced but fixed message content. However, an advantage of in-vehicle systems over traditional information providers is that they can allow the user to input message preferences. This can be accomplished through standard controls (e.g. button presses) or via 'smart card' technology, in which system preferences are stored on an electronic card.

The present study attempted to examine some of the above issues through semi-structured interviews with a group of drivers. Many previous studies (e.g. Barfield et al, 1991; Streff & Wallace, 1993) have used mail surveys to reach large samples of up to a few thousand drivers. Although the person-person approach is time-consuming and allows only small sample sizes, subjects can be more carefully pre-selected, and a more flexible and in-depth question format used. However, a problem of all questionnaire- or interview-based research is that the results usually do not lend themselves to statistical analysis (due to the data being nominal or ordinal). Therefore, the interview/questionnaire in the present study was designed to yield descriptive results, but at the same time examining driver informational requirements and preferences, and the reasons for these preferences.

Method

Subjects

Forty subjects (27 males and 13 females, mean age 36) were tested. They were made up of a cross section of different types of driver, including HGV drivers, taxi drivers, business and leisure drivers, and were selected as 'regular motorway drivers'. For the purposes of the study, this was defined in terms of specific criteria; either they drove short journeys on the motorway of at least three junctions at least weekly, or they drove longer journeys on the motorway of at least 100 miles at least monthly. Sixteen gave their most common purpose of travel as business, fifteen as leisure, and nine as commuting (i.e. driving to and from their place of work). Subjects were paid for their participation.

Experimental design

The study used a combination of questionnaires and interviews, with the experimenter present at all times. This allowed dialogue between the experimenter and subject, leading to fuller and more meaningful answers. Altogether, the process took between 40 and 60 minutes.

The first section consisted of a simple self-completion questionnaire addressing the subject's driving profile and driving habits. This included questions relating to current uses of traffic information and circumstances for diverting from an intended route.

The second section presented a series of hypothetical motorway scenarios to drivers. A schematic of a stretch of the M1 motorway was laid down, and card representations of the 'subject's vehicle' and 'event' were placed at certain locations on the motorway. The traffic events included fog, accident, congestion, ice and roadworks, and the distance from the vehicle to the event was varied. Drivers were asked to imagine that their car was fitted with a system able to provide information about the problem ahead. They were then required to indicate what information they would ideally like to receive about the event, and when they would first like to receive it. Prompts were given to ensure that the information was at the lowest level possible (for example, subjects would be encouraged to state how they wanted an event location represented, rather than simply "I want to know where it is!").

The final section returned to the questionnaire format and considered additional features that an in-vehicle system might be able to provide; for example, messages relating to service stations or local tourist attractions.

Results

Subjects were initially asked questions concerning their general driving habits. In terms of what factors were important to drivers when travelling on a motorway, most (53%) cited saving journey time as most significant. This was followed by increasing journey safety (32%), reducing distance travelled (10%) and increasing journey comfort (5%).

Questions about diversion situations were presented on a five point scale (*always, often, sometimes, seldom, never*). The majority of subjects (65%) felt they would sometimes change routes or stop at services in the case of poor weather and traffic conditions, and only 7% would never divert. Only a small proportion of drivers would usually divert in cases of poor visibility (e.g. fog), a slippery road (e.g. ice), or fairly slow moving traffic (50-30 mph); 15%, 18% and 15% respectively gave *often* or *always* as their answers. However, when traffic speed dropped below 30mph, more subjects would consider diverting; situations of slow moving (<30mph) or stationary traffic would cause 38% and 55% of drivers respectively to *often* or *always* divert.

When asked what traffic information they used on the motorway, a number of currently-available sources were cited. All subjects had used the simple dot-matrix displays available in the central reservation of motorways, and most (65%) had seen the new, larger variable message signs in operation. The majority listened to national radio (88%) or local radio (63%) in their cars, at least *sometimes*. Radio Data System (RDS) radios were used, at least *sometimes*, by only 23% of respondents, and mobile phone services by 13%. Many subjects also had access to Teletext (70%) or route-planning computer software (40%) to help them plan journeys before beginning travel.

Problems with existing information-providers on the motorway included messages arriving too late, or relating to problems which no longer existed. For radio reports, drivers did not always know the times of day when traffic information would be given, or did not know which station to tune to for local news in an unfamiliar area. Some subjects with RDS radios commented that they tended to be swamped with irrelevant information.

The last questionnaire section considered additional information that an in-vehicle system might provide; subjects rated certain types of information on a five-point scale (*very useful, useful, fairly useful, not very useful, useless*). Most subjects (80%) stated that they would find information about the next junction, such as its number and destinations, *useful* or *very useful*. Service station messages, such as the distance to services, was also rated highly (62%). However, subjects were not so positive about information concerning parking in the nearest town, park and ride facilities, or local tourist attractions (only 35% of subjects in each case rated such messages *useful* or *very useful*).

Some of the results from the scenarios section are given in table 1 below. This considers four possible scenarios;

• fog a number of junctions away from the current position.
• fog a short-range ahead such that it was not possible to avoid the problem by diverting or stopping at services.
• an accident a number of junctions ahead from the current position.
• an accident a short range ahead.

The table gives the percentage of subjects who stated that they would want each piece of low-level information in each scenario. Only the most requested message elements are shown.

Table 1. Drivers' information requirements across various scenarios

Information element	Scenario			
	Long-range fog	Short-range fog	Long-range accident	Short-range accident
Event type (e.g. "fog")	90%	90%	88%	93%
Location start of event	80%	60%	78%	65%
Extent of event	75%	63%	n/a	n/a
Diversion information	70%	30%	63%	n/a
Speed of traffic	58%	50%	38%	35%
Visibility distance	50%	43%	n/a	n/a
Extent of tailback	48%	25%	73%	40%
Lane closures	n/a	n/a	53%	70%
Est. problem duration	3%	3%	30%	18%
Estimated delay time	13%	10%	30%	30%

Some of the patterns emerging from the above results are discussed below.

Discussion

The first result of interest is that saving journey time was rated more important to drivers than safety or comfort. This is in accordance with the findings of Spyridakis et al (1991) and shows that in-vehicle messages should be designed with time saving in mind. In order to save time, the vast majority of drivers were quite prepared to divert from their chosen route, particularly when they encountered slow moving or stationary traffic. Again, this supports the findings of previous surveys (Streff & Wallace, 1993; Spyridakis et al, 1991). It is therefore likely that driving behaviours such as choice of route can actually be influenced by advanced traffic information. However, it must be proven to drivers that the benefits of such advanced information include a reduction in journey time.

As expected, most drivers used radios and variable message signs as their main sources of on-road information. Common problems experienced with these sources include the timing of messages and the lack of information at all times. In-vehicle systems should be able to provide more timely information, and the use of roadside beacons will allow messages to be more relevant to the current location. However, it is the responsibility of the traffic control centre, or other information provider, to ensure that they take full advantage of the potentials of new technology. Based on the problems with current information providers, it is also recommended that the capability to interrogate an in-vehicle system at any time be included. This could involve a simple "repeat last message" function, or more complex interactions to find specific message details.

The ALERT-C protocol (RDS-ALERT, 1990) defines five basic message elements; event description, location, extent, estimated duration and diversion advice. Although reducing journey time was stressed by subjects in the questionnaire, in the scenarios section not many requested an estimation of the problem duration. The only two pieces of information consistently mentioned by subjects across all scenarios were the event description and the location of the start of the problem. The present study also highlighted the fact that the extent of a problem and diversion advice are only applicable to certain events.

The information required by drivers was clearly different between cases when they were and were not able to divert and avoid the problem. When avoidance was possible, subjects were less interested in knowing about the extent of any tailback and the duration of the problem. Also, subjects generally did not consider diversion information to be appropriate in such short-range situations, even though an event such as fog may extend beyond the next junction or services.

It was also apparent that drivers required different information according to the type of event. A message element such as visibility distance is clearly only applicable to poor weather situations, whereas single lane closures, for example, are unlikely to be caused by poor weather. Drivers preferred to know about the actual speed of the traffic for a weather event compared to an accident scenario. For accidents, they were more interested in an estimation of how long the problem was expected to last, and what delay time it would add to their journey.

Previous research (e.g. Labiale, 1990; Graham & Mitchell, 1994) indicates the importance of keeping the length of in-vehicle messages to a minimum, perhaps around 4-6 elements. From this and the current study, it is possible to recommend a concise message content for each of a number of scenarios. For example, a message for a long-term accident warning, taking into account driver preferences from the scenarios section might read "Accident. M1 junction 23-24. 2 mile tailback. Left lane closed. Recommended diversion M69." Of course a number of other factors have to be taken into account in optimising the design of messages. The order of message elements has not been considered in this study; neither has the issue of particular wordings or abbreviations. Moreover, any message has to be designed with the technical capability of the display in mind. The results emphasise the need for flexibility in the content of traffic messages, but this could contradict the common ergonomics recommendation of consistency in the position or location of elements within a message, and a careful trade-off is necessary.

It is also clear from the present study that their are large individual differences in preferred message contents; where one driver might be most concerned with saving travel time and require an estimation of the problem duration, another might prefer safety information such as which lanes to avoid or a recommended speed of travel. The only robust solution to this problem is to allow customisation of various system parameters. Certain route planning software, for example, allows drivers to indicate whether they are most interested in reducing journey time, reducing distance travelled or taking only certain types of road. General system preferences, along similar lines, might also be appropriate to a real-time information system.

References

Barfield, W., Haselkorn, M., Spyridakis, J. and Conquest, L. 1991, Integrating commuter information needs in the design of a motorist information system, Transportation Research, **25A** (1), 71-78.

Department of Transport 1993, *Press Notice 489, 2nd December*, (Dept. of Transport, 2 Marsham Street, London SW1P 3EB).

Färber, B. 1993, Determining information needs of the driver. In A.M. Parkes & S. Franzen (eds.), *Driving Future Vehicles*, (Taylor & Francis, London), 69-76.

Graham, R. and Mitchell, V.A. 1994, An experimental study into the ability of drivers to assimilate and retain in-vehicle traffic information, *Proceedings of the Vehicle Navigation and Information Systems Conference - VNIS '94*, (IEEE, Piscataway, NJ), 463-468.

Labiale, G. 1990, In-car road information: comparison of auditory and visual presentations, *Proceedings of the Human Factors Society 34th Annual Meeting*, (Human Factors Society Inc, Santa Monica, CA), 623-627.

RDS-ALERT 1990, *ALERT-C Traffic Message Coding Protocol Proposed pre-Standard*, DRIVE I Project V1029, (DRIVE Central Office CEC, Brussels, Belgium).

Spyridakis, J., Barfield, W., Conquest, L., Haselkorn, M. and Isakson, C. 1991, Surveying commuter behaviour: designing motorist information systems, Transportation Research, **25A** (1), 17-30.

Streff, F.M. and Wallace, R.R. 1993, Analysis of drivers' information preferences and use in automobile travel: implications for advanced traveller information systems, *Proceedings of the Vehicle Navigation and Information Systems Conference - VNIS '93*, (IEEE, Piscataway, NJ), 414-418.

AN EVALUATION OF AN IN-VEHICLE HEADWAY FEEDBACK SYSTEM

C. Carter, A. May, F. Smith and S. Fairclough

HUSAT Research Institute,
The Elms, Elms Grove,
Loughborough, Leics LE11 2AF

The aim of the study was to test the effect of continuous headway feedback on driving behaviour and user acceptance issues. An in-vehicle radar gave driving subjects visual and auditory feedback on the time headway (TH) to the vehicle ahead. The effects of headway feedback on driving behaviour were evaluated. Results revealed that the proportion of time spent at below one second headway was reduced by the headway feedback system, but only when the vehicle was in a 'steady following scenario'. Mental workload did not appear to be increased, speed and steering control remained unaffected and subjective comments towards the system were generally favourable.

Introduction

Measuring inter-vehicle separation on motorways is a technically feasible and attractive means of encouraging drivers to keep 'safer' headway's. However little is known about the behavioural implications of providing drivers with feedback on headway information. The purpose of this study was to evaluate a real-time TH feedback system and to address the implications of introducing a system.

Following the vehicle in front too closely (tailgating) is cited as the primary contributory factor in rear-end shunts, which account for 12.13% of all vehicle involvement in road traffic accidents (Wagstaff, 1992). Postans and Wilson (1983) observed that on the Bedfordshire stretch of the M1 motorway the most frequent cause of personal road accident injuries between 1976-1979 was the rear-end shunt. It has been proposed that larger headway's will reduce the number of road accident injuries incurred. The UK Department of Transport publishes stopping distances for vehicles in the Highway Code in feet, metres and car lengths (HMSO 1992). Another criteria for a 'safe' following headway, specified in literature by the Police and the Department of Transport is that drivers should maintain a two second time gap between their vehicle and the vehicle in front (HMSO, 1985, 1988). This measure is independent of vehicle speed, road and weather conditions.

Many studies however have shown that a large percentage of drivers follow at much less than the recommended two second headway. Sumner and Baguley (1978) reported from two different motorway sites that 31% of vehicles followed with less than 2s TH and 15% less than 1s TH. Postans and Wilson (1983) observed that of close-following incidents on the motorway, 20% involved a TH of less than 0.5s.

There seems to be a difference between the recommended safe headway for the set of driving conditions and the actual headway adopted by drivers. In a study carried out by Wagstaff (1992) subjects indicated the headway that they would normally adopt when travelling at 30 and 50 mph in a vehicle, by standing at that point behind a stationary vehicle. The mean response for 30mph equates to a TH of 0.7 seconds and for 50 mph a TH of 0.9s in a dynamic situation.

The human operators' ability to detect relative motion is very sensitive, ie. relative closing velocity between two vehicles (Evans and Rothery, 1974). However this ability only comprises part of the overall task of headway monitoring. Distance judgement is also an important part of the task; drivers must be able to judge distances in order to keep a certain headway. Rockwell (1972) observed that experienced drivers estimated distances with a 20%-100% error and inexperienced drivers with up to 300% error in estimation. Furthermore Wagstaff (1992) showed that most drivers overestimated the distance to the car in front when placed at the recommended following distance in a static estimation task. When placed at 46 m for 50 mph, 63% of drivers exceeded the correct response and nearly a third over estimated the distance by 100%. Drivers experience difficulty in estimating distances accurately and generally believe themselves to be further away than they actually are when asked to quantify distance. Michon commented that because of perceptual deficiencies in driver performance, instrument aids are required in order to help the driver make more accurate judgements (1973).

Drivers are unaware of minimum stopping distances and safe following distances. Wagstaff (1992) asked drivers whether they knew what the Police recommended following time gap was; 17% of the subjects responded 'yes' and only 7% of these correctly.

The present study is concerned with the impact of headway information, in the form of continuous time-headway feedback, on driver behaviour. The study sought to examine whether providing drivers with feedback about their headway to the vehicle in front would affect their headway-keeping. It also examined the effect of high and low traffic density on drivers' headway-keeping, with and without the provision of headway feedback.

Method

Sixteen subjects, eight male and eight female, were recruited from a subject listing held by HUSAT and were paid for their participation in the experiment, which continued over five consecutive days.

The study was carried out in an instrumented Vauxhall Cavalier 2.0 GLi. A PC-driven microwave radar was installed above the front bumper of the car; this measured the range to the vehicle in front at a rate of 2 Hz and logged this data. This radar enabled the provision of visual feedback of real-time TH on a colour CRT small screen TV, mounted centrally at dashboard height in the car. TH information was displayed to the driver in the form of a variable length bar, divided into three equal segments: top segment coloured red (indicating a TH of 0-1 seconds), middle coloured amber (1-2 seconds headway) and bottom segment coloured green (2 or more seconds headway). Time headway feedback was also presented as a repeating intermittent (0.7Hz) auditory tone when the TH was less than one second ie. when only the red segment (or part of it) on the visual display was visible.

The experimental measures taken during the study were as follows: (a) Driver glance allocations and durations, recorded by an in vehicle video camera, directed at the drivers' face. (b) Car control parameters (vehicle speed, steering wheel deviation, brake pedal depression, longitudinal acceleration) recorded at 2 Hz by an in-vehicle 'datalogger'. (c) Driver ECG (0.1 Hz quantification of HR variability: psycho-physiological indicator of mental effort), sampled at 100 Hz and recorded on a Maclab and a Macintosh Powerbook. (d) Instances of overtaking by the driver, recorded by the experimenter using a time stamp on the ECG file. (e) Subjective mental workload measures, obtained using a modified version of the NASA RTLX. (f) Subjective questionnaires (completed by subjects): A short session questionnaire assessed the prevailing traffic density, and the degree to which the driver felt at ease with the car. A more detailed feedback system questionnaire determined an overall driver evaluation of the TH feedback system.

A repeated measures experimental design was employed. CONDITION (feedback device or control) and TIME (early or late) acting as a manipulation of traffic density were within-subject factors and GENDER was a between-subject factor. Each subject undertook five sessions on consecutive weekdays. The first was a vehicle and headway feedback training and acclimatisation session, lasting approximately an hour. This session also included a measurement of the drivers' normal "safe and comfortable" driving headway at speeds of 35, 45, 55 and 65 mph. The remaining four sessions were experimental sessions: (i) with headway feedback in high density traffic; (ii) with headway feedback in low density traffic; (iii) without headway feedback in high density traffic; and (iv) without headway feedback in low density traffic. The four drives were performed on four consecutive days at either 8.00-9.00 a.m. (high density traffic) or 11.00-12.00 a.m. (low density traffic). The order of presentation of CONDITION and TIME was counterbalanced across subjects and the order of GENDER was alternated. After the completion of the final motorway session, the measurement of the drivers' normal "safe and comfortable" driving headway was repeated.

During the four experimental sessions, the subjects were asked to drive as they normally would, without talking to the experimenter, who sat in the rear offside passenger seat. When performing a feedback session, subjects were asked to use the headway feedback system as personally preferred. Instances of the driver overtaking were recorded as they occurred. The NASA RTLX and the session questionnaire were presented immediately after each of the four experimental sessions. The feedback system questionnaire was completed after the second feedback session.

Results

For reasons of brevity, only the main significant results relating to the effect of the TH feedback are discussed. ANOVA tests were carried out to investigate the effects of the presence of TH feedback (CONDITION) and time of day (TIME), representing traffic density, on the dependent variables.

The TH data collected from the radar sensor was quantified in seconds [TH= (Inter-vehicle distance)/(Following car speed)]. This data was post-processed to obtain data which was appropriately representative of motorway driving. This involved filtering out data where the radar target confidence rating (a binary measure) was zero and filtering out data where speeds fell below 45 mph (eg. due to traffic queues). TH data was converted into a frequency distribution ranging from zero to five seconds in ten steps of 0.5 seconds (HEADWAY INTERVAL). The total time spent in each bandwidth of TH was quantified as a percentage frequency (to allow for lost data due to the filtering process) and subjected to a square root transformation for the purposes of statistical analysis.

A significant interaction occurred between CONDITION and HEADWAY INTERVAL [$F(9,126)=3.25$, $p<0.05$]. See Figure 1 below.

A comparison of means (students f-test) revealed significant differences at interval class 0.5-1.0 seconds [$p<0.01$]. Subjects spent less time in this bandwidth due to the introduction of TH feedback and more time in the bandwidth 1.0-1.5 sec. [$p<0.05$].

The TH data was categorised according to the relative velocity (Rv) between the two vehicles and analysed in more detail. Steady following behind the lead car (SF), Rv between -1.5mph and 1.5 mph; dropping back from the lead car (DB), Rv>1.5 mph; closing on the lead car (CL) Rv<-1.5 mph.

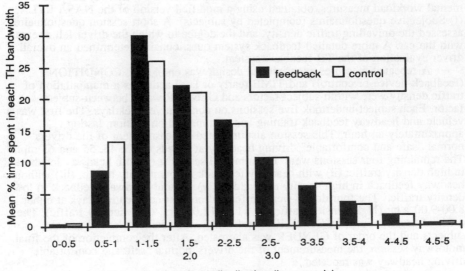

Figure 1. Time headway distributions in control and feedback conditions (n = 16).

Subjects spent approximately twenty minutes per journey in both CL and SF compared to approximately ten minutes in the DB scenario. The effect of TH feedback within each following scenarios was analysed. The ANOVA results from the CL scenario and the DB scenario revealed no significant effects due to the presence of feedback. The analysis of the SF scenario revealed a significant interaction between CONDITION x HEADWAY INTERVAL. These differences were identical with those found during the overall analysis which included all headway data, i.e. subjects spent less time in the TH bandwidth 0.5 - 1.0 sec. and more time in the following bandwidth 1.0 - 1.5 sec. [$F(9,126) = 6.42$, $p < 0.01$)].

Pre and post-test driving headway analysis indicated no significant differences in TH's employed during steady following at various speeds.

The ANOVA analysis on the 0.1 Hz HR variability revealed no significant effects due to the presence of TH feedback.

Over 70% of the drivers felt generally positive towards the TH feedback system. The information provided by the display was perceived to be useful and a majority of subjects looked at the headway display fairly frequently. Although most subjects found the information presented onscreen to be easy to interpret, a large number criticised the central position of the CRT screen. A number of subjects felt that the auditory warning was important but they also found the tone to be distracting (not at all distracting n=8, a little distracting n=7, very distracting n=4). Those subjects who found the system distracting cited the auditory warning as the reason. However most subjects agreed they would prefer some form of auditory feedback included in the system display. The subjects indicated that a discrete tone would have been equally preferred to provide headway warnings.

The majority of subjects felt that they drove "a little differently" when using the headway feedback system. The subjects indicated that they were "more aware" of headway separation and consequently were more cautious drivers. Several subjects reported that their driving style was altered simply to avoid activation of the auditory warning. When asked about their perceptions of traffic safety, approximately half the subjects felt "a little safer" when using the headway feedback system than they did normally. Fourteen subjects claimed to trust the screen information "quite a lot", however they also recognised that this high level of accuracy was not consistent.

Discussion

Evidence was found that providing real-time TH information to drivers did produce a change in driving behaviour. This was reflected in an overall shift of about 5% of the journey time from a TH of between 0.5 to 1.0 seconds to a TH of between 1.0 to 1.5 seconds. However, there was much less of a shift from the 1.0 to 1.5 second band to the 1.5 to 2.0 second band; this indicates that subject behaviour is principally a reaction to changes in the system at 1.0 seconds (the appearance of the auditory tone and the change in colour of the display from red to amber), and is not a general trend for driving at an increased TH. Since the tone is intermittent (and not a single discrete tone), it is possible that subjects drove just outside the 1 second mark in order to avoid the auditory warning. (Driving continuously at less than one second headway resulted in a repeating tone). This change in TH behaviour due to the feedback system occurred only during periods of steady following behind the car in front.

As expected, no change in driving behaviour was found when the lead car was pulling away, even when the inter-vehicle range was small, for example when a faster car overtook, cut in front of the experimental vehicle, and pulled away. In these cases, subjects fulfilled a relatively passive role, initiated by the increasing TH to the vehicle in front.

When closing on the lead vehicle, the presence of TH feedback produced no behavioural changes in either high or low traffic density. During periods of approaching and overtaking, the drivers' attention is focused on the overtaking manoeuvre, and TH probably perceived to be of secondary importance.

There were no apparent learning effects due to using the system ie. TH's judged as "safe and comfortable" were not altered after having used the system. This could have been due to lack of ability of subjects to relate displayed TH's with actual on-road distances. An additional study (Fairclough, May, Carter & Smith, in preparation) has confirmed the apparent inability of drivers to correlate inter-car distances with TH's (or even distance estimations). This lack of any training effect is also likely to be due to the relatively limited use of the feedback device - only one training session and two experimental sessions, approx. 2.5 hours total.

When overtaking, the provision of TH feedback produced an initiation of the overtaking manoeuvre at a greater inter-vehicle TH, but only during periods of low traffic density. This is probably due to the increased social pressure on drivers during high traffic conditions to maintain prevailing inter-vehicle separations.

The physiological index of mental effort used (0.1 Hz HR variability) indicated no increase in mental workload associated with use of the system. The subjective determination of workload (NASA RTLX) confirmed that minimal extra demands were imposed on the driver by the TH feedback system.

The generally positive attitudes towards the TH feedback system are based on the use of the system for only 2.5 hours and have to be interpreted with care due to the likely novelty factor. The subjectively slightly more cautious driving style employed is also likely to be influenced by the experimental set-up.

In terms of the driver interface, the visual display was well liked, and the visual representation of TH found to be easy to interpret. The only real criticism regarded the size and positioning of the screen, with a preference for a smaller display located in the vehicle dashboard. The negative feelings expressed towards an auditory verbal warning are probably based on exposure to poor implementations of speech synthesis, but may also indicate a preference for auditory feedback perceived as advisory as opposed to mandatory.

It is apparent that the subjects placed a high degree of trust in the accuracy of the information provided by the system. Subjects were also aware of the technical limitations of the system, for example, the spurious TH's produced on curved sections of road and when approaching large vehicles in adjacent lanes.

Conclusion

The provision of TH feedback to the driver did result in a reduction in the percentage of journey time spent below 1.0 second TH. This behavioural change only occurred during periods of steady following behind a vehicle. This quantitative finding was confirmed by subjective assessments of driving a little more cautiously when using the system. However this reduction in chosen TH only appeared to occur when the driver felt unpressured by external cues, and the use of such a system may be limited during periods of overtaking or high traffic density.

Based on the results of this experiment (with limited exposure to the system), the training effects in terms of reducing normal driving headway's of such a system appear to be limited, with a difficulty in translating an artificial representation of TH to actual on-road separation between vehicles.

Subjective and physiological measures of mental workload indicate minimal additional demands placed on the driver when using the system. Overall, the system was highly trusted and the visual interface found to be effective and easy to interpret. The auditory warning was found to be useful but a little distracting.

The findings of the study suggest that while the use of a TH feedback system may decrease the TH adopted by a driver, these behavioural changes are likely to be a reaction to the system implemented, and not so much a general adoption of increased TH's. In addition, although such a system could be well accepted, in practice, its value may be limited in traffic conditions where focused attention occurs or a degree of social pressure is imposed upon the driver.

Acknowledgements

This work was funded by the CEC under the DRIVE programme (DETER V2009). The authors wish to acknowledge the technical assistance of Philips Research Laboratories, and in particular Dr Andrew Stove.

References

Evans, L. and Rothery, R. 1974, Detection of the sign of relative motion when following a vehicle, *Human Factors*, **16**, 161-173

HMSO 1985, Road craft: The police driver's manual. London: Her Majesty's Stationery Office.

HMSO 1989, Driving: The department of transport manual. London: Her Majesty's Stationery Office.

HMSO 1992, The highway code. London: Her Majesty's Stationery Office.

Michon, J.A. 1973, Traffic participation-Some ergonomic issues. Soesterberg, The Netherlands: Institute for Perception, TNO, Report No. IZF 1973-14.

Postans, R.L., and Wilson, W.T. 1983, Close following on the motorway. *Ergonomics*, **26**, 317-327.

Rockwell, T. 1972, Skills, judgement, and information acquisition in driving. In T.W. Forbes (Ed.), *Human factors in highway trafic safety research*, (New York: Wiley Interscience) 348-379.

Sumner, R. and Baguley, C. 1978, Close following behaviour at two sites on rural lane motorways. Transport and Road Research Laboratory, Report 859, Crowthorne, Berks.

Wagstaff, G.F. 1992, What constitutes reckless driving? A psychological study of motor vehicle following distances. In Losel et al (ed.), Psychology and law

A ROAD-BASED EVALUATION OF DIFFERENT TYPES AND LEVELS OF ROUTE GUIDANCE INFORMATION

Louise Dicks[1], Gary Burnett[2] and Sue Joyner[2]

[1] BT, BT Laboratories, Human Factors Unit,
Martlesham Heath, Ipswich, IP5 7RE
[2] HUSAT Research Institute, The Elms,
Elms Grove, Loughborough, LE11 1RG

A road-based evaluation of different types and levels of route guidance information was conducted. Sixteen subjects each undertook two conditions each on a different route, one with visual information and one with visual and auditory information. Half the subjects were presented with simple visual information (directional arrows), half with complex visual information (accurate road layout). In addition, street names were randomly presented to all subjects. Few differences were found between the simple and complex visual conditions. However, visual plus auditory information significantly reduced the impact on visual behaviour, resulted in fewer navigational errors and was consistently preferred to visual information alone. Subjects made significantly longer glances to the display when the street names were presented, but were more confident in making the correct manoeuvres.

INTRODUCTION

'Automotive' navigation at present depends almost entirely on road signs and road and city maps. If a driver is required to travel in an unfamiliar area he will, at best, plan the trip and base his navigation through the unfamiliar area on his expectations of names, road numbers, etc. to be found during his progression. In urban areas the navigation task is made more difficult by the short distances between decision points and the need to perform several conflicting tasks. This is confounded by the fact that there is normally no time or opportunity for the driver to consult the map during driving, if only for safety reasons, and most drivers hesitate to stop and park up to examine a map. The resulting navigational confusion can result in loss of orientation, nervousness and reduced safety.

Electronic route guidance systems that guide drivers from A to B along efficient routes using visual, and often verbal instructions, will soon be available to the everyday driver. While there is potential for these systems to aid and support drivers in the navigation task, such systems also have the potential to overload the driver with additional information and distract him/her from the primary task, that of driving safely.

Previous studies have primarily concentrated on how to present route guidance information to the driver (e.g. visually (maps, symbols or alphanumeric), aurally, or by a combination of the two modalities). As an example of findings, there is now a convergence of opinion (e.g. Ashby, Fairclough and Parkes (1991), Walker, Alicandri, Sedney and Roberts (1990) and Verwey and Janssen (1993)) that symbol-based information presenting step-by-step instructions relevant to each decision point along the route is 'better' than map-based information.

There has been relatively less work carried out regarding the types and amount of navigation information to present to the driver. In fact, it could be argued that this is the crucial issue to address. A system may present timely information in an easy-to-understand manner, but unless the system presents enough of the right sort of information, then driver uncertainty will be high. Such uncertainty may then have consequences for the safe control and manoeuvring of the car, since drivers may become focused on navigational task elements and may neglect or fail to process important road and traffic information.

An important point to consider is how route guidance systems will be implemented within vehicles. If, as is predicted, these systems become popular accessories to the vehicle, then it is inevitable that motor manufacturers and suppliers will look towards pushing down the costs. An obvious way of doing this would be to use small low-cost displays, especially since this would have the additional benefit of requiring limited space on the already cluttered car cockpit. However, such displays would only be able to contain limited road layout information, which may reduce driver performance and confidence.

The aim of this experiment was threefold. Firstly, to investigate the effects on visual behaviour and driver confidence of two different levels of visual route guidance information, simple (which could be presented on a low resolution screen) and complex (which could be presented on a high resolution screen), Secondly, to investigate how the presentation of auditory information interacted with the two levels of visual information. Thirdly, the presence or absence of street names information was also investigated to determine the effect on driver behaviour and confidence when driving in an unfamiliar area.

METHOD

The experiment was a field study carried out in the suburbs of Leicester. Two matched experimental routes were developed and two short familiarisation routes (to allow to subjects to experience the method of presenting route guidance information before carrying out the condition). The 'Supercard' program was used to create a stack of cards for each combination of conditions for each route used during the experiment. Four experimental conditions were developed: simple visual (directional arrows with limited road layout information), simple visual plus auditory, complex visual (accurate road layout information) and complex visual plus auditory. Figure 1 shows the two levels of visual information.

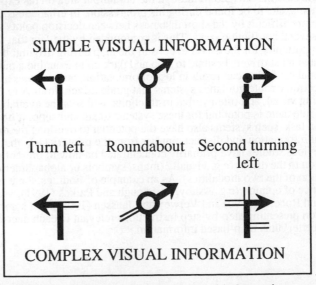

SIMPLE VISUAL INFORMATION

Turn left   Roundabout   Second turning
                                   left

COMPLEX VISUAL INFORMATION

Figure 1. The two levels of visual information

The verbal instructions were recorded using the 'Hypercard' program and imported into the trial stacks onto the relevant cards. The presentation of the route guidance information was controlled by a Macintosh computer operated by the experimenter. The experimenter advanced the cards at the specified changing point by pressing the space bar on the keyboard.

The route guidance information was presented to the subjects on a 5.75" TFT, monochrome liquid crystal display. This was mounted in a fixed position on the dashboard above the central column of instruments. The auditory messages were recorded digitised speech prompts, presented following an auditory beep, and simultaneous with the start of the visual presentation. The verbal information was presented via a loudspeaker positioned slightly behind the driver, close to the hand brake. The test car, a Saab 9000i, was equipped with cameras and recording equipment to capture subjects' visual behaviour and comments throughout the trial. Each subject was given the same information about the route guidance system, and was presented with the route guidance information at the same points on the routes (approximately 200 yards before the manoeuvres where possible). This distance was considered to be sufficient to allow subjects to prepare for the manoeuvres, based on the fact that the routes were mainly residential. Pilot trials confirmed that this distance was appropriate.

Sixteen subjects were recruited; thirteen male and three female, with ages ranged from 41 to 60. Subjects were randomly assigned to conditions and a repeated measures trial design was adopted enabling each subject to experience two conditions, one level of visual information both with, and without auditory information, the order being counterbalanced. Subjects also experienced street names randomly at approximately half of the manoeuvres for each condition.

The following objective data was collected; visual glance behaviour to the route guidance display (glance duration, frequency and allocation) and navigational errors (defined as points at which subjects strayed from the experimental routes). Glance durations were measured to an accuracy of 1/25th. of a second and were defined as the time between the eyes leaving the road to view the LCD screen to the moment they returned to the road ahead. Subjective data were collected via specially developed questionnaires to investigate usability, acceptability and preferences between the two conditions each subject carried out. In addition, subjective ratings of perceived task difficulty were collected via the NASA Raw Task Load Index (Hart and Staveland (1988)) to investigate subjective feelings of 'effort' or attentional demand from the standpoint of the actual user. Statistical analysis of the data was carried out using the 'SuperAnova' and 'StatView' packages.

RESULTS

Method of presenting route guidance information

Table 1 shows that subjects spent more time looking at the route guidance display when presented with only visual information. No significant differences were found between the simple and complex visual information for any aspect of glance behaviour. However, the visual plus auditory conditions significantly reduced the impact on visual behaviour: glance duration was reduced by 10%, glance frequency was reduced by 27% and subjects spent 1.4% less of their time in motion glancing at the route guidance system (all p < 0.01).

Table 1. Visual behaviour of subjects

		Mean glance duration (sec.s)	Mean glance frequency (per manoeuvre)	Mean % of time looking at the display
Simple visual information	Visual information only	0.81	1.96	10.07
	Visual and auditory information	0.74	1.59	8.18
Complex visual information	Visual information only	0.83	2.28	11.91
	Visual and auditory information	0.74	1.48	7.23

Subjects consistently preferred visual plus auditory information, for ease of understanding, for deciding which turn to make, for using whilst driving and for confidence in where to turn. Subjects consistently reported that they relied more on the voice component of the route guidance information than on the visual component. No significant differences were found between the simple and complex visual route guidance information for either perceived task workload or any of the components as rated on the NASA RTLX perceived task workload scale. However, subjects experiencing the complex visual information rated their mental demand, distraction, stress level and overall perceived task workload to be significantly lower with visual plus auditory information than with visual information alone ($p < 0.05$).

Street names
When street names were presented subjects made significantly longer glances ($p < 0.001$) towards the route guidance system (on average 12%). In addition subjects spent a significantly greater proportion of time in motion, an extra 1.7%, ($p < 0.001$) looking at the route guidance system when they were presented with the street name. With the exception of the complex visual and auditory condition, subjects were significantly more confidence about making a manoeuvre when they were presented with the street name than when they were not ($p < 0.05$).

Navigational Errors
The majority of navigational errors (87.5%) occurred when subjects experienced the visual route guidance information conditions only, and of these 62.5% occurred with the simple visual information and 37.5% occurred with the complex visual information. All navigational errors occurred either at roundabouts (simple visual information) where subjects reported that the route guidance information did not clarify which exit to take, or on right turnings (complex visual information) where the subjects misjudged which turning to take when prior turnings not required were involved.

DISCUSSION

When only visual information was presented, there was a disparity between the subjective and objective findings. Subjects experiencing the complex visual information consistently rating their confidence higher than those experiencing the simple visual information. However, the simple visual information was found to impact less on subjects visual behaviour. This is very likely to be due to the amount of information incorporated in the complex visual information, i.e. the symbols presenting complete road layout information were more detailed and required longer uptake times, leading to longer glance durations.

When auditory information was incorporated fewer glances were made and less time was spent glancing towards the route guidance display. This supports previous findings (Burnett and Parkes (1993), Verwey and Janssen (1993)) indicating that the auditory mode should be the primary source of information since it has the obvious advantage of not requiring a driver to take their eyes away from the road ahead. Many subjects reported that they liked having the verbal instructions because it meant that they could concentrate on the task of driving without having to take their eyes off the road, using the visual information solely for confirmation or as a back-up. There is now converging evidence that route guidance systems should present oral instructions as the main source of information backed-up by simple visual instructions. The visual back-up is needed for two main reasons:

1) oral information is system paced, so information may be missed in critical situations. The visual information can be used for confirmation,
2) auditory messages may not be appropriate for complex manoeuvres, such as roundabouts where an auditory message may become too complicated to comprehend quickly

The fact that the vast majority of navigational errors occurred in the visual only conditions could suggest that they were due more to the absence of auditory messages than to the composition of the visual information. All of the errors with the simple information occurred at roundabouts. It is felt that this occurred because the simple information only consisted of spatial directions relating to the point at which the subject stopped at a roundabout junction. Once the subject moved onto the roundabout the initial angle related to their exit was lost, and subjects had to recall their starting position in order to choose the correct exit. The auditory information provided the number of the exit (e.g. "Take 3rd exit") which allowed drivers to successfully carry out the roundabout manoeuvre.

With the complex visual information the navigation errors occurred at right turnings when there were prior turnings which were not required. Subjects felt that they missed the turns due to glancing at the screen (i.e. when they looked at the screen they had driven past and hence missed one of the prior turnings).

When street names were presented there was a disparity between the subjective feelings of confidence, and the effect on visual behaviour. Subjects said they preferred the presence of the street name when driving in an unknown area because it assured them they were on the right route. This was the case regardless of the level of visual information that they experienced. The presence of street names increased the glance duration (presumably since it takes a finite time to read the name on the display) and increased glance frequency since subjects looked at the screen whilst manoeuvring to check the street name against that on the screen presumably to confirm that they had taken the right route. It must be noted, however, that the percentage of journey time in motion spent glancing towards the information when street names were provided was approximately 11%. This level of visual demand compares very favourably to the findings from studies of other systems. For example, Burnett and Joyner (1993) found that, on average, drivers glanced towards a map-based system for 20% of their journey time.

CONCLUSIONS

The primary aims of this study were to investigate the effectiveness of combining visual and auditory route guidance information, and to see whether it would be of value when used with a low resolution display (simple) or a high resolution display (complex). The findings provide evidence to suggest that visual information alone is not the most effective method of presenting route guidance information. Since auditory information alone seems inadequate, a route guidance system should preferably present both visual and auditory information to make use of the benefits of each type. If it is only possible to incorporate a visual route guidance system, then full road layout information on a high resolutions display is required for optimum performance and driver confidence. If the display has poor resolution and the visual information is consequently degraded, then street names become more important, as they help to restore the confidence lost from not having the road layout information. A suggestion made is that the presence of street name information is made available as an option for the driver to select if required.

If a lower resolution display screen is incorporated into a route guidance system, then degrading the visual information can be compensated for if complimentary auditory information is provided. However supplementary visual information is required at complex junctions, such as roundabouts, to reduce the probability of navigational errors, and to give the driver confidence.

REFERENCES

Ashby, M., Fairclough, S.H. & Parkes, A.M. 1991, A comparison of two route information systems in a urban environment, Report no.47, DRIVE Project V017 (BERTIE)

Burnett, G.E. and Parkes, A.M. 1993, The benefits of "pre-information" in route guidance systems design for vehicles. In Lovesey, E.J. (ed), *Proceedings of the Ergonomics Society's 1993 Annual Conference*, (Taylor and Francis).

Burnett, G.E, Joyner, S.M. 1993, An investigation into the man-machine interfaces of existing route guidance systems, In D. Hugh and M. Reekie (ed.), *Proceedings of international conference of vehicle information systems*, October 12-15 1993, Ottawa, Ontario.

Hart, S.G. and Staveland, L.E. 1988, Development of the NASA TLX (Task Load Index): Results of empirical and theoretical research. In Hancock & Meshlcati (ed.), *Human Mental Workload* (North Holland).

Verwey, W.B. and Janssen, W.H. 1993, Further evidence for benefits of verbal route instructions over symbolic spatial guidance instructions. In D. Hugh and M. Reekie (ed.), *Proceedings of the IEEE-IEE Vehicle Navigation and Information Conference*, October 12-15 1993, Ottawa, Ontario.

Walker, J, Alicandri, E, Sedney, C. and Roberts, K. 1990, In-vehicle navigation devices: effects on the safety of driver performance. *SAE Paper No. 912793*.

VEHICLE OVERTAKING IN THE CLEAR-OUT PHASE AFTER OVERTURNED LORRY HAS CLOSED A HIGHWAY

Tay Wilson* and Charlotte Neff**

*Psychology Department
**Department of Law and Justice
Laurentian University
Ramsey Lake Road
Sudbury, Ontario, Canada
P3E 2C6
tel (705) 675-1151
fax (705) 675-4823

Immediately upon recommencement of traffic flow after an overturned truck (lorry) blocked Highway 69 from Toronto to Sudbury, Canada for over an hour, the experimenters entered the traffic stream, proceeded in a northerly direction at the speed limit (90 km/h) and recorded all instances of overtaking and being overtaken over a 139 kilometre section of the highway ending at Sudbury. A significant difference in the distance of overtakings from the site of the accident was noted with over half of the overtakings occurring within the first 20 kms. Three of 34 overtakings were judged to be dangerous. Pro-rating the number overtakings in specially designed passing lanes was by the proportion of the total distance occupied revealed a significantly greater number of overtakings in the passing lane.

Wilson, Lonero, and Brezina (1973), Wilson (1983, 1991) have identified the desirability of assessing actual driving practices as the basis of the development of driver improvement programmes. Driving practices have been examined at pedestrian crossings (Wilson & McCardle, 1992; Wilson & Godin, 1994), in north-London commuting (Wilson & NgAndu, 1994) and in lane merging situations (Wilson & Godin, 1993).

One particular dangerous aspect of driving involves overtaking on a highway. Studies of actual overtaking in Briton have, for instance, been carried out by Postans and Wilson (1983), Wilson, Postans and Garrod (1983), and Wilson and Best (1982). Although overtaking generates a considerable number of serious collisions, it is difficult for ordinary drivers to assess the actual overall frequencies of various types of safe and unsafe overtaking manoeuvres on highways as compared with those personally experienced and so to adjust their driving practices if indicated. It is suggested here that, because of their limited personal experience and idiosyncratic interpretations of this experience, drivers are likely to develop all sorts of odd notions about the nature of general or modal driving practices in various situations which might lead to the adoption of dangerous driving practices. It is further suggested that individual models of actual driving practice might be most idiosyncratic in the case of relatively rare situations. One such situation is the case of traffic clearing after a traffic mishap has blocked a highway. There is a clear need for data in such situations. Unfortunately, all too seldom, do experimenters get to observe events as they unfold in such circumstances. However, on Thursday, May 26, 1994, a lorry (truck) overturned across highway 69 from Toronto to Sudbury, just up the hill from where the experimenters were refuelling. Traffic was blocked for most of an hour in both directions. Here was a golden opportunity for observation of drivers overtaking practices.

Method

Immediately upon recommencement of traffic flow after an overturned truck (lorry) blocked Highway 69, the two-lane major link from Toronto to Sudbury, Canada and the major Trans-Canada Highway, for over an hour, the experimenters entered the traffic stream, proceeded in a northerly direction at the speed limit (90 km/h). One experimenter drove and the other recorded all instances of overtaking and being overtaken over a 139 kilometre section of the highway ending at Sudbury. This section of highway was punctuated by the addition of 12 specially designed passing lanes each of which widened the north-bound direction by one full lane of the road. The approximate length of these passing lanes was as follows: eight at 2.5 km, two at 3 km, two at 1.5 km, and one at 0.5 km. Account was taken of whether overtaking occurred within or outside of these designed passing areas. Finally overtakings judged to be unusual, potentially dangerous or illegal were noted. When the experimenters entered the traffic cue, the tailback of stopped cars from the accident site measured 1 km. Platoons of about 50 cars were allowed past the accident site on an north-south alternating direction basis. Observation of overtakings began at 3:37 p.m. when the experimenters' car was finally flagged past the truck rollover site.

Results

The experimenters took 95 minutes and averaged 88 km/h for the 139 km trip from the roll-over site to the outskirts of the city. The average speeds were 90, 80, 93, and 88 for successive 27, 35, 54, and 23 km segments respectively. Given that the second segment involved a section of highway construction including a brief stop of one minute and that there were two separate 0.5 km slow speed zones averaging 80 and 75 km/h respectively, it can be concluded that the experimenters achieved their aim of driving at a constant speed of 90 km/h, the posted speed limit for the road.

A summary of the 34 observed overtakings at or not at passing lane locations by distance from the site of the truck rollover which blocked the highway appear in table 1. Consider the column headings. In column one is recorded the distance in kilometres travelled from the site of the over-turned truck. In column two the road section is identified as either a designated passing lane (yes) or not (no). In column three appears the length in kms of each designated passing lane site. In column four appears the number of vehicles overtaking the experimenter at that juncture. In column five appear coded comments regarding attributes of the juncture or unusual events. In particular, "DBL-YELLOW" means overtaking on a double yellow line section of the highway; it is used to indicate a location where insufficient overtaking distance is available. Overtaking on a double yellow line section is expressly forbidden in the driving code and hence by definition dangerous. "ROAD CONST PASS ON RT SHOULDER" means that a vehicle overtook the experimenters vehicle

Table 1. Frequency of 34 overtakings (otake fr) in and outside designated passing lanes (pass lane), by length of passing lane (leng) and distance from truck roll-over road blockage site (kms).

KMS	PASS LANE	LENG	OVERTAKE	COMMENT
1	Y	2.5	4	
10	Y	2.5	5	
14	N		1	
19	Y	2.5	5	
23	N		1	DBL-YELLOW
30	N		1	RD-CONST PASS ON RT-SHOULDER
39	N		1	
42	Y	2.5	0	
55	N		1	DBL-YELLOW
56	Y	2.5	0	
69	Y	3	1	
87	Y	2.5	5	
91	N		3	
94	N		1	
96	Y	2.5	0	
107	Y	3.5	1	
117	N		1	
122	Y	2.5	0	
131	N		2	
135	Y	0.5	0	
137	Y		1	4-LANE STARTS
139				SUDBURY
total	12	28	34	

in a construction zone by using an unmade-up gravel shoulder. "SUDBURY" means the southern outskirts of the city of Sudbury at Richard Lake where overtaking observations were terminated.

The number of overtakings of the experimenters' vehicle by distance from accident site in 20 km blocks were 15, 3, 1, 1, 9, 2, and 3 respectively. This distribution of overtakings by distance from roll-over site was significantly non-uniform ($X^2 = 33.9$, df = 6, p< 0.01) with almost half of the overtakings occurring within the first 20 km and another large block occurring at the 80-100 km distance from the mishap. The proportion of passing lanes in which some overtaking took place (7/12) was not significantly different from the proportion of passing lanes in which no overtaking took place (5/12) ($X^2 = 2$, df = 1, n.s.). The proportion of overtakings in a passing lane (22/34) was not significantly different from those not occurring in a passing lane (12/14) ($X^2 = 2.64$, df = 1, n.s.). However, when the number overtakings in passing lanes was pro-rated by the ratio of the total distance not occupied by passing lanes and that occupied by passing lanes (109/28 km), there was seen to be a significantly greater number of overtakings in the passing lane (86:12) ($X^2 = 55.8$, df = 1, p< 0.01). Furthermore, there was a significant difference between the number of overtakings judged to be dangerous (3/34) and the number judged to be not dangerous (31/34) ($X^2 = 23.1$, df = 1, p< 0.01). It might be noted that, of the three overtakings judged to be dangerous, two occurred on double yellow lines where there was insufficient sight line to carry out the overtaking safely and one involved a small truck overtaking on the right road shoulder in a construction zone where there was no markings or pavement.

Discussion

Four points about the traffic might be made. First, the tailback of 1 km when the experimenters joined the cue might represented 100-200 vehicles (depending upon an estimate of 10-5 metres space occupied per vehicle). A traffic count taken at the location some months later but at the same time gave an estimated flow of 300 vehicles per hour at the location. Second, although the speed limit was 90 km/h, it was generally reckoned by drivers that speeding tickets would not be issued below 100 km/h. Thus, most traffic drove in a 10 km/h speed band between 90 -100 km/h. In the approximately 1.5 hour trip to Sudbury, it would further be expected that all 100 km/h vehicles up to six-seven minutes or 15 kilometres behind the experimenters at the start of recording would overtake them. Combining a time value of 6.5 minutes with the 34 overtakings gives a traffic volume estimate of about 310 vehicles/hour which accords with the above traffic count; if the figures are correct, it further suggests that most traffic was travelling near the 100 km/h zone. Third, the experimenters overtook only two vehicles, both in passing lanes. The first vehicle was overtaken at the first passing lane, occurring immediately after traffic was flagged through the accident site. This vehicle re-overtook the experimenters at the next passing lane. Fourth, that there was, as usual on this road, police monitoring of traffic speed was indicated by the fact that the last vehicle overtaking the experimenters was a police car.

It is of interest to consider overtaking in the specially designed passing lanes. Overtaking in passing lanes was almost double other overtaking, although passing lanes covered only about 1/4 of the trip distance. All non-passing lane road was grouped together in this assessment since overtaking even on supposedly clear sections of two-lane road is so fraught with driver error. It might be further noted that traffic was encouraged to use passing lanes for the most part by the posting of signs advertising their presence 2 km before each occurrence. The aim seems to have been met of encouraging traffic to wait for passing lanes to overtake in the post-traffic mishap clear-out situation studied by the experimenters.

Finally, the number of judged dangerous overtakings was three. This might surprise some drivers who expect, in part, because of their own driving practices, a higher frequency of such overtaking particularly when a major blockage of the highway has delayed drivers for most of an hour. One of the three judged dangerous overtakings occurred in a construction zone only thirty km from the traffic mishap site. While one swan does not make a summer, it might be hypothesized that the second unexpected traffic delay played a causative role in inducing this unusual and dangerous overtaking manoeuvre. Future studies might be aimed at studying the effects of multiple unexpected traffic delays on driving practice. Furthermore, what is now needed is a further comparative study of overtaking on the same road during normal (non-collision) circumstances.

References

Robertson, Sandy, 1993. Drivers at traffic Signals: a qualitative analysis. In Contemporary Ergonomics, Praeger, London, pp 381-386.

Postans, R. L. and Wilson, W. T. 1983. Close Following on the Motorway. Ergonomics, vol. 26, no. 4, 317-327.

Wilson, W.T., Lonero, L and Brezina, E A ,1973. Skills Improvement Approach to Collision reduction. Canadian Psychologist, vol 14, no 1, pp 8-16.

Wilson, Tay, 1983. Undesirable aspects of British driving. In Road Safety, Praeger, London, pp. 206-9.

Wilson, Tay, 1991. Locale Driving Assessment - A Neglected Base of Driver Improvement Interventions. In Contemporary Ergonomics, Praeger, London, pp 388-393.

Wilson, Tay, and Best, W., 1982. Driving Strategies in Overtaking. Accident analysis and Prevention. vol 14, no. 3, pp 179-185.

Wilson, Tay and Godin, Marie, 1993. A Study of Cooperation extended to trapped Merging drivers. In Contemporary Ergonomics, Praeger, London, pp 387-391.

Wilson, Tay and Godin, Marie. 1994. Pedestrian/Vehicle crossing Incidents Near Shopping Centres in Sudbury, Canada. In Contemporary Ergonomics, Praeger, London, pp 186-192.

Wilson, Tay, and McArdle, G, 1992. Driving Style Caused Pedestrian Incidents at Corner and zebra Crossings. In Contemporary Ergonomics, Praeger, London, pp 388-393.

Wilson, Tay and Postans, Rod and Garrod, Glen. Lorry and Coach Overtaking on the Motorway. Traffic Engineering and Control. June/July 1983, pp. 311-314.

Wilson, Tay and Ng'Andu. 1994. Trip Time Estimation Errors for Drivers Classified by Accident and Experience. In Contemporary Ergonomics, Praeger, London, pp 217-222.

MUSCULOSKELETAL TROUBLES AND DRIVING: A SURVEY OF THE BRITISH PUBLIC

Diane E Gyi and J Mark Porter

Vehicle Ergonomics Group,
Department of Human Sciences,
Loughborough University of Technology,
Epinal Way,
Loughborough ,
Leics LE11 3TU.

In order to explore the relationship between car driving and musculoskeletal troubles, interview data were collected from 600 members of the general public based on the Nordic Musculoskeletal Questionnaire. The results clearly showed that exposure to car driving was significantly related to reported sickness absence due to low back trouble. There was also a higher frequency of reported discomfort as annual mileage increased. Drivers of cars with a flexible driving package had fewer reported musculoskeletal troubles. It seems from the results that those who drive as part of their job appear to be more at risk from low back trouble than those whose jobs primarily involve sitting (not driving). The results indicate an urgent need for the training of managers in the importance of measures to reduce this problem, for example, selection of the car with respect to postural criteria.

Forward

This research was carried out as part of the Brite Euram European Initiative (Project 5549). Loughborough University was one of several European based partners in the consortium (which also included car manufacturers and seat designers) whose joint objective was to improve car seat design. The overall purpose being to support national and international standards relating to issues about designing high quality seat systems.

Introduction

It is nothing new to say that low back discomfort frequently accompanies driving. Porter, Porter and Lee (1992) in a study of 1000 drivers at Motorway Service Stations in England, found that 25% of all drivers and 66% of all business drivers were suffering from some low back discomfort at the time of the interview. Furthermore Pietri, Leclerk, Boitel, Chastang and Morcet and Blondet (1992) in their study of commercial travellers found a significant relationship between car seat comfort (in drivers who drove between 10 and 20 hours a week) and the incidence of low back trouble . Kelsey and Hardy (1975) also carried out a well documented study and found that men who had ever had a job where they spent more than half their time driving were nearly three times as likely to develop an acute herniated lumbar disc. These and other studies indicate that the relationship between musculoskeletal trouble and driving warrants further investigation.

Epidemiological studies examining the relationship between driving and musculoskeletal troubles however are scarce, perhaps indicative of the difficulties of identifying causal factors. As driving is now so much a part of our culture, it is difficult to advise 'giving up' driving and due to costs, to advise 'changing a vehicle' just in order to investigate if driving a particular vehicle was causing the problems. Also the driving task involves prolonged sitting, postural fixity and vibration any of which individually could lead to musculoskeletal trouble. Variables such as gender, lifestyle, work tasks and motivation may also have an effect on reports of symptoms in the lumbar region. It is probable however that symptoms arise from multiple relationships and influences. Pheasant (1992) hypothesised that the pattern of occurrence of musculoskeletal troubles could be described like a pyramid, with a large proportion of people (prevalence 70-90%) at the bottom who suffer task related musculoskeletal trouble but do not complain very much and usually do not develop serious clinical conditions and a few people at the top who suffer severe pathological effects, but between these extremes are a continuum of people with problems many of which could be prevented by redesign of the work task.

An interview survey was carried out in August 1993 of 600 members of the British public in order to investigate the relationship between driving and the prevalence and severity of musculoskeletal troubles including reported sickness absence. The general public were selected at random (roughly within the strata of age and gender) from public places throughout England for example shopping malls, motorway service areas and holiday resorts. They were not told the exact purpose of the survey thus avoiding selection bias. The sample contained individuals who were non-drivers, low mileage drivers, high mileage drivers and people who drive as part of their job. A complimentary study of 200 police officers was also carried out, the results of which are reported elsewhere (Gyi and Porter 1994).

Methods

The survey was based on the standardised format of the Nordic Musculoskeletal Questionnaire or NMQ (Kuorinka, Jonsson, Kilbom, Viterberg, Beiring-Sorenson, Andersson and Jorgensen 1987). The NMQ consists of a general questionnaire for the analysis of the prevalence and severity of musculoskeletal trouble in different anatomical areas and optional, more detailed, question sheets which concentrate more thoroughly on the common sites of musculoskeletal troubles i.e. neck, shoulders and low back and the severity of the impact of this trouble on work and leisure activities. The questionnaire was not intended to be used for the diagnosis of musculoskeletal disorders and it is accepted that a medical examination is required for this, instead the term 'musculoskeletal troubles' is used to mean aches, pains, discomfort or numbness experienced in the different body areas. Firm diagnosis of low back pain is difficult anyway for example a disc protrusion seen on a CT scan may be asymptomatic (Conte and Banerjee 1993).

The NMQ has been tested for reliability and validity both in Scandinavia (Kuorinka et al 1987) and more recently by the Health and Safety Executive in England (Dickinson, Campion, Foster, Newman, O'Rourke and Thomas 1992). It has also been used in several published studies, for example Beiring-Sorenson and Hilden's (1984) study of low back trouble in the general population; Anderson, Karlehagen and Jonsson's (1987) study of Swedish bus drivers and shunters and Burdoff and Zondervan's (1990) study of low back pain in crane-operators. The NMQ is short, can accommodate different workforces and individuals and has shown itself to be non threatening and accepted.

Questions were added to the NMQ based on work by Hildebrandt (1987) regarding other possible risk factors for low back pain such as age, prolonged sitting, lifting and previous back complaints. A list of occupational task demands was taken from Pheasant (1992). A list of sports felt to be high risk for neck and back ailments was taken from a study by Porter and Porter (1990) of the views of physiotherapists, osteopaths and chiropractors. Kelsey, Githens, O'Conner, Weil and Calogero (1984) have found an association between cigarette smoking and back pain; a question was therefore included regarding cigarette smoking. Scales for measuring factors like job satisfaction and motivation were considered but were felt to be too lengthy and invasive for this type of interview study. A single question about job satisfaction with a 5 point scale was therefore included. Finally a series of questions regarding the age, type and the adjustment features of any vehicles driven regularly were included, with details of their exposure to driving in terms of annual mileage, distance driven as part of work, length of journey to work etc. These questions were asked at the end of the interview to avoid any clues being given as to the reason for the study. The authors are aware of the difficulties of obtaining quality data about many of these factors without the backup of objective measures.

Results

Personal details

Table 1. The age distribution of the sample by gender.

Group	Mean (SD)	Age range
Males (n=303)	38.48 (13.09)	17-73
Females (n=297)	38.47 (13.65)	17-74

The sample of the general public consisted of a wide range of age groups (Table 1), annual mileage, vehicle types, heights, Body Mass Indices etc. All the results reported in this paper refer to driving a car.

Exposure to driving

The results clearly indicate that exposure to driving does have a potential effect on reported sickness absence due to low back trouble. Figure 1 shows that the mean number of days ever absent from work with low back trouble was 22.4 (SD 111.3) for high mileage drivers who drove more than 25,000 miles in the last 12 months, compared with 3.3 days (SD 14.7) for low mileage drivers who drove for less than 5,000 miles. Male drivers with longer journeys to work, perhaps representing regular daily exposure, also experienced significantly more low back trouble in the last 12 months ($p<0.001$). This figure was approaching significance for females.

Considering those whose job involved driving as part of their job, the results again clearly showed that the number of occasions and days ever absent with low back trouble was higher in those with the greatest exposure to driving. Figure 2 shows that the mean number of days ever absent with low back trouble was 51.4 days (SD 192.9) for those who drove more than 20 hours a week as part of their job, compared with 8.1 days (SD 34.2) for those who drove less than 10 hours as part of their job . Also the mean number of days ever absent from work with low back trouble was nearly 3 times higher for those who drove for more than 500 miles a week as part of their job compared with those who drove less than 200 miles .

Discomfort was reported in at least 1 body area by 54% of car drivers. There was an increased frequency of reported discomfort with higher annual mileage with 20% of high mileage drivers (over 25,000 miles a year) 'always' or 'often' having discomfort with their vehicle compared with only 7% of low mileage drivers (under 5,000 miles). The most frequently reported discomfort areas were the low back (26%) and the neck (10%), which is comparable with the work carried out by Porter et al (1992) where the figures were low back (25%) and neck (10%).

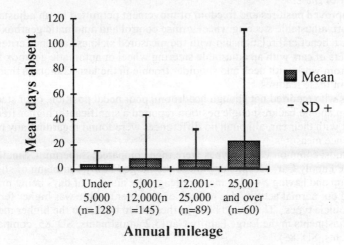

Figure 1. Number of days ever absent from work with low back trouble for car drivers according to annual mileage (n=422)

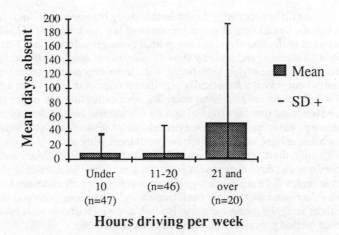

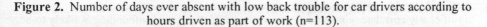

Figure 2. Number of days ever absent with low back trouble for car drivers according to hours driven as part of work (n=113).

Comparison of driving with other working postures

Comparisons were made between those whose job involved driving a car for more than 20 hours a week with 3 other separate groups; those whose work involved sitting (not driving) for > 4 hours a day, a group whose job involved standing for > 4 hours a day and a group whose job involved lifting for a large part of the day. Surprisingly it was found that the driving group experienced more low back trouble in the last 12 months than the sitting and standing groups, however no significant differences were found with the lifting group. For example 34% of the driving group compared with only

16% of the sitting group and 26% of the standing group experienced low back trouble for more than 8 days in the last 12 months ($0.1 > p > 0.05$).

Adjustability of the car

The improved postures and freedom of movement permitted by an adjustable lumbar support, adjustable steering wheel, cruise control and automatic gearbox were found to have a beneficial relationship with the measured sickness absence criteria. For example drivers of cars with an adjustable steering wheel or automatic gearbox had less days absent from work with neck and shoulder trouble in the last 12 months than those drivers without these features.

Drivers who reported not enough headroom, poor pedal position, poor steering wheel position and no backrest angle position reported a significantly higher frequency of discomfort with their car, although no differences were found regarding any of the sickness absence measures.

The 3 most common vehicle types were then compared; Supermini, Small Family Car and Large Family Car. Despite drivers of the Large Family Car being of slightly older age group and having a higher mean mileage, the number of days being prevented from carrying out normal activity due to neck or shoulder trouble was higher for both of the smaller vehicle types. This could be hypothesised to be due to the higher mean number of adjustments in the Large Family Car (3.2 adjustments, SD .85, compared with 1.18 adjustments, SD .85).

Confounding variables

Having shown a clear association between driving and low back trouble, it was important to look more closely at the influence of some of the possible confounding factors.

No significant differences were found in this study between males and females for any of the prevalence or sickness absence measures of low back trouble. However the differences between males and females for reported point prevalence (last 7 days), period prevalence (last 12 months) and severity (last 12 months) of neck, shoulder, upper back and wrist/hand trouble were highly significant, with females reporting more trouble.

Our data did not reveal a statistically significant relationship between age and low back trouble, nor did exposure to driving correlate with increasing age in our sample. It can therefore be assumed that the effect of age on driving and low back trouble is minimal. There was however an increased prevalence of musculoskeletal troubles of the large joints such as the hips, ankles and elbows with increasing age.

For males only, there was a positive correlation between Body Mass and the number of occasions and days ever absent from work with low back trouble. Although, as Body Mass in males did not show a significant correlation with exposure to driving, it is unlikely to be the main cause of low back trouble in high mileage drivers. As may be expected due to its weight bearing capacity, Body Mass was found to be related to the point prevalence, period prevalence and severity of knee trouble.

Significant positive correlations were found between sports activity and low back, neck and shoulder sickness absence criteria. However sports activity did not significantly correlate with exposure to driving and therefore confounding is likely to be minimal.

No significant correlations were found between driving and cigarette smoking or job satisfaction.

Discussion and Conclusions

The results of this study implicate that exposure to car driving should be taken seriously with regard to reported sickness absence, particularly with individuals who drive as part of their job. Employers need to be made aware of the potential 'hidden costs' incurred if individual postural comfort is not considered. The adoption of a good posture with efficient movement patterns is essential to delay the onset of discomfort and to help avoid more serious health problems. Affordable, highly adjustable driving packages are needed which can be adjusted with minimum effort (even during a journey if necessary) with guidance provided on how to adjust the seat and controls for optimum postural comfort.

People who drive as part of their job appear to be more at risk from sickness absence due to low back trouble than those whose jobs primarily involved sitting (not driving) and standing tasks. Literature, posters and other media are available warning of the dangers of lifting or sitting for long periods at computers, but nothing exists to inform drivers of the importance of ensuring that the cars they drive allow them to select comfortable and efficient driving postures. Managers with the responsibility for purchasing cars for use by others also need training in the importance of careful selection of the car with respect to postural criteria and the demands of the job, time allowed for exercise regimes and active participation in back care programmes. The potential benefits of such training of managers and their employees then needs to be fully researched and evaluated. It is hoped that as awareness increases, car manufacturers will be under increased pressure to offer suitably adjustable driving packages or risk a fall in their market share.

References

Andersson, K., Karlehagen, S. and Jonsson, B. 1987, The importance of variation in questionnaire administration, *Applied Ergonomics*, **18**;3, 229-232.

Beiring-Sorenson, F. and Hilden, J. 1984, Reproducibility of the history of low back trouble, *Spine*, **9**;3, 281-286.

Burdoff, A. and Zondervan, H. 1990, Epidemiological study of low back pain in crane operators, *Ergonomics*, **33**;8. 981-987.

Conte, L.E. and Banerjee, T. 1993, The rehabilitation of persons with low back pain, *Journal of Rehabilitation*, **Apr/May/Jun**, 18-22.

Dickinson, C.E., Campion, K., Foster, A.F., Newman, S.J., O'Rourke, A.M.T. and Thomas, P.G. 1992, Questionnaire development: An examination of the Nordic Musculoskeletal Questionnaire, *Applied Ergonomics*, **23**;3, 197-201.

Gyi, D.E. and Porter, J.M. 1994, Musculoskeletal troubles and driving in police officers. In *Proceedings of the IVth Annual Conference on Safety & Well-being at Work 1994: A Human Factors Approach*, Loughborough University 87-94.

Hildebrandt, V.H. 1987, A review of epidemiological research on risk factors of low back pain. In P. Buckle (ed.), *Musculoskeletal Disorders at Work,* (Taylor & Francis, London) 9-16.

Kelsey, J.L. and Hardy, J. 1975, Driving of motor vehicles as a risk factor for acute herniated lumbar invertebral disc, *American Journal of Epidemiology*, **102**;1, 63-73.

Kelsey, L., Githens, P.B., O'Conner, T., Weil, U. and Calogero, J.A. 1984, Acute prolapsed lumbar intervertebral disc. An epidemiological study with special reference to driving automobiles and cigarette smoking, *Spine*, **9**;6, 608-613.

Kuorinka, I., Jonsson, B., Kilbom, A., Binterberg,, H., Biering-Sorensen, F., Andersson, G. and Jorgensen, K. 1987, Standardised Nordic questionnaires for the analysis of musculoskeletal symptoms, *Applied Ergonomics*, **18**;3, 233-237.

Pheasant, S, 1992, *Ergonomics, Work and Health*, (Taylor & Francis, London) 57-76.

Pietri, F., Leclerk, A., Boitel, L., Chastang, J.F., Morcet, J.F. and Blondet, M. 1992, Low back pain in commercial travellers, *Scandinavian Journal of Work and Environmental Health*, **18**, 52-58.

Porter, J.M. and Porter, C.S. 1990, Neck and back ailments: A survey of practitioners views, *Vehicle Ergonomics Group*, Loughborough University (unpublished report).

Porter, J.M., Porter, C.S. and Lee, V.J.A. 1992, A survey of driver discomfort. In E.J. Lovesey (ed.), *Contemporary Ergonomics 1992*, (Taylor & Francis, London) 262-267.

PERCEIVED RISK, OBJECTIVE RISK AND THE COMPENSATORY PROCESS

Di Haigney, Thomas W Hoyes, A Ian Glendon and Ray G Taylor

Psychology Group
Aston University
Aston Triangle
Birmingham B4 7ET

In this report, a principal hypothesis of Risk Homeostasis Theory - that compensation occurs according to *perceived* rather than *objective* risk is tested through use of the Aston Driving Simulator. The data are statistically treated so that perception of risk and movement in 10 possible driving-related variables may be observed whilst intrinsic risk is held constant. Accident frequency was determined as being insignificant across treatment conditions whilst differences in speed-related behaviours were established as being significant - possibly indicating a 'behavioural pathway' through which compensation may occur. Implications of the findings with respect to the future development of RHT are discussed.

Introduction

A utility-based theory of behavioural compensation in drivers has been developed in the last decade which has been the subject of much controversy (Trankle and Gelau, 1992). The model is 'Risk Homeostasis Theory' (RHT) which was formerly referred to as the 'Theory of Risk Compensation'.

RHT was initially proposed by Wilde to account for the rate of accident loss observed per time unit of exposure, per capita and per kilometre driven (Wilde, 1989) and as a framework from which interrelations between these measures can be explained. RHT has also been posited as being a potential 'general theory of behaviour' (Wilde, 1985, 1989), although this paper is concerned with the theory as it relates to driver behaviour.

In RHT, the population temporal accident rate (the summed cross-products of the frequency and severity of accidents per time unit of road user exposure) is regarded as arising not from a general attempt to minimise perceived risk levels as is postulated in some theories (e.g. Naatanen and Summala, 1976), but rather to match the perceived against the 'target' or 'desired' degree of risk.

The 'target' level of risk is determined by relative utilities associated with the riskier or safer behaviours which could be adopted as alternatives to the road-user's current activity - and so it varies relative to the situation the user currently deems him/herself to be in. According to this model, road-users engaged in a driving task will experience a participative level of risk which is generally not central in the driver's awareness, and this is frequently compared with the 'target' level of risk, at a 'pre-attentive' level of cognition (Wilde, 1982, p.210).

Should a difference between these two variables become great enough for the discrepancy to be perceived, each road-user will then behave in a manner aimed at reducing this difference to a subjective zero, below their own individual perceptual thresholds (the just-noticeable-difference or jnd), the sensitivity of which varies both between individuals, and with the speed with which risk conditions are altered (Wilde, 1982). Such compensation is hypothesised as occurring through whatever behavioural means are available to the individual.

This study attempts to induce differences in perceived risk whilst objective risk is held constant, so that compensatory changes in specific behavioural pathways responsible for any increase or reduction in accident frequency may be isolated and identified. The means by which this may be achieved is through exposure of participants to a simulated driving environment in which they are required to drive in a 'British' condition (i.e. to drive on the left hand side of the road) and in a 'Continental' condition (i.e. to drive on the right hand side of the road) - conditions which whilst indistinguishable in terms of objective risk may realise differential levels of perceived risk (Wilde, 1982). In addition, Adams (1988), argues that engineering initiatives may influence driving behaviour indirectly - through influencing the amount of attention allocated to the driving task. Mental workload during the task was therefore also measured to determine whether it may function as a pathway.

'Compensation' was operationally defined as any significant differences within the variables logged by the Aston Driving Simulator (ADS) or in the performance of the secondary task over the experimental treatments of the 'British Condition' and the 'Continental Condition'.

Method

The ADS consists of a car seat, a steering wheel, accelerator and brake pedals placed so as to mimic operating conditions within an automatic car. The 'windscreen view' is represented by a computer graphic output of a road image, displayed on a terminal placed in front of the participant. When the simulation program is initiated, the 'view' also incorporates images of 'other traffic' on a single carriageway road, travelling on both carriageways. The 'other traffic' is capable of 'intelligent action', such as overtaking the user.

Participant responses, through the the use of pedals and steering wheel, are not only used to update the screen output (so that moving the steering wheel caused the 'car' to move towards one edge of the 'road', for example), but are also stored as 10 driving performance variables (mean speed, minimum speed, maximum speed, speed variance, kerb collisions, collisions with others, mean braking, braking variance, mean acceleration, acceleration variance).

After a 'practice session', participants were randomised between two conditions on the ADS - half initially 'driving' the simulator for ten minutes under a British Condition, then for a further 10 minutes under a Continental Condition. All participants were read standard instructions which informed them of the experimental conditions, the equal 'seriousness' of any collisions experienced and of the secondary task to be performed.

This task consisted of participants verbally pairing the letters of the alphabet with a numeric representing that letter's position in the alphabet (i.e. A1, B2, C3, etc.) repeating the exercise again from 'A1' after the letter 'Z' had been reached. This task was scored for incremental errors (for both letters and numbers) and the number of 'fresh attempts' (use of the 'A1' pairing).

When a participant had completed both conditions on the ADS, they completed the Risk Homeostasis Awareness (RHA) questionnaire.

Results and Discussion

Repeated measures ANOVAs for all the dependent variables provided by the ADS, related t-tests for the Secondary Task data and Pearson correlations between all ADS variables and responses to the RHA, were calculated.

Although the ADS logged performance data every half second, in order to simplify the analysis, data were condensed into 'blocks' of information, which represented mean values for the dependent variables measured. Blocks were formed for each consecutive 120 seconds of data - there were therefore 5 blocks of data for each participant in each condition. When some analysis is described as being by 'block', it refers to the comparison of some value against these 5 measures of performance. Using 'blocks' allows some variation in behaviour to be observed within conditions, and is thus a more sensitive measure than a single 'aggregate' value. The term 'Risk', refers to any significant differences between dependent variables over the two conditions.

Speed Variables

Speed variance was associated with a significant difference between conditions ($F[1,160]=9.59$; $p=0.004$). As significantly higher speed variance was observed in the Continental Condition relative to the British Condition, through interactionally significant levels of driver behaviours such as acceleration (a 'risk acceptance' behaviour - Matthews, Dorn and Glendon, 1991) and braking (a 'safety behaviour' - Wagenaar and Reason, 1990), data appear to indicate greater uncertainty as to the environmental contingencies in the former condition. As the majority of participants rated the Continental Condition as 'more risky' than the British Condition on the RHA, the findings concerning the speed variance variable support RHT in as much as they serve to illustrate that individuals adopt different strategies of driving behaviour in response to perceived risk between conditions when objective risk is constant.

The Maximum Speed variable only had Risk as a main effect ($F[1,40]=5.59$; $p=0.023$).

Differential levels of certainty (and so perceived risk) associated with the experimental conditions, may also serve to provide some means for interpreting the significantly greater maximum speed values recorded in the British Condition. Senders, Kristofferson, Levison, Dietrich and Ward (1964), describe speed behaviours in the driving task as being related to the limited 'sampling rate' of which a person is capable - that is, the number of 'bits of information' per time unit which the individual is able to process cognitively. Where a participant has experienced a scenario a number of times so that s/he becomes familiar with various subjective event-probabilities, greater speeds may be exploited with no perceived increase in risk.

Given that participants may be able to import massive amounts of practical driving experience into the simulation task, a participant may well be expected to drive at greater maximum speeds in the simulated British Condition relative to novel situations (e.g. the Continental Condition). This supposition agrees with the basic postulates of RHT.

The values obtained for mean speed gave rise to a significant result when the Block by Risk interaction was calculated ($F[4,160]=7.09$; $p<0.0001$).

The fact that mean speed was highly interactionally significant, indicates that some distinct condition-specific behavioural trend across time occurred - and so may possibly represent the process of compensation along a speed-based behavioural pathway. The discussion of the fluctuations observed in other ADS variables (see below), each of which influences the value of the mean speed in some fashion, would support this interpretation, although the exact nature of each variable's relation with any other cannot be deduced with any certainty due to both the aggregation and 'blocking' of results.

Accident Variables
Two measures of accident loss were logged by the simulator -'Kerb Collisions' (where a participant has effectively driven the car off the 'road'), and 'Collisions with Others', in which the participant's simulated vehicle comes into contact with another car displayed on the screen. RHT maintains that the homeostatic effect will be realised through total accident loss (i.e. aggregated accident severity by accident frequency), in which the 'severity' variable remains an elusive concept to quantify (Adams, 1988). As participants were informed that all accident types were equally serious and given that physical risk is not really an issue in simulated environments, severity was considered to be controlled for as far as was practicable, and so treated as a constant value. In effect then, accident loss was considered equivalent to accident frequency.
Neither Risk (F[1,160]=1.64; NS), nor Block by Risk (F[4,160]=1.43; NS) were significant when calculated for Kerb Collisions.
The other measure of accident loss - the absolute number of collisions with other vehicles - was found to be nonsignificant by Block, Block by Risk and by Risk alone. Whilst this nonsignificance in terms of accident loss could also be explained in terms of RHT through the 'complete compensation' phenomenon (and in a similar manner for Kerb Collisions), the use of non significant findings as grounds for support for a theoretical proposal must be borne in mind (McKenna, 1985). Further, some compensatory 'learning' process would be expected throughout the sessions (i.e. by block), and there is no evidence for this.

Braking Variables
Neither the Risk factor (F[1,160]=1.64; NS), nor the Block by Risk factor (F[4,160]=1.43; NS) had any significant relationship with the Mean Braking variable. Block by Risk had significant effects on Brake Variance (F[4,160]=4.36; p=0.002), which is taken as indicating a 'learning process' throughout the conditions. The Risk factor when considered alone, was found to be non significant (F[1,40]=2.02; NS).

Accelerator Variables
None of the factors investigated for Accelerator Mean were found to be significant (F[1,160]=3.14; NS) for Risk and the interaction between Accelerator Mean, Block and Risk were also non significant (F[4,160]=1.08; NS).
The Accelerator Variance by Block by Risk interaction was found to be significant - although only weakly so (F[4,160]=2.48; p=0.046). Risk (F[1,160]=0.00; NS) was non significant for this variable, again taken to indicate 'learning'.
As greater variance was recorded in the British Condition, it could be argued that RHT is supported through greater acceleration variance occurring in the more familiar scenario (the British Condition) in order to increase perceived risk (and possibly by extension, objective risk - which would be realised in nonsignificant differences in accident frequencies over the conditions). This would seem to be supported to some extent through the responses given on the RHA (see below). This finding remains consistent with that relating to higher speed variance in the Continental Condition because of higher braking variance in the latter case.

Attention
The performance of the secondary task was evaluated by calculating the absolute number of errors made (defined as some inconsistency in the secondary task - either of verbal or numerical elements), the number of fresh attempts made (taken to be the occurrence of the 'A1' pairing when not preceeded by 'Z26'), and the percentage of task errors arising in both conditions (the task as a means of testing mental loading has been validated by Hoyes, 1990).

The number of errors made gave a non significant result when a paired t-test was used (t=1.04, NS). The percentage of secondary task errors was also not significant (t=1.22, NS). The number of 'fresh attempts' at the secondary task was also non significant when subjected to a t-test (t=0.81, NS).

It is possible that the insignificant results recorded stemmed from the simple nature of the simulated driving task itself. Participants were only required to drive forwards (the driving task with the lowest mental loading), on a track where only the road and other traffic were represented - and then in simple 'blocked graphics'. The 'information density' of the 'road', and so the potential for mental loading may have allowed the use of additional attentional abilities not available in the normal driving task. Although simulator data suggest that 'attention' did not form a compensatory pathway, participants indicated differential levels of attention between conditions on the RHA (see below), which would be consistent with 100% compensation occurring via this pathway.

The Risk Homeostasis Awareness Questionnaire

The RHA questionnaire was designed by Hoyes (1992) to evaluate whether participants are aware of any differences in perceived risk between the experimental conditions and their compensatory reactions to them (if any were manifest).

The significant correlation between Question 1 ('Do you think that you concentrated more when the environment was less safe?'), and 'attention' in the Continental Condition (r=-0.3358, p=0.034) is discussed above.

Three of the RHA questionnaire items were significantly related to the ADS variables used. Question 2 on the RHA, for example ('During the experiment when you felt less safe, did you drive more slowly?'), demonstrated negative values for 'r' for Mean Speed, and Number of Overtakes performed by a participant (r=-0.4183, p=0.006; r=-0.4240, p=0.01 respectively). Responses to Question 3 ('During the experiment, when you were driving on the "right hand side" of the road, did you feel less safe?'), were significantly related to Accelerator Variance (Continental Condition, r=0.4223, p=0.005), and Question 6 ('When I can see that road conditions are worsening I tend to drive more carefully') was significantly correlated with Overtakes in the Continental Condition (r= -0.3535, p=0.022).

The significant correlation between Question 2 ('During the experiment, when you felt less safe did you drive slower?') on the RHA and the speed variance achieved between conditions, however, indicates that participants were aware that they compensated for perceived differences in risk through alteration of their speed behaviours. As participants maintained that speed would be reduced when the environment was perceived as becoming 'more risky', the condition with the greater speed variance could be taken as the situation which participants associated with a greater accident probability - i.e. the Continental Condition (Garber and Gadirajo, 1989). This proposition would be supported by the significant relation between Question 3 ('During the experiment, when you were on the "right hand side" of the road, did you feel less safe?'), and Accelerator Variance (see 'Speed Variables').

Whether the correlations deemed to support this aspect of RHT actually do so, depends both upon the empirical verifiability of RHT as well as the construct validity of the questionnaire. There would seem to be grounds on which both may be questioned although they are useful in risk behaviour research.

The non significant differences in accident loss achieved on the ADS (i.e. the 'Collisions with Other Vehicles' and 'Kerb Collision' variables), would not refute RHT in as much as differences in subjective risk could be said to have been recognised by the sample and compensated for through the use of speed behaviours.

As no specific pathways are defined in RHT (Wilde, 1989, p.277), the model has in the past been criticised on the grounds that it may not necessarily be considered as being refuted when findings are non significant (McKenna, 1985). As a result of this, several authors, such as Adams (1988), whilst recognising the process of compensation, maintain that RHT cannot be utilised as 'an explanatory model of causation' (Wilde, 1988, p.463), until a time when analysis techniques are sufficiently advanced to enable derivation of functional relations between variables. Since this was not possible on the ADS, the significant differences in behaviours observed may only be taken as possibly implying some compensatory mechanism occurring through speed behaviours - it is beyond the scope of this study to establish whether such behaviours were initiated in order to maintain a homeostatic level of accident loss, or whether the non significant variation in the accident loss variables are more attributable to other, unrecorded factors.

However, further research and model development are required in which the direction and strength of relations between variables are known, so that the status of RHT as an explanatory framework of driver behaviour as it relates to accident loss may be more comprehensively established.

References

Adams, J.G. 1988, Risk homeostasis and the purpose of safety regulation, *Ergonomics,* **31(4)**, 407-428.

Garber, S. and Gadirajo, J. 1989, Factors affecting speed variance and its influence on accidents, In *Human Performance and Highway Visibility, Design and Safety,* Methods Transportation Research Board, National Research Council, Washington D.C., **TRR 1213** pp. 67-71.

Hoyes, T.W. 1990, Risk Homeostasis in a simulated air-traffic control environment, *Unpublished Master of Science Thesis,* Hull University.

Hoyes, T.W. 1992, An examination of Risk Homeostasis in a simulated driving environment, PhD Thesis, Aston University.

Matthews, G., Dorn, L. and Glendon, A.I. 1991, Personality correlates of driver stress, *Personality and Individual Differences,* **12**, 535-549.

McKenna, F.P. 1985, Does Risk Homeostasis Theory represent a threat to ergonomics? In D.J. Oborne (ed.), *Contemporary Ergonomics,* (Taylor and Francis, London).

Naatanen, R. and Summala, H. 1975, A simple method for simulating danger-related aspects of behaviour in hazardous activities, *Accident Analysis and Prevention,* **7**, 63-70.

Senders, J.W., Kristofferson, A.B., Levison, W.H., Dietrich, D.A. and Ward, J.L. 1964, The attention demand of automobile driving, *Highway Research Record,* **195**, 15-33.

Trankle, U. and Gelau, C. 1992, Maximisation of subjective expected utility or risk control? Experimental tests of risk homeostasis theory, *Ergonomics,* **35 (1)**, 7-23.

Wagenaar, W.A. and Reason, J.T. 1990, Types and tokens in road accident causation, *Ergonomics,* **33 (10-11)**, 1365-1375.

Wilde, G. J.S. 1982, The theory of Risk Homeostasis: Implications for safety and health, *Risk Analysis,* **2(4)**, 209-225.

Wilde, G.J.S. 1985, Assumptions necessary and unnecessary to risk homeostasis, *Ergonomics,* **28**, 1531-1538.

Wilde, G.J.S. 1989, Accident countermeasures and behavioural compensation: The position of Risk Homeostasis Theory, *Journal of Occupational Accidents,* **10(4)**, 267-292.

HUMAN FACTORS IN THE DESIGN AND DEVELOPMENT OF ROAD SIGNS AND MARKINGS FOR BRIDGES

M. Halliday[1], T. Horberry[2] & A.G. Gale[2]

[1] BR Scientifics, Faraday House, London Road, Derby. DE24 8UP.
[2] Applied Vision Research Unit, University of Derby, Mickleover, Derby. DE3 5GX

An account is presented of the application of human factors to reduce and mitigate the incidence of high sided vehicles striking rail-over-road bridges. This paper focuses on the design and development of both bridge markings and intermediate bridge warning signs together with the subsequent methodology used to evaluate them. This research was conducted in two phases.

Introduction

In the UK, there are numerous rail-over-road bridges of varying height above the road surface. There are also a large number of high vehicles travelling on these roads. Consequently, on average, two bridges are struck every day by high vehicles. This results in delays to rail operations whilst the bridge is closed for damage inspection, potential damage to the vehicle, injury to the driver, and in rare circumstances track distortion capable of derailing a train. "Bridge bashing", as it is termed, is a multi-dimensional problem encompassing a large number of factors which may directly or indirectly contribute to a bridge-strike.

The signing on the approach to bridges is often in three stages. The first, the primary route indicator, informs the driver of a bridge height restriction somewhere ahead. The sign may also provide appropriate diversionary information. The height restriction is displayed as numeric information within a warning triangle or mandatory roundel. This type of sign is referred to as 'Headroom at hazard ahead'. Research by Mackie (1967) and Cooper (1989) has shown this to be an effective method of presenting height information, although understanding of the intrinsic meaning of the different sign shapes is less conclusive.

Second stage signing, termed here as 'intermediate' bridge warning signs, is found closer to the hazard. The information on the sign provides details about the bridge hazard, how far away it is and its height. This information is essential for drivers of high vehicles to make decisions about whether or not they are able to pass under the bridge ahead.

Third stage signing, categorised here as those signs immediately prior to and on the bridge. These signs re-emphasise the height restriction using the 'Headroom at hazard ahead sign' but may also present more specific information, such as the width across an arch bridge which the headroom restriction applies to.

The two critical information components required by a driver to make an informed judgement about whether a bridge can be passed under or not are the bridge height and the vehicle height. Facilitating the integration of these two pieces of information by displaying the vehicle height in the driver's cab should improve drivers' judgement. Research by Galer (1981) supports this. A survey of lorry drivers' perception of their vehicle height showed that a large proportion of estimates were inaccurate compared to actual measures.

The present research focuses on two key factors regarding the display of visual information, concerning low bridges, to drivers of high vehicles. The first is the information content of road signs necessary to inform a driver about the hazard ahead and how to manage it. The second is the need to mark the low bridge itself in a manner that emphasises its height restriction.

This study presents a comparative evaluation of the two current standard signing methods against various alternative sign designs for the intermediate bridge warning sign. The alternative designs retain the key information elements of the current intermediate bridge warning signs ie. bridge height and distance ahead.

The alternative designs in this study may be described as forming two generic types. These are termed 'New Worded' and 'Symbolic'.

The 'New Worded' signs use different vocabulary compared to the two current intermediate signs and an alternative format for their presentation. The vocabulary used for the 'New Worded' sign designs is based on Sanders & McCormick's (1993) definition of the minimum content of a warning message and hazard levels.

The use of 'Symbolic' signs is based on research supporting the use of symbols over words to convey information. Jacobs, Johnston & Cole (1975), for example, suggests that symbolic signs are superior to verbal ones in terms of their conspicuity, legibility and ease of comprehension. MacDonald & Hoffman's (1991) research also shows symbolic signs to be superior to verbal signs based on an index of similar measures. Unlike the "New Worded' signs, the 'Symbolic' signs also include pictographic representations of the possible consequences of ignoring the bridge warning, namely, a bridge strike.

All of the four sign types were tested with and without a yellow border. This follows research by Cooper (1988) in which a yellow backing board was found to increase conspicuity more than sign size. Two other border types, using two colours to form a hatched patterns, were also evaluated (red & yellow and black & yellow). Red and yellow is a well known hazard warning to most HGV drivers since it denotes a long vehicle. Black is used in the other border type since it provides the highest contrast value to yellow and these two colours form the current standard for bridge markings.

The design of bridge markings themselves are evaluated in this study in order to establish whether it is possible to reduce the perceived height of a bridge by altering the markings placed on it. Where the mis-match of information is based on marginal differences in the relative perceived height of a bridge and the vehicle this should result in a more cautious judgement by drivers regarding their ability to pass under a bridge or not. The alternative designs produced here for evaluation draw on ideas from the BR Bridge Strike Task Team, European standards and established psychological principles regarding perceptual manipulation.

Phase One

Twenty designs for both bridge markings and intermediate warning signs, including the current DoT standards, were developed. These were then reduced to a smaller experimental set through the judgement of the project team and from extended informal interviews with drivers of high vehicles. Primarily signs that were not understood were discarded, although in both groups for signs and bridge markings, designs deemed similar in design but poorer in effectiveness were also discarded.

The design criteria by which warning signs were selected were the speed and accuracy of their comprehension and for bridge markings it was those resulting in the perception of bridge height as lowest. The drivers' opinions were sought regarding perceived effectiveness for both bridge markings and signs.

Thirty two subjects took part (sixteen car drivers and sixteen drivers of high vehicles). All had corrected far visual acuities of 6/12 or better, and only one subject had some degree of colour defect.

Bridge Markings

Several photographs were taken of bridges which were regularly struck by high vehicles. From these a photograph of an arch bridge was selected as an exemplar and digitised on a PC. A graphic designer then modified this and replaced the original bridge markings with experimental markings. The images were then printed on high quality paper using a thermal transfer printer. Eight pictures of the bridge with different bridge markings were presented to all the subjects in a paired comparison study where they were asked to indicate which of the pair of bridges shown looked lower. Following this, each subject was then presented simultaneously with all eight pictures and asked to pick out the three which they considered to be the most effective. They were then asked to explain their reasons for selection.

Analysis of the data showed a marked difference in the perception of bridge height depending on the marking type. Notably, markings which made a bridge appear lowest were considered the least effective by drivers. Most comments regarding why marking was considered effective referred to its 'visibility' rather than whether it made the bridge look lower.

Warning Signs

The sixteen warning signs were developed as computer images. The warning signs were divided into four trial sets and the warning signs from each group were then embedded in a set of twelve other digitised signs taken from the Highway Code. The presentation order of all signs was randomised. Subjects were divided into four groups and each group presented with a set of stimuli.

Car drivers viewed these as video projected stimuli in a laboratory and drivers high vehicles viewed them on a PC monitor at a transport cafe. All signs were presented so that the subtended angle of vision was the same in each case.

The time taken for a subject to initiate the display of and acknowledge comprehension of the signs, by pressing the space-bar on the PC keyboard, was recorded. Following each sign's presentation, subjects were asked to explain the meaning of the sign. Only answers for the bridge warning signs were recorded and classified by a key-word set. After this, subjects were asked to pick the two bridge warning signs, from the signs seen, they considered most effective and to explain why.

None of the measures showed any statistically significant differences between signs. It is possible that confounding factors were present such as a 'ceiling-effect' in the comprehension time measures where a subject had understood the sign but failed to acknowledge this quickly enough. It was decided to progress this element of the project by further analysing a sample of the existing Sign-set, together with additional variations, using a larger subject group.

Phase Two

Bridge Markings

A dynamic (animated) PC display of bridge markings was developed against which subjects' perception of bridge height were measured. For this, flat top bridges rather than the previously used arch bridges were utilised. The results of further analyses made of a Railtrack Bridge Strike database revealed that 'girder' style (flat-topped) bridges rather than arch bridges are struck more often and the consequences are more severe.

The PC animation was then transferred to video and viewed on a high quality monitor. The scene comprised a brief pre-test route to familiarise the driver with the animation, followed by a specified test-route template (based on driving around a block) featuring four bridges of different heights. The speed of travel was kept constant. The viewpoint, through which the video was seen, was set at an approximated HGV driver's real world eye height. The animation comprised five templates in total. These represented the four types of marking taken forward from Phase One and a 'null' template incorporating no bridge markings. Each template presented the bridges in a random order. Subjects were asked whether they thought they could pass under or would strike the bridges. A confidence rating of their decisions was also taken which was forced at a set point before each bridge. This point was also specified from a pilot trial.

Warning Signs

The methodology for sign evaluation used relative comprehension levels. The modified Sign-set additionally comprised a 'null' sign. This was a sign designed in contravention of recognised ergonomic practice. These 13 signs were embedded in the set of the 12 Highway Code signs used in Phase One. Each subject viewed all signs, in a random order, on a PC. The signs were presented for a short, fixed time period. The subjects viewed a mask prior to and following each sign presentation. Subjects fixated on a cross on the pre-stimulus mask then they pressed the space bar on the PC keyboard to reveal the sign. Following each sign presentation the subject had to state the sign's meaning. Verbal responses to the alternative designs within the Sign-set were compared and scored against the original key-word set. Subjective measures regarding hazard awareness with respect to the signs were used to support the objective data.

Discussion

This paper concentrates on the study methodology and design whilst final detailed results will form the subject of a separate paper. The various data analyses can be considered as follows:-

◆ The study had two main areas of research, both of these examine comparative data between the current standards and proposed alternative bridge signs and markings.

◆ The measures obtained were designed to 'benchmark' the current standards and to enable the relative effectiveness of the alternative designs to be assessed.

◆ The effect of bridge markings on drivers' decision making, with respect to whether bridge can be passed under safely or not, was compared.

◆ The relative effect of bridge signs on subjects' comprehension levels was analysed with respect to inaccurate and acceptably accurate information recall. The effect of signs on the subjects perception of hazard awareness was also studied.

The results from this study will form the basis of further work to formalise recommendations for the optimal design of both intermediate bridge warning signs and bridge markings. It is intended that other ergonomic issues related to the display of visual information to drivers of high vehicles will be incorporated in future work.

Acknowledgements

We acknowledge the guidance received from members of the Steering Group. We thank Richard Jones - BRB, the project sponsor and a member of the Steering Group, for permission to publish this paper and Noel Douglas for his graphic design input.

References

Cooper, B.R. 1988, A comparison of different ways of increasing traffic sign conspicuity. Department of Transport, TRRL Report RR 157.
Cooper, B.R. 1989, Comprehension of traffic signs by drivers and non-drivers. Department of Transport, TRRL Report RR 167.
Galer, M. 1981, A survey among lorry drivers about the striking of low bridges. Department of the Environment, Department of Transport, TRRL Supplementary Report 633.
Jacobs, R.J., Johnston, A.W. & Cole, B.L. 1975, The Visibility of Alphabetic and Symbolic Traffic Signs. Australian Road Research Board. Vol 5, Part 7 68-86
MacDonald, W.A. & Hoffman, E.R. Drivers' awareness of traffic sign information. *Ergonomics* **34**, 5, 585-612
Mackie, A.M. 1967, Progress in learning the meanings of symbolic traffic signs. Ministry of Transport, RRL Report LR 91.
Sanders, M.S & McCormick, E.J (1993) *Human Factors in Engineering and Design*, 7th edn, (McGraw-Hill, Inc, USA).

Anthropometry

ANTHROPOMETRY OF CHILDREN 2 TO 13 YEARS OF AGE IN THE NETHERLANDS

L P A Steenbekkers

Delft University of Technology
Faculty of Industrial Design Engineering
Jaffalaan 9, 2628 BX Delft
The Netherlands

During the design of products for children it might be useful to have a computer child-model to evaluate the design in an early phase of the design process. Results of a national study on development of children, aged between 2 and 13 years, are used to develop such a child-model. Diversity in characteristics of children, relevant to the development of such a computer model are described.

Introduction

In order to be able to design products for children designers of daily life products need data on development of children. Much is known concerning growth and development of children, especially in auxology and medical sciences (see for example Tanner, 1990). For designers, however, this information is hardly suitable to be used during the design process. Therefore a study was set up to get normative data on developmental characteristics of children in the Netherlands based on empirical assessments and suitable to be used in the design process. These characteristics were: body dimensions, force exertion, motor performance, physical flexibility, technical comprehension and temperament. A second goal of the study was to investigate whether mutual relationships could be found between these developmental aspects and between each of them and the liability to have accidents. A third goal was to provide suggestions for design and evaluation criteria that will lead to safer daily-life products for children (Steenbekkers, 1993). In this paper results on the anthropometric variables will be discussed.

Method

In this study 40 dimensions were chosen to be measured, see figure 1. Determination of the method of measurement of the dimensions was based on existing methods in order to be able to compare our results with data from other sources. These sources were: the Dutch Standard NEN 2736 (1987), the German

Standard DIN 33402 (1981), Anthropometric Source Book (1978) and Snyder et al. (1977). When no definition could be found in these sources, we defined our own method of measurement.

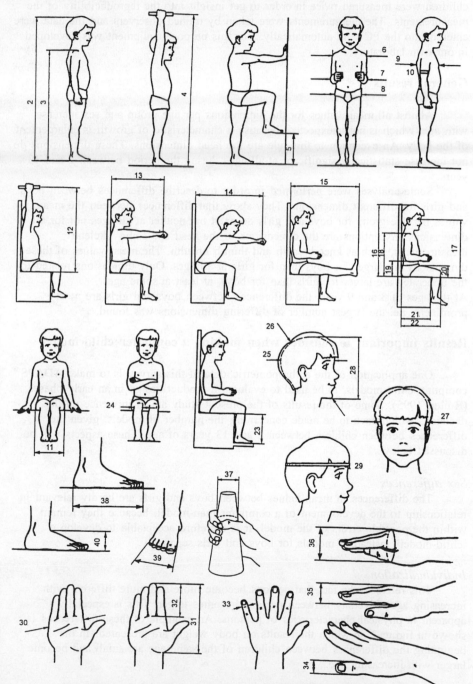

Figure 1. Dimensions measured in the KIMA project.

A national representative sample of 2245 children aged between 2 and 13 years was measured at health care centres for infants and toddlers (children between 2 and 4 years of age) and at primary schools all over the country. Some 206 children were measured twice in order to get insight into the reproducibility of the measurements. The measurements were taken by trained observers and the data were entered into the PC semi-automatically. For this purpose equipment was developed in our own laboratory.

General results

Almost all mean values for the dimensions per age group and sex increase with age, which is not unexpected, because a characteristic of growth is enlargement of the body. An exception to this rule are the head dimensions. These dimensions do not increase statistically significant at all ages during the period between 2 and 13 years.

Some analyses were performed in order to describe differences between boys and girls for different dimensions. They show that differences between the mean values per age group for boys and girls do exist but neither at all ages nor for all dimensions. Exceptions are the dimensions of the head and some 'skeletal' measurements, such as knee breadth and thumb breadth. The mean values of these dimensions are larger for boys than for girls at all ages. Only dimensions related to the upper leg are larger for girls than for boys, at least at some ages.

At the ages of 3 and 9 years the differences between boys and girls are most prominent, i.e. the largest number of differing dimensions was found.

Results important to consider when making a computer child-model

One application of the anthropometric data of this survey is to make ADAPS computer child-models, to be used to evaluate a product design in an early phase (Ruiter, 1995). Some of the results of the present study are of influence on the decisions which have to be made concerning the number of models, given the differences between children between 2 and 13 years of age. These aspects will be discussed below.

Sex differences

The differences in mean values between boys and girls are hardly relevant in relationship to the development of a computer man-model, because they remain within the possible errors of the model. It is therefore acceptable to develop a 'child-model' instead of models for boys and girls separately.

Individualization

This refers to the fact that persons become more and more different with increasing age. Diversity between persons becomes larger. This is especially apparent in physical characteristics of persons. An example of these differences is shown in figure 2, in which the results on body weight are presented. In the beginning the differences between children of the same age are small and become larger with increasing age.

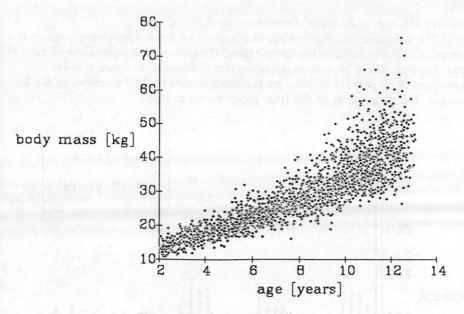

Figure 2. Body mass according to age.

Overlap between age groups

This is related to the previous point. An increase in diversity implies an overlap between age groups. Children of the same age have very different body dimensions and on the other hand, children with the same, say, stature may differ very much in age. For example: the largest 6-year-old child is as tall as the smallest 12-years-old child in our sample. This is illustrated in figure 3, in which for some age groups the number of children per stature group is presented. In the stature group 130-140 cm. children of 6, 9 and 12 years of age are present.

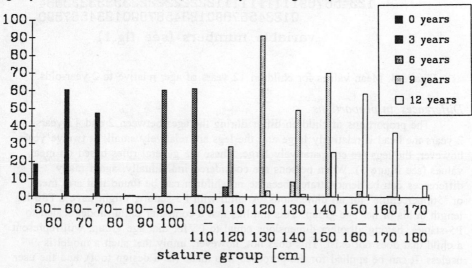

Figure 3. Number of children of different ages according to stature group.

Relative differences between dimensions at different ages
      The differences as described above are not in all dimensions clearly present. It
is especially demonstrable in the length-related dimensions. The dimensions of the
head, however, hardly increase implying smaller differences. In figure 4 body
dimensions of 12-year-old children are presented relative to the dimensions of the 2-
year-old. The mean values of this latter group are set at 100%.

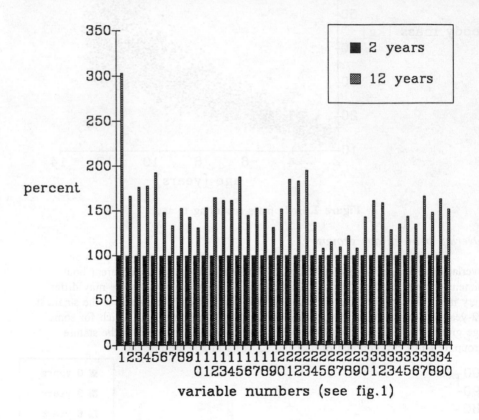

Figure 4. Mean values for children 12 years of age, relative to 2-year-olds.

Differences in proportions
      The proportions of children differ during the ages between 2 and 13 years. At
2 years the head is relatively large and the legs are relatively small; at twelve years,
however, the legs are comparatively large. These are general rules based on mean
values (see figure 5). When persons are considered individually, again many
differences can be demonstrated because no children can be found that are 'mean'
or '5th percentile' for all dimensions. A child with P5-stature, might have a P30-leg
length in his or her age group. This implies that a computer child model scaled to
P5-stature, having all other dimensions equal to P5 for that age group, will represent
a child that does not exist. This does not, however, imply that such a model is
useless. It can be applied for the purposes it is made (as a design tool), and the user
has to realise that differences between children in the age group concerned are
larger than the model suggests.

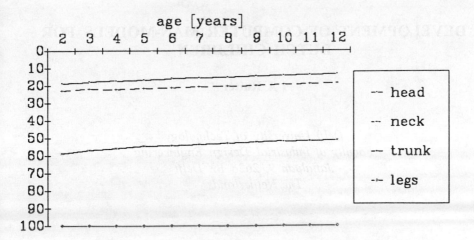

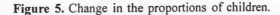

Figure 5. Change in the proportions of children.

Discussion

The results of the anthropometric part of the study on development of children in the Netherlands (KIMA) are currently being used to develop computer child-models. The result will be a number of child-models, each for a different age group.

When using such a model one has to remember the differences between individual children in the age group the model represents. Of course models are a simplification of reality, but nevertheless they are very useful in certain phases of the design process, when data in, for example, tables do not give appropriate information.

A future development of ADAPS child-models might be a feature to adjust body proportions of the child-model used. This would give the user the opportunity to evaluate a design with a model which is more according to reality.

References

Anthropometric Source Book 1978, Volume 2, NASA Reference Publication 1024, (National Technical Information Centre (NTIS), Springfield, Virginia, USA).
DIN 33402, Körpermasse des Menschen 1981, (Berlin).
Nederlands Normalisatie Instituut 1987, *Nederlandse vertaling en bewerking van ISO/DIS 7250, document nr. 20* (Translation in Dutch of ISO/DIS 7250), Commissie 30113 Antropometrie en Biomechanica, Delft.
Ruiter, I.A. 1995, *Development of computer man-models for Dutch children*, this volume.
Snyder, R.G. et al. 1977, *Anthropometry of infants, children and youth to age 18 for product safety design*, (Society of Automotive Engineers, Inc. Warrendale).
Steenbekkers, L.P.A. 1993, *Child development, design implications and accident prevention*, (Delft University Press, Delft).
Tanner, J.M. 1990, *Fetus into man, Physical growth from conception to maturity*, Revised and enlarged edition, (Harvard University Press, Cambridge, Massachusetts).

DEVELOPMENT OF COMPUTER MAN-MODELS FOR DUTCH CHILDREN

I A Ruiter

Delft University of Technology
Faculty of Industrial Design Engineering
Jaffalaan 9, 2628 BX Delft
The Netherlands

ADAPS (Anthropometric Design Assessment Program System) can be used to evaluate a 3D workspace or product with man-models. This paper describes the process of development of a series of models for Dutch children with the emphasis on the determination of the optimum number of models.

Introduction

This paper presents the first results of a combination of two research projects: KIMA and ADAPS.

The aim of the KIMA project (Steenbekkers, 1993) was to collect data on normal, healthy Dutch children (age 2-12). One of the intentions was to provide designers of daily life products with useful data on children. Six groups of variables have been measured, one of them being body dimensions.

ADAPS (Anthropometric Design Assessment Program System) is a computerprogramme that enables the user to assess 3D workspaces and products with the use of computer man-models. The available models are: Dutch man and woman, Dutch elderly (man and woman) and a 4-year-old boy.

The model of the 4-year-old boy is an ad hoc model based on a combination of sources from different origin. The data on body dimensions collected during the KIMA project provided a good basis to update the existing child-model as well as to develop new child-models.

Data

May the collected data on body dimensions of children be used to develop computer child-models? To answer this question the following should be taken into consideration.

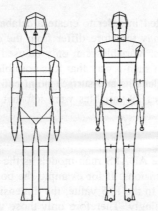

Figure 1. Model of a 4-year-old boy scaled up to P50 male adult height.

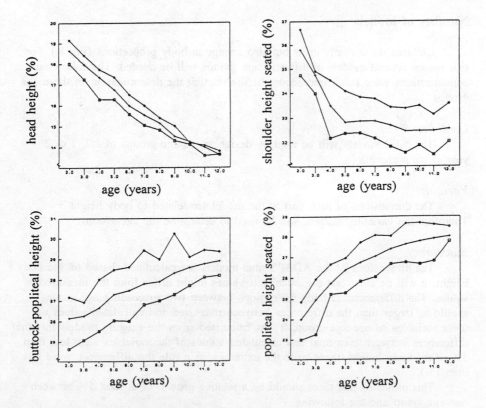

Figure 2. Segment length as percentage of body height (□ =P3,+ =P50,◊ =P97).

Target group
        Data were collected from a representative sample of Dutch children, so this will give no restrictions to the use of the data.

Purpose
        The data were collected in order to create a databank for designers. The selected body dimensions may therefore differ from the dimensions required for the development of man-models. Variables can only be used to create an ADAPS man-model if they are defined in such a way that it is possible to describe them as a combination of model link lengths and surface point coördinates.
Out of 40 measured variables 29 variables were left that could be used to develop a child-model.

Correlation
        A characteristic of the ADAPS man-models is the relationship between 'stature' and the other dimensions. If for example the body height changes from the 50th percentile (P50) value to the P95 value, the dimensions of each part of the model will be scaled accordingly. Therefore only those variables that show a sufficient correlation with body height may be used to develop a model. An arbitrarily chosen correlation of 0.5 reduces the number of variables to 19.

Number of models

        Children do not only grow but also change in body proportions (Fig 1.). For that reason several models of different age groups will be needed. The following considerations were taken into acount before starting the determination of these age groups.

Usability
        The child-models will be used by designers, so age groups of 1/2, 1 or 2 year(s) are preferred.

Variable
        The dimensions of each part of the model are related to body height. Therefore the variable 'stature' will be used to determine the age groups.

Inaccuracy
        The dimensions of the ADAPS man-models are calculated, based on the body height. It will be clear that the calculated values might differ from the measured values. The differences in body dimensions between two succeeding age groups should be larger than the differences between measured and calculated values of these variables of one age group. It was calculated from the existing models that the differences between measured and calculated values of the variables were less than 5% of the body height (these were the extremes, as a rule the differences were less than 1%).
        This indicates that there should be a relative growth of at least 5% between one age group and the following.

Table 1. Absolute and relative growth of boys and girls per 1-year age group.

age	stature P50(cm)	absolute growth(cm)	relative growth (%)
2	93.3		
3	101.3	8	8.6
4	108.4	7.1	7
5	116.4	8	7.3
6	122.6	6.2	5.3
7	128.7	6.1	5.0
8	134.0	5.3	4.1
9	140.5	6.5	4.9
10	146.6	6.1	4.3
11	150.9	4.3	2.9
12	156.4	5.5	3.6

Table 2. Absolute and relative growth of boys and girls per 2-years group.

age	stature P50 (cm)	absolute growth (cm)	relative growth (%)
2	93.3		
3+4	104.5	12.3	13.3
5+6	119.9	15.4	12.8
7+8	131.0	11.1	9.3
9+10	143.8	12.8	8.9
11+12	152.5	8.7	5.7

Table 3. Comparison between dimensions of 4-year-old model and children.

variable	model	children	difference
stature	105.4	108.4	3.4
head height	18.0	18.8	0.8
shoulder height seated	33.6	36.3	2.7
buttock-popliteal height	33.6	35.6	2.0
popliteal height seated	23.4	27.9	4.5

Boys and girls

Boys and girls show the same growth patterns. Girls are a little smaller on the average but the differences in body length stay within the inaccuracy of the model. The decision was made to develop child-models instead of separate models for boys and girls.

To get a first impression of the changes in body height, growth per year (absolute as well as relative) was calculated (Table 1.). It will be clear from Table 1 that growth speed slows down when children get older.

To get an insight into the changes in body proportions we looked at the changes of four variables: head height, shoulder height seated, buttock-popliteal length and popliteal height seated. For each of those four variables the P3, P50 and P97 values were calculated as a percentage of the body height. The results (Fig. 2.) show that the legs get relatively longer, head and trunk smaller.

The KIMA results showed that the standard deviation of 'stature' increases with increasing age, so the absolute differences in body height between small and large children of the same age group get larger. The relative differences however (compared to average body height) stay nearly constant. The relative differences of the P3 and P97 value (compared to the P50 value) of the four segment lengths show a tendency to stay constant or to get smaller.

When s (standard deviation) is calculated as percentage of the mean of the connected variable, the values calculated for the variable 'stature' are smaller than for the other variables. This means that 'stature' diverges relatively less than the other variables.

The increasing standard deviation of the body height combined with a decreasing or constant standard deviation of the four segment lengths leads to an increasing accuracy of the calculated values of the dimensions of the man-model with increasing age. Calculation of the differences between measured and calculated body dimensions proved this to be true.

Considering the facts that the models should show at least a 5% relative growth and that the models tend to be more accurate with increasing age, we gave a closer look to a division in 2-year age group (Table 2.). Calculating the differences between measured and calculated body dimensions showed no difference between the results of 1-year and 2-year groups. Based on these results we choose to develop 8 child-models: 2, 3, 4, 5, 6, 7-8, 9-10 and 11-12 year old.

Development of a model for 4-year-old children

The model for 4-year-old children will be based on the existing model of a 4-year-old boy. The process will be nearly the same as described for the development of the models for Dutch elderly (Ruiter, 1990). For 19 variables the values calculated for the 4-year-old boy were compared to the measured values of these variables for 4-year-old children. Results of this comparison for 5 variables are given in Table 3.

Next step will be enlarging of the P50 boy-4-model thus far that its height equals the measured P50 height of 4-year-old children and to compare the measured and calculated values of the variables. Based on the differences of these values the P50 model will be changed (by changing link lengths and coördinates of surface points) until these differences are as small as possible. Then the P3 and P97 values will be calculated and compared to the measured ones. If necessary the P50 model will be changed until the differences for P3, P50 as well as P97 for each variable are as small as possible. When the 'child-4-model' is finished the same procedure will be followed for the other seven models.

Discussion

The main issue of this paper is the determination of the number of models that should be developed. The division into age groups is based on several arguments, most important one the usefulness for designers. Therefore the group division might not be as statistically sound as could be.

An important shortcoming of the model is that no use was (could be) made of the weight related variables. So the model makes no distinction between thin and fat children; within the same age group the model for small children will be thin, the model for large children fat, due to the fact that all body dimensions are related to 'stature'.

Another shortcoming of the model is that its ranges of motion are copied from the adult model. During the KIMA project no ranges of motion of seperate joints have been measured. When all eight child-models are finished it might be possible to adjust some ranges of motion by using data on the 'reach' postures (postures nr. 4, 5, 12 and 14 of Figure 1, Steenbekkers, 1995).

A problem with anthropometric data is that measurements are usually taken in standard anthropometric postures instead of in functional postures. This means that although the model might match the measured body dimensions closely, we still have no insight into the way the model behaves in functional postures.

In spite of all these shortcomings we still think the child-models will be useful design tools, like the other models proved to be.

References:
Ruiter, I.A. 1990, Development of man-models for the Dutch senior citizen. In E.J. Lovesey (ed), *Proceedings of the Ergonomics Society's 1990 Annual Conference*, (Taylor & Francis, London) 223-227.
Steenbekkers, L.P.A. 1993, *Child development, design implications and accident prevention*, (Delft University Press, Delft).
Steenbekkers, L.P.A. 1994, Anthropometry of children 2 to 13 years of age in the Netherlands, *this volume*

Domestic Ergonomics

EVALUATION AND DESIGN OF THE DOMESTIC GAS BILL

by

Armelle Schaad and Ken Parsons

Department of Human Sciences
Loughborough University of Technology
United Kingdom

Two experiments were conducted to evaluate the domestic gas bill and to determine an improved design. In the first, eight subjects evaluated the existing gas bill in an interview questionnaire involving tests of comprehension and subjective measures of satisfaction. It was found that although some information was well understood some important information was not. Subjects were generally not satisfied with the bill design. The second experiment evaluated the original gas bill and three proposed designs based upon the results of experiment 1. Eight subjects evaluated each bill. As before the original bill was found to be unacceptable. Two of the remaining three bills were acceptable and bill C received unanimous support.

Aim

The aim of the experimental work presented in this paper was to evaluate the appropriateness to users of the existing British domestic gas bill and to provide a proposal for an improved design.

Method

The project was conducted in two parts. The first part identified the user requirements for a gas bill and evaluated how effectively customers could use the bill in terms of its functionality and comprehension and also how satisfied they were with its design and presentation. This led to three proposals for alternative designs. These designs were evaluated along with the existing bill in a usability trial.

Experiment One: User evaluation of the existing domestic gas bill.

The aim of this experiment was to determine the comprehension and expectations of British Gas customers concerning the current gas bill.

Method

Eight subjects each completed a usability trial of the existing East Midlands gas bill. Four male and four female subjects were used, one each from the age range 20-35, 36-50, 51-65 and >65 yrs. Subjects were presented with an example of the bill and completed tasks concerned with their understanding and ability to use it. These included tasks concerned with who the bill was for and for what period it covered, total amount, what the components of the bill and symbols used represented, how calculations of energy used were carried out and how to obtain further information. Information was also obtained concerning overall design and satisfaction and suggestions for improvement.

Results.

Questions concerning amount to be paid, who the bill was for, how to get further information, customer reference number and period for standing charges and how they were calculated, were all answered correctly by subjects. Major problems of comprehension occurred with numerical information in particular where calculations were involved. How to pay was not always understood. None of the subjects understood the transformation of energy used between cubic feet to cubic metres and cubic meters in kilo Watt hours. Subjects also did not understand the period for the gas used and basic information about how much gas was used. The customers overall impressions of the bill were divided into ability to find information, its ability to facilitate understanding and the bills graphic design. The bill was generally rated poorly on these criteria and in particular on its ability to facilitate understanding in terms global information and phrases and symbols used (kiloWatt hour, cubic feet, calorific value etc.). All eight subjects said that the bill was not pleasant to use and all suggested improvements that could be made. Almost all subjects ignore the information on the back of the bill and had not read the booklets explaining the bill. These are essential to understand fully what is presented. The main requirement was for further information regarding how calculations were made. Some subjects thought that less information was required. No effect of gender was found however older people above the age of 50 years had lower understanding than those younger than 50 years.

Experiment 2: Usability trials of four gas bill designs

The aim of this experiment was to evaluate in user trials, three proposed improved gas bill designs and the original design.

Method

Thirty two subjects completed a usability trial of four gas bill designs. The bill designs were the existing East Midlands gas bill (bill A), and three other bills (B,C and D) varying in information content and layout according to the results of experiment 1. Four male and four female subjects were used to evaluate each bill, one each from the age range 20-35, 36-50, 51-65 and >65 yrs. Subjects were presented with an example of one of the bills and completed tasks concerned with their understanding and ability to use it. These were similar to those used in experiment 1 and included tasks concerned with who the bill was for and for what period it covered, total amount, what the components of the bill and symbols used, represented, how calculations of energy used were carried out and how to obtain further information. Information was also obtained concerning overall design and satisfaction and suggestions for improvement. After all information had been collected, the three remaining bills (of which the subjects had no knowledge) were presented to the subjects and they provided a rank order of their general impression of all four bills. The experiment was conducted by interview questionnaire in the waiting room of the local railway station and took approximately thirty minutes. Subjects were volunteers who had previously paid a gas bill.

Results

Figure 1 presents the results concerning bill comprehension for each of the four bills. For bill A (original) questions concerning amount to be paid, who the bill was for, how to get further information, customer reference number, period for standing charges and how they were calculated and the date of the bill (questions 1, 2, 3, 4, 6, 7, 8, 11, 12, and 25,) were all answered correctly by subjects. Major problems of comprehension occurred with numerical information in particular where calculations were involved. How to pay was generally understood (7 out of eight subjects). None of the subjects understood the transformation of energy used between cubic feet to cubic metres and cubic meters in kilo Watt hours. Subjects also did not understand the period for the gas used (only two understood) and some (2 out of 8) did not understand basic information about how much gas was used(questions 10, 14, 15, 16, 21, 22, and 24,)

The customers overall impressions of the bill were divided into ability to find information, its ability to facilitate understanding and the bills graphic design. An average of positive responses over eight subjects provided a score of positive aspects. The average values obtained for bill A were 0.66, 0 and 0.77 respectively on the criteria presented above. The value of 0.66 above for example means that on average only 0.66 subjects responded positively concerning ability to find information. A summary of information over all four bills is presented in Table 1.

Bill B (changed layout and information) scored highly across subjects on all questions. There were a few comments regarding it to be too technical but no major problems. Although the bill was considered easy to understand some subjects thought that it was not pleasant to read.

Bill C (logical layout for calculations and changed information) performed highly on all aspects of use and satisfaction. So much so that all subjects could use it and found it pleasant to read.

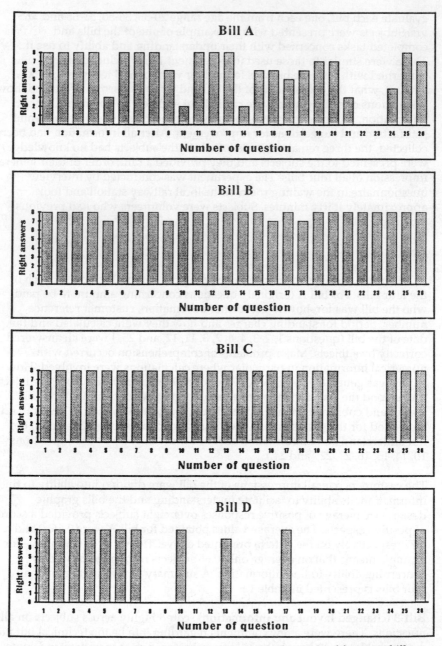

Figure 1: Results concerning bill comprehension for each of four gas bills.
       Bill A - existing bill
       Bill B - changed layout and information
       Bill C - logical layout for calculations and changed information
       Bill D - changed layout and reduced information
       Zero scores imply that the question was not applicable for that bill.

British Gas
East Midlands

CONTACT:

PAYMENT:	British gas PO Box 58 De Montfort Steet Leicester LE1 9DF

ENQUERIES:

About the bill:	Leicester (0533)559666
Others:	Look in the telephone directory under "gas"

Customer reference number: 610-3690-02400-1-1

Mrs A.SADE
FAULKNER/EGGINTON COURT
LOUGHBOUROUGH LEICESTER
LE 12 8NN

DATE:

Bill(Tax point):	07/01/93
Meter reading:	06/01/93

PERIOD:

12/10/92 to 06/01/93
(86 days)

METER READING:

Present:	2530E
Previous:	2306E
	E=Estimated*

GAS USED:

2530-2306=224 hundred cubic feet
=6740 KWh

	CALCULATION	CHARGES Debit	Credit
GAS	6743 KWh X 1,477p. per KWh	99,59	
STANDING CHARGES	90 days X 10,10p. per day	9,09	
V.A.T	0,00% on £ 108,68(99,59+9,09)	0,00	
PREVIOUS DEBIT	... to 12/10/92	21,29	
PAYMENT	2 X £ 24,00		48,00

AMOUNT   £ 81,97

*
If the present meter reading is an estimate (E) and you would like us to use your own meter reading, please write your reading on the back of the bill and send it to us as soon as possible.Or telephone and tell us your rereading.Thank you.

Figure 2: Gas bill C - logical layout for calculations and changed information. This bill performed 'best'. In the form tested a traditional blue background was used.

Bill D (changed layout and reduced information) performed well on a number of functional aspects but was generally considered unacceptable because of a lack of information.

Table 1. Average number of positive statements out of eight subjects over three criteria.

Criteria	Bill			
	A	B	C	D
Find information	0.66	4.66	7.0	4.33
Facilitate understanding	0.00	6.50	7.0	6.00
Graphic design	0.70	6.00	7.3	5.00

Comparison of bills

The rank order of bills was consistently found to be bill C (best), B, D and almost all subjects ranked bill A as worst. Bills C and B performed consistently better than bills A and D. Bill A was considered to require further information, was too technical and was unacceptable. Bill D was considered to require further information and half of the subjects found it unacceptable. Bill B was a little technical but overall bills B and C performed relatively well and were acceptable. Bill C performed best overall and is shown in Figure 2.

Conclusion

The two experiments presented in this study and other considerations lead to the conclusion the existing gas bill required improvement in design and that improvements could be made by adopting the design used in bill C from experiment 2.

Acknowledgements

The authors acknowledge the help and information provided by British Gas at Loughborough and Leicester.

PLEASURE IN PRODUCT USE: BEYOND USABILITY

Patrick W. Jordan and Mirjam Servaes

Philips Corporate Design
Building SX, P.O. Box 218
5600 MD Eindhoven
The Netherlands

Much of the current work in human factors concentrates on the usability of products. This tends to focus on utilitarian, functional product benefits. However, a usability based approach does not necessarily take account of the range of hedonic or experiential benefits associated with products. This paper reports on an interview based study that was a 'first pass' at investigating those emotional benefits ('pleasure') in a more systematic way. The interview was designed to investigate the feelings that are induced by using particularly pleasurable products and the properties of design that influence whether a product is particularly pleasurable.

Feelings associated with pleasurable products included security and assurance, confidence, pride, excitement and satisfaction. Results suggested that appropriate functionality, usability, aesthetics, performance and reliability were the principal factors in determining the pleasurability of a product.

The data gathered from the interview support the idea that product pleasurability involves more than usability alone. It is suggested that, in order to optimise the experience of product use, those involved in user-centred design should look both at and beyond usability — moving towards products that are a positive pleasure to use.

Introduction

The importance of user-centred design is increasingly being recognised. So far, the main approach to user-centred design has been to address the issue of usability. Designing for usability and evaluating for usability tend to focus on ensuring that products support efficient and effective task completion and that the user doesn't suffer any discomfort when using the product. The International Standards Organisation's (ISO) definition of usability underlines this. The ISO define usability as being:

> "... the effectiveness, efficiency and satisfaction with which specified users can achieve specified goals in particular environments." (ISO DIS 9241-11).

Effectiveness is referred to as the level of accuracy and completeness with which tasks are completed. Often this will be measured in terms of error rates during interaction or the quality of the output from a task. Efficiency is to do with the amount of effort that the user must exert in order to complete a task. Time on task and mental workload are examples of measures of this. Satisfaction, meanwhile, refers to the acceptability of the product to the user and the degree of comfort associated with product use.

There are many cases where satisfaction may be the most important aspect of usability. In particular, when a user has a choice of whether or not to use a product. After all, however objectively effective and efficient a product is in use, people will avoid it if they have negative attitudes towards it. Issues connected with satisfaction seem particularly salient for those involved in the design of consumer products. If users develop negative attitudes towards such products, they may simply stop buying them, or start buying a competitor's products.

Pleasure

Traditionally, when considering the attitudinal aspects of usability, the emphasis has tended to be on avoiding negative attitudes rather than on striving to induce positive attitudes. The ISO definition mentions "comfort and acceptability". Similarly, an earlier definition from Shackel (1986) recognises the importance of product use being "... within acceptable levels of human cost in terms of tiredness, discomfort, frustration and personal effort." But what of the more positive aspects — the aspects that make a product a real pleasure to use. Perhaps such considerations should also play a role in user-centred design.

This paper reports on a study, aiming to give a 'first pass' at a systematic investigation of pleasure in product use. Within the area of consumer behaviour pleasure is a central concept in the literature on experiential products and hedonic consumption (see for example Hirschman and Holbrook 1982; Holbrook and Hirschman 1982). The experiential or hedonic approach focuses on the emotions, fantasies, images etc. related to one's experience with a product. Product usage is explained by the feelings of pleasure and enjoyment it offers.

The study reported here used a semi-structured interview, which was designed to help achieve the following:

- Identify the emotions that are engendered by using particularly pleasurable products.
- Identify the properties of a design which can contribute to making a product particularly pleasurable to use.

Method

Participants
A semi-structured interview was conducted with 18 respondents (10 female, 8 male). The vast majority of these were students of less than 25 years. All participants were residents of Glasgow in Scotland.

Interviewees had been contacted beforehand. It was requested that, before the interview, they thought of a particularly pleasurable product which they owned or used (or used to own or use). The questions that were asked in the interview centred around this product.

The Interview
The interview contained a series of open-ended questions. Users were asked to consider the product which they found particularly pleasurable (question 1). They were asked to say what it was and to give a general description of it (2). They were then asked about the aspects of the product that made it particularly appealing (3) and about the feelings engendered by using the product (4). Finally, they were asked when they experienced these positive feelings (5) — was it only during product

usage, or would, say, anticipation or remembrance of using the product induce positive emotions.

Results

Products chosen

The products that participants selected as being particularly pleasurable were as follows: stereo equipment (8 participants), video cassette recorders (VCRs) (3), TV sets (2), hairdryers (2), electric guitar (1), food processor (1), personal stereo (1).

Question 2 of the interview was asked in order to see if users would volunteer any information about design factors or emotions engendered without being specifically asked about these issues. Little of interest emerged from these questions however, with respondents usually giving straight factual descriptions of the products.

Emotions associated with pleasurable products

Question 4 of the interview asked about the emotions associated with pleasurable products. In analysing responses, interpretations were made of the semantics used by participants. For example, if respondents had said that they "found it thrilling" to use a product, this would be interpreted as 'excitement'. There were also cases where the analyst regarded different responses as referring to the same emotion. These were added together for the purposes of the analysis. For example, security and comfort were regarded as representing the same concept. Most respondents mentioned more than one emotion in connection with each of their products.

Figure 1 summarises the responses that participants gave when asked about the feelings associated with pleasurable products.

EMOTION

Security / comfort										
Confidence										
Pride										
Excitement										
Satisfaction										
Entertainment										
Freedom										
Sentiment / nostalgia										
Number who mentioned	1	2	3	4	5	6	7	8	9	10

Figure 1. Emotions associated with pleasurable products (n = 18).

The single most common aspect mentioned was security and comfort. Users liked to know that the product was there when needed. For example, the owner of a stereo said that it gave him the security of knowing that he would be able to enjoy his leisure time listening to music. Confidence was another important factor. This could refer to confidence with respect to the quality of a product's performance or an effect of the product that made the user feel self-confident after use. For example, the user of a hairdryer said that she had confidence in her appearance after using the dryer to style her hair.

Respondents often indicated that they felt pride in themselves for having chosen a good or stylish product, or felt that the ownership of a particular product gave them recognition within a particular group. For example, the owner of an electric guitar was proud as he felt that fellow musicians would recognise its qualities.

Some products were rated as pleasurable because they were exciting to use or were exciting to anticipate using. For example, a radio owner felt excited at the prospect of tuning in to an hour or so of talk radio after he had finished his day's work. Similarly, some products were rated as pleasurable on the grounds of providing the user with entertainment. For example, a user found her VCR pleasurable, as she enjoyed watching her favourite films on it. Conversely, satisfaction seemed a more 'background' type of emotion and appeared to be chiefly associated with products that were pleasurable in that they caused the user no bother, rather than products that were positively cherished by the user. For example, a respondent said that she was satisfied with her TV as she was able to take it for granted — knowing that it would work satisfactorily — and found it to be unobtrusive.

A feeling of freedom was associated with some pleasurable products. For example, the user of a compact disc (CD) player enjoyed listening to CDs whilst doing household chores. He said he gained a sense of freedom from the drudgery that would normally be associated with these chores.

Other pleasurable products had a 'history' attached to them. This induced sentimental or nostalgic feelings. For example, a user said he was "attached" to his stereo as he remembered the care he'd taken in purchasing it and was pleased that he'd made such a good choice.

Nine other concepts were also mentioned. These included a number of emotions that were mentioned on a one-off basis, which were apparently different from those mentioned above. 'Surprise' and 'escapism' are examples of these. Other responses included references to concepts that didn't really appear to be emotions, such as 'informativeness' and 'homeliness'.

When are these emotions experienced?

The vast majority of respondents said that they experienced these emotions when using the products concerned. However, a significant number of respondents also experienced feelings connected with the product before and after use. For example, excitement is an emotion that was sometimes felt before using a pleasurable product.

Responses also indicated that the emotions associated with products were felt at other times too. For example, pride could be taken in a product simply when the owner saw it.

Design properties associated with pleasurable products

Question 3 of the interview asked about the design properties which made a product particularly pleasurable. As with the question regarding emotions, analysing the data sometimes involved making interpretations of participants responses, as well as sometimes combining responses that were regarded as being similar. So, for example, if respondents said that they "liked the appearance" of their products, this would be interpreted as 'aesthetics'. Similarly, 'features' and 'functionality' are examples of two words which were thought to refer to the same aspect of design. Most respondents mentioned more than one design property in association with each product.

Figure 2 gives a summary of the design properties mentioned in connection with pleasurable products.

DESIGN PROPERTY

Design property	Number who mentioned
Features / functionality	10
Usability	9
Aesthetics	7
Effective performance	7
Reliability	6
High status product	4
Convenience	3
Size	3
Brand name	1

Figure 2. Design properties connected with pleasurable products (n = 18).

Having an appropriate range of functionality was the most commonly mentioned issue. For example, a stereo user was pleased to have a turntable, tape player and compact disc (CD) player in the same unit. Pleasure was also often gained through a product simply performing its central functions well (effective performance). For example, a number of stereo users remarked that they derived particular pleasure from their systems due to the good sound quality.

Seven of the respondents mentioned the importance of aesthetics. Both style and colour were important to users. For example, one stereo owner insisted he would not buy a stereo unless it was black. Similarly, another stereo owner was pleased with how well it complimented the layout of her room. Another important aspect of product form was the appropriateness of its size. For example, one respondent mentioned the advantages of having a small, compact stereo, whilst another mentioned that she was pleased with her TV as it had a large screen.

Usability was mentioned by eight of the respondents. People enjoyed products that were easy to use. For example, a VCR user was pleased because it was easy to understand what each of the buttons on the machine did and a stereo user was pleased because the controls were laid out in a helpful way. Convenience was also mentioned. For example, the owner of a CD player found CDs to be more convenient than cassette tapes.

Some products were rated as pleasurable because of their reliability. Some respondents said that they had become attached to products that had given them years of good service. For example, the user of a hairdryer had been using it for 11 years and had become very attached to it. Four users mentioned the appeal of having a prestigious product with associated status. For example, the owner of a stereo was pleased with its upmarket image. Having a product with a respected brand name was regarded as important by two of the respondents.

Another six properties got one-off mentions. These included, for example, 'individuality' and 'installability'.

Discussion and conclusions

The results reported represent a first pass at tackling two aspects of the issue of pleasure in product use — the emotions associated with using pleasurable products and the properties of a design which can contribute to making a product pleasurable to use.

The study reported has a number of limitations. Firstly, the participants represent a fairly narrow sample and there is no guarantee that these results would generalise to a wider population.

Another limitation comes from the way the data were analysed. The analyst categorised responses based on his own 'view of the world', rather than according to any particular model of emotions or of design properties. As the analyst (the first author) is a human factors specialist based in a design department, it might be hoped that the classification used in terms of design properties is fairly satisfactory. However, the classification in terms of emotions may be far less so. Indeed, there appears to have been little systematic study of emotions by psychologists (Plutchik 1994). So, because of the possible inaccuracies in classification, it would probably be unwise to treat the results as giving a full and definitive list of the emotions and design properties connected with pleasure in product use.

A third limitation comes from the open ended nature of the questions in the interview. When asked about emotions and design properties, participants may have found some potential responses more easy to articulate than others. Equally, some responses may have come to mind more easily than others. Thus, it may not be valid to treat the number of mentions that each emotion and design property received as being an index of their comparative importance.

However, despite these limitations, the results still indicate that the issue of pleasure with product use goes beyond usability. Responses concerning the properties of a product that make it pleasurable indicate that, whilst usability is important, it is only one of a number of factors influencing pleasurability. Similarly, satisfaction — the emotion associated with usable products — appears as only one aspect of the emotions associated with pleasure. It seems, then, that if the goals of user-centred design are to include facilitating pleasurable experiences for the user, factors other than usability alone must be considered.

Implications for user-centred design
    Data gathered from the studies suggest that users are able to identify design issues that are important to them with respect to creating pleasurable products. Designing for usability appears to be very important, but so too does (amongst others) including appropriate features and functionality, aesthetics, effective performance and reliability.
    The design issues that emerged from these interviews refer to 'high level' properties of the design. Whilst they give a broad idea of the design qualities that need to be achieved, there is no specification of the lower level design issues that need to be addressed. For example, the high level property 'usability' could be broken down into a number of lower level properties such as consistency, feedback, and error prevention. These can form the basis for guiding design decisions. Before pleasurability can be engineered into a product, more concrete design criteria affecting these higher level issues will have to be established. This may be the subject of future studies. In the meantime, the high level criteria reported may give a basis for steering designers towards creating products that are a pleasure to use.

Implications for product evaluation
    The emotional dimension associated with usability is satisfaction. However, the results of this study indicate that pleasurability is associated with a wider range of emotions such as security, confidence, pride, excitement and nostalgia. Just as tools exist for measuring satisfaction (e.g. Kirakowski 1987), it might be equally possible to design tools for measuring these other emotional aspects, giving the possibility of quantifying pleasure in product use. Another issue for future investigation is the possibility of identifying behavioural correlates to product pleasurability — these might include, for example, frequency of usage, or future purchase choices. This could give the prospect of making more objective estimates of product pleasurability.
    As with designing for pleasurability, the outcomes of these studies are useful in terms of steering evaluators' thoughts, rather than being authoritative statements about the set of emotional elements to analyse. Again, it was the high level issues that were addressed, not the low level detail which might be needed for comprehensive evaluation. It seems clear however, that checking whether or not a product is satisfying to use is only a part of predicting whether or not that product will be positively pleasurable to use.

Acknowledgements

Thanks are due to our colleagues Irene McWilliam, Mark Hartevelt and Hans Kemp for their advice and comments on this topic.

References

Hirschman, E.C. and Holbrook, M.B. 1982, Hedonic consumption: emerging concepts, methods and propositions, *Journal of Marketing*, **46**, 92-101.
Holbrook, M.B. and Hirschman, E.C. 1982, The experiential aspects of consumption: consumer fantasies, feelings and fun, *Journal of Consumer Research*, **9**, 132-140.
ISO DIS 9241-11, *Ergonomic requirements for office work with visual display terminals (VDTs)*:- Part 11: Guidance on usability.
Kirakowski, J. 1987, The Computer User Satisfaction Inventory, IEE Colloquium on Evaluation Techniques for Interactive System Design, II, London.
Plutchik, R. (1994), *The Psychology and Biology of Emotions*, (Harper Collins, London).
Shackel, B. (1986), Ergonomics in design for usability. In M.D. Harrison and A. Monk (eds.), *People and Computers*, (Cambridge University Press, Cambridge).

HCl

GENTLY DOES IT: DOES GRADUAL
ADAPTATION IMPROVE USABILITY?

R.E. Griffin, C.M.C. Dowlen, D.K. Williamson

S.E.S.D., South Bank University,
103 Borough Road,
London, SE1 0AA.

eMail to: r_griffin@cosmos-uk.org

Adaptive systems have been reported in the literature for some years now.
Although they have become accepted in some areas, in the field of interaction
design/HCI, adaptive interfaces have been viewed by many practitioners with
distrust. We examine some possible reasons for this and argue that good
adaptive design *is* possible through examples from outside of the HCI arena.
Gradual adaptation is then discussed together with the design of a gradually
adaptive menu. Finally we describe the preliminary results of this system and
pay particular attention to describe the algorithms employed.

Introduction

Adaptive systems adapt themselves to their environment; "the idea of an adaptive
interface is straightforward; simply, it means that the interface adapts to the user rather than
the user adapt to the system"; Norcio and Stanley (1988). Adaptive U.I.s have not been
embraced by all. One reason for the poor uptake of adaptive interfaces is the confusion that
has resulted from the many different classes of adaptation that have been proposed; many
of these have been applied to many different applications. Further-more, many researchers
have reported that users are suspicious of adaptive interfaces; the typical user fear is stated
as "loss of control" (from the user to the product) Norcio and Stanley (1988) and widely
reported elsewhere. There are isolated cases where it is reported that users reacted
favourably towards adaptive interfaces: Burford and Baber (1994) reported that "Subjects
would respond positively (towards an adaptive interface) if forewarned". Similarly Chin
(1990) found that "Adaptable menus... were not rated less consistent than the fixed menu".
Because the majority of research has reported negatively in respect of users' responses,
there is a widespread view that adaptive interfaces are neither needed nor wanted by users.

Another reason for the lack of enthusiasm shown toward adaptive interfaces is that
much of the published research to date has reported problems concerned with maintaining
the user's cognitive model (UCM). This results in confusion, errors and consequently mal-

adaptation: Chin (1989), Norcio and Stanley (1988) and Malinowski, Kuhme, Dieterich and Schneider-Hufschmidt (1993).

There are many adaptive user interfaces around us, but by their very nature we are not always aware of them. For over ten years Bang & Olufsen televisions have automatically adjusted picture contrast according to ambient lighting conditions. Vehicles have had adaptive systems built-in for some time: automatic gearboxes, automatic braking systems and speed sensitive power steering are all examples of adaptive interfaces.

While we do not propose that the incorporation of adaptive interfaces into *other* products is justification for adaptive computer user interfaces, it does go some way towards rebuking the argument that people neither need nor want them.

We propose that, just as it is accepted that there is varying quality of design in non-adaptive systems (i.e. consumer goods), adaptive U.I.s also possess variable quality of design.

Gradual adaptation

We propose that by designing an interface such that it adapts *gradually* to its users, user acceptance may be increased and mal-adaptation should be reduced. The idea of gradual adaptation is not new; Mitchell and Shneiderman (1989) concluded that by giving users control over the dynamic aspects of the U.I., many of the problems that they reported may be overcome.

By gradual, we imply that the adaptation become "viscous". The degree of viscosity will be dependant upon the nature of the U.I., and in particular on the functionality that it is applied to. A frequently used function may require an increase in viscosity because the user will have built-up a strong cognitive model of the function's behaviour. A function that is seldom used may be more rapidly adapted on account that the user will be less aware of its behaviour. Furthermore, more viscosity will be required if there are several functions "fighting" for supremacy. This scenario is typical of adaptive menus, where menu selections are logged to determine an order list. We expect a number of observations to be borne out of experimental results, providing that the viscosity of the interface is correctly set:

- The number of changes (adaptations) made will be small. This we believe should minimise the number of mal-adaptations.
- Users may not be aware of changes, or at least may not be aware of the number of changes. Characterised by consistent under-estimation of the number of adaptations
- There should be little difference between the selection times for a gradually adaptive system and a similar non-adaptive system.

Experiment Design

The experiment is intended to validate whether gradually adaptive menus have any advantages (selection time, error and user preference) over instantly adaptive and non-adaptive menus. It is not intended to validate the *need* for adaptive menus.

The experiment was designed around three stages, or chapters. During each chapter, subjects would use gradual, instant or non-adaptive menu programmes.

General considerations
In order to make the trials less abstract, it was decided that the test-bed to validate the hypothesis would be based around a real product. It was felt necessary to chose a product that most subjects would have had experience with. This was to try and ensure that subjects would be familiar with the tasks being set and quickly recognise the menu names and contents. A calculator was chosen for this reason.

The time taken for each menu selection is recorded. Timing starts when the user indicates that they have read the instructions and stops once they have made a menu selection. In order to proceed to the next instruction it is not necessary for the subject to make the correct menu selection. Subjects indicate that they have read the instruction by moving the screen pointer below a line drawn on the screen (by the programme) and "click" the mouse button. This method of instruction is used to reduce the requirement for dextrous accuracy of the subject, whilst ensuring that timings should not be affected, even if the subject makes a selection from the wrong menu. After clicking below the line, subjects then make a menu selection, after which a new instruction is displayed. This process continues until the chapter is finished.

Although the programmes were not capable of performing calculations, instructions were presented in a logical manner: number, operator, number, operator... The instructions were ordered such that during the use of the gradual-adaptive programme twelve changes would be made. Subjects were asked to make selections from four menus. Only the "Operators" menu was adaptive, when appropriate.

Each adaptive menu item has an associated counter. The values of each item counter are restricted within a range determined by two variables: mincval (minimum value) and maxcval (maximum value). Each counter may have a value between these limits. Items are ordered by sorting on value. The highest counter value corresponds to the top menu position, the lowest to the bottom position. Each time an item is selected its value is incremented by the variable cinc and the values of all other counters decremented by the variable cdec. Together, the counters' maximum, minimum, decrement and increment values determine the rate of adaptation; viscosity. The adaptation algorithm was designed this way to enable changes to the adaptation rate to be made quickly and easily. If required, one or more menu items may be "weighted" such that their increment and/or decrement values are greater/smaller than other menu items. This may be useful when a menu contains items which are known to be more frequently used than others. By making the numerical difference between each counter large the duration of the learning period may be extended.

- When an item changes its position it swaps place with the item immediately above it.
- When menu items exchange positions the values of their counters are compared. If they are equal the counter values are exchange, otherwise the selected item's counter is incremented and the other item's counter is decremented.
- When a counter is decremented, a comparison is made to ensure that following this operation, it will not be equal to another counter. Should this be the case the counter is not decremented.

Experimental Procedure

Each trial was divided into three chapters: adaptive gradual, adaptive instant and non-adaptive. Each subject performed all three chapters in one sitting and the average time to complete a trial was around twenty minutes. The order of presentation of chapters was varied, as illustrated in diagram one, to combat statistical dependence.

Diagram 1. Organisation of trials.

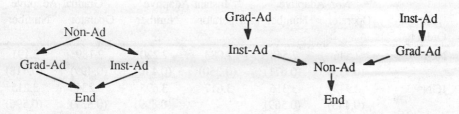

Before starting any of the chapters, each subject read a set of printed instructions; these explained the structure of the trial, explained what they would be asked to do and gave a very brief explanation of menus. The difference in behaviour between each of the programmes was explained. The instructions strongly emphasised that it was the three programmes that were being tested, not the subjects. Before each chapter was started subjects were told which programme they would be using: instant, gradual or non-adaptive and were given time to "play" with it in order to familiarise themselves with it. Once the subject was satisfied that they were confident with using the programme, the trial was begun.

During the trials, on-screen instructions presented the subjects with selection tasks to follow. After each menu selection the programme automatically, and covertly, recorded the selection task, the menu selection made, time taken and the order of the adaptive menu contents. A message informed subjects when each chapter had been completed. They were then asked to make any comments about the programme. The cycle was repeated until all three chapters had been completed.

Results

A very small initial pilot study with six subjects is described. The subjects (three male, three female) had an average age of twenty five and had a wide range of computing experience: three subjects had used a computer on very few occasions (less than four times during the past year) and three used computers regularly. Subjects were not paid to take part in the trials. Although the size of this study is extremely small, Virzi (1990) has shown that with four or five subjects eighty percent of usability problems will be detected.

A very small initial pilot study with six subjects is described. The subjects (three male, three female) had an average age of twenty five and had a wide range of computing experience: three subjects had used a computer on very few occasions (less than four times during the past year) and three used computers regularly. Subjects were not paid to take part in the trials. Although the size of this study is extremely small, Virzi (1990) has shown that with four or five subjects eighty percent of usability problems will be detected.

The mean times for each subject are shown in Table 1. All subjects show a large difference between the first and second chapters, while the difference between the second and third is substantially less marked. Only subject three demonstrates a slower mean time for the third chapter (gradual) when compared to the second (instant). All other subjects display "practising effects". Because of the wide variation in times it is difficult to draw

Table 1. Mean subject times for the three adaptation types.

Menu -> Order I	Non-Adaptive		Instant-Adaptive		Gradual-Adaptive	
	Operator	Number	Operator	Number	Operator	Number
NGI	2.546	2.540	2.083	2.220	2.189	2.191
	(0.715)	(0.671)	(0.350)	(0.440)	(0.499)	(0.418)
IGN	2.942	3.316	3.617	3.695	3.234	3.215
	(0.445)	(0.569)		(0.899)	(0.537)	(0.596)
NIG	6.561	7.015	4.178	4.178	4.332	4.412
	(3.094)	(2.515)	(1.084)	(1.540)	(1.510)	(1.284)
GIN	2.562	2.575	2.571	2.901	3.100	2.792
	(0.513)	(0.385)	(0.441)	(0.442)	(0.847)	(0.575)
NGI	3.050	3.052	2.434	2.467	2.672	2.306
	(0.792)	(1.039)	(0.388)	(0.534)	(3.380)	(0.349)
IGN	2.870	3.638	4.196	4.339	3.380	3.338
	(0.793)	(2.526)	(1.215)	(1.580)	(0.749)	(0.956)

comparisons between the subjects' mean times for each adaptation type.

Comparison of selection times for the (non-adaptive) "Numbers" menu and the "Operators" menu reveals very little difference in the results. Thus there is little difference in selection times for the adaptive menus after taking practising effects into consideration. Subjects one, four and six demonstrate marginally faster times for the instant-adaptive menus than for the fixed Numbers menu. Selection of items from gradual adaptive menus for subjects four and five was slightly slower than selection of items from the Numbers menu. It must be considered, however, that the Numbers menu contains ten items, while the Operators menu contains only five.

The number of selection errors made while using the Operators menu is very low. Three subjects made no errors at all. Subject two made one error (grad-ad). Subject three made one error (non-ad) and two errors (inst-ad). Subject six made most errors (five in total: three (non-ad), and one each inst-ad and grad-ad.

Two subjects preferred gradual-adaptive menus to instant adaptive while three preferred instant-adaptive to gradual-adaptive. One subject could not express a preference. Without any prompting, three subjects said that they were not aware of many changes occurring when using the gradual-adaptive programme. When subsequently asked about the number of instant-adaptive changes subjects responded "lots" or "all the time". Two of the subjects (one and two) that preferred the instant-adaptive menus said this was because it was easier (than with the gradual-adaptive) to predict where the items would move to. All subjects preferred non-adaptive menus overall. Subject five thought that an adaptive font menu would be of great benefit.

Conclusion

It would be wrong to draw too many deep-rooted conclusions from the preliminary study presented here. However, we now have some evidence to suggest that subjects may underestimate the number of changes made during gradual adaptation.

Although response to gradual-adaptive menus was mixed, this was based upon an adaptation with only five menu items. With more items it would be difficult for those subjects who use think-ahead strategies to predict the order of menu items. Under these circumstances gradual-adaptive systems may be preferred. Future work will:

- investigate the effects of increasing the number of dynamic items.
- investigate viscosity effects.
- incorporate the parsing of keyboard short-cuts to pull the associated menu item down the menu.

After further testing, the next task will be to build a standalone programme incorporating adaptive systems. The user will be able to switch the adaptive system(s) on or off easily and "undo" changes. System invoked changes will be pre-warned and users will be given the choice of accepting, rejecting or trying the change for a trial period. After the trial period has expired, the user will again be asked to select, reject, try, and so on. An adaptive menu may be inserted "on top" of the normal menu, or as a sub-menu leading off it. We also intend to investigate using adaptive interfaces as a perverted way of enabling users to program the U.I.; by deliberately making a mistake a user can train the interface to behave the in way s/he wants it to.

References

Burford, B.C. and Baber, C. 1994, A User-Centred Evaluation of a Simulated Adaptive Autoteller. In S.A. Robertson (ed.), *Contemporary Ergonomics 1994, Proceedings of the Ergonomics Society's 1994 Annual Conference*, (Taylor & Francis) 46-51.

Chin, J.P. 1990, Fixed vs. Dynamic User Adaptable Menus: A Comparison of Menu Structure, Search Performance and Subjective Satisfaction. In Karwowski and Rahimi (eds.) *Ergonomics of Hybrid Automated Systems*, (Elsevier, B.V.), 553-560.

Kurtenbach, G. and Buxton, W. The Limits of Expert Performance Using Hierarchic Marking Menus. In S. Ashland, K. Mullet, A. Henderson, E. Hollnagel & T. White (eds.), *InterCHI '93 - Bridges Between Worlds, proceedings of the Conference on Human Factors in Computing Systems*, (Addison-Wesley) 482-487.

Malinowski, U., Kuhme, T., Dieterich, H. and Schneider-Hufschmidt, M. 1993, Computer-Aided Adaptation of User Interfaces with Menus and Dialog Boxes. In G. Salvendy and M.J. Smith (eds.), *HCI: Software and Hardware Interfaces*, (Elsevier, Amsterdam) 122-127.

Mitchell, J. and Shneiderman, B. 1989, Dynamic vs. Static Menus: An Exploratory Comparison, SIGCHI Bulletin, **20** [4] 33-37.

Norcio, A.F. and Stanley, J. 1988, Adaptive Human-Computer Interfaces. NRL Report 9148 Naval Research Laboratory (September 30th 1988).

Virizi, R.A. 1990, Streamlining the Designing Process: Running Fewer Subjects. In *Proceedings of the Human Factors Society's 34th Annual Meeting*, 291-294.

THE PROBLEMS THE SOFTWARE DEVELOPER FACES

P.E. Waterson, C.W. Clegg and C.M. Axtell

Institute of Work Psychology,
University of Sheffield,
Sheffield, S10 2TN.

In this paper we outline some of the problems which are involved in building large scale computer programs, our arguments are partly based upon an analysis of some recent software "catastrophes", as well as two case studies. In the first case study, we describe the implementation of a set of CASE tools within a software project within a major UK bank. In the second, we describe an attempt to use a new methodology within the public sector to increase the extent of user involvement within systems development. Both cases were viewed as successful and worthwhile by those involved, nevertheless they show symptoms of the many difficulties faced software projects. We conclude the paper with a summary of the problems facing the software developer, as well as suggestions as to how some of these problems can be overcome.

Introduction

The nature of software development has changed considerably in the last ten or so years, the advent of many technological changes to the work of software engineers and programmers have resulted in a great many pressures and problems for those involved. Recent reports of software "catastrophes" have highlighted these difficulties and brought to the attention of the wider public the problems inherent in building large scale software systems. The findings from various surveys have shown that around 30-35% of all major UK projects overrun in terms of time and budget. Similarly, Griffiths and Willcocks (1994) cite evidence which shows that as many as one-fifth of all projects costing in excess of half a million pounds result in systems which are inappropriate, or unusable.

A number of explanations can be given for the difficulties involved in writing large scale software. One of the most obvious is the fact that the task itself is intrinsically difficult, large scale programs often involve tens of thousands, if not millions, of lines of code which may take many years to write, test and eventually implement. In order to cope with this complexity, software projects involve a number of parties with widely differing skills and knowledge. A further problem arises when one considers the fact that very few individuals will be in a position to hold a complete understanding, or mental model, of the program, let alone be able to comprehend the relationships and interdependencies which exist amongst its various parts. Co-ordinating and managing the activity of these individuals is therefore an important, but also problematic task.

Whilst it is fair to say that these problems have always been around it is also true that the process of building software has become more difficult in the last decade or so. These difficulties in themselves are unlikely to be attributed to one single cause, but crucially involve a number of difficult explanations which cut across a number of levels of analysis including developments in IT strategy, planning and management, as well as the increasing drive toward using new technology in the form of programming tools and methodologies. In what follows we concentrate specifically on some of the possible psychological explanations which can be given for the problems which confront software projects. Our arguments are partly speculative, the main intention being to examine software development from a number of perspectives, including cognitive and organisational approaches to the analysis of complex work domains which involve distributed problem solving and processes such as communication and knowledge sharing.

The next section considers evidence drawn from reports which have sought to explain why some prominent software catastrophes occurred. A further section describes two empirical case studies drawn from our own experience. Final sections of the paper consider factors which are shared between these examples, our overall intention being to raise a set of issues which stimulate further debate and research within this area.

Recent Software Failures and Catastrophes

One of the most widely publicised software catastrophes which has occurred in recent years is that of the Taurus system developed for the London stockmarket. Taurus was intended to replace many of the activities of the stockmarket leading eventually to "paperless trading". Currie (1994) outlines some of the major problems which led up to the failure of the project, these include the fact that the program had no overall architecture and software development was characterised by its ad-hoc and incremental nature. Taurus was also beset throughout with problems relating to poor project management, this affected development tasks such as quality control and testing of the program, as well as compounding problems such as the fact that teams working on the program were fragmented and communication links between them were minimal. In addition, the use of external consultants with limited experience in the finance area combined with senior managers underappreciation of the complexity involved in building the system lead to many problems in themselves. The Taurus program took over three years to develop before eventually being abandoned, the estimated cost of the project was around £400m.

Taurus, although costly in financial terms, did not lead to injuries or fatalities, unlike the program which was designed to cope with emergency calls to the London ambulance service, Despite the differences in terms of the application domain of the two projects, they share some similarities in terms of the symptoms which led up to eventual systems failure. In the report which describes the failure of the LAS system for the London Ambulance service, Page, Williams and Boyd, (1993) point to a number of issues which led up to the failure of the system. Some of the most prominent difficulties included the use of inexperienced contractors and inadequate project management. Despite repeated warnings the system was widely perceived as over ambitious and too complex to implement against what proved to be an impossible timetable. Added to these difficulties was the fact that the resulting software was not properly tested and the system had brought about changes to the work carried out by staff which clashed with existing practices. The LAS program was finally abandoned and staff went back to using a manually based system in late 1992.

A final example is drawn from the Field system which was designed to provide IT support to local Training and Enterprise Councils (TEC's) in the early 1990's. Bourn (1993) outlines a number of problems which eventually led to the cancellation of the project. Once again, one of the main problems related to project management - managers did not possess the relevant technical skills with which to deal with the day-to-day running of the project. Likewise, the system did not provide for adequate user involvement, training or support. These problems were compounded by the use of external consultants whose activities were not closely monitored or coordinated, despite the fact that in 1992 £11m had been spent on this part of the project. The combined effect of these problems led the Department of Employment to abandon plans to provide local TEC's with IT support.

Although the examples described in this section represent to a large extent the worst extremes of the problems which beset software projects, there is reason to believe that these problems are by no means uncommon. In the next section we describe two case studies of software development projects, both were regarded as successful in terms of the final systems which were produced, however, they share many similarities with the systems described above.

Case Studies

Case Study 1 - The Charging project

The study took place in the IT department of a large UK finance company. The Charging project (as we have called it) involved approximately 100 people from the IT and finance sections of the company. The program being developed was intended to cope with the management of information relating to corporate (ie. business) accounts. During our visits to the organisation we conducted 30 interviews with key personnel on the project including managers, analysts, programmers, as well as users seconded to the project for its duration. In particular, we chose to examine the organisation of work, the knowledge and expertise of project members, the use of programming tools, and the major problems which occurred over the course of the programs development.

One of the most striking features of the project related to the way in which work was organised according to principles of formal and emergent design. Roles and responsibilities were subject to negotiation and discussion rather than being fixed. As a result of patterns of communication which evolved over time collaboration between developers took place according to the knowledge and skills of individuals. The presence of key individuals such as senior analysts and programmers allowed knowledge to be shared and distributed across team boundaries. Smooth and successful coordination throughout the project was largely due to the activity of these "boundary spanners", particularly when technical problems occurred (eg. a programming tool was not working properly) or information was required regarding the program (eg. clarification of the function as a sub-procedure). No direct procedures governing the activity of senior project members had been formalised on the project; rather project members used a "support network" which had evolved over time and depended upon the sharing of expertise and knowledge regarding the program.

The types of knowledge which were required to build the program ranged from information concerning the data structures in the program (ie. computational knowledge), aspects of the programs intended application domain (eg. inputs to the system and work activities associated with its use), as well as more specific skills associated with project management (eg. timescales and budgets) and software engineering (eg. data modelling, use of programming tools). Collaboration within the project was very much the norm, and rather than being centred solely around the technology to be used, was determined by people and their skills. The precise form of what might be termed the "division of cognitive labour" changed according to the type of tasks to be completed and the availability of knowledge and expertise. In one important case decisions were dependent upon the knowledge and influence of one individual in particular. Within the project a senior analyst acted as an "overseer", that is, he was frequently called upon to clarify problems and educate other personnel in the details of the program. Part of the reason why this individual was seen as vital to the project related to his experience on other projects, as well as the fact that he had worked at every stage of the program's development including the writing of the original functional specification.

During the course of the program's development a decision was made to invest in a set of CASE (Computer Assisted Software Engineering) tools, these were used to cover aspects of the analysis, design and coding of the final system. Whilst initially there was much enthusiasm associated with the introduction of the tools, it was clear that they were causing serious delays to the progress of the project. Part of the problem with the tools was that they caused confusion about which parts of the program developers were working on. A major complaint was that many tasks were now being inefficiently carried out by a number of people in parallel. An unexpected consequence of the introduction of the tools was that the working patterns which had evolved within the project were upset. One of the main reasons this had occurred is reflected by the "technology-led" nature of the implementation (Symon and Clegg, 1991). Most effort was placed on re-organising the project around the demands of the tools, rather than vice-versa.

A number of other problems occurred during the development of the program. One of these relates to the involvement of external contractors who were responsible for the bulk of the final coding of the program. Because the contractors were physically remote from the project communication links between project members and contractors were severely limited. One consequence of the decision of senior managers to contract out the later stages of the program was that differences in standards and methods for coding the program caused difficulties. Within the project, a major concern was that code should be easy to maintain and enhance rather than being overly complicated. Unfortunately, large sections of the contractor's code had to be abandoned mainly because of misunderstandings regarding the design of the program, but also because it was difficult to comprehend. Likewise, the pressure of work and the need to meet deadlines meant that project members had to compromise a number of aspects of the program, one major consequence being that the testing of the program had to be limited to sections which were for immediate use rather than sections which would be used later. Despite these difficulties the program was successfully implemented, however, many project members were aware that maintaining the system was going to be a major challenge once it had been released. In other words, the program, although being on time and within budget, was very much a compromise, the danger being that this in itself might lead later on to problems which in hindsight could have been avoided. The second case study is in many ways like the Charging project in that it shows evidence of similar difficulties.

Case Study 2 - *User involvement in the Public sector*

This example is drawn from the IT section of a public sector organization, the system being developed was intended to deal with the daily administrative requirements of several thousand users in local offices. One of the most innovative aspects of the project was an attempt to increase the extent of user involvement within systems development by using a new method for user participation which had been developed within the organisation. Over the course of the study we employed a number of research methods including interviews with developers and end-users, as well as questionnaires and an observation study which was video taped.

One aspect of the participation methodology was an attempt to bring users and developers together in order to take part in the design of screens and, at a later stage, user acceptance testing. Although this part of the methodology initially received much support it ran into problems later on during the development of the system. One of the main reasons these "cooperatives" proved to be difficult to run effectively was that development deadlines took precedence over aspects of user involvement. As developers approached completion of a section of the program they found that they had little time to take part in discussions with users, as a result of these pressures this component of the methodology was dropped and only resumed once programming had been completed, or was more relaxed. Developers often felt that the cooperatives were time consuming and users varied a great deal in terms of their ability to supply requirements and/or provide criticisms of the system. In addition, because of the importance of development tasks the links between the cooperatives and the construction of the system became uncoupled, developers had to resort to including user decisions piecemeal as no time was available for user decisions to keep up with changes which had already been made to the system.

The second component of the methodology was also difficult to run effectively. User groups were primarily set up so that users could gather together and collectively evaluate parts of the system which had been just been written. One of the main problems with the methodology was that it did not scope the activities of the user groups. Preliminary analysis of video tapes of these groups shows that users engage in a variety of activities including, but not limited to, system design and usability evaluation. In other words, user roles were difficult to constrain and, as a result, the exact function of this part of the methodology became ambiguous. Handover between users and developers also became increasingly difficult as the system neared completion since time was limited and more likely to be given over to programming tasks rather than a consideration of feedback from user groups. Again the links between users and developers became out of synch. with one another, eventually leading to a situation where interaction between the two parties was increasingly limited.

Added to the difficulties of ensuring that the combined work of users and developers was running in parallel was the fact that part of the system was being constructed by internal consultants. The involvement of consultants partly caused some resentment amongst developers, largely because contact with the consultants was not encouraged, and once again, because tasks which were shared between the two groups became uncoupled. In the face of these difficulties the original methodology was thrown off course and became less effective as time went by. Although user involvement was quite extensive at the beginning of the project it later became minimal since taking into account the views of users was seen as time consuming and holding up the progress of development in what was already a very hectic and stressful environment.

Some Common Problems

In this section we consider some of the common factors which are shared across the examples described above. The list of problems which software developers have to deal with is not meant to be exhaustive, rather the intention is to consider some of the most persistent difficulties, as well as some of the most recent trends which have occurred. These include:

(a) Lack of experience/background of managers and personnel - in a number of the examples senior managers were unfamiliar, or lacked any background, in IT and software development. This is an important problem as most of the important decsions which can throw projects off course are made at these levels;

(b) Use of consultants - all of the examples have involved at one stage or another the involvement of external consultants, whilst this is not problematic in itself, it is the case that very often contracts are placed against unrealistic timescales, In addition, consultants are often brought into projects at a late stage and often disturb communication patterns and channels which have evolved within projects;

(c) Overambition and complexity - a characteristic of many systems is the increasing drive toward systems which are complex and overly ambitious. Instead of simplifying the process using techniques as BPR (Business Process Re-engineering) in order to evaluate new requirements, systems are often built (and fail) because of attempts to unnecessarily automate work activities or add extra complexity to an already difficult domain;

(d) Lack of user involvement - even where active user participation is attempted it often occurs against unrealistic objectives which are not properly thought through;

(e) Technology over people issues - typically at lest 95% of resources are invested in technical concerns, when issues relating to the availability of expertise and skills within projects for example, are perhaps more important and likely to directly affect productivity;

(f) <u>Time pressures and trade-offs</u> - as we have seen the increasing drive for projects to be "on time and within budget" means that comers are cut. In addition, it may cause "knock on" problems in that user involvement and testing become less of a prioity with resulting consequences in terms of the adequacy of final systems;

(g) <u>Inadequate testing</u> - validation of software is frequently given a low priority, particularly in the case of non-safety critical systems. The program described in case study 1 for example, although perceived as successful, was so large that adequate testing was impossible;

(h) <u>Poor management and control procedures</u> - almost all cases involve difficulties in terms of project management, even when tools such as PRINCE are used they are often not fully exploited or adhered to.

Conclusions and Recommendations

From the examples we have described above it should be obvious that even when a software project is viewed as successful it shows symptoms of the problems which are characteristic of projects which fail or are abandoned. A major concern which should be addressed relates to the fact that the problems we have described are not seen as important, mainly because the context in which software development takes place means that other goals such as economic and technological imperatives take precedence. There are, however, signs that things are changing. Recent attempts to model the process of software construction for example (eg. Curtis, Kellner and Over, 1992), are addressing issues of personnel allocation and the importance of expertise, as well as the influence of organisational constraints. In addition, our examples have underlined the importance of the training of individuals across differing roles and responsiblities (ie. from managers to programmers), this should take the form of education in strategy and planning, as well as supporting activites within projects such as collaboration and the appropriate use of planning and programming tools. All too often attempts to automate aspects of development using the latest technology are poorly thought out and result in confusion and disruption within projects. Finally, we recommend that more studies of the kind we have described are carried out, partly to counter the bias toward laboratory studies of programmers which dominate the literature, but also to validate and build upon the problems we have set out above.

References

Bourn, J. 1993, Computer systems for training and enterprise councils: The department of employment's management of the field system. *National Audit Office,* (London: HMSO).

Currie, W. 1994, The strategic management of a large scale IT project in the financial services sector. *New Technology, Work and Employment,* 9, 1, 19-29.

Curtis, B., Kellner, M.I. and Over, J. 1992. Process modelling. *Communications of the ACM,* 35 (9), 75-90.

Griffiths, C. and Willcocks, L. 1994. Are major information technology projects worth the risk? (*Oxford Institute of Information Management, Templeton College, Oxford*).

Page, D., Williams, P. and Boyd, D. 1993, Report of the inquiry into the London Ambulance Service, (*South West Thames Regional Health Authority*), February 1993.

Symon, G. and Clegg, C.W. 1991, Technology-led change: A study of the implementation of CADCAM. *Journal of Occupational Psychology,* 64, 273-90.

EFFECTS OF TECHNOLOGY ON THE WAY PEOPLE WORK

K. Brownsey and M. Zajicek

*School of Computing and Mathematical Sciences,
Oxford Brookes University, Oxford OX3 OBP, UK,
Tel: 0865 483683, Fax: 0865 483666,
Email: MZAJICEK@uk.ac.brookes*

This paper addresses the issue of the effect of technology on problem solving. It covers the historical effect of general technology on the way people work, problem solving and organisational structures and then proceeds to look more specifically at the effect of computer technology on certain forms of problem solving. We examine in particular the way in which the process of problem solving can become more concurrent when information relevant to the problem is made readily accessible. We also discuss ways in which possible solutions to problems can be explored more creatively when easy backtracking is made available.

Introduction

This paper seeks to identify ways in which the use of technological tools has affected the process of using facts to solve problems.

We first look at ways in which low level technology such as the use of diagrams has enabled the process of problem solving, and then explore the affects of computer technology on organisational structure, and scientific discovery.

The paper then reports the results of experiments that were designed to explore the way in which the provision of readily available, computer based information can affect the strategies people use for solving complex problems. We use as our experimental platform a resource allocation system in which the problem is to build a satisfactory modular degree program out of a set of modules covering different topics. This has to be done within the constraints represented by the rules of the modular degree.

We look particularly at the way in which the process of problem solving can become more concurrent when information relevant to the problem is made more accessible. This is discussed in the context of memory retention and ordering of facts We also discuss ways in which possible solutions to problems can be explored more creatively when easy backtracking is made available.

Historical effects of technology on problem solving

There are several senses in which technology affects the way people work in large organisations, in areas of scientific discovery, and at the level of human problem solving.

Low level or narrowly utilised technology can impede the full use of human potential in problem solving as shown by Mey (1992) and Gorayska and Cox (1994). For example the limited use of early command line interfaces obliged users to adapt their task to fit in with the technology available.

Conversely the appropriate development and use of technology particularly in the field of human-computer interaction can substantially enhance the functionality and efficiency of task execution.

Diagrams as a form of technology

Here we cite the use of diagrams as a form of technology, in the same way as some of today's most useful software is technology. Certainly the latter rests upon technology of a more 'traditional' type - printed circuits and so on - but we are concerned with technology for reasoning. In this sense an important strand of the new generation of software, utilising all the technical and theoretical advances made in the field of human computer interaction, is the development of diagram technology.

There are clear examples of early diagram technology and its importance for problem solving in many areas of human endeavour - maps, plans for buildings and machines etc. - followed by the use of diagrams for more dynamic or abstract problems such the critical path analysis diagrams in project planning. In the field of logic there is a tradition of the use of diagrams from Aristotle, Gardner (1982) onwards. Unfortunately such use seems to be looked down upon by some of the more 'formal' logicians. For example, the American pragmatist C. S. Peirce, Peirce and Ketner (1993), developed a diagrammatic method for predicate logic, which has been subsequently shown to be correct and complete. Yet the idea that a diagrammatic approach to logic could be valid was stoutly resisted by Frege.

Using structures as deductions, as Peirce advocated, seems to be worth investigation, given all the potential for diagrammatic work we now have through new hardware and software, supporting graphical interfaces. In the field of human computer interaction it opens up possibilities of interactive reasoning which could transcend practical and theoretical limitations of fully automated reasoning.

Effects of computer technology

We have identified three types of effect.

The first effect is the way in which technology has transformed the nature of an organisations activity, such that the day to day activity of its workforce is greatly changed from that of a generation before. This effect is the most familiar, impinging on our everyday activity through banks, travel agencies, shops etc.

It is an interesting point to consider quite how much more is being achieved. Joseph Weizenbaum commented that it was not the case that the banks would not be able to do what they were doing without computers, they would just be doing it in a different way, Weizenbaum (1976). However we are concerned here with the effects on how people work and not with the achievement.

The second effect is the way in which powerful computational facilities have enabled scientists to spot connections and theorems which could not have been anticipated using solely the power of the human brain. For example, a great stimulus to the development of chaos theory was the availability of cheap and rapid computational power, Gleick (1993).

This came first in the form of hand held electronic calculators, allowing many more calculations per working hour than paper and pencil or slide rule. Subsequently cheap personal computers allowed the use of tools such as spreadsheets and third generation programming languages, with graphics capability. These were used to give global views and pictures of the phenomena being studied. Spreadsheets are fairly easy to use, for most literate and numerate people.

Most people in the wide scientific community seem to be capable of learning and using third generation programming languages, given the motivation. As computer professionals we have to underline the point that this does not make them software engineers but certainly capable of using these tools in a rough and ready experimental fashion to 'see what happens'. The explanation of the subject matter, for educational purposes, is difficult to imagine without the use of simulation and illustration by computer, Becker and Dörfler (1989).

The third effect is the way in which the provision of 'information' to hand and readily accessible enables people to solve problems in a more concurrent and interleaving way than previously. We ascertain that the capacity of the human brain to hold information dictates the use of a sequential type of problem solving without a computer. This discussion is the subject of the main part of the paper.

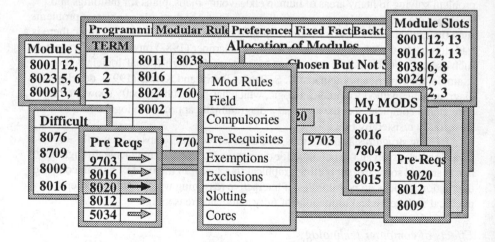

Figure 1. Typical Interface for modular Building

Concurrent versus sequential approaches

The problem solving application

The particular application providing problem solving opportunities was that of the Modular Degree Program Builder for Oxford Brookes University. Using this computerised system, students are required to select suitable modules, or units of study, out of approximately 2000 modules.

The allocation of modules is performed under a wide range of constraints such as the necessity for prerequisite modules, compulsory modules for particular fields. The problem to be solved is build the most appropriate degree program subject to the constraints operating. There are potentially several solutions which have varying degrees of suitability.

A typical interface for this form of problem solving is shown in Figure 1. The state of module allocation so far is presented centrally with information to support the allocation ranged in windows around it.

Sub tasks and constraint checks associated with module allocation

The main task of building a suitable degree program can be broken down hierarchically into sub tasks. It is impossible given the state of human cognitive processing to 'see' the allocation in one step. Program builders must break the main problem or task into manageable recognisable sub tasks. We carried out an experiment with twenty students from a variety of disciplines in the Sciences and Humanities. The first task for our module program builder was to hold a group discussion in which they were asked to identify the main tasks involved in program building. These were found to be:

A - Allocate all compulsory modules for the field you are following
B - Start with modules you would enjoy most and then put in their prerequisite modules.
C - Put in sufficient acceptable modules to make up the number of modules required with a view to changing individual modules later.
D - Allocate modules to particular years to balance the learning effort.
E - Swap chosen modules for as yet unchosen modules to improve termly balance.

Our experimental program builders were also asked to list the most common constraint checks they were likely to perform in order of complexity. These were ordered as follows, most likely first.

1 - Check that module prerequisites are selected.
2 - Check slotting clashes do not exist.
3 - Check all compulsory modules for a field are selected.
4 - Check for more complicated constraints such as 'you must take two out of three modules for a particular field'.
5 - Check termly balance of modules.

In addition it may be that the program builder finds after several allocations that they have built a program that will not work because certain constraints have not been satisfied. It is then necessary to deallocate back to a point where the constraint problems have been removed and allocation can proceed again. This is called backtracking.

We then proceeded to look at the ways the subject builders used sub tasks, constraint checks, and backtracking to build the optimum degree program.

Structure of experiments

The aim of experimentation was to analyse how the two groups of subjects, those using a paper based system, and those using a computerised system, organised their sub tasks and constraint checks.

We wanted to look at the difference in the ways in which program builders ordered and interleaved their tasks. Those using the paper based system had to look up information in the modular program book and write intermediate decision information on scraps of paper. Those using the computerised system had supporting information on constraint violation, and available modules etc. This was readily available in easily accessible windows as shown in Figure 1.

Ten subjects used the paper based system and ten used the computerised system to build degree programs. We asked them to identify by the codes above when they were engaged in any of the sub tasks or constraint checks listed. They were also asked to insert U for 'undo' when they found themselves backtracking to a previous position when the allocation of modules had become untenable.

Therefore a typical module program building session might look like

AAA1A1UA3A3UCC1.....

Equipped with this empirical evidence of what program builders were doing we were able to work out coefficients to establish how much interleaving of tasks, and backtracking was happening and if there appeared to be a significant difference between the computer based and paper based builders.

We used the following formulae

Task change coefficient = (no of task changes) / (no of task actions - 1)

Backtracking coefficient = (no of backtracking events) / (no of constraint checks) where a backtrack consists of one or more undos in a row.

We worked out the two coefficients for each subject program builder and then averaged all task change coefficients for each group and all backtracking coefficients for each group. In this way the results were not affected by the way in which subjects chose to solve the problem or how many different steps were needed for each subject.

The table below shows the two average coefficient scores for both groups

	Paper based	Computer Based
Task change coefficient	.25	.83
Backtracking coefficient	.43	.79

Performing a statistical test on the results enabled researchers to claim that there was a significant difference at the 5% level between the two groups in their use of task interleaving and concurrent tasking.

We can also say that there was more backtracking activity among computer based builders than paper based. What is not shown by the figures is the depth of backtracking. computer based builders backtracked more often but to less depth. In all they made fewer backtracking steps, from which we can conclude that they were making better decisions.

Qualitative results

In addition to recording instances of sub tasks, constraint checks and undos, subject program builders were also asked to talk through their decisions and reasons as they built a program, and to evaluate the program they had built. They also answered questions about their own perception of their problem solving. These exercises provided researchers will some qualitative data which we summarise below.

We found that in a paper based system, with access to 'background information' limited by human memory and concentration , the builder is forced into a sequential mode of planning for problem solving. Subject builders using the paper based system frequently stayed with the same sub task until it was completed and used only constraint checks 1 and 2 which left them vulnerable to constraint violations checked by 3, 4, and 5. They often found themselves involved in long and complex backtracking processes.

The ability to backtrack effectively using a computer was found to provide a significant advantage. It was difficult to quantify this but researchers felt that computer based builders who had all details of constraint violation to hand could backtrack earlier and therefore more effectively.

With a computer system for resource allocation, one can maintain for some time a global view of an inconsistent solution but with the possibility of a consistent solution emerging. In this way inconsistencies can be viewed globally, alongside each other, or the details of individual inconsistencies readily accessed. This is also discussed by Liu, Fuld, and Wickens (1993).

Conclusions

By controlling the number of sub tasks attached to the resource allocation task, we have been able to provide a form of quantitative measure of the amount of concurrency and interleaving of tasks which is made available by the use of a computer system.

We have also been able to show that the task of resource allocation has been approached in a substantially different way by those using a computer.

Researchers at Oxford Brookes University intend to explore further the use of diagram technology and its application to graphical interfaces.

References

Becker, K. and Dörfler, M. 1989, *Dynamical Systems and Fractals,* (Cambridge University Press) 19-22

Gardner, M. 1983, *Logic Machines and Diagrams*, 2nd edn (Harvester Press, London) 28 -30

Gleick, J. 1993, *Chaos*, (Abacus Press, London) 56-59

Gorayska, B. and Cox, K. 1994, *Hidden Meaning in Computer Applications, Proceedings IFIP/GUI'94*

Liu, R., Fuld, R. and Wickens, C. 1993, *Monitoring behaviour in manual automated scheduling systems*, International Journal of Man-Machine Studies **39**, 1015-1029

Mey, J. 1992, *Adaptability,* AI & Society **6**, 186-191.

Peirce, C. S. and Ketner, K. L. (ed.), *Reasoning and the Logic of Things*, (Harvard University Press) 68-72

Weizenbaum, J. 1976, *Computer Power and Human Reason*, (W. H. Freeman, New York) 34 -35

EMPHASIZING USABILITY IN THE DESIGN OF MULTIMEDIA SYSTEMS

Hilary Johnson

Department of Computer Science
Queen Mary and Westfield College
University of London.
Mile End Road
London
E1 4NS

Multimodal/Multimedia (M4) systems are very much in vogue at the moment due mainly to hardware achievements which are presently outstripping theoretical and empirically led research on how to use the different media. There is an underlying assumption that more media is better. Moreover, the allocation and combination of media used has wide ranging implications for the ultimate usability of the system. This paper gives a brief introduction to the research issues related to the design of M4 systems. The assumed benefits to the users, of interacting with multimedia systems are outlined, followed by a consideration of current research in allocating and combining media. The paper concludes with a brief research program related to improving the usability of M4 systems.

Introduction

Multimedia/multimodal (M4) systems are very much in vogue at the moment with a burgeoning number of journals, books, conferences and workshops devoted to research related to providing M4 systems. M4 systems present media in various different formats, from text, graphics, voice, video and so on, to be interpreted by our different senses. Bernsen in a recent paper (1994), concerned with the relationship between modality theory and supporting multimodal interface design, provides a taxonomy of 28 generic unimodal modalities thus highlighting the range of modalities on offer. M4 systems relate to both media and modality. Despite having different meanings, the two terms "multimedia" and "multimodal" are often used interchangeably. Crudely speaking, from a user perspective the term multimodal is often assumed to be related to the sensory modality used for interpreting the information, for example, the auditory modality will be used for interpreting auditory output, for example, voice output. Multimedia, on the other hand is often taken to relate to the different media that might be used to represent information, for example, graphics might represent pictorial information that is interpreted by the visual sensory modality.

From a system perspective, multimodality can be characterized by the capacity of the system to interpret raw inputs to a high level of abstraction, for example, that of the task domain, or to render information starting from high level internal representations. Although multimedia includes interpretation and rendering capabilities, these processes function at low levels of abstraction. Thus, as from the users' perspective, the distinction between modality and media is also significant from the technology viewpoint. Given these definitions it can be argued that all interactive computer systems are multimedia and multimodal and that none are unimodal/media.

With respect to recent research, there have been significant advances in technology, including networking, which make it possible to provide and support a variety of different media at the human computer interface. However, the minus side is that hardware achievements are presently outstripping theoretical and empirically led research on how to apply and use the different media at the interface, and even research into establishing if there are benefits to users from using multiple media. Hollnagel (1991) has argued that it is often the case in HCI that technological advances are ahead of corresponding theory development. Although in most cases it certainly is preferable for theory to lead the design of better interactive systems, there could also be a bottom up case where technology leads to advances in theory. This is possible because technology can in some cases provide an environment where the *motivation* for advances in theory is overwhelming.

It is argued that such is the case in relation to M4 systems, and this is because there is an implicit assumption that more media is better for the user. The problem is that inappropriate use of media may do nothing to enhance HCI. There is a distinct possibility that M4 systems, rather than benefit, actually hinder the user due to indiscriminate use and application of media and media combinations. Given the costs of constructing M4 systems there is a need for theories to guide the investigation, testing and evaluation of the "more media is better" assumption. Moreover, it is important to conduct theory led research into the combination and allocation of different media with usability of the system at the forefront.

Fundamental questions that need to be addressed are: what is the appropriate media for representing information and how can this representation be optimized; what is the best media combination for representing information; and, how can M4 systems be evaluated? In addition, the allocation and combination of media must take into account the context in which the task(s) are to be performed. Therefore, the problem is not just one of establishing optimum solutions, although this is complex enough, but for those solutions to be tailored to the tasks that the users must accomplish using the system, i.e. the tasks that the system has been designed to support.

The next section is concerned with the potential benefits of using multiple media/mode systems.

Benefits of M4 systems.

Over the years many benefits of using M4 systems have been claimed. The claims listed below are the result of a detailed discussion at an AAAI Symposium on Intelligent Multimedia-Multimodal Systems; Johnson et al, 1994). The claims have not been systematically investigated and many are implicit. Below each claim an attempt has been made by the author of the paper to discuss the basis of the claims. An underlying common theme is that because we use and interpret multiple media in our everyday lives that this leads to a corresponding need for computers to mimic that rich sensory experience and that such systems benefit the user. Each of the following claims should provide the basis for research hypotheses which could be tested in relation to the benefits of M4 systems. They should also provide a view of what might constitute a usable M4 system. Additionally, it is important to note the interdependent nature of the claims.

*M4 systems increase user comprehension.

The assumption here is that providing information in more than one media will enable people to perceive and interpret more information. There are two senses in which this statement can be understood; first, that more information is provided using different media to provide and therefore distinguish that information, so the total information displayed is greater. Secondly, no more information is provided by the use of additional media, but more of the information provided is understood, for instance, if there is confusion and misunderstanding as a result of perception of information in one media, then this can be corrected or otherwise supplemented by providing the same material in another media.

However, there is a debate about the benefits of using multiple media on user comprehension with Willows and Houghton (1991) arguing that multiple media is no better than unimedia and Peeck (1991) arguing that there are benefits. The problem is that it is hard to judge the validity of either argument without knowing minute details about the nature of user tasks, quality of the user interface and the allocation and combination of media in the different research reported.

The claim about benefits to user comprehension relates very much to the research issue concerned with redundancy, i.e. is it a good thing to replicate the same information in more than one media, and does the notion of redundancy extend to information that is in some way similar?

Measures of usability related to this claim would include testing user comprehension either by questionnaire, conceptual maps, etc. and also independent judgements of task output quality after using the M4 system.

*M4 systems result in enhanced human computer interaction.

This assumption relates to the argument that when we interact with computers our input and the corresponding computer output is considerably restricted in comparison with interaction in an everyday human-human sense. There will always be differences in how we both interact with and through computers since we have different expectations of the computer and because the computer has limited capacity for understanding for example, natural language. However, the argument states that we can considerably enhance our interaction with the computer if we are able to use our senses in a fuller and more orchestrated manner.

Measures of usability here might relate to users' subjective impressions of using the different media. Additionally, it would be expected that there would be both comprehension and user performance benefits which were over and above those expected with systems employing less media.

*M4 systems result in improvements in speed, accuracy and naturalness.

This assumption is hard to substantiate, at the moment there is not enough evidence to say that M4 systems are better in terms of speed and accuracy than other interactive systems which employ fewer media. In fact, given the difficult task of coordinating and synchronising the presentation of a variety of media, the converse seems more likely. The claim with respect to naturalness appears to relate to employment of our senses in a fuller and more orchestrated manner. Measures of usability would relate to both users' acceptance of the system, but also tests of the speed and accuracy of the system in comparison to other systems employing less media.

*M4 systems provide greater flexibility and expressive power.

There are two aspects to this claim. First, the basis for the claim to greater flexibility is that there might be aspects of the context in which roles are played and tasks conducted, which necessitate the use of different media tailored to specific aspects of the work context. For instance, Oviatt (1994) in researching the use of new pen/voice systems for use on future portable devices found that in certain circumstances one media/mode was used over another media/mode. For example, "voice" was not used in situations where the material to be discussed was confidential and must not be overheard or where there was a risk of disturbing others. Therefore,

M4 systems give the users choice and flexibility in how they use computers which could enhance and change for the better the way in which they work.

Second, there are two aspects to the claims about expressive power. One relates to the gestalt assumption that the whole is more than the sum of the parts, that using a combination of media gives something more than use of uni- or less media. The other aspect is concerned with the argument that some concepts or attributes of concepts are better represented by some media than others, see for example, Larkin and Simon (1987). Therefore, greater expressive power means representing that media at the interface.

Measures of usability here might relate to changes for the better in work practices, subjective judgements by users and greater comprehension of users as a result of optimizing expression of information in a representation.

*M4 systems lead to reductions in tiredness, heightening of awareness and directing of attention.

The basis for this three-pronged claim is that use of different senses will avoid users becoming tired and that changes in the nature of stimuli (media of representing that stimuli) will heighten awareness and direct attention. An example of heightening awareness and directing attention might be where the system uses a particular media to highlight or filter information for the user. A warning sound, might be one currently employed, crude example of different use of media to direct the users' attention. All these related arguments rely on research in psychology, for instance, perceptual theories of invariance and also theories of selective attention. With respect to usability measures we would expect if such claims had any foundation that there would be more information that was understood, perceived and attended to.

In this section, claims about the benefits of using M4 systems have been outlined along with possible bases for such claims. Since this paper is primarily concerned with emphasizing the usability of M4 systems there has also been brief discussion as to the measures that might be taken to establish whether these claims are realistic in improving the usability of interactive systems.

Allocation and combination of media in M4 systems

The previous two sections have raised a number of important issues to be considered in emphasizing the usability of M4 systems, such as the benefits to users of using multimedia systems, suggestions as to measuring usability and also the redundancy of information provided. A further problem with the design of M4 systems relates to the topic of this section, allocation and combination of media, which has fundamental consequences for the ultimate usability of the system.

It is not possible in this short paper to review all of even the limited research related to the allocation and combination of media presently available. Four papers only have been selected to give a representative flavour of different aspects of this research. One is concerned with providing a research agenda related to the allocation and combination of media in M4 systems, another describes an architecture for a M4 explanation system and the final two are interesting in that they attempt to provide rules and guidelines for the problems of allocating and combining media in designing M4 systems.

In the first paper, Alty (1991) rather than providing rules for the design of M4 systems, reviews research from a number of disciplines relating to the allocation and combination of different media. The paper concludes with a list of questions which comprise a research agenda for the design of M4 systems. These include: How do humans use multimedia approaches in everyday life? When do media in combination improve bandwidth? What are the strengths and weaknesses of particular media in combination? What is media interference, when does it occur and why?

Feiner and McKeown (1992) in the second paper, present an architecture they have developed for an explanation system that generates explanations for equipment repair and maintenance. One of the components of the architecture is a media coordinator which is responsible for decisions about media allocation and combination. They argue that informal experiments have been carried out which distinguish various types of information and categorizes them according to the most appropriate presentation medium, for instance text or graphics. Unfortunately, the authors repeated requests for the results and design of the informal experiments have produced nothing.

The last two papers to be briefly mentioned are concerned with rules and guidelines. Hovy and Arens (1990) provide an initial set of guidelines to be followed in designing M4 systems. They argue that the basic idea is for multimedia communication systems to assign a particular modality to each data type to be presented. Their approach is based on a two-step generalisation of the basic scheme. The first generalisation is to assign the modality not to each data type, but instead to each feature that characterizes data types. The second generalisation is to assign characteristics of data not to modalities but instead to characteristics of modalities. Examples of the two step generalisation rules are:
*Data duples (e.g. locations) are presented on planar modalities (such as graphs, tables and maps).
*Data with specific denotations are presented in modalities which can convey the same denotations.
*If more than one modality can be used, and there is an existing presentation, prefer the modalities already present as exhibits in the presentation.
*If more than one modality can be used, and there is additional information to be presented as well, prefer modality that can accommodate the other information as well.

There is however, no mention in the paper about any empirical validation of the guidelines and no discussion about the subsequent effects on usability of the system. In addition, it remains to be seen how easy M4 designers find the rules to follow.

Andre and Rist provide the other set of rules for media allocation and mode preferences to be followed in designing M4 systems:
* Use graphics over text for concrete information (shape, colour, texture and also events and actions if there are visually perceptable changes).
* With respect to spatial information (location, orientation, composition and also physical actions and events) for speed use graphics, for accuracy use text.
* For temporal relations between states, events and actions, encode information about sequence(s) by using spatial or temporal layout, use text to express temporal overlap, temporal quantification (e.g. "always") and temporal shifts (e.g. three days later).
* Present semantic relations (e.g. cause/effect, action/result, problem/solution, condition, concession) where there is ambiguity in picture sequences of temporal sequences vs cause/effect, problem in text, condition and concession, graphically with verbal comments.
* Use text for quantification, especially most, versus some, versus any, versus exactly-n.
* Express negation graphically (e.g. overlaid crossing bars) unless scope is ambiguous then text is to be used.

As with the Hovy and Arens paper there is no empirical validation of the rules, no direct emphasis on improving or testing usability and no guidelines for application of the rules. However, since Andre and Rist have constructed a system which follows these rules, they have provided an environment in which theoretically led experiments into system usability can take place.

Conclusion

This paper has very briefly outlined some of the research into the design of M4 systems. Two questions are considered fundamental to the ultimate usability of M4 systems; the redundancy of the information provided by the systems, and the allocation and combination of media to support users in conducting the tasks associated with their work roles. Our aims for future work in this area are to undertake a research programme which will consider in detail each of the following:
i) identification of a range of representative work tasks to be analysed;
ii) identification of the task knowledge that people possess in order to successfully undertake those tasks;
iii) map the knowledge components of those tasks in the media in which they are represented in the external world to appropriate media for computer representation. This will involve conducting experiments, applying and testing the validity of the rules given in the previous section, and with full consideration of the literature. In addition, a theory of task knowledge and behaviour should indicate where errors and misconceptions often occur in understanding system output during tasks and these occasions should have implications for the use of media redundant information.
iv) Iteratively evaluate both the use of media and the computer support for the users' tasks.
A series of experiments are currently under way. The intention is that ultimately a contribution can be made to improving the usability of M4 systems.

References.

Alty, J. 1991, *Multimedia - What is it and how do we exploit it?* Keynote address, HCI91 Conference, Edinburgh. LUTCHI Report No. 0110.

Andre, E. & T. Rist. 1993, *The design of illustrated documents as a planning task.* In M.Maybury (ed) Intelligent Multimedia Interfaces, AAAI Press Menlo Park, 94-116.

Bernsen, N.O. 1994, *Modality theory in support of multimodal interface design.* In P. Johnson et al. Proceedings of AAAI Symposium on Intelligent Multimedia Multimodal Systems, 37-44.

Feiner, S. & K.McKeown. 1990, *Coordinating text and graphics in explanation generation.* Intelligent Interfaces, 442-449.

Hollnagel, E. 1991, *The influence of artificial intelligence on human-computer interaction: Much ado about nothing.* In J. Rasmussen, H.B. Andersen and N.O. Bernsen (eds) Human computer interaction: Research directions in cognitive science. Volume 3. London, LEA Press.

Hovy, E. & Y.Arens, 1990, *When is a picture worth a thousand words? - Allocation of modalities in multimedia communication.* In Proceedings of AAAI Symposium on Interfaces, Stanford, March 1990.

Johnson, P., Feiner, S., Marks, J., Maybury, M. & J. Moore. 1994, *Proceedings of AAAI Symposium on Intelligent Multimedia Multimodal Systems.* Stanford, March 1994.

Larkin, J.H. & H. A. Simon, 1987, *Why a diagram is (sometimes) worth ten thousand words.* Cognitive Science, 11, 65-99.

Oviatt, S. 1994, *Toward empirically-based design of multimodal dialogue systems.* In P. Johnson et al. Proceedings of AAAI Symposium on Intelligent Multimedia Multimodal Systems, 30-36.

Peeck, J. 1987, *The role of illustrations in processing and remembering illustrated text.* In Willows, D.M. & H.A.Houghton, (eds)The Psychology of Illustration. Springer-Verlag.

Willows, D.M. & H.A.Houghton, 1987, *The Psychology of Illustration.* Springer-Verlag.

Iterative Development of a Free-Text Retrieval Tool

N. I. Beagley & R. A. Haslam

Human Sciences Department,
Loughborough University,
Loughborough,
Leics.,
LE11 3TU

Improved scanning and data storage methods have opened the door to document archiving. The value of this method of storage can only be realised when archived material has been effectively retrieved. This paper describes the evaluation of a computer based tool designed to provide a small user group with desktop access to internal reference documents. The tool's interface presents a condensed list of all words contained within the document. Evaluation of the system as part of the iterative development cycle was based on a direct comparison between computer-based and paper-based searching. The results of the evaluation illustrated the separate search strategies adopted for the different media. Whilst overall search performance was similar between the two methods, use of an optimised computer search strategy offers equivalent performance for unfamiliar documents.

Introduction

This paper reports the development and evaluation of a text retrieval module of the Database Application in Vehicle Ergonomics (DAVE). The stated aim of the DAVE system is "To assist the work of a group of experts in the field of vehicle ergonomics" (Beagley et. al. 1992). A modular approach has been taken providing access to discrete sections of data through specifically designed access programs. Previous modules developed for the DAVE system include a module to access the group's Anthropometric database (Beagley et. al. 1993) and a module to formalise the acquisition and display of vision assessment data. The aim of the text retrieval module was to provide on-line access to text resources relevant to the work of the group.

The application of computers to text retrieval has been in many cases based on the card catalogue approach in which information is condensed to accommodate restricted storage space. In these systems a requirement is placed on a person to draw relevant text and keywords from a document to form the basis of the index. This is a large workload overhead that raises the effective cost of maintaining an index. The increased power of personal computers has opened up text retrieval to a lower entry level market (Holloway 1991) There have been significant improvements in the reliability and cost of optical character recognition techniques (Andon 1990). Advances in storage technology have reduced the cost of storing large volumes of data. This opens up an alternative strategy to traditional paper based filing for document archiving. There has been a significant growth in computer based library systems that have access to the entire text of a document (Dunlop & Van Rijsbergen 1993). A paper document translated to recognised text permits searches based on the entire content of the text. Rather than requiring an individual to undertake the task of distilling keywords from the document, the indexing task can be passed to the computer. Computer systems are now able to effectively manage the complete text of documents. This allows documents to be added to an archive without the task overhead of indexing. The result however, is a very different approach to searching the document archive. Removing the restriction of searches based on selected keywords, the user is able to search for a word or combinations of words throughout the entire document archive. Dumas (1988) compared searches using controlled language index terms with searches on all terms contained in the full text of document abstracts. The results support a move towards the full text approach to document archiving. Glavash (1994) found that on-line access of the full text of reference documents was preferable to traditional hardcopy retrieval. However, the value of rapid access to the text had to be balanced against the client's requirements for graphics.

The approach taken for the DAVE text retrieval module was to provide the user with the means of accessing the complete text of the chosen reference documents. By producing a complete list of words contained within the retrieved document, the module provides the user with an unrestricted choice of keywords when searching the document. Processing limitations of the desktop computer systems employed are overcome by pre-indexing of the stored documents. When a document is opened, an alphabetical list of all words contained in the document is presented (Figure 1a). Words in this list may be found by visual scanning or by entering a text search string. The word of interest is selected from the list by clicking on it. The computer then presents the user with a list of occurrences of the selected word. The words are presented in context. This context list may be narrowed by boolean filtering for words surrounding the keyword. Where the list contains a small number of occurrences, visual sorting may be more appropriate. Once a suitable word occurrence is found it can be selected by clicking on it. This changes the display from the index screen to the document screen in which the selected occurrence is highlighted as it appears in the reference document (Figure 1b).

The goal of replacing the paper document with computer access was to improve the speed and accuracy of information retrieval. It was initially considered sufficient to simply provide the reference document in the form of indexed text. Early evaluation of the initial text-only prototype highlighted the importance of the graphical figures contained within the chosen reference documents. Consequently the prototype was evolved to include access to photographic representations of all pages contained within the document. Once in the document screen, the user may choose to view a photographic representation of the selected page. The system makes no attempt to index the graphical elements of the document other than to include the textual descriptions accompanying the figures in the recognised text database. The recognised text is arranged in a page format allowing the user to flip through the document page-wise. The module indexes, stores and retrieves text across a network allowing the central maintenance of the document and image store. This also has the consequence of reducing the storage space required on each machine for operating the software.

The approach taken for the development of the DAVE modules has been user centred, iterative prototyping. This approach has evolved to match the characteristics of the development environment (Beagley, et. al. 1993). The development is in-house, facilitated by good lines of communication between developer and the user population. This close association has been used as the basis for the development approach. Luqi (1989) observed that users require some experience of a software product before they are able to produce robust specifications. Graphically based representations of specifications have been proposed as a suitable medium for communicating requirements (Spence 1991). Rapid prototyping goes a stage further by adding functionality to the proposal

model. A functional prototype provides a platform for testing program feasibility
and assisting the users to determine the system specification.

The software development has been carried out using SuperCard, a high level,
object oriented, programming environment. SuperCard's flexibility supports the
evolution of the software prototype towards the implemented system.

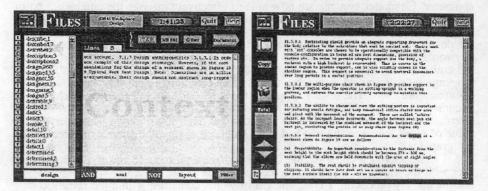

Figure 1a Index Screen   Figure 1b Document Screen

Aim

The aim of the evaluation was to provide guidance for the further iteration of
the electronic file module. The task set for the user group, investigates the ability of
the Electronic file module and the paper source to provide the user group with access
to vehicle specific reference data. The capability of the module to replace the
requirement for the paper source is investigated by a direct comparison of the two
media.

Procedure

Organisational constraints limited the time available to the users for the
purpose of assessment. It was necessary to minimise the group's time cost for
completion of the evaluation task. The target user group consisted of six individuals,
each of whom took part in the evaluation. The stated development requirement of
tailoring the system to the specific needs of a small specified group dictated the
participation of all members of the group in the evaluation task. The task was
constructed to allow all stages to be completed within one hour. This estimation was
based on the varied experience of the user group in both in terms of computer
utilisation and use of the reference document in its paper form.

The subject was seated at a desk in a quiet office. Positioned on the desk was
the computer displaying the DAVE entry screen and a complete copy of the
Reference Document. The experimenter demonstrated the completion of two
example questions. The examples were demonstrated using both the Electronic File
module and the paper based Reference document. The subject was then asked to
complete the six questions listed on their task sheet, in order of presentation. Each
user answered three questions with the aid of the paper document. The other three
questions were answered using the Electronic File module. The questions required the
subject to write a single line answer followed by the reference page on which they
located the answer. The subjects were asked to complete the questions at their own
pace with particular emphasis on accuracy as this most closely matched the normal
requirements for reference searches within the organisation. The subjects were
required to complete the questions in the order of presentation. This order was
balanced to overcome the potential learning effect. If assistance was required, the

subject was directed towards DAVE's built in help system. Each subject completed the task within office hours.

Measures taken included error rate & completion time. An action log based on observation was compiled by the observer. Valid task actions were predefined in a pilot test. Following completion of the task, each subject was interviewed. In each semi-formal interview the user was asked to highlight adverse and favourable features of both functionality and interface with reference to their particular working requirements.

A telephone enquiry log generated by the section was used to identify six categories in which the section has previously received enquiries. The questions were generated by an individual experienced in vehicle ergonomics from outside of the designated user group. The questions were considered to be of equal difficulty by the questioner.

Results

Completion time was measured from the opening of the document to the writing of the answer for the paper based task and from the first key/mouse press to the writing of the answer for the computer based task. Errors were recorded for incorrect answers. Three errors were recorded for answers based on a paper search. Two of the errors related to insufficient detail in the answer. The third related to a complete failure to find the page reference. Two errors were recorded for questions completed by computer searching. Both computer based errors related to insufficient detail in the answer.

Table 1 Completion time (seconds) & Task Actions

	Average time	Min. time	Max. time	Average actions	Min. actions	Max. actions
Paper	219.39	77	702	12.22	5	32
Computer	225.89	52	507	12.39	5	33

User stated problems

* Would like to see more text in the context window.
* Trouble recognising whether the document was in text or photo mode.
* Confused by the movement between the index and document screens.
* Would like to be able to flick between photos quickly.
* Would like the text to include cross-references to related texts.

Observed problems

* Use of Index scrolling in place of faster index search function.
* Occasional visual searching in place of faster and/not filtering.
* Filtering of complete words rather than a wider, partial word, filter.
* Unsuccessful use of document page flipping, mimicking skimming.
* Non-computer user had initial confusion with scroll bars.
* Failure to recognise the full functionality of scroll bars.
* Double click to select a line when single was sufficient.

Discussion

Observation of the task showed two distinct search strategies. The computer tool demands a structured approach in which a successful search is dependent on he distillation of appropriate keywords. Success using the paper source involved a high proportion of skimming through the document. This approach hinged on the individual's previous experience of using the reference document. Comparison of the two media in this evaluation is immediately biased in favour of the paper medium. All of the subjects required the reference document for their work and therefore had experience in searching the paper document for relevant references. It is important to note here that the enhanced ability of individuals to skim through the document was not evident with subjects who had not used the document recently despite considerable previous experience. The subjects had no previous experience of free text searching. Prior training in the use of the computer tool was minimal, as it was designed to support occasional use. The tool must present an intuitive interface to be effective. The evaluation aimed to highlight features of the system which were not obvious to a naive user.

Completion time and error rate provided the best performance measures. The subjects showed a wide spread of performance for both media, demonstrating the loose nature of the task for which the computer tool was designed. The average completion times (Table 1) show a slightly lower average completion time for the paper source. The action list used to record user strategy was based on observable actions. Whilst showing the degree of task complexity, it does not provide a clear measure of performance. The action list serves rather as a record of the adopted procedure. Task analysis based on the action list showed the deviation of the subjects from the optimal computer approach. Training of the subjects in parallel with functional and interface iterations highlighted in this evaluation provide scope for improvement in the access times for all subjects tested.

Development guidelines dictated the inclusion of a help system for users. Despite the novice level of the users for this evaluation the help feature remained unused. Consideration of the task actions carried out by each subject showed that subjects used the functions of the system appropriately, demonstrating an understanding of the interface. Slow computer search times generally resulted from the failure to choose appropriate keywords. Guidance is therefore required in developing the user's strategy when using the system as opposed to help in accessing the functions of the system which caused few problems for the users.

The strategy of page navigation was used by one subject. This related most closely to the page skimming approach favoured by all subjects when using the paper source. Failure to find the reference using the document's index resulted in skimming. It was possible to quickly flick through pages in the text mode. The picture mode, however, had a system delay of 3.9 seconds making picture skimming unfeasible. When skimming, using the paper source, the subject derived a fast impression of context through diagram recognition. This relates more closely to picture skimming in the computer tool. Effective application of the skimming strategy for the computer tool would require a solution to the update delay encountered when picture skimming. In the meantime the optimal search strategy when using the computer module remains text based filtering.

The computer based document was split into chapters to narrow search areas. The selection of a chapter required a computer processing time of 11.6 seconds. When added to the time required by each subject to select from the chapter list, this delay represented an important proportion of the access times. A recommendation for the iteration of the module was to remove the choice of chapters providing the complete document. In addition to speeding up the direct access to the document's text, this removes the necessity for the user to judge the content of the separate chapter headings based on their titles alone. Removing chapter divisions allows the system to retrieve related references that may be present in other chapters of the document, further increasing the depth of the search.

Providing the reference module on the desktop computer of each user removes the need to overcome the inertia of the user. Using the reference document requires the user to move from their seat, in most cases to another room. When using a computer system an individual makes inherent cost/benefit assessments that dictate the extent to which the functionality of a system is used (Eason 1984). The decision whether to move from one's seat or to choose the computer tool is an extension of this assessment. If desktop access to reference documents reduces the "costs" to the user of accessing the reference document it is expected that use of the reference document will increase.

Iteration Recommendations

* Increase the effective viewing area of the context list
* Improve text/photo mode indication
* Improve index/document mode indication
* Educate the users in the optimal search strategy
* Broaden the document database

Conclusion

The computer generally matched the paper source for the speed and accuracy of specific reference location when used by novice subjects. Whilst matching the access rate of paper, the computer source is unlikely to enhance the access rate for individuals intimately familiar with the paper source. The computer does however offer enhanced access to users less familiar with the paper source. The adoption of a suitable strategy should improve the access times for the user population. As the user group gain experience in applying the computer tool, improved access time for all documents is anticipated. The value of the computer tool to the user group should increase with the growth of the tool's document database allowing users to maintain a high level of searching performance between unfamiliar documents.

References

Andon, M., 1990, Great Powers. The price of Optical Character Recognition, *Personal Computer Monthly,* December, 133-138

Beagley, N.I., Haslam, R.A. & Parsons, K.C., 1992, Designing an Information System for Experts in Vehicle Ergonomics, *Contemporary Ergonomics 1992* (London: Taylor & Francis) 322-326.

Beagley, N.I., Haslam, R.A. & Parsons, K.C., 1993, A Computer Based Tool for Accessing Anthropometric Databases, *Contemporary Ergonomics 1993* (London: Taylor & Francis) 142-147.

Beagley, N.I., Haslam, R.A. & Parsons, K.C., 1993, Hypermedia for the In-House development of Information Systems, *Human-Computer Interaction: Applications and Case Studies,* 374-379

Dumais, S. T., 1988, Textual Information Retrieval, *Handbook of Human Computer Interaction,* Helander (ed), Chapter 30, 673-700

Dunlop, M. D. & Van Rijsbergen, C. J., 1993, Hypermedia and Free Text Retrieval, *Information Processing & Management,* Vol 29, No. 3, 287-298

Eason, K. D. , 1984, Towards The Experimental Study Of Usability, *Behaviour And Information Technology,* Vol. 3, No. 2, 133-143

Glavash, K., 1994, Full-Text retrieval for document delivery - A viable option?, *Online,* May, 81-84

Holloway, H., 1991, Text Retrieval in the Office, *Information Media & Technology,* Vol. 24, No. 3, 106-108

Luqi, 1989, Software Evolution Through Rapid Prototyping , *Computer,* May, 13-25

Spence, I. T. A. & Carey, B. N. , 1991, Customers Do Not Want Frozen Specifications, *Software Engineering Journal,* July, 175-181

THE PORTABLE USABILITY LABORATORY: A CASE STUDY

T.J. Hewson and P.V. Marsh
t.j.hewson@lut.ac.uk, p.v.marsh@lut.ac.uk

HUSAT Research Institute,
The Elms,
Elms Grove,
Loughborough, Leicestershire LE11 1RG

HUSAT has undertaken many usability evaluations over the last ten years, but recently new demands have been placed on the evaluations by clients. Major drawbacks of evaluations in the laboratory are the artificial situation, and the difficulty of recruiting representative users to participate in the studies. These difficulties are somewhat overcome by the development of a portable laboratory. This paper discusses through a case study those problems that are specific to the creation of a portable laboratory, and demonstrates that a laboratory set up in the field can yield results every bit as valid and useful as a full scale evaluation conducted on a permanent site.

Introduction

At HUSAT a fully equipped laboratory has been in existence for nearly ten years, for both research and consultancy work, and has undergone an evolution of technological advances since its inception.

Recently new challenges have faced the usability evaluation team. Clients now readily acknowledge the usefulness and importance of evaluations of their products. They demand however, full evaluations, not only at their own sites but at remote sites potentially anywhere in the World. Although the evaluation procedures are required to be mobile and more flexible, there must be no trade off in quality. Usability data have to stand comparison with information gathered in the main laboratory. Tapes of edited highlights must also be of the highest possible standard.

These seemingly irreconcilable requirements have stimulated a change of approach to usability evaluation, and an in depth look at the potential for high fidelity field trials.

Advantages of a portable laboratory
Usability evaluations at clients' sites hold a number of clear advantages. The first, and most important is the realistic environment in which the testing is conducted. No matter how well one equips a usability laboratory, the testing environment will always be artificial. If the evaluation is of a product likely to be used in an office situation then it would make sense to evaluate the product in that environment. Evaluating in the laboratory requires all of the possible variables to be simulated. These problems in reproducing the context of product usage are well documented, particularly by the ESPRIT project MUSiC which placed special emphasis on these issues (Maissel et al

1991). This project Deliverable lists over 30 environmental factors alone which contribute to the usability of a product. Bevan and Macleod (1994) reiterate these findings by stating that 'It is not meaningful to talk simply about the usability of a product, as usability is a function of the context in which the product is used.'

One of the major problems we have found in conducting laboratory trials is persuading a representative sample of a specific subject population to visit the laboratories at HUSAT. This is a particularly problematic issue when the product under evaluation is intended for a specific user population. This can manifest itself in a number of ways, the geographical location, the work commitments, or the scarcity of the intended users to name but three.

Disadvantages of a portable laboratory
There are also disadvantages to usability evaluation in the field. The first is the setting up of the laboratory equipment. This is a technical procedure, which can intimidate many human factors practitioners. In the case of a permanent usability laboratory equipment is less likely to fail, and if it does, there is likely to be a technician on hand. A portable laboratory however, requires setting up and configuring, often being closely integrated with the client's own equipment. A clear understanding of the total configuration is imperative if faults have to be diagnosed. There is much greater potential for connections to fail in a temporary set-up. Wires may inadvertently be dislodged or incorrectly connected during initial set-up.

The traditional approach of pointing a camcorder at a computer screen has led to a common misconception that the quality of data capture from portable laboratories is of low quality. This practice is unacceptable to both us and our clients, who expect professional quality recordings of the evaluation, with a high quality highlights tape. At HUSAT we have found that we can recreate the fidelity of a fully equipped laboratory with one ergonomist and a rationalised set of equipment.

Finally the quality of the set-up of a usability laboratory in the field is always to some extent, dependent on external factors. Fundamental problems of lighting, cable management and available space can however be minimised by clearly specifying requirements beforehand.

This paper centres on a case study which describes how none of these disadvantages should stand in the way of usability evaluation in the field, and that the advantages of evaluation within the context of the product's eventual use are clear. The paper will also assure that portable usability evaluations do not necessarily have to be low fidelity evaluations, and that simply using a camcorder to make the best of a poor situation is no longer acceptable to many commercial clients.

The HUSAT Permanent Laboratory

The first laboratory at HUSAT was set up in 1987, it has been evolving ever since and now is in regular demand for both research and consultancy projects. The main laboratory consists of two rooms separated by a one-way mirror. We have two remote controlled cameras as well as a static camera. The image of the computer screen is placed onto video through a scan converter, which can then be mixed with the images from the cameras.

A recent acquisition has been a comprehensive video editing suite, which facilitates the production of highlights tapes for the client. This has repeatedly shown to be one of the most effective ways of conveying the findings of the trial to the system designers, but only if produced in a professional manner. The importance of producing a good quality highlights tape has become paramount in the work of HUSAT and is well described by Dumas and Redish (1994).

Case Study —The Brief

The client is a provider of specialist financial information to a specific user population. The user groups are concentrated in several, specific geographical locations. The users have tremendous pressures on their time, as their jobs are stressful but with high rewards.

A new version of the user interface of the client's product had been developed. End users stressed that the new interface should be a tool to ease their workload, without creating additional effort to learn it. An improved version of a rival product had just been released, making the trials more urgent.

HUSAT were asked to evaluate the new version, in a variety of locations around the World with representatives of the target population. These demands precluded the use of the usability laboratories at HUSAT. The solution was the portable laboratory.

The Considerations

Factors to be considered for a portable trial are:

How much equipment do you really need?
The amount and type of equipment selected is dependent on a number of factors. The first is a trade off between the location of the trial and the resources available for transportation. Additionally one must consider the size and layout of the room in which the evaluation is to take place. Different configurations are described by Wiklund (1993) but typically, a usability test laboratory includes one room for test participants and another for observers. The available budget for the evaluation is nearly always an important factor.

Which equipment do you really need?
The most basic list of recording equipment is a camcorder, tape recorder and notebook, but these rarely satisfy quality requirements and their limitations are quickly apparent. The presentation of findings is one of the most critical aspects of the evaluation procedure. As described before there is no better method than the highlights video, and so the quality of recorded material is paramount.

Rubin (1994) lists an inventory of equipment required for setting up a travelling or mobile laboratory, costing $10,000 and $15,000 including a Hi-8 video camera, colour video monitor, Hi-8 video cassette recorder, VHS recorder, for recording the Hi-8 videos onto VHS for distribution, microphone, tripod, and a trolley.

Which video format?
There are a number of video formats currently available including VHS, Video 8mm, Hi-8, S-VHS, U-Matic and Betacam each has its advantages and disadvantages. Rohn (1994) lists considerations including the length of the tape, image quality, and cost. Hi-8, for example is more convenient to store because of its compact size., but its narrow tape width makes it more prone to drop out problems and tape fragility. S-VHS provides the robustness of a half inch wide tape along with the high resolution of Hi-8. In addition the production of VHS tapes from S-VHS is more straightforward as the formats are interchangeable on S-VHS machines. Whilst U-Matic, especially in SP form, offers excellent quality, the bulkiness of the equipment and limited tape length (approximately an hour) preclude it from serious consideration for field use. Betacam's only clear disadvantage is its cost. We would certainly endorse Uyeda's (1994) list for choosing S-VHS as the preferred video format:

- has superior characteristics to Hi-8 in retaining video quality across multiple generations of edits
- allows the use of the same decks to dub down to VHS (for demos and distribution of tapes to product teams
- allows the use of the most common video format (VHS) if there is not time to dub source tapes
- is cost effective relative to other formats
- is reasonably portable for field work.

Practical Problems

Drawing up a contingency plan is vital. It is helpful to consider a number of different equipment failure scenarios and how the trials might be continued with alternatives.

What to take as a backup
Equipment fails. No matter how much preparation is carried out in advance, there is always a chance that some of the equipment transported for a portable evaluation, will not work when it arrives. In practice it is usually not possible to duplicate every piece of equipment. Therefore one has to assess each piece of equipment in terms of the likelihood of failure and the convenience of its contingency plan. These problems will range from not getting an image from the scan converter, through to batteries running out if a piece of equipment is left on.

Take or hire locally
With the range of equipment taken it soon becomes bulky. It is relatively simple to pack the equipment in the back of a car, but when the lab has to become truly portable rather than transportable e.g. in the case of taking the laboratory abroad, the option of hiring equipment locally should be considered.

   The logistics of organising the equipment to be transported, set up and configured in countries other than the UK are problematic. There are a number of practical issues which arise when setting up electrical equipment which rely on compatibility with each other, when so many standards and formats prevail in the industry.

Shipping equipment
On organising the equipment to be transported abroad, we soon discovered that customs procedures are detailed and complicated. In order to export and subsequently import sophisticated equipment of this nature a Carnet is required from the London Chamber of Commerce. This has to be stamped at every customs station on both outward and inward legs of the journey. Failure to do so may lead to serious complications. Clearly, baggage constraints for the airlines is another concern which should be addressed.

Powering the equipment
The most obvious first consideration is the local conditions regarding mains electricity. The compliance of pieces of equipment may be ensured but those items running off mains electricity must be passed through a transformer. In the schematic representation of a laboratory set-up (fig. 1) the VCR, monitor, lap top computer, mixer and scan converter must all pass through the transformer.

Video standards
The compatibility of video standards is the next important issue. PAL, NTSC and SECAM are the three main World-wide standards. The video standard is dictated by the site chosen for the eventual video analysis and production of highlights tape, not by the location of the trial itself. Standards conversion invariably leads to a deterioration in quality.

Our Solution

Laboratory Set-Up
The evaluation was to be conducted in one room approximately 5m x 5m, with a temporary screen separating the control side to the evaluation side.

   The full range of equipment consisted of a video camera, a scan converter, a video mixer, an S-VHS video recorder, a 14 inch high resolution video monitor, a lap top computer, and ancillary equipment including lapel microphone, video tapes, spare microphones and batteries etc.

   We decided to use S-VHS video format for the reasons cited earlier. Our choice of Hitachi studio camera was in retrospect unnecessarily bulky and complicated. A more compact, (and less intrusive alternative) would be a smaller desktop type camera such as a Flexicam™.

In this case study, budget was not a major factor and so for the purposes of producing the highlights tape, we took the step of incorporating the scan converter into our inventory. These expensive pieces of equipment translate the video monitor output from a computer or work station to a video standard such as PAL that can be recorded. Because of the cost it was hired just for the duration of the evaluation.

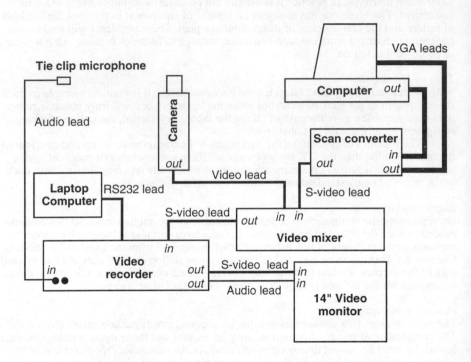

Figure 1 Laboratory System Diagram

The other equipment that we considered to be invaluable were a lap top computer and a video mixer. The computer was for logging of incidents and real-time analysis of the evaluations and the mixer combined the image from the computer screen via the scan converter with the view of the user's face. This gave context to the users' interaction with the software.

In this case two experimenters were available, one to conduct the testing, and one to operate the equipment, including the data logging device. If real time data logging is not required then it is possible to have just one experimenter. Likewise if it is not necessary for an experimenter to sit in with the test subject, then one supervisor/equipment operator will suffice.

One of the key differences between the portable set-up and the permanent usability laboratory is the absence of a one-way mirror. In most cases this is not crucial, as a high quality video monitor will portray all that is happening on the user's side of the partition, in fact this partition can be less intimidating and distracting than a one-way mirror.

There is also no remote camera control in this portable set up. This is a feature of our permanent laboratory, but seldom used. We have found that with careful camera set-up prior to testing, it is possible to obviate the requirement for a remote control camera .

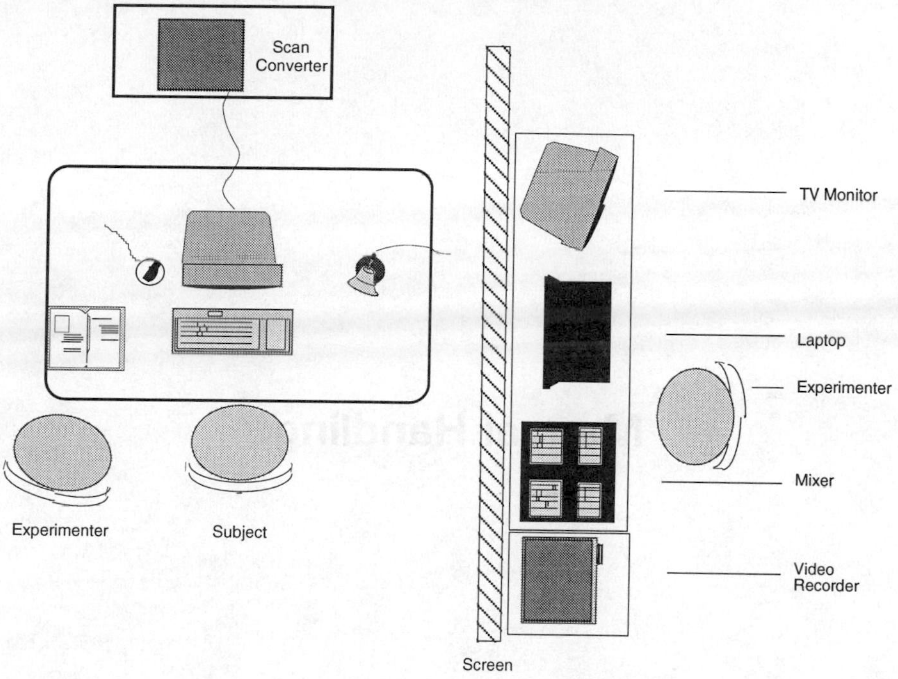

Figure 2 Set up of the Laboratory

Conclusion

In conducting evaluations on site, we found that there was no compromise in the quality of data and video capture.

In attempting to overcome the problems presented to us in this case study, we at HUSAT have found that it is often preferable to conduct usability evaluations in the field, so that context of use can be considered when investigating usability. We found that the practical problems of setting up a laboratory on a client's site although plentiful, are easily overcome, whilst the advantages in conducting an evaluation in the context of the intended working environment are considerable.

References

Bevan, N. and Macleod, M. 1994, Usability Measurement in Context, Behaviour and Information Technology **13** (1 and 2), 132-145(Taylor and Francis, London.)

Dumas, J. S. and Redish, J. C. 1994, *A Practical Guide to Usability Testing* , (Ablex Publishing Corporation, Norwood, NJ)

Maissel, J., Dillon, A., Maguire, M., Rengger, R. and Sweeney, M. 1991, Context Guidelines Handbook, MUSiC Project Deliverable IF2.2.2

Rohn, J. A. 1994, The Usability Engineering Laboratories at Sun Microsystems, Behaviour and Information Technology **13** (1 and 2), (Taylor and Francis, London.)

Rubin, J. 1994, *Handbook of Usability Testing* (Wiley and Sons, New York)

Wiklund, M. E. 1993, Building a Usability Test Laboratory, Medical Device and Diagnostic Industry, March, pp 68-77.

Manual Handling

ASSESSING RISK IN NURSING ACTIVITIES - CAN PATIENT HANDLING ASSESSMENTS BE IMPROVED?

RJ Graves, A Morales, and A Seaton

Department of Environmental & Occupational Medicine,
University of Aberdeen

The introduction of the Manual Handling Operations Regulations requires employers to assess risk of manual handling. Previous research showed that senior nurses had poor performance when assessing manual handling risks in nursing. This was due possibly to inadequate training and the process of risk assessment. Subsequently a methodology for risk assessment and a training programme for senior nurses was developed using risk assessment worksheets containing accepted criteria and risk factors for six patient handling activities. Twenty senior nurses (a 21% sample) used the worksheets to assess risk from videos of patient handling activities. Independent ratings were used as an expert standard for comparison. Overall, the study showed that the worksheet provided senior nurses with a means to carry out assessments of patient handling activities which was less dependent on their technical ergonomic knowledge.

Introduction

According to statistics published by the Health and Safety Executive (HSE), medical, veterinary and other health services have a high incidence of manual handling injuries. During 1990/91 55% of reported over-three-day injuries in these services were caused by manual handling. In the Health Services 70% of manual handling accidents were related to patient handling. According to several studies nurses are the main group at risk within the Health Services. This finding could be explained by the wide range of manual handling tasks involved in their profession and particularly by the complexity of handling patients (live loads). Lifting and assisting patients in particular have been identified as the main cause of back injury in nursing.

Under the Manual Handling Operations Regulations (MHOR), employers have to assess the risks to their employees from manual handling of loads and, having made that assessment, to take action to reduce those risks. Complying with these duties in the Health Services could be a difficult task. Nursing, in particular, could present special demands for the assessment. The characteristics (size, weight, unpredictability) and medical conditions of patients, the need to use special techniques to handle them and constraints of the working environment in the wards (size of beds, personal lockers, tables, medical equipment) are conditions for which the MHOR do not provide enough information to allow a straightforward application of the regulations.

Another practical issue in nursing is determining how the Health Services will carry out the duties imposed by the MHOR. The MHOR states that employers should be able, in the majority of cases, to carry out the assessment themselves or delegate it to others within their organisation. What are the implications of delegating the employer's duties? According to the legal position, delegation does not discharge legal responsibility and the employer remains legally responsible for his duties. In conjunction with the employer, line managers are also legally responsible for the implementation of regulations. Determining who will carry out the employer's duties has therefore practical and legal implications. In order for the employer's duties to be successfully delegated, in-house risk assessor candidates will be required to understand the nature of the handling operations to be assessed and the risks involved.

In addition, risk assessors will require detailed knowledge of the demands of the regulations, training and practice in risk assessment. Senior nurses (Sisters and Charge Nurses) are likely candidates to carry out in-house assessments in nursing. They know the practicalities of manual handling in their profession and the restrictions of the environment. Applying the MHOR will involve them in the process of risk identification, assessment and control and because of their role as line managers, staff will be more inclined to participate in these activities and to comply with future measures of risk control. However, there is an important question. What training is needed to enable senior nurses to perform the duties of in-house assessors?

A pilot study carried out by Morales (1992) looked at senior nurses' performance on risk assessment of manual handling activities. A checklist of four pages, based on the draft manual handling regulations and guidance was used to assess video recordings of manual handling activities. These included patient and non patient handling activities. A 6% sample of senior nurses from a large teaching hospital received individual training and participated in trials of the document. The results of the study showed that senior nurses found the checklist difficult to use, and their agreement on risks involved in the activities was good in only 20% of cases. Two possible causes of this poor performance were inadequate training and the checklist was too complex. It was concluded that senior nurses could not identify and assess risk reliably under these circumstances. One outcome from the study was a recommendation that further research be carried out into improving senior nurses' performance, especially in relation to a better risk assessment approach and training method.

It is clear that in order to satisfy the MHOR, there was a need to develop a practical methodology for senior nurses to carry out their risk assessment duties. As patient handling is a key issue in manual handling in nursing, there was a need to develop a risk assessment approach that takes into account the complexities of such handling. Examination of the literature showed there was a consensus on techniques considered to be good practice. The accepted techniques provides criteria that need to be met in order for the patient handling procedure to be carried out correctly. In some cases these techniques have been evaluated scientifically. Intuitively, if it were possible to encapsulate accepted practice in the form of a risk assessment method, then this would overcome some of the problems associated with the requirements of a complex assessment process. This was the approach that was adopted in the study described below, where a methodology for risk assessment and a training programme for senior nurses were developed (Morales et al, 1993).

Approach

The principal aim of the study was to develop a methodology to help senior nurses apply the MHOR reliably. Specific aims were:

a) to design a worksheet containing specific criteria to identify, assess and propose control measures for risk of patient handling activities in nursing.
b) to develop a prototype training package to provide background knowledge to apply the MHOR and information on how to use the worksheet.
c) to evaluate the effectiveness of both in a study with controlled trials.

A worksheet was developed to provide senior nurses with a list of 'good practice' criteria for patient handling and a list of causes of risk associated with them. Criteria on accepted practice were obtained from the literature although it was accepted that the scientific basis of these criteria was weak. This is discussed later. Causes of risk related to accepted practices were obtained from the literature and covered those; arising from patients (live loads), wards (working environment) and nurses (individual capabilities). As the MHOR require that a record of the assessment be made, the risk assessment approach needed to provide senior nurses with means of recording their findings.

It was decided that senior nurses would require training in the following areas in order to be able to carry out their risk assessment duties:

a) basic background on ergonomics and biomechanics;
b) the MHOR and a definition of the role of in-house assessors; and
c) the use of a worksheet approach to risk assessment.

Based on this approach to risk assessment and training a study was undertaken to evaluate senior nurses' performance on the use of a prototype assessment approach for patient handling. The following activities were selected among those for which a literature review provided criteria on accepted practice;

a) a shoulder lift in bed,
b) a cradle lift in bed,
c) turning a patient in bed,
d) a rocking pivot transfer to a chair,
e) a shoulder lift transfer to a wheelchair,
f) a side transfer to a trolley, and,
g) attending a patient with a plastered leg.

The worksheet was designed to fit on a single A4 page, formatted in three columns and using landscape orientation. The first column contained a still frame from video recording of the activity under analysis, and a list of conditions under which the activity (TASK) would be safe to perform (accepted practice criteria). The second column listed selected risk factors related to patients (load), wards (working environment), nurses (individual capabilities) and other factors, to support risk identification and assessment. Selection of these risk factors was based on their relevance to the activity under analysis. All questions in the worksheet required 'YES/NO' answers. The third column of the worksheet was divided into two sections; one allowed the user to suggest measures of risk control and the other to suggest dates for compliance with control measures and when reassessment should be done.

Drafts of the risk assessment worksheets for the seven activities were evaluated and modified prior to the trials. Two of the authors used the worksheets to make an independent assessment of what could be expected from an expert evaluation of the activities. This formed the basis for a consensus on conditions for safe performance and on causes of risk. It was used as a baseline to compare senior nurses' performance.

Twenty senior nurses, 3 males and 17 females, comprising a 21% sample participated in the study. Selection of the sample was based on selecting wards after a review of accident records over a period of five years. Participants in the study came from wards with a high frequency of manual handling accidents.

A training package with written guidance was prepared to standardise the training given to senior nurses. The first part outlined the magnitude of the problem of back injuries in nursing, provided basic concepts of biomechanics of manual handling, and introduced the ergonomic approach to risk identification, assessment and control proposed by the MHOR. The second section outlined the duties imposed on employers by the MHOR, the role of in-house assessors, and provided information on the use of the risk assessment worksheet designed for the study.

Trials were arranged within two weeks of the training. During each session, a senior nurse carried out a series of assessments using the worksheets. Six worksheets were used to analyse a video recorded sequence of patient handling activities. During the session each nurse had to observe the activity and check if the conditions in the first column of the worksheet were met. A diagram of the activity was provided to assist the analysis.

Any negative answer was considered to be evidence of risk and nurses had to assign the risks identified to the generic causes of risk outlined in the second column of the worksheet. The third column of the worksheet was used to record control measures, based on the risk factors and causes of risk.

After the trial each participant assessed his/her experience using a self administered questionnaire. The questionnaire was divided into three sections. The first two sections obtained their opinions on the contents of the training sessions, the worksheets and the videos used during the trials, using a five point scale (1 very easy; 5 very difficult). The third section contained three questions designed to analyse nurses retention of knowledge from the training sessions.

Results and Discussion

To allow comparisons with the previous study (Morales, op cit.) senior nurses' agreement among themselves and with the standard 'expert' rating were analysed using the Kappa statistic (chance corrected inter-observer agreement). Kappa values were classified according to Altman's criteria (Altman, 1991). The binomial distribution was used to analyse the answers to questions in columns one and two. Frequency distribution was used to analyse the third column of the worksheet and the results of the self administered questionnaire.

Kappa statistic analysis showed that the worksheet for risk assessment provided senior nurses with a valid methodology to carry out their duties of in-house assessors reliably. However, the worksheet had some limitations that could have influenced senior nurses' performance. One of these was the amount of data that could be fitted onto one page. This excluded some definitions and criteria that could have increased senior nurses' performance. This limitation could be solved in future studies by using an annexe to the worksheet where further information could be included.

The influence on senior nurses' performance of the use of video recordings is more difficult to assess. In some activities, videos did not provide adequate information for risk assessment. However, studies from the literature have highlighted the benefits of video recordings in risk assessment. One of these benefits is being able to rewind the tape to obtain a more detailed observation of activities under analysis. During the trials senior nurses rewound the tape an average of five times for each activity. This facility could have improved their risk assessment performance in this study but raises questions about applying the worksheet to 'real life' observations where the possibilities of looking at the same actions again will be limited. On the other hand, this limitation could be solved by using 'generic' assessments. This alternative, proposed by the MHOR, uses a broadly similar range of manual handling operations for the assessment, and provides a better idea of the risks involved.

Overall, senior nurses' answers to questions on accepted practice criteria showed that in the majority of cases (80%) senior nurses could use them to identify risk reliably. However, the scarcity of scientific evidence to support criteria of accepted practice for patient handling limits the use of a worksheet. Although most criteria of accepted practice may have an intuitive appeal, their scientific validity have not been demonstrated. The need for this data was evident in the tasks involved in rocking pivot transfer and transfer to a trolley. The use of accepted practice for these activities required nurses to bend their backs and handle loads away from their bodies, or stretch across the trolley to grab the easy slide, both of which go against accepted MHOR practice.

Similar findings have been highlighted by other review articles in the literature. For example, it has been stated by Pheasant et al (1991) that even with techniques such as the shoulder lift, which is considered as 'low risk,' there is still a risk of injury in 20% of occasions it is used. Furthermore, the patient's weight limits used to define the 'safe' level would expose 50% of handlers to risk. Under these conditions this technique would not comply with the MHOR requirements of providing reasonable protection to nearly all (95%) of handlers.

Overall, senior nurses' answers to questions on identification of causes of risk showed that in the majority of cases (91%) they could use them to assign causes of risk reliably. Kappa statistic analysis of the assignment of risk to causes showed that senior nurses were able to assign causes of risk more accurately for risks caused by factors related to Individual Capabilities and the Load than to those caused by factors related to the Task and the Working Environment. Difficulties in the assessment of these areas were described in a previous study Morales (op cit.) and have also been reported by other studies.

Analysis of control measures suggested by senior nurses highlights the key role assigned by them to training on patient handling techniques and to the availability of mechanical aids. Evidence on the benefits of training on patient handling techniques to reduce back injuries are controversial. However, some recommendations have been published in the literature in relation to the use of a training programme as a risk control measure for back injuries. Bearing these recommendations in mind, health managers should be advised against the easy option of buying 'off the shelf' packages, including videos on how to apply the MHOR, to discharge their duties.

The benefits of mechanical aids for handling patients have been researched extensively in the literature, but it is widely recognised that they are under used for a variety of reasons (e.g. they are time consuming, patients don't like them, they are not readily available, lack of skills on how to use them). In order to overcome these problems, the role of mechanical aids in patient handling and in preventing back injuries in nurses, along with the provision of adequate criteria for their selection will have to be addressed as part of the training programme on manual handling.

Analysis of the self administered questionnaire showed that the worksheet and the video were rated 'easy' to use but retention of knowledge was low. The use of a single day for the training sessions, due to time constraints on senior nurses' availability, could explain this result.

The use of open questions to assess retention of knowledge could have been an additional factor which influenced the results. Open questions require better knowledge of the subject than multiple choice questions and are therefore more difficult to answer. The high percentage of senior nurses that could recall three types of equipment for manual handling supports this hypothesis. This type of equipment is more familiar to senior nurses than the risk factor categories of the ergonomic approach and therefore, more easy to remember.

Finally, whether low retention of knowledge is a relevant factor for senior nurses' performance is debatable. It could be argued that what is needed is not to improve retention or knowledge as much as to improve the definitions and criteria used in the worksheet. If senior nurses can use the worksheets reliably, it seems less important practically whether they can remember the ergonomic definitions behind them.

Time constraints on senior nurses did not allow them to attend two separate training sessions. Condensing both sessions in one could have affected retention of knowledge and effectiveness of training. Based on the results of the study it also seems necessary to improv the training package to provide more information on the assignment of risk and more practical training exercises.

Conclusions

Overall, the study showed that, within certain limitations, the worksheet provided senior nurses with a means to carry out assessments of patient handling activities which was less dependent on their technical ergonomic knowledge. Limitations of the worksheet were:- lack of definitions and criteria from the amount of information on a one page worksheet and difficulties in obtaining adequate information from some of the videos for risk identification Further development of the worksheets and educational methods for delivering the training could increase the good level of senior nurses' performance reported by this study.

References

Altman D.G. 1991 *Practical Statistics For Medical Research*. (Chapman and Hall, London).

Morales A. 1992 Assessing manual handling risk in nursing. MSc dissertation in occupational medicine. Department of Environmental and Occupational Medicine. Medical School - University of Aberdeen: Aberdeen.

Morales, A. Graves, R. J. Seaton, A. 1993 Evaluation of the use of the HSC Draft Regulations and Guidance on Manual Handling to assess risk in nursing activities. Colt Foundation Grant. Final Report. Department of Environmental and Occupational Medicine University of Aberdeen: Aberdeen.

Pheasant S, Holmes D, Stubbs D. 1991 Back pain in nurses: some ergonomic studies. In E.D. Lovesey, *Contemporary Ergonomics 1991; Proceedings of the Ergonomics Society's 1991 Annual Conference*: (Taylor and Francis, London) 323-327.

The Ergonomics of Digging and the Evaluation of a Novel Design of Spade

RS Bridger*, P Sparto** and WS Marras**

*   Dept of Biomedical Engineering, UCT Medical School
    and Groote Schuur Hospital, Observatory, 7925, South Africa

** Biodynamics Lab, Dept of Industrial and Systems Engineering,
    The Ohio State University, Columbus, Ohio

A laboratory investigation of digging was carried out to evaluate the risk of low back injury and to compare a conventional spade with a prototype 2-handled spade based on designs reported in the ergonomics literature. Digging was found to carry a high risk of back injury. The prototype reduced the risk by approximately 8%. Bending was reduced by 40% when the protoype was used but this was partly offset by an increase in twisting. From a fundamental point of view, the new concept merits further evaluation.

Introduction

Although a great deal of research has been carried out on the ergonomics of manual handling, the main emphasis has been on industrial applications where load size and shape are known, or can be specified. Other forms of manual handling have received less attention. A great deal of research has also been carried out on certain types of hand tools, particularly powered tools or small manually powered tools used in industry. Digging and shovelling are examples of manual handling tasks using hand tools which, on a priori grounds, would appear to impose considerable stress on the musculoskeletal system of the user. However, little is known either about the stresses imposed by these tasks or about the scope for amelioration through ergonomic redesign of the tools,

Research on digging and shovelling dates from the turn of the century with the work of FW Taylor. Amongst a variety of findings, Taylor specified an optimal shovel load (approximately 9 kg.) and pointed out the need to use different shovel designs for materials of different density and consistency. Frievalds (1986a) has reviewed the literature on the ergonomics of shovel design and shovelling. A shovelling rate of 18-21 scoops per minute with a load of 5 to 11 kg has been found to be most efficient. Lighter shovels increase efficiency especially for low loads. A throw height of 1 to 1.3 metres is acceptable. Low ceilings (as found in mine stopes) constrain posture and increase energy expenditure. Frievalds (1986b), in an experimental study, presents the following recommendations for shovel design; a lift angle of approximately 32 degrees, a long handle, a large square point blade for shovelling, a large round point blade for digging, hollow back construction to reduce weight, as light a weight as possible without sacrificing too much strength and durability. Frievalds and Kim (1990) found the minimum energy cost of shovelling to be obtained when the ratio of blade size to shovel weight was $0.0676 \ m^2/kg$.

The work described above is an attempt to optimise the design of shovels and spades by manipulating existing tool features and observing their effects on various performance parameters. An alternative, and possibly complementary, approach is described by Sen (1984) who reported that shovel design can be improved by fitting a second handle at the neck of the shovel, thus reducing the need to stoop. The second handle pivots at the point where it is connected to the shovel. Frievalds (1986b) reported that the second handle was more of a hindrance than a help due to usability problems - the hand on the second handle tended to hit the main handle during fast throwing movements. Neither of these investigators provide objective data to support their claims for and against the use of the second handle, however.

Degani et al. (1993) evaluated a modified shovel design with two perpendicular shafts. The secondary shaft could be positioned along the main shaft, towards the blade end of the shovel, to accommodate user preferences and anthropometry. Its handle was able to rotate around the secondary shaft to reduce wrist stress when throwing a load. In an experimental study, a significant reduction in lumbar paraspinal EMG was observed when the secondary shaft was used. In a field evaluation, ratings of perceived exertion were significantly lower (up to approximately 20%). Subjects commented that less bending was required with the modified shovel but that is was less suitable for digging in narrow trenches.

By bringing the handle closer to the user, the second shaft would be expected to reduce the need to stoop when digging. This may reduce both the musculoskeletal and the physiological load of the task. A fundamental study of digging, using the approach of Marras, was carried out to determine whether this was the case. 10 subjects were videotaped in the laboratory while digging with a conventional spade both with and without a second handle. Forces and moments at the feet were measured using a force platform and lumbar motions with a lumbar motion monitor (see Marras et al., 1992). Marras and his colleagues have presented an abundance of evidence which demonstrates the importance of a three-dimensional approach to the evaluation of occupational loading of the spine (Marras et al, 1993). In particular, no evaluation can be complete without an analysis of trunk motion characteristics (Marras, 1992, Marras et al, 1994). This is because many of the forces acting on the spine at work arise from the motions required by the task rather than just the posture or the load characteristics. For example, Marras and Mirka (1993) have demonstrated that increases in trunk velocity are accompanied by greater co-activation of trunk muscles with a corresponding increase in spinal loading.

Since usability issues are always embedded in a particular context, in this case that of a laboratory investigation, no attempt was made to gather data about them.

Method

Procedure
A repeated measures design was used in which 10 subjects transferred sand using a conventional spade and a spade fitted with a second handle. Two secondary handle conditions were used giving a total of three digging conditions. Data were captured for each digging trial which consisted of transferring 5 spadeloads from a sandbox to an adjacent container. The trial was repeated once for each spade type.

Spade Designs.

A commercially available gardener's spade was used. A second handle was designed whose shaft could be bolted to the neck of the spade via a pivoting plate (Figure 1). The second handle could be rotated through 90 degrees and fixed such that subjects either dug with the wrists pronated or supinated.

Apparatus

Subjects dug while standing on a force platform which recorded forces and moments in 3 axes. A lumbar motion monitor was used to record lumbar dynamics in three dimensions and a video camera was used to record body posture in 2 dimensions. The sampling rate was 60 Hz (Figure 1).

Results and Discussion

The data were analysed using ANOVA to generate F-ratios for the main effects due to spade type, trial number (first trial or repetition) and subjects and the interactions. Only the main findings are presented here due to space limitations.

Digging Performance

The time per trial and the weight of sand transferred were similar for the 3 spade types. Digging time per 5 dig trial was greater when the second handle was used (18.9 and 18.6 compared with 17.5 seconds, $F = 5.34$, $df = 2$ and 18, $p< 0.05$).

Trunk-Thigh Angle

The data were obtained from the automatic analysis of the joint markers captured on the video recordings. The mean trunk-thigh angle per trial is the mean of all the trunk-thigh angles sampled throughout the 5 digging cycles. Table 1 gives mean and standard deviation trunk-thigh angles when digging with the conventional and two handled spades. Data are combined across first and second repetitions as this effect was not statistically significant ($F = 3.25$, $df = 1$ and 9, $p>0.05$).

Table 1. Mean and Standard Deviation Trunk-Thigh Angles Under Three Digging
Conditions (radians)

	Mean	Standard Deviation
Conventional spade	1.335	0.258
Handle 1 (wrist pronated)	1.857	0.183
Handle 2 (wrist supinated)	1.868	0.241

With the conventional spade, the mean trunk-thigh angle was low throughout the digging cycle (1.335 radians or 76 degrees). With both prototype spades, the mean angle was larger (1.857 radians is approximately 106 degrees).This effect was statistically significant ($F = 64.4$, $df = 2$ and 18, $p<0.001$). Thus, use of the prototype spade was accompanied by approximately 30 degrees less bending of the trunk. Figure 1 illustrates this finding with photographs taken at those points in the digging cycle with the most and least bending of the trunk. To further assist visualisation of these findings, it can be said that with the conventional spade, the level of both hands is below that of the knees for a large part of the digging cycle. With the second handle, both hands can be kept above the level of the knees for a large part of the digging cycle - a reduction in bending of about 40%.

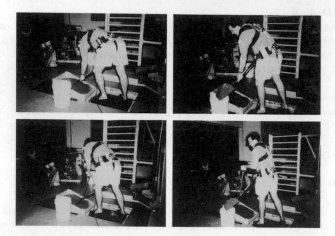

Figure 1. Laboratory evaluation of spade designs

Force Plate Data

Forces and moments about three orthogonal planes were collected as described in the methods section. There was great variability in all forces and moments throughout each trial. Further, the variability was cyclic, tracking subjects' movements during the 5 digs. Analysis of mean forces would have been unlikely to yield meaningful data and was abandoned in favour of an analysis of peak forces and moments. For each subject/trial combination, 5 peak forces or moments were manually extracted from the force plate data. These were analysed by ANOVA to test the significance of any effects due to the spade. Only statistically significant differences are reported here.

The moment about the x-axis of the force platform was significantly lower when the second handle was used (mean peak moment was 133.17 nm and 127.56 nm with the second handles and 159 nm when the conventional spade was used). This was statistically significant ($F = 10.63$, df = 2 and 18, $p<0.001$). The x axis of the force platform runs perpendicular to the direction of digging, therefore this result indicates more of subject and load weight being above the platform when the second handle was used (rather than being displaced anteriorly) and is compatible with the previous finding of reduced bending when digging with the second handle.

The peak forces in the y axis (running parallel to the direction of digging) was greater when the second handles were used (60.76 n compared with 73.86 n and 67.88 n respectively). This was statistically significant ($F = 3.70$, df = 2 and 18, p <0.05). The positive value of the force suggests it reflects the force reaction when the spade was inserted into the sand. This may be explained partly by the increase in weight of the spade when the second handle was attached to it (an increase of 0.9 kg).

The peak forces in the z axis were greater when the conventional spade was used (1029.85 n compared with 1008.12 n and 1012.85 n respectively). This was statistically significant ($F = 3.65$, df = 2 and 18, $p < 0.05$). The z axis is the vertical axis through the force platform and subject's body. A possible interpretation of this finding is that, at an early stage in the lifting of the dug sand, the acceleration of the spade or body is slightly lower when the second handle is used, which would imply lower peak forces acting on the subject.

The peak moment about the z axis was greater when the second handles were used than with the conventional spade (- 14.14nm versus -15.91nm and -15.90nm respectively). This was statistically significant (F = 4.62, df = 2 and 18, p < 0.05). Any twisting movements of the load or body would be transmitted to the force platform via the subjects feet causing a moment about the platform in the z axis. The present finding implies that there was less twisting of the load, or of the body, when the conventional spade was used.

Risk Analysis Using Lumbar Motion Data

The data obtained from the lumbar motion monitor (LMM) were analysed using the model of Marras et al. (1994). The model states that a combination of 5 trunk motion and workplace factors can predict medium and high risk low-back disorder. These factors are lifting frequency, load moment, trunk lateral velocity, trunk twisting velocity and trunk sagittal angle. The first two factors were estimated indirectly by the experimenters and the remaining factors were provided by the LMM. Table 2 presents the output of the model which is in terms of probabilities - i.e. the probability that the component belongs to a high-risk group and the overall probability after the components have been combined.

Table 2. Biomechanical Risk of Low Back Disorder Associated
with 5 Components of Digging

Spade	Lateral Velocity	Sagittal flexion	Load Moment	Average Twisting	Lift Rate	Overall Probability
Convent-ional	0.364	0.959	0.558	0.069	0.980	0.586
Handle 1	0.264	0.857	0.272	0.164	0.980	0.507
Handle 2	0.252	0.863	0.270	0.140	0.980	0.501

As can be seen, the riskiest component in conventional digging is sagittal flexion. This supports the present attempt to reduce bending in digging by means of a second handle. A reduction of almost 10% has been obtained. The load moment is also reduced when the second handle is used. This is likely to be a direct consequence of the more upright posture. A 10% reduction in lateral trunk velocity was also found when the second handle was used. However, use of the second handle was accompanied by an increase in the average twisting velocity of the trunk (which is also suggested by the force plate findings described above). The overall reduction in risk (from 0.586 to 0.507 and 0.501) is therefore not as large as might have otherwise been expected (digging with the second handle causes an increase in the average twisting velocity which nullifies to some extent some of the other benefits).

Conclusions and Recommendations

Digging with a conventional spade is a task associated with high risk of low back disorder, even when the task is evaluated in ideal (laboratory) conditions using easily dug material (soft sand).
Addition of a second handle to the spade in the laboratory digging task produced large reductions in factors known to be associated with high risk of low back disorder. An overall reduction in risk of about 10% was observed. These findings provide a firm foundation for future ergonomics research aimed at reducing the risk of back injury in activities such as gardening and small scale farming and in industries such as construction. In particular, the concept of adding a second handle to a spade seems to be a valid one from a fundamental point of view. Further work, to investigate detailed design and usability issues, would seem to be justified. In particular, it would be worthwhile to carry out user trials of a variety of prototypes to determine the circumstances, tasks and soil types under which the greatest benefits can be obtained.

Finally, the findings demonstrate clearly the inadequacy of static, sagittal approaches to the evaluation of occupational trunk stress. Although the addition of a second handle greatly reduces easily observed variables such as trunk-thigh angle and sagittal flexion, it increases other, less easily observed variables such as average twisting velocity. Without the use of appropriate instruments to characterise trunk motions in three dimensions, the present investigation would have overestimated the potential advantages of the two handled spade.

References

Degani, A., Asfour, S.S., Waly, S.M. and Koshy, J.G. 1993, A comparative study of two shovel designs, Applied Ergonomics, **24**,306-312.
Frievalds, A. 1986, The ergonomics of shovelling and shovel design - a review of the literature, Ergonomics, **29**,3-18.
Frievalds, A. 1986, The ergonomics of shovelling and shovel design - an experimental study, Ergonomics, 29,19-30.
Frievalds, A. and Kim, Y.J. 1990, Blade size and weight effects in shovel design, Applied Ergonomics, **21**,39-42.
Marras, W.S. 1992, Toward and understanding of dynamic variables in ergonomics, Occupational Medicine, state of the art reviews, **7**,655-677.
Marras, W.S., Fathallah, F.A., Miller, R.J., Davis, S.W. and Mirka, G.A. 1992, Accuracy of a three-dimensional motion monitor for recording trunk motion characteristics, International Journal of Industrial Ergonomics, **9**:75-87.
Marras, W.S. and Mirka, G.A. 1993, Electromyographic studies of the lumbar trunk musculature during the generation of low-level trunk acceleration, Journal of Orthopaedic Research, **11**:811-817.
Marras, W.S., Lavender, S.A., Leurgans, S.E., Rajulu, S.L., Allread, W.G., Fathallah, M.S. and Ferguson, S.A. 1993, The role of three-dimensional trunk motion in occupationally-related low back disorders, Spine, **18**:617-628.
Marras, W.S., Lavender, S.A., Leurgans, S.E., Fathallah, F.A., Ferguson, S.A., Allread, W.G. and Rajulu, S.L. 1994, Biomechanical risk factors for occupationally-related low back disorders, Ergonomics, In press.
Sen, R.N. 1984, Application of ergonomics to industrially developing countries, Ergonomics, **27**:1021-1032.

TRUNK MUSCLE STRENGTH AND MANUAL HANDLING SKILLS: THE EFFECTS OF TRAINING

Diana Leighton and Thomas Reilly

School of Human Sciences
Liverpool John Moores University
Mountford Building, Byrom Street
Liverpool, L3 3AF

Injuries to the back may result from repetitive lifting and manual handling, particularly if trunk muscle strength is inadequate. A regimen of physical training could increase the lifting capability of the worker and so reduce the load on the spine. This study aimed to determine the effects of physical training on trunk muscle strength and lifting capability. Fifteen females participated in a 10 week physical training programme. Two weight-training sessions comprising progressive resistance exercises and one circuit-training session were performed each week. Four physical assessments were performed during training; a control group (n=14) only completed these assessments. Psychophysical tests of lifting capability performance improved with training.

Introduction

Exposure to repetitive lifting has been associated with a high prevalence of back pain and injury (Frymoyer et al., 1983). In addition, inadequate strength of the trunk musculature has been identified as a risk factor in the incidence of low back pain (Smidt et al., 1983). If trunk muscle strength is inadequate and stabilisation of the ligaments insufficient, injury or loss of optimal function may result.

Numerous studies have been performed to investigate muscular strength in relation to the incidence of low back pain. The majority of these may be differentiated according to distinct objectives: Elnaggar et al. (1991) examined the effectiveness of physical exercise in reducing the severity of low back pain; prospective and retrospective investigations have aimed to identify predictive variables associated with the risk of experiencing back pain (Pope et al., 1985; Burton et al., 1989), and the implementation of training programmes has stimulated research investigating the relationship between muscular strength and lifting capability in manual handling operations (Asfour et al., 1984). The efficacy of training programmes applicable to personnel within various occupations is less well documented and has important implications for the prevention of low back pain.

In order to reduce the stress placed on the back during manual handling activities, it is advisable to reduce the load lifted and/or the number of repetitions of the lift, or to increase the lifting capacity of the worker. The latter recommendation was the subject of

397

this study which aimed to determine the effects of physical training on trunk muscle strength and lifting capability.

Method

Subjects

Fifteen females participated in a physical training programme of ten weeks' duration (training group). The subjects were of mean age 20.6 (± 1.8) years, mean height 165.6 (± 6.30) cm and mean body mass 65.4 (± 5.7) kg. A control group of females (n=14) did not participate in the training but performed physical assessments at identical time intervals to the participants in the training programme. The control group subjects were of mean age 20.4 (± 3.0) years, mean height 165.4 (± 4.3) cm and mean body mass 63.2 (± 8.1) kg.

Physical assessments

Physical assessments were performed before the start of training, during weeks 4 and 7, and at the end of training (week 10). Each assessment consisted of eight test parameters, details of the test procedures are given below.

Subjects performed a "one repetition maximum" vertical lift (1-RM). A square box with handles was lifted from 87 cm to 30 cm and returned to the starting position. The weight of the box was selected by the subject from a number of visually identical lead-filled bags, weighing between 0.25-2.5 kg. The weight of the box, unknown to the lifter, was recorded.

The maximum acceptable weight of lift (MAWL) was identified using the box and weights utilised for the 1-RM. Subjects were asked to select the load they perceived as a comfortable weight to lift at a rate of 6 lifts per minute for 10 min without becoming out of breath or fatigued. The weight of the box, unknown to the lifter, was recorded

Maximal isometric lifting strength (MILS) in a leg lifting technique was assessed on an isometric lift dynamometer as described by Birch et al.(1994). The equipment consisted of a platform in front of two 2 m vertical bars. Handles protruded perpendicular to the vertical bars, the height of which could be adjusted. Strain gauges were mounted on each bar to detect force applied vertically to the bars. Subjects adopted a squat lifting posture, grasping horizontal bar handles either side of the knees; handles were positioned at the level of the upper border of the patellae when the knees were straight. Subjects were instructed to exert a maximum force, lifting the bars vertically. Three maximum lifts were performed, each over a three second period. The peak force from the three trials was recorded.

Endurance of the leg muscle was assessed using the isometric lift dynamometer detailed above. Similarly, the leg-lift position was adopted and subjects were required to exert a vertical lifting force on the horizontal bars corresponding to between 45% and 55% of the MILS. Lifting force was displayed on a computer screen and the 45-55% limits marked. An audible signal indicated when the force exerted by the subject was not within the set limits. Subjects were instructed to maintain the lift for as long as possible. Endurance time was recorded in seconds.

Isokinetic strength of the lumbar flexors and extensors was assessed from standing and at a velocity of 2.09 rad s⁻¹ (120 deg s⁻¹). Tests were performed on a computer-controlled isokinetic dynamometer (Lido Active, Davis CA). Three submaximal movements through the range of motion (maximum flexion - maximum extension) were performed to familiarise subjects with the test. The peak torque from four maximal trials was recorded.

Skinfold measurements were obtained using skinfold calipers (Harpenden). The sites of measurement were: triceps, biceps, subscapular and supra-illiac. Percentage of bodyfat was predicted from the sum of the skinfolds (Durnin and Womersley, 1974)). Body mass was recorded at each session of testing (kg).

Physical training

The duration of the training programme was 10 weeks. Three supervised training sessions were undertaken each week. In addition, exercises were performed at home during the weekends.

Two of the weekly supervised sessions involved progressive resistance exercises performed on circuit-weight training equipment and activities involving exercise against body weight. A circuit-training session constituted the third supervised class. At the start of each exercise class a standard whole body warm-up regimen was performed and stretching exercises concluded each session.

The progressive resistance exercises were based upon the subject's one repetition maximum for each exercise. The 1-RM was re-evaluated during weeks 3,5 and 8 and subsequent exercise intensities were based upon the revised 1-RM.

Three circuits comprising ten exercises were performed during each weight training session. Table 8.2 lists the exercises performed during the weight training sessions. The intensity of the workload increased from the first to third circuit (between 50-80% of 1-RM).

Table 1. The exercises comprising the weight training circuits.

*Seated pulley row	*Lateral pull-down
*Reverse grip lateral pull-down	*Hamstring curl
*Knee extension	*Bench press
Back extension	Swimmers kick
Inclined sit-up	Leg-raise (bent leg)

*exercises performed against variable resistance

Analysis of data

Statistical analysis aimed to detect any changes in back muscle strength and lifting capability as a result of the physical training. Two-way analysis of variance for repeated measures was performed on the test parameters for the two groups during the four physical assessment sessions. In addition, Bonferroni T Tests localised significant differences ($P<0.01$) among the four test sessions for the experimental group, only where the ANOVA had revealed an overall change in the test results.

Results

The results of the tests performed by both groups in the four physical assessments are displayed in Table 2. Significant increases in the load selected by the training group for the 1-RM task were observed during the course of the physical assessments ($F_{3,81}=23.97$, $P<0.001$); increases were significant between tests 1-2, 1-3, 1-4 and 2-4 ($P<0.01$). The submaximal load selected by the same group for the MAWL task also

Table 2. Mean values (±SD) achieved by the training and control groups during the four physical assessments

Test	Test 1		Test 2		Test 3		Test 4	
	Training	Control	Training	Control	Training	Control	Training	Control
Max. Isometric Lifting strength (N)	637 (103)	743 (204)	679 (113)	831 (232)	706 (108)	803 (232)	734 (113)	837 (276)
Endurance (s)	38 (21)	33 (19)	47 (23)	29 (13)	61 (57)	38 (21)	51 (30)	35 (17)
Max. Acceptable Lift (kg)	15.5 (4.6)	20.5 (5.2)	17.4 (4.1)	21.4 (5.5)	18.5 (4.3)	19.3 (4.5)	20.7 (5.1)	18.5 (5.6)
Submaximal lift (kg)	10.7 (2.7)	12.2 (3.7)	12.3 (3.1)	14.5 (5.2)	14.3 (3.2)	13.7 (4.9)	16.3 (3.3)	12.9 (4.8)
Extension Peak Torque (Nm)	96.9 (21.0)	121.5 (46.7)	91.5 (16.3)	126.7 (45.8)	106.0 (23.5)	132.1 (53.1)	101.8 (15.6)	142.1 (62.5)
Flexion Peak Torque (Nm)	147.7 (21.0)	147.4 (33.0)	135.7 (20.5)	154.4 (35.3)	152.8 (18.4)	159.1 (31.6)	159.9 (21.7)	159.3 (33.0)
Body Mass (kg)	65.4 (5.7)	63.2 (8.1)	64.7 (5.3)	64.4 (8.4)	64.9 (5.1)	63.7 (8.3)	65.2 (4.9)	63.3 (8.1)
Body Fat (%)	27.9 (3.3)	27.5 (3.9)	27.5 (3.2)	27.5 (4.0)	27.7 (3.3)	27.5 (3.9)	26.8 (3.4)	27.8 (4.2)

demonstrated consistent increases following the second physical assessment (P<0.01). Maximal isometric lifting strength increased significantly over the physical assessments ($F_{3,81}=10.79$, P<0.001); the greatest improvement in performance was observed in the training group (P<0.01). Leg strength endurance times were significantly greater in the training group compared to the control subjects ($F_{3,81}=19.13$, P<0.001). Peak torque values for the trunk flexors increased in both groups over the four physical assessments ($F_{3,81}=2.57$, P=0.06). The training group demonstrated an increase in trunk flexor peak torque of 8.3% from the start to the end of training; the change in the control group was not significant (P>0.10). The mean trunk extensor peak torque increased significantly in both the training and control groups over the physical assessments ($F_{3,81}=5.42$, P<0.005). Body mass did not alter significantly during the ten week period in either group of subjects ($F_{3,81}=0.68$, P=0.569). In the training group, the percentage of body fat was significantly reduced between tests 1-4 and 2-4 (P<0.01).

Discussion

Training for muscular strength resulted in a significant increase in the two psychophysical measures of individual lifting capability. The increase in the maximum weight that subjects were willing to lift increased significantly after three weeks of physical training. By the end of the training programme MAL had increased by 33.6%. In addition, the load individuals considered themselves capable of lifting once every 10 seconds for 10 min increased overall by an average of 52.3%. The MAL results support the findings of Asfour et al. (1984) although their test consisted of three lifts of different heights compared to the one task assessed in the current study. Asfour et al. (1984) attributed the significant improvements in the maximum amount of weight lifted to both an increase in muscular strength and an improvement in lifting technique. In the current study improvements in perceived lifting capability exceeded the magnitude of increase in muscular strength and were greater in the trained group.

The two parameters of psychophysical lifting capability did appear particularly sensitive to physical training; however, the static and dynamic measures of muscular strength appeared to be influenced partially by familiarisation with performing the tasks; this was demonstrated by the results of the control group. Maximal isometric lifting strength at knee height and peak torque for the trunk flexors and extensors increased over the four assessment sessions in the group who did not undertake the physical training programme. However, for MILS, the magnitude of the increase was greatest in the training group. For the isokinetic assessment of the trunk flexors and extensor muscles at 2.09 rad s^{-1} it may be implied that variation took place under test-retest conditions. The subjects performed the physical assessments at the same time of day on each testing occasion, therefore controlling for the circadian variation in muscular strength (Atkinson, 1994). Literature relating to reliability data of isokinetic trunk assessment in the standing position is lacking and there is a need to examine sources of variation in performance of such assessments. Several implications arise relating not only to the screening of individuals for muscle function, but to the application of the methodology for pre-employment screening of individuals in occupations requiring manual handling. Present observations suggest a familiarisation effect which persists over repeated tests and that test frequency itself induced a training response.

Griffen et al. (1984) reported a test-retest correlation for MILS at knee height in a population of eighty females aged 29 years or less (r=0.855); the assessment consisted of the initial test followed by a re-test 5-7 days later. The statistical method of applying correlation coefficients to test-retest data is subject to criticisms which do not apply to techniques such as coefficients of variation or limits of agreement. Although the training group in the present study demonstrated an increase in MILS over the 10 week period, it seems that some of this improvement may be attributable to the effect of repeated testing.

Conclusions

This study demonstrated that a physical training programme of 10 weeks' duration induced an increase in the lifting capability of the participants. Improvements in the maximum weight that could be lifted were apparent following three weeks' of training (equivalent to 9 exercise sessions). Improvements in MILS and dynamic peak torque of trunk extensors and flexors were observed in the control group of non-training individuals, demonstrating an effect of repeated testing. Nevertheless, the greatest improvements in MILS were observed in the participants of the training programme. The results demonstrate the beneficial effects of physical training programmes for personnel involved in occupations demanding manual handling.

References

Asfour, S.S., Ayoub, M.M. and Mital, A. 1984, Effects of an endurance and strength training programme on lifting capability in males. *Ergonomics*, **27**, 435-442.

Atkinson, G. 1994, *Effects of Age on Human Circadian Rhythms in Physiological and Performance Measures*. Ph.D. Thesis, Liverpool John Moores University.

Birch, K., Sinnerton, S., Reilly, T. and Lees, A. 1994, The relation between isometric lifting strength and muscular fitness measures. *Ergonomics*, **37**, 87-93.

Burton, A.K., Tillaston, K.M. and Troup, J.D.G. 1989, Variation in lumbar saggital mobility with low back trouble. *Spine*, **14**, 584-590.

Durnin, J.V.G.A. and Womersley, J.j 1974, Body fat assessed from total body density and its estimation from skinfold thickness: measurements on 481 men and women aged from 16 to 72 years. *British Journal of Nutrition*, **32**, 77-97.

Elnaggar, I.M., Nordin, M., Sheikhzadeh, A., Parnianpour, M. and Kahanovitz, N. 1991, Effects of spinal flexion and extension exercises on low back pain and spinal mobility in chronic mechanical low back pain patients. *Spine*, **16**, 967-972.

Frymoyer, J.W., Pope, M.H., Clements, J.H., Wilder, D.G., MacPherson, B. and Ashikaga, T. 1983, Risk factors in low-back pain. *The Journal of Bone and Joint Surgery*, **65**, 213-218.

Griffin, A.B., Troup, J.D.G. and Lloyd, D.C.E.F. 1984, Tests of lifting and handling capacity: their repeatability and relationship to back pain symptoms. *Ergonomics*, **27**, 305-320.

Smidt, G., Herring, T., Amundsen, L., Rogers, M., Russell, A. and Lehmann, T. 1983, Assessment of abdominal and back extensor function. *Spine*, **8**, 211-219.

Snook, S.H. 1978, The design of manual handling tasks. *Ergonomics*, **21**, 963-985.

MANUAL HANDLING IN A DISTRIBUTION WAREHOUSE: THE IMPORTANCE OF TASK SPECIFIC TRAINING

Rebecca J Lancaster

**Horton Grange Cottage, Horton, Skipton,
N. Yorks., BD23 3JT**

**as part requirement for the degree of Master of Science (Engineering) in Work
Design & Ergonomics
School of Manufacturing and Mechanical Engineering, University of Birmingham,
Birmingham, B15 2TT**

In response to the 1992 Manual Handling Operations Regulations, a thorough
assessment was made of all the manual handling tasks in the distribution
warehouse of a multinational medical company. The level of risk was
established using a taxonomy of methods including; body part discomfort
ratings, posture analysis, NIOSH evaluation and the company's own manual
handling assessment checklist. Recommendations for reducing the risk were
made in terms of workplace redesign, changing work practices and task
specific training. The outcome identifies the lack of transfer of basic manual
handling training to the real work environment and illustrates the importance
of task specific training in reducing the risk of work related injury. Training
is required which provides employees with methods for handling loads other
than 'boxes', in environments other than the 'classroom'.

Introduction

According to official injury statistics, warehouse workers belong to a risk group
with a high frequency of reported back injury (Ljungberg et al 1989). The nature of the
work and the design of the warehouse environment present an ongoing risk of manual
handling injury. Implementation of the Manual Handling Operations Regulations (1992)
requires a consideration of the following; the task, the load, the working environment and
individual capabilities.

In the distribution centre employees are involved in a whole range of handling tasks
including lifting, pushing, pulling and carrying. They are required to handle a variety of
loads including boxes, pallets, barrels, and packaging sleeves. The technique for the
handling of these loads is often restricted by the workplace environment; for example,
unloading 40 ft containers, picking products from inside the racking, and transporting
loads from different levels using steps. Furthermore employees represent a varied
capability, both male and female, from the age of 16 to 50 yrs, with varying physical
capability and fitness.

In addition to the above, the regulations require employers to assess the adequacy of manual handling training provided for their employees. Ultimately the aim of the investigation was to implement the regulations, in doing so short comings in the existing basic manual handling training were identified. It is these shortcomings that will be the focus of this paper.

Terminology

A number of tasks performed in the warehouse are referred to throughout the paper, these will be described to give clarity of understanding to the reader;

Unloading 40 ft containers: Containers come to the warehouse from all over the world filled edge to edge and floor to ceiling with boxes, these boxes must be manually handled one by one onto pallets ready to be placed in the racking system.

Loading vans: the vans must be loaded with boxes manually. The task involves passing boxes from pallets at the top of the loading bay down to the bottom of the bay and then into the van.

Picking orders: Boxed products are picked from two tier racking and placed on a pallet to be transported, using a palletiser, to the despatch area. Employees randomly take boxes from the pallet, a technique known as 'pyramid picking'. The technique made in the recommendation , 'systematic picking' involves taking boxes from the pallets layer by layer. This technique allows the boxes at the back of the pallet to be pulled to the front of the pallet by sliding it on the layer of boxes below, thus enabling the load to be brought close to the body before lifting it.

Export packing: The process of preparing orders to be transported around the world by air freight or cargo. The products are either placed in large boxes and sealed or placed on a pallet and then covered by a packaging sleeve and strapped.

Pallet handling: The handling of empty pallets is inherent in habitual tasks.

Method

The level of risk for each of the manual handling activities was identified using a taxonomy of methods.

Body Part Discomfort Ratings provided an insight into the level of experienced discomfort for specific tasks. Employees were required to record their tasks in a daily log book, any discomfort experienced when performing these tasks was recorded on a body part discomfort map and rated from mild to extreme discomfort.

Posture analysis was conducted for each of the tasks using the OWAS (Ovako Working Posture Analysing System) method and the percentage of postures categorised in order to identify those requiring further attention.

Adequacy of the existing manual handling training, given to employees, was made using the checklist below. The checklist was devised by the author using the following publications as a guideline; HSE (1992) and Ayoub & Mital (1989).

In addition NIOSH evaluation of Maximum Acceptable Weight Limits for Lifting was calculated to demonstrate the importance of task specific training for one particular task - that of order picking.

Assessment of the risk was made using the company's own assessment checklist, derived from the 'Guidance on Regulations' (HSE 1992).

Results

Body part discomfort

The results show that there are more incidents of reported discomfort to the lower back than any other area of the body. The results were further analysed to identify those tasks that were reported to cause moderate/severe discomfort ever time the task was performed. These tasks were; loading vans and unloading containers.

Posture analysis

The percentage of postures in each of the four action categories of the OWAS system were calculated for each task. Two tasks were found to require attention; loading vans and unloading containers.

NIOSH Evaluation

The maximum acceptable weight limits were calculated for the order picking task. The evaluation was made for both the 'Pyramid' and 'Systematic' picking techniques. The measurements were taken for the picking of two boxes on the pallet; one from the front of the bottom layer (FL), and the second from the front of the top layer (FH). A comparison of the maximum acceptable weight limits for the two techniques is presented in figure 1.

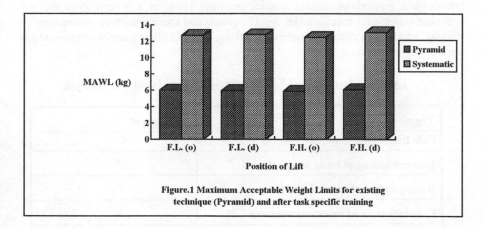

Figure.1 Maximum Acceptable Weight Limits for existing technique (Pyramid) and after task specific training

Assessment outcome

Four out of five of the manual handling tasks were calculated to be of medium to high risk. The results of the analysis methods appear to be consistent in categorising the level of risk of manual handling injury posed by the handling tasks.

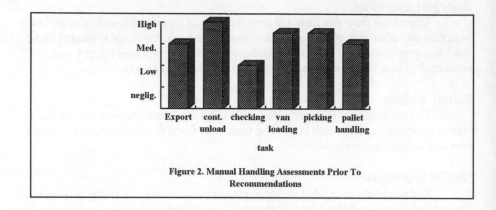

Figure 2. Manual Handling Assessments Prior To Recommendations

Evaluation of training

From Table 1. it appears that the existing training is effective in providing knowledge of general safe manual handling practices. However, it is insufficient in providing training for specific tasks, specific people and specific work environments. The training, which takes place in a classroom setting, does not allow transfer of knowledge to the working environment.

Table 1. Assessment of Existing Basic Manual Handling Training

Training Factor	Assessment
Safe handling technique	✓
Understanding of basic anatomy	✓
Basic physics of manual handling	✓
Consideration of work environment	X
Realistic Training Setting	X
Recognising potentially hazardous handling operations	X
Handling unfamiliar loads	X
The proper use of handling aids	X
The proper use of PPE	X
The importance of good housekeeping	X
Factors affecting individual capability	X

Discussion

The risk of injury posed by the medium and high risk tasks can be reduced by improving the manual handling training given to employees. Training which addresses the individual tasks specifically will be described to demonstrate the role of task specific training in reducing the risk of manual handling injury.

Unloading containers

This task largely involves repetitive lifting and alternative methods are being sort including slip sheet technology. The risk of injury can however be reduced by tailoring the training to the task. The outcome of the posture analysis revealed that many of the poor postures that were observed were due to poor handling technique. In some instances this technique was inherent in the task and therefore could not be improved by training, however, many of the postures which fell into action categories three and four were exerted during the handling of pallets. It is believed that because of the shape and size of pallets very few employees know how to handle them safely. The technique used to handle pallets varies extensively among employees, often resulting in; holding the load away from the body, stooping, twisting and holding the load above shoulder height. Training is required which demonstrates the safe lifting technique for the range of loads that are handled by employees and not just boxes.

Loading vans

The risk of injury when loading vans was largely due to poor handling technique. This poor technique is caused by incorrect layout of the task by the employee. For example, employees were observed leaning over pallets to lift boxes instead of positioning the pallet in such a way that boxes could be accessed form all sides of the pallet. The importance of workplace layout for the safe handling of loads should be an essential part of manual handling training. Due to workplace layout, employees are often prevented from using good handling techniques taught.

Picking

The advantage of systematically picking boxes from the pallet in layers is supported by the outcome of the NIOSH Evaluation. The maximum acceptable weight limits for this task are increased by approximately 100% when adopting this technique. The load can be brought close to the body before lifting and stooping into the racking is reduced. It may be necessary to give instruction on specific work practices for the safe handling of loads.

Export packing

The importance of being aware of individual capabilities was emphasised when observing this task. Many very large and heavy loads must be handled, some of which are unfamiliar loads including sacks and packaging sleeves. Training should provide good handling techniques for these unfamiliar loads as well as the importance of knowing individual capabilities and the advantage of team lifting.

Conclusion

The outcome of the current investigation identifies the importance of task specific training in reducing the risk of manual handling injury. Basic manual handling training, presented to employees in an unrealistic work setting, does not appear to be transferred to the real working environment. Training is required which provides employees with methods for handling loads other than 'boxes' in environments other than the 'classroom', and an awareness of individual capabilities.

References

Ayoub, M. M. and Mital, A. 1989, *Manual Materials Handling*, Taylor & Francis, chapter five, 265 - 273.

Health & Safety Executive (HSE), 1992 *Manual Handling: Guidance on Regulations,* (HMSO, London)

Ljungberg, A., Kilbom, A. and Hagg, G. M. 1989, Occupational lifting by nursing aids and warehouse workers, *Ergonomics*, **32**(1), 59-78.

Posture

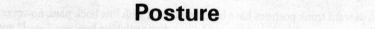

POSTURE TOLERANCE

Gerald Weisman, Kathryn Coughlin, James Fox

The Vermont Rehabilitation Engineering Research Center
The University of Vermont
One South Prospect Street
Burlington, Vermont 05041
USA

Awkward trunk postures have been associated with low back pain, however there are no standards and few guidelines distinguishing between "good" and "bad" postures. This paper describes a work-in-progress which is studying the effects of awkward trunk postures on discomfort levels, cognitive and manual task performance, activity of the erector spinae muscles assessed via RMS EMG, and fatigue of the erector spinae muscles assessed via a shift in the median frequency of EMG. Preliminary data collection on one subject in six postures, revealed discomfort, measured with a visual analog scale, changes with posture. It was also noted that task performance changed with posture. Forty-eight subjects will be recruited for the purpose of studying twenty-two postures in a block design experiment. We will compare the same postures between subjects and different postures within subjects.

Introduction

Twisting and bending have been identified as risk factors for low back pain (Bauer, 1989; Kelsey and Golden, 1988). The more a posture deviates from upright (neutral) posture, the less well it is tolerated. Biomechanical explanations for discomfort in non-neutral postures include muscle and ligament loading (Kumar, 1989; McGill and Norman, 1986; Potvin, McGill, and Norman, 1991; McGill, 1992). Few investigators have assessed the relationships among postures, posture tolerance and task performance and injury rates: thus, there are no widely accepted workplace guidelines for acceptable postures. In this study the relationships among specific back postures and duration for which they can be tolerated without discomfort will be investigated.

It is hypothesized that discomfort, assessed via a visual analog scale, varies with postural biomechanical characteristics. The following are the specific hypotheses of the laboratory study: (1) trunk postures can be analyzed in terms of tri-axial trunk angle and duration; (2) tolerance (discomfort as measured by VAS) of postures correlates negatively with the magnitude, complexity, and duration of the posture; (3) muscle fatigue correlates positively with the magnitude, complexity, and duration of the posture; (4) measures of muscle activity (RMS EMG) correlate positively with magnitude, complexity, and duration of the posture; and (5) task performance, measured by cognitive and physical measures, correlates negatively with the magnitude, complexity, and duration of the posture.

This paper presents work-in-progress. As of this writing preliminary data has been collected on one subject. Data collection and analysis will be complete and available by April, 1995.

Methods

Subjects

Forty-eight subjects between the ages of 18 and 50, who have been free from back pain in the previous 24 months, will be recruited to participate in the study. In this paper, we will report the results of the pilot study.

Materials and Methods

Postures were characterized in terms of trunk angles and categorized to reflect low-, medium-, and high-stress postures. The postures were separated into three sets, and the order of the postures within the sets will be randomized for each subject (see Table 1). Each subject will hold 13 of 22 static postures for a duration of 10 minutes or until intolerable. These postures include single plane and complex postures, each at three angle magnitudes, all within mild to moderate ranges as defined by Keyserling (1986) and Corlett and Manencia (1980). The dependant variables measured included: trunk angle, RMS EMG and median frequency of the erector spinae muscle, task performance, and discomfort.

Table 1. 13 Test Conditions. [Flexion (F), Lateral Bend (L), Rotation (R)].

Set A	Set B	Set C
0° F, 0° L, 0° R	0° F, 0° L, 0° R	0° F, 0° L, 0° R
20° F, 10° L, 10° R	20° F, 10° L, 10° R	20° F, 10° L, 10° R
35° F, 20° L, 20° R	35° F, 20° L, 20° R	35° F, 20° L, 20° R
50° F, 30° L, 30° R	50° F, 30° L, 30° R	50° F, 30° L, 30° R
20° F, 0° L, 0° R	20° F, 0° L, 0° R	0° F, 10° L, 0° R
35° F, 0° L, 0° R	35° F, 0° L, 0° R	0° F, 20° L, 0° R
50° F, 0° L, 0° R	50° F, 0° L, 0° R	0° F, 30° L, 0° R
0° F, 0° L, 10° R	0° F, 10° L, 0° R	0° F, 0° L, 10° R
0° F, 0° L, 20° R	0° F, 20° L, 0° R	0° F, 0° L, 20° R
0° F, 0° L, 30° R	0° F, 30° L, 0° R	0° F, 0° L, 30° R
20° F, 0° L, 10° R	20° F, 10° L, 0° R	0° F, 10° L, 10° R
35° F, 0° L, 20° R	35° F, 20° L, 0° R	0° F, 20° L, 20° R
50° F, 0° L, 20° R	50° F, 30° L, 0° R	0° F, 30° L, 30° R

Volunteers wear a three-axis electrogoniometer (BackTracker, Isotechnologies, Hillsborough, NC, USA), and four pair of electrodes (two mounted over the left and right erector spinae at L3 and two mounted over the multifidus at L5). The raw EMG is sampled at 512 Hz, filtered (band pass 20-500 Hz), stored on hard disk, and displayed on the screen in real time for visual inspection during the experiment. EMG signals are recorded through a Synergy™ EMG System (Fasstech, Burlington, MA, USA). The Synergy is configured to accept four channels of EMG and four A/D channels of unipolar voltage signals. The signals from the BackTracker are input into three of the A/D channels of the Synergy resulting in a comprehensive data collection system.

The subject is positioned into a posture using a posture guide. This device allows the subject to maintain a specified posture by providing non-supportive tactile feedback using a system of lightweight rings, the position of which defines the posture. If the subject attempts to change his or her posture during the experimental period, the subtle resistance of the jig provides sufficient feedback to maintain the posture.

Twenty-four of the subjects will complete a dexterity task assessment three times throughout the duration of the posture. The task assessment is based on Fitts' Law reciprocal tapping movement. A digitizer is used to record the number of hits and misses in a predefined time period. The tapping task utilizes pairs of targets, each pair consisting of two parallel lines 4cm long, printed on templates which fit on the digitizer. All combinations of W=1, 2, and 3cm (W=distance between the two parallel lines comprising a target), and A= 10, 20, 26 cm (A=distance between the centers of the two targets) are used to provide nine configurations. The configurations are printed three to a sheet on three separate sheets, so that no single W or A appeared more than once on a single sheet (Fowler, White, Holness, Wright, and Ackles, 1982). Each of the 9 configurations per task corresponds to a unique index of difficulty, a parameter based on the width of the targets and the distance between the target centers.

The other 24 subjects will complete a cognitive task three times throughout the duration of the posture. The cognitive task is a modified version of the INSPECT task created by Reynolds at the University of Buffalo. The subject searches for a specific character on the computer screen by moving a trackball to highlight different areas of the screen. The performance score is based on the average time to completion, and the number of errors.

The subjects complete a pain location drawing and a visual analog scale with endpoints labelled "No Discomfort" and "Extreme Discomfort", at two minute intervals during the test to elicit subjective measures of discomfort.

Data Analysis

RMS EMG and median frequency are obtained by post processing the raw EMG signal.. These parameters are used as indicators of trunk muscle activity and muscle fatigue, respectively. RMS EMG was normalized three ways to allow inter- and intra-subject comparisons. The normalization methods consisted of a five second MVC (maximum voluntary contraction), a five second Sorensen test, and five seconds of quiet standing in the upright posture with hands hanging straight down.

The following data are collected for each of 13 postures:
- Discomfort (VAS ratings) at 2 minute intervals;
- Task Performance Score at t=0 minutes, t=4 minutes, and t=8 minutes,
- RMS EMG at 20 Hz;
- Median frequency of EMG at 1 Hz;
- Angular position (via BackTracker) at 20 Hz.

The discomfort score is calculated by measuring the distance from "No Discomfort" to the mark, and dividing this number by the total length of the scale. The dexterity score is represented by the slope of the movement time versus difficulty index regression line. The raw EMG signal is post-processed, yielding four channels of RMS EMG and four channels of median frequency. Three channels of angular position are also recorded. This method provides 11 continuous and two discrete scores for each posture and subject combination.

For each combination of subject and posture, 6 lines of data are recorded for analysis, corresponding to 0, 2, 4, 6, 8, and 10 minutes into the posture. Thus, there are no missing data points unless a posture is not held for the full 10 minutes. Plots of the data are made for each combination of subject and posture, and average scores for each posture.

Results

Preliminary data has been collected on one subject performing a shortened version of the protocol, where he held six postures for 10 minutes or until intolerable. The subject completed the dexterity task up to three times while holding the specified posture. The following six postures were investigated:

P01:	0° flexion,	0° lateral bend,	0° rotation
P09:	0° flexion,	20° lateral bend,	0° rotation
P21:	35° flexion,	20° lateral bend,	20° rotation
P04:	50° flexion,	0° lateral bend,	0° rotation
P14:	20° flexion,	10° lateral bend,	0° rotation
P02:	20° flexion,	0° lateral bend,	0° rotation

The discomfort ratings are shown in Figure 1. It can be seen that the neutral posture (P01) produced the least discomfort, while the most complex posture (P21) is associated with the largest increase in discomfort.

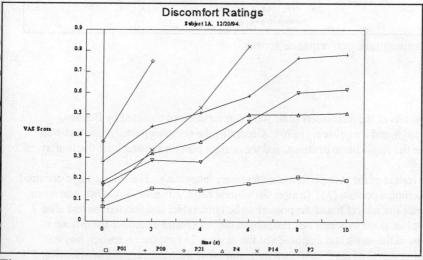

Figure 1. Discomfort ratings

The task performance scores were evaluated if the number of misses did not exceed five percent. Assuming a normal distribution of hits, 95% of the hits will fall within two standard deviations. Thus the maximum allowable error is five percent. The score reported is the slope of Movement Time versus Index of Difficulty regression line. A high positive score indicates that as the Index of Difficulty becomes larger (more difficult), the time to move the stylus between two targets becomes slower. The results show that the greatest

slopes were generated while the subject was in the neutral posture (P01) and in single axis
bending (P02). The least slopes were generated when the subject was in the single axis
bending (P09), or single axis flexion (P04), although the task slope for the 50° flexion
posture (P04), did increase during the second test. The slope of the Movement Time versus
Index of Difficulty was the smallest for the most complex posture investigated (P21). The
results are shown graphically in Figure 2.

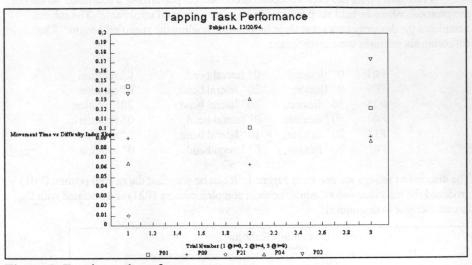

Figure 2 Tapping task performance scores

Discussion

The results of the discomfort scale were similar to those obtained by Bonney,
Weisman, Haugh, and Finkelstein (1990). Generally, the complex postures are less well
tolerated than the single plane postures, and the upright posture showed little discomfort, as
we expected.

The results of the task performance test were interesting. The lowest slope occurred
in the most complex posture (35° flexion, 20° lateral bend, 20° rotation). Only one score
was noted since the subject found the posture to be intolerable, and aborted the test after 2
minutes. The low slope indicates that the subject did not change speed for the different
configurations of the width and amplitude of the targets. In fact, the regression line was
calculated using only 3 of the 9 configurations, since 6 of the trials resulted in more than 5%
error indicating the subject was more likely to err while in the most complex posture. The
greatest slopes occurred when the subject was in the neutral posture, and when the subject
was in a single plane posture (20° flexion). Perhaps the subject was more willing to slow
his speed for the more difficult configurations in these less extreme postures. This
explanation is supported by the discomfort ratings for the neutral (P0) and complex (P21)
postures, however, the discomfort ratings do not support this hypothesis for the single plane
posture (P02).

Conclusions

A laboratory experiment is being conducted to determine the effect of awkward trunk postures on discomfort, cognitive and manual task performance and activity and fatigue of back musculature. Preliminary data supports the hypothesis that extreme and/or complex postures correlate negatively with task performance and tolerance. Data collected from the full study may be helpful in providing guidance for the design of workplaces to avoid awkward trunk postures, thus reducing the risks of low back pain and injury.

Future Research

The work reported here is part of an ongoing laboratory project, which will involve 48 subjects. The results for the pilot study are reported here. The laboratory study will be followed by a field study, where postures will be evaluated in various workplace environments with respect to the percent of total workday spent in the postures defined in the laboratory study.

Acknowledgements

This work was funded by a grant from the National Institute of Disability and Rehabilitation Research, US Department of Education (H133E30014-95).

References

Bauer, W.I., 1989, Scope of Industrial Low Back Pain. In Wiesel, Feffer, Borenstein, and Rothman (eds.), *Industrial Low Back Pain*, (The Michie Company, Charlottesville, Virginia).

Bonney, R., Weisman, G., Haugh, L.D., Finkelstein, J., 1990, Assessment of Postural Discomfort. In *Proceedings of the Human Factors Society 34th Annual Meeting*, 684-687.

Corlett, E.N. and Manencia, I., 1980, *The Effects and Measurement of Working Postures*, Applied Ergonomics, **11**, 7-16.

Fowler, B., White, P.L., Holness, D.E., Wright, G.R., and Ackles, K.N., 1982, *The Effects of Inert Gas Narcosis on the Speed and Accuracy of Movement*, Ergonomics, **25**, 203-212.

Kelsey, J.L. and Golden, A.L., 1988, *Occupational and Workplace Factors Associated with Low Back Pain*, Occupational Medicine: State of the Art Review, **3**, 7-16.

Keyserling, W.M., 1986, *Postural Analysis of the Trunk and Shoulders in Simulated Real Time*, Ergonomics, **29**, 569-583.

Kumar, S., 1989, *Cumulative Load as a Risk Factor for Back Pain*, Spine, **15**, 1311-1316.

McGill, S.M. and Norman, R.W., 1986, *Partitioning of the L4-L5 Dynamic Moment into Disc, Ligamentous, and Muscular Components During Lifting*, Spine, **11**, 666-678.

McGill, S.M., 1992, *A Myoelectrically Based Dynamic Three-Dimensional Model to Predict Loads on Lumbar Spine Tissues During Lateral Bending*, **25**, 395-414.

Potvin, J.R., McGill, S.M., and Norman, R.W., 1991, *Trunk Muscle and Lumbar Ligament Contributions to Dynamic Lifts with Varying Degrees of Trunk Flexion*, Spine, **16**, 1099-1107.

EFFECTS OF TASK POSITION AND TASK DIFFICULTY ON SEATED WORK POSTURE

Guangyan Li and Christine M Haslegrave

Department of Manufacturing Engineering
and Operations Management
University of Nottingham
Nottingham NG7 2RD UK

An experiment has been carried out to test the effects of manual task distance, task height and task difficulty on seated work postures. The task simulated was a manual assembly task with two levels of difficulty. The results indicated that the postures of head/neck, trunk and arm were all influenced by task distance and task height, with task distance being the more important factor. Manual task difficulty had a significant effect on elbow angle and upper arm posture, but did not show a significant effect on head or trunk postures. Within the range of locations tested, the best posture was found with this manual task located at 50-70 cm distance and at 20 cm below the horizontal eye level.

Introduction

It has been widely recognised that a poor and prolonged postural loading is a contributory factor in the development of some musculoskeletal problems (eg. Westgaard and Aarås, 1984; Aarås et al., 1988). With continuing technological advances, sitting has become the most common posture in today's workplace. Many tasks performed on production lines are characterised by low force exertion, but with high speed manipulation, frequent repetition of hand movement and long duration of continuous work under a constrained body posture, such as the tasks in electro-mechanical assembly, sewing-machine operation, VDU work, and so on.

An awkward sitting posture can be influenced by the interaction of several factors, including workplace layout (eg. seat and table height and inclinations), task location (task height, distance and direction), manual and/or visual requirements of the tasks, and the anthropometric characteristics of the seated worker.

Much effort has been made to improve seated working postures through the modifications of the workplaces, with studies which mainly include the design of work table and seat, and the provision of arm or back support, in the hope of reducing muscular stress in the shoulder and the back. These studies have sought to minimise the ill effects of constrained posture and have also provided incidental but convincing evidence for the association between postural loading and discomfort (eg. Mandal, 1981; Bendix, 1986).

However, despite the studies on table or seat design, it is still not clear how seated postures can be improved for many industrial work situations or, in other words, how to design all seated workplaces so that the operator can perform manual and/or visual tasks with a less stressful working posture. Among the workplace design and task factors which may affect a seated operator's posture, task distance, task height and manual difficulty levels of the tasks might be considered to be the most important, and these were tested in the present study for their effects on working postures.

Experimental method

A manual assembly task was designed to accommodate two levels of manual task difficulty. Four holes (25mm deep) were positioned equally around a disc (50mm in diameter, measured from the centre) with letters A, B, C and D arranged clockwise marking the corresponding positions of the holes. The size of the letters A-D was large enough so that no extra visual effort was needed as the task distance was varied during the experiment. The easy manual task was to insert a smooth plug into the hole (with their axes horizontal in the sagittal plane); the second more difficult task was to insert a plug made with three keys into the hole with three keyways. The position of the three keys/keyways appeared to be almost equally located around the plug/hole but with 1-2 degree deviation between each pair so that the plug could not be put into the hole unless each key matched its keyway (the possibility being 1/6). When the assembly task was completed (ie. the plug reaches the bottom of the hole), a signal light was automatically presented. To minimise the body twisting, the task was performed two-handed in the sagittal plane.

The experiment was designed with three independent variables: task distance (3 levels: 50cm, 70cm and 90cm from the subject's hip joint when sitting upright); task height (3 levels: 0cm, 20cm and 40cm downwards relative to the subject's horizontal eye level when sitting upright) and manual task difficulty (2 levels: easy and difficult). The experimental conditions were randomly presented to each subject.

Ten subjects (5 male and 5 female) participated in the experiment. All were right-handed. Their mean age was 27.8 years (SD=3.29, Range=21-33); their mean stature was 175.3cm (SD=6.8, Range=165.4-185.2). All subjects were paid for the trial. The subjects' eyesight was tested before the experiment to make sure that this met the standard visual acuity of 20/20 (with normal corrections if required).

After the anthropometric characteristics were taken, each subject was seated in an upright posture on a height-adjustable seat (no backrest) with its height adjusted to the subject's preference with thigh set horizontal and feet resting on the floor (with foot support if required). A specially designed equipment was used to maintain this upright position during the calibration, and to measure the posture changes during the trials. The task height and task distance were than set corresponding to the subject's horizontal eye level and hip joint. The seat height and seat position, subject's hip position relative to the seat, and the foot position were maintained the same during the whole experiment. All changes in posture therefore were movements of the upper body.

Head/neck flexion, trunk flexion, and right elbow angle were recorded by electronic goniometers (Penny+Giles). These postural data were loaded into a portable data logger and then transferred into a computer where the postural recordings were given as continuous curves. The head/neck posture was measured using goniometers attached on the back of the head and the back at C7 (with a special hat used to fix the goniometer). The head/neck posture recorded was therefore relative to the back. Trunk posture was measured using goniometers placed on the right horizontal thigh and upright trunk, with the hip joint in the middle corner of the goniometers. The trunk posture was recorded as the posture change from the upright position.

The subjects' right upper arm flexion was recorded by a video camera positioned on the right-hand side normal to the frontal plane and at the shoulder height. Upper arm flexion/extension was measured from the video records using markers which were placed on the shoulder joint, the hip joint and the lateral side of the elbow. The upper arm flexion was defined from a line drawn through the marker on the elbow and shoulder joint and the line through the shoulder joint and hip joint.

Subjects worked on each experimental condition for 3 minutes during which at least 4 recordings were made of head/neck, trunk flexion/extension, elbow angle and upper arm flexion. The mean value of the first three recordings was used to represent the posture adopted at the test condition. After completing each test condition, the subject rested for 2 minutes, and there was a 15 minute's break in the middle of the 90 minute experiment. The lighting condition was kept constant throughout the experiment.

Results

Postural parameters were analysed with ANOVA on Minitab. The effects which were shown to be statistically significant for each of four parameters are indicated in Table 1. The main significant effects are illustrated in Figures 1–4. The results indicated that both task distance and task height had significant effects on head/neck, trunk, elbow and upper arm postures.

Table 1. The significance in ANOVA analysis of the effects of independent variables on postural parameters

	Task distance	Task height	Task difficulty	Task distance x task height	Task height x task difficulty
Head/neck	p≤0.001	p≤0.001	NS	NS	NS
Trunk	p≤0.001	p≤0.025	NS	NS	NS
Elbow	p≤0.001	p≤0.001	p≤0.01	NS	NS
Upper arm	p≤0.001	p≤0.001	p≤0.001	p≤0.001	p≤0.001

NS: not statistically significant (p>0.05).

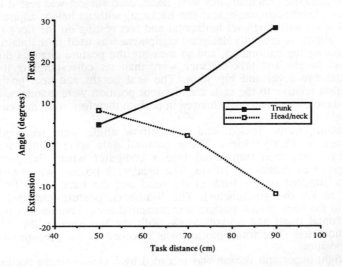

Figure 1. The effect of task distance on head/neck and trunk postures

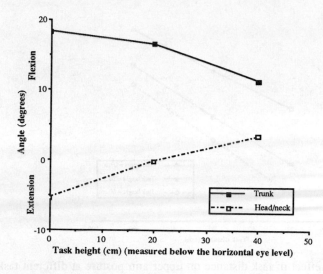

Figure 2. The effect of task height on head/neck and trunk posture

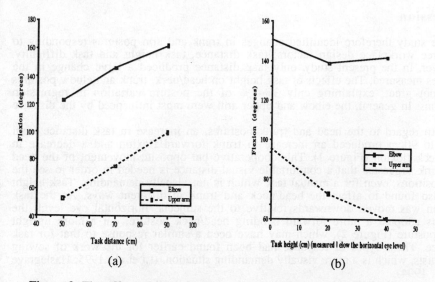

(a)   (b)

Figure 3. The effect of task distance and task height on elbow and upper arm postures

Manual task difficulty had a significant influence on elbow and upper arm postures, but not on head and trunk postures. Subjects tended to adopt a smaller elbow angle (5 degree smaller on average) and less upper arm flexion (3 degree less on average) when performing a relatively difficult manual task than when performing an easier manual task.

There was a significant interaction in the effects of task distance and task height on upper arm posture, as shown in Figure 4. In addition, there was a significant interaction between task height and task difficulty, and both these interactions suggest that postural responses of the arm can be quite complex when performing manipulative tasks.

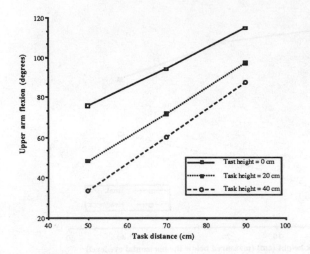

Figure 4. The effect of task distance on upper arm posture at different task
heights

Discussion

The study therefore identified changes in trunk and arm postures responding to
the three workplace design factors, task distance, task height and task difficulty.
However, in the present study, only task distance produced a large change in the
postures measured. The effects of task height on head/neck, trunk and elbow postures
were not great, explaining only 3%-7% of the posture variation in regression
equations. In general, the elbow and upper arm were most influenced by the distance
factor.

With regard to the head and trunk postures, an increase in task distance from
50cm to 90cm produced an increase in trunk forward flexion and a decrease in
head/neck flexion (Figure 1). This cooperative but opposite movement of the head
and trunk suggested that a comfortable visual distance is needed in order to see the
task positions, even for a manual task which is not visually demanding. Task height
was also found to affect the head/neck and trunk in different ways. As the task
position was moved downwards relative to the subjects' horizontal eye level, the
subjects adopted a more forward bending head/neck posture, but a more upright
trunk posture (Figure 2), which may have been a similar response to that for task
distance. These types of response had been found earlier for the work of sewing
machinists, which is a more visually demanding situation. (Li et al., 1995; Haslegrave
and Li, 1994).

Manual task difficulty was shown to have statistically significant influence on
elbow and upper arm postures, but task difficulty showed almost no effect on head or
trunk postures. The result might have suggested that a manually more demanding
task needs to be performed with the arms either closer to the trunk, or more
abducted. In doing so, whether the subjects can use their fingers more easily or they
can see the task better, is still not fully understood. As the effect of task difficulty on
elbow angle and upper arm flexion was independent of the effect of task distance,
and the trunk was not significantly affected by the task difficulty factor, a different
upper arm abduction must have been involved. Furthermore, a significant interactive
effect of task height and task difficulty on upper arm flexion (Table 1) might have
also indicated that to perform an easy or difficult manual task, different wrist
positions relative to the forearm as well as to the task are needed, and so the arm is
adjusted to obtain the wrist posture. In other words, if the arm position is not free to

adopt the task-required wrist posture, (if, for example, the elbow or forearm is rested on table or on an arm support, or the arm abduction is constrained by a limited work space), an awkward wrist posture may have to be adopted in performing the manually demanding tasks.

It is shown in Figure 4 that the effect of task distance on upper arm flexion varies significantly according to task height. Clearly, to maintain the upper arm close to the trunk, a low task height and near task distance are preferable. Furthermore, a positive correlation between upper arm flexion and trunk flexion found in this experiment (r=0.63, p≤0.05) was consistent with previous studies (Li et al., 1995) suggesting that reducing task distance can be beneficial for both arm and trunk postures. Nevertheless, this should be considered together with the effects on head/neck postures in order to determine the situation which is best overall.

Conclusion

The study has shown how a seated working posture can be influenced by several work factors and by the interactions between these factors, and is useful for better understanding the posture behaviour in particular tasks. To maintain a more upright trunk and less extended arm postures, a near and low task location is preferred. However, a more upright head/neck posture requires a relatively more distant and higher task position (close to the horizontal eye level). Based on this study, the task located within the distance of 50-70 cm and at the height about 20 cm below the horizontal eye level appears to be beneficial overall for the assembly task tested.

As has been suggested by the current experiment and the previous studies, a close manual task almost always involves visual effort in a greater or less extent, and this may play an important role in posture adoption. Task difficulty has been found to affect arm posture, but not the head/neck and trunk, suggesting that to perform different manual tasks, arm posture should be freely adjustable, which should be considered especially when an arm support is used.

References

Aarås, A. et al., 1988, Postural angles as an indicator of postural load and muscular injury in occupational work situations. Ergonomics, 31, 6, 915-933.
Bendix, T., 1986, Chair and table adjustment for seated work. In: The Ergonomics of Working Postures, eds. N. Corlett, J. Wilson and I. Manenica. (London: Taylor & Francis), 355-362.
Haslegrave, C M. and Li, G., 1994, Influences on posture adopted for a skilled manual task. In: Proceedings of the 12th Triennial Congress of the International Ergonomics Association. August 15-19, Toronto, Canada.
Li, G., Haslegrave, C. M. and Corlett, E. N., 1995, Factors affecting posture for machine sewing tasks: the need for changes in sewing machine design. Applied Ergonomics (in press).
Mandal, A. C., 1981, The seated man (Homo Sedens), the seated work position, theory and practice. Applied Ergonomics, 12, 1, 19-26.
Westgaard, R. H. and Aarås, A., 1984, Postural muscle strain as a causal factor in the development of musculo-skeletal illness. Applied Ergonomics, 15, 3, 162-174.

INVESTIGATION OF POSTURES ADOPTED BY BREAST SCREENING RADIOGRAPHERS

JL May, AG Gale, CM Haslegrave[1],

Applied Vision Research Unit, University of Derby, Derby, UK
Manufacturing Engineering, University of Nottingham, UK[1]

The aim of this study was to determine which postures are adopted by radiographers during breast screening. Video recordings of radiographers performing mammography at two screening centres were made and posture analysis performed on these data. The results indicated that breast screening radiographers do adopt potentially harmful postures while performing mammography. Possible causal factors in the development of these postures were then identified by linking their occurrence with a certain stage in the screening cycle.

Introduction

During the screening process the radiographer has to position appropriately each woman being screened against the mammography unit so that an X-ray image (mammogram) of each breast is obtained. Usually two different radiological views of each breast are taken and each view entails different, precise and careful positioning of the woman. The radiographer must also maintain a certain throughput of women through the centre (roughly one every five minutes) and so the procedure has to be performed efficiently within a minimum amount of time. Positioning the woman may often involve the radiographer in adopting unusual postures so leading to the possible development of musculoskeletal disorders.

Method

Observational Study

An observational study was undertaken at two Breast Screening Centres which encompassed a wide cross section of radiographers of different stature (1.52m-1.73m) and mammography experience. Full ethical approval, as well as the consent, of the women concerned had first been obtained. Video recordings were then made of eight radiographers as they screened women using two different mammographic views, the medial-lateral oblique and the cranial caudal. These recordings were made for each of the four different mammography units used at the two centres.

Task analysis

A task analysis was constructed to establish the relevant sequence of radiographer operations during screening. This was based upon the appropriate literature, published in accordance with the breast screening programme, which illustrated how the radiographer should undertake the process of obtaining a high quality mammogram. In order to establish if this procedure was that which actually was performed by the radiographers in practice several radiographers then were observed while performing mammography and the sequence of actions which they performed and the manner in which they undertook them was noted.

Results

Observational Study

A list of postures generally considered to contribute to muscular discomfort was constructed from data from Van Wely (1970), who produced a list of 'bad postures' thought to be connected to certain sites of pain, and Corlett's (1988) 'ten principles for the arrangement of workplaces' . The frequency and length of time that each radiographer adopted one of these unsuitable postures throughout the screening cycle was recorded. This was performed a minimum of five times for each radiographer while they took the two mammographic views. It was also related to various stages of the screening cycle (obtained via task analysis) in order to identify possible causal factors for the adoption of the harmful postures recorded. A list of these postures, and why they occurred within the screening cycle are given in Table 1.All the radiographers adopted some of these postures with every mammogram taken. The postures adopted for the longest amount of time were:-

Posture	Time(s)
Bent/twisted posture	33.4
A single fixed/cramped posture	18.5
Static work above heart level	32.6
Upper arm hanging unsupported out of vertical plane	52.8
Mean Screening Cycle Time	*78.1*

The frequency and duration of these postures depended upon; the mammographic view taken, which breast was X-rayed, which mammography unit was used and which radiographer took the mammogram. This data was analyzed using ANOVAs and significant differences found in; the cycle time to take a mammogram between radiographers, which view was taken, and which breast was imaged (Table 2).

Table 1. Unsuitable postures adopted by radiographers when performing mammography and reasons for their occurrence.

Postures adopted		*Reasons*
Bending/twisting		-Bending under X-ray tube head
		-Avoiding reflections from X-ray compression plate
		-Observing breast when positioning due to dimensions the women and the unit.
		-Locating the foot pedal to operate the unit
		-Locating and positioning woman's arm.
		-Observing rotational guideline of unit
Unequal distribution of weight on both feet		-Using foot pedal to apply breast compression
Joints at extreme range of movement	*Leg*	-Using foot pedal which may be located out of reach
	Arm	-Adjusting height and rotation of unit particularly for tall women and if rotating in a direction away from the radiographer
		-Lifting breast when behind woman, particularly if women of heavy build
		-Positioning breast, reaching across woman
	Wrist	-Positioning and removing X-ray film
		-Lifting breast onto compression plate
Muscle force not co-linear with limbs		-Inserting and removing X-ray film
Static work performed above the level of the heart		-Bending under X-ray tube head
		-Height of woman screened
Repetitive force by one hand/foot		-Use of foot pedal
		-Altering height/rotation of unit
Upper arm hanging unsupported from vertical		-Having to work reaching across the woman to position breast
Restricted controlled movements		-Precise positioning of breast to quality standards

Even when the data were controlled for the number of left and right breasts imaged and the number of cranial caudal and medio-lateral views taken, a significant difference was still observed between the cycle times of different radiographers. The type of mammography unit used also had a significant effect on the cycle time when the cranial caudal view was taken. This is shown in Table 3.

Table 2. **Results of ANOVAs showing the mean cycle times for the view taken, the breast imaged, the radiographer and the unit used.**

	Variable	Cycle Time(s)		
View	Cranial Caudal	62.9)	Significant	
	Medio-lateral	87.3)	Difference	p<.001
Breast	Left	85.1)	Significant	
	Right	72.1)	Difference	p<.001
Radiographer	1	83.5)		
	2	74.5)		
	3	66.5)		
	4	76.6)		
	5	99.2)		
	6	73.4)		
	7	65.5)	Significant	
	8	83.7)	Difference	p<.001
Mammography Unit	1	75.3)		
	2	79.7)		
	3	86.5)	No Significant	
	4	76.9)	Difference	p>.01

Table 3. **Results of ANOVAs showing the significance levels for the differences in mean cycle time between radiographers and the type of mammography unit used when data is controlled for the view taken and which breast is imaged.**

	View		Breast	
	Cranial caudal	*Medio-Lateral*	*Left*	*Right*
Radiographer	p=.01	p=.01	p=.02	p=.09
Type of Mammography Unit	p=.03	N/S	N/S	N/S

ANOVAs were conducted on the data to examine whether the mean time for which particular postures were adopted varied significantly with the unit being used. An example of the results is given in Table 4 which shows that there are significant differences in the presence of extension/flexion of the wrists depending upon which mammography unit is used.

Table 4. **Results of ANOVAs showing the significance levels between the type of unit used, the radiographer and the time poor wrist posture is adopted.**

	Extension/Flexion of Left Wrist	Extension/Flexion of Right Wrist
Mammography Machine	p=.0001	p=.0001
Radiographer	p=.001	p=.0008

The above table also shows that there is also a significant difference between radiographers in the amount of time they adopt certain postures throughout the process of taking a mammogram. Further work is being undertaken to extend these results, and confirm the reliability of the method used in this study for recording data.

Task analysis

The task analysis revealed that the radiographers studied followed the same method of working on different mammographic units. There was a difference in working patterns between radiographers. The less experienced were slower on all units and needed to re-adjust the position of the unit and the women's breasts several times throughout the screening cycle. This resulted in more frequent use of the controls and more frequent walking from one side of the unit to the other more often. Many radiographers deviated in some way from the constructed ideal theoretical method, but this tended to be very slight, such as standing on the opposite side of the woman when taking the cranial caudal view, or taking mammograms in a different order.

Discussion

While occupational factors are not the only cause of musculoskeletal problems, the video analysis performed here revealed that breast screening radiographers do adopt postures which are believed to contribute to muscular discomfort and injury. It is possible that some of the reports of pain previously recorded from such radiographers may therefore originate from work related activities. The postures adopted for the longest period of time were found to be; bending/ twisting, a single fixed/cramped posture, performing static work above the level of the heart and extension of the upper arm while working.   The amount of time radiographers were forced to adopt many of these postures was significantly increased while the medial lateral view was taken. This was partly due to the fact that all radiographers took significantly longer to perform this view rather than the cranial caudal, (87.3 versus 62.9 seconds) and also because of the position of the tube head when this view was taken.

There was a significant difference in the amount of time some postures were adopted between different radiographers. This still remained true when the data were controlled for the number of mammograms each radiographer took of the left and right breast from the two different views.   This is possibly due to slight differences in technique, and differences in mammography experience - with radiographers of less experience taking longer to position a woman.   The anthropometric dimensions of the radiographer, stature for example would also play a part in the adoption of some postures and research is continuing to investigate this.

The style and design of the mammography units themselves are also important as demonstrated by significant differences in the time certain postures are adopted and the use of the mammography units.   For example the extension/flexion and deviation of both wrists occurred significantly more often when one particular unit was used due to the fact that the final breast compression had to be applied manually rather than by use of a foot pedal. For a similar reason the unequal weight distribution and extension of the legs was also therefore of shorter duration for the latter unit. The dimensions of some units and the layout of controls resulted in radiographers having to adopt a more severe or awkward angle to perform a function than on other units. As no measurement of angle is taken this difference is missed by the present recording system.

As observed breast screening in this study radiographers are often subjected to frequent; bending and twisting, static working postures, forceful movements, and repetitive work - factors which have all been related to muscular problems (Stubbs and Buckle, 1984). It is feasible that this may have a cumulative effect of wear and tear on their musculoskeletal system, aggravating possible previous injuries or paving the way for new ones. The main postures which may contribute to the back, neck and shoulder pain reported by radiographers in a previous study (May et al, 1994) are; stooping, bending and static muscular loading of the shoulders and upper arms. These postures are caused primarily by the basic design of the mammography units which fail to adequately address the radiographers visual and postural requirements while being used.

Conclusions

Whilst it may not be possible to eliminate all contributory factors in the development of musculoskeletal pain, it should be possible to reduce the amount of bending and twisting that the radiographers have to perform throughout the working day. This is important not only for the health of the radiographer but also for the quality of current breast screening and potentially the future compliance of women attending for screening. This research is working towards the proposal of recommendations to further the ergonomic design of these mammography units.

Acknowledgements

This research is funded by the National Health Service Breast Screening Programme.

References

Corlett, E.N. 1988, The investigation and evaluation of work and workplaces, *Ergonomics*, **31**, 5, 727-734.
May, J.L. Gale A.G., Haslegrave C.M., Castledine J., Wilson A.R.M. 1994, Musculoskeletal problems in Breast Screening Radiographers. Contempory Ergonomics. Ed. S.A. Robertson. Taylor and Francis.
Stubbs, D.A. and Buckle, P.W. 1984, The epidemiology of back pain in nurses, *Nursing*, **32**, 935-938.
Van Wely, P. 1970, Design and disease, *Applied Ergonomics,* **1**, 5, 262-269.

Health and Safety

ERGONOMICS ASSESSMENT OF PASSENGER CONTAINMENT ON FAIRGROUND RIDES

John Alan Jackson

Health & Safety Laboratory

(a Division of the Health & Safety Executive)

Broad Lane

Sheffield S3 7HQ

Recent research has shown that many accidents occurring on fairground rides can be attributed to poor design of the passenger containment. This paper describes a structured method for ergonomics assessment of passenger containment systems. The method has been used successfully by the Health & Safety Executive to identify problems and help in assessing solutions.

Introduction

Recent research (Jackson, 1993) has shown that poor design of passenger containment systems on fairground rides may be a contributory cause to many of the accidents reported to the Health and Safety Executive (HSE). The HSE has a large field force of Inspectors covering the whole of Britain who, although multi-skilled professionals, are unlikely to have a specialist knowledge of ergonomics. A simple, yet reliable, method for making ergonomics assessments of containment systems needed to be developed for use by HSE field Inspectors and other ride examiners.

Passenger containment systems

Passenger containment is more than lap bars and seat belts, it is everything which contributes to the safe containment of the passengers such as seating and footwells. The passenger can be considered to be at the centre of a containment system, interacting with the various system components. Three types of interaction have been identified, static interactions - the physical size relationships; dynamic interactions - the effects of force; and psychological interactions - passengers' perceptions and behaviour. To simplify matters, these can be assessed separately but as they are part of the same system the final analysis must recognise their interdependence.

429

Static interactions

In assessing static interactions, one should consider aspects relating to the physical fit of passengers in the ride. It was found that most containment systems could be broken down into a number of common components. The five main components are seating, lap bars, hand rails, footwells and over shoulder restraints. Each of these components has an associated series of critical body dimensions. As an example, consider the height of the seat pan as shown in figure 1. The critical body dimension is popliteal height. If the seat pan is too high shorter passengers will not be able to brace themselves with their feet. Thus it could be suggested that the seat pan height should be no more than the 5th percentile value for the lowest age range of the passengers. Considering such dimensions can help in setting a suitable minimum height restriction for the ride.

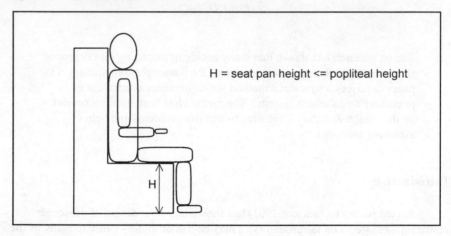

Figure 1. Seat pan height

A proforma has been developed to ensure that the important dimensions of the five main components are all measured, an example of part of this is shown in figure 2. This is supported by tables of critical body dimensions associated with these measures along with suggestions for percentile ranges. Actual figures are not given as several sources of anthropometric data are available for many populations. (eg. Pheasant, 1990). When designing a ride, the range of dimensions in the population for which it is intended needs to be considered. It may be necessary for the designers to stipulate maximum and / or minimum height restrictions, or other passenger selection criteria, for the design to function safely. When assessing existing containment systems, the examiner would measure those parts of the containment components which have critical body dimensions associated with them. The use of the proforma ensures consistency in collection of data. The ride dimensions would then be compared to data for the associated body dimensions and any mismatches identified for further investigation or action.

LOCATION _____   **DATE** _____

RIDE NAME_____   **RIDE TYPE**_____

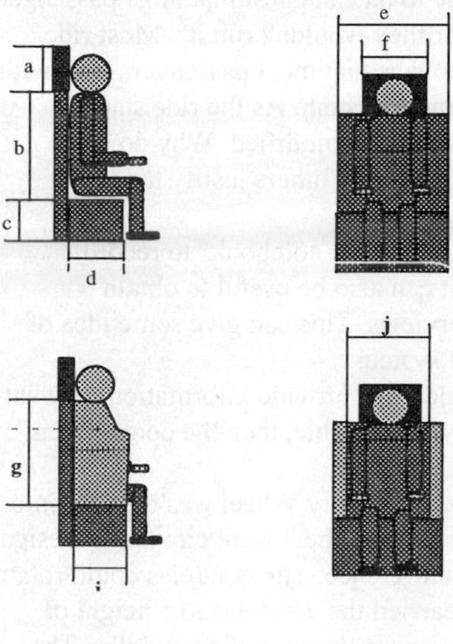

SEATING

a_____

b_____

c_____

d_____

e_____

f_____

g_____

h_____

i_____

j_____

Figure 2. Section from proforma

Dynamic interactions

Fairground rides are designed to impose forces on passengers and this is why there is a need for containment. The passenger is not a fixed part of the ride and they will attain their own momentum. If the passenger is moving in the same direction as the ride at the same velocity there is little danger of them being ejected. However, when the ride changes direction or velocity the passengers will tend to carry on moving in the direction they were originally travelling unless they are acted upon by a containment component such as the seating or the lap bar. Some knowledge of the size and direction of ride forces is useful to determine the placement of containment components and how substantial the containment needs to be.

Objective measurements of forces require expensive equipment and access to rides for instrumenting them. The analysis of the results is not straightforward as the motion has to be resolved into three dimensional vectors and then interpreted in terms of passenger momentum and orientation at any one time. Some degree of subjective assessment of the forces can be made by using direct observation and video recordings. This method, combined with an understanding of the overall ride motion, gives some idea of direction of forces which is often sufficient to determine whether a particular containment system will be effective.

Psychological interactions

Psychological interactions tend to be more difficult to measure. The problem is to assess the passengers' perception of the containment system and how this affects their behaviour in the system. Observations made to date suggest that most passengers will make the assumption "well it must be safe or they wouldn't run it". Most ride operators will make the same assumption. This means that most passengers start with an optimistic view of the security of the containment system. As the ride starts to impose forces on the passenger their perceptions may be modified. Why do some people ride roller coasters with their arms in the air while others justify the term 'white knuckle' ride?

Again, the video camera can be used as a "dynamic notebook" to record passengers actions whilst the ride is in motion. It can also be useful to obtain passengers' opinions of the ride by informal interviews. This can give some idea of pre and post ride perceptions of the containment system.

Observation of passenger behaviour on rides will provide information on what is predictable misuse. If the misuse is reasonably foreseeable, then the containment system should be designed to take account of it.

An example of this problem was seen when a Liberty Wheel was brought into the UK from elsewhere in Europe. The gondolas on the wheel were circular in design with moulded plastic seats around the inner circumference. The gondolas could rotate and swing backwards and forwards. The wheel carried the gondolas to a height of approximately 27m. There was no form of passenger restraint in the gondolas. The dynamics of the ride are such that they do not pose a risk of ejection if the passengers remain seated. However, it is common for passengers to stand up on this type of ride to get a better view. If this occurs the risk of falling out of the gondola is greatly increased. The HSE insisted that some form of restraint, to stop passengers standing up, was fitted before allowing the ride to operate. The criteria for this was that it was reasonably foreseeable that passengers would stand up.

Field trials

The methods outlined in this paper have been applied to actual investigations by HSE ergonomists. Using this structured method it was found that gathering the relevant information on a particular ride was achieved in a thorough and rapid manner and that the interpretation of the data was guided by the general principles identified in the research. Suggested solutions could also be assessed by considering the containment system dimensions as they relate to the anthropometry of the population in question and the ride dynamics before modifications to the ride are made.

One example was a study of bench type fairground rides. Several accidents have occurred where passengers have been ejected from this type of ride. Video recordings of one particular model showed that the passengers' legs were swinging from side to side, due to lateral motion, and that there was a twisting of the hips and the body was forced downwards, due to a vertical component in the motion. Measurements taken of the ride showed that the seat pan was 500mm above the foot rest. This suggests that British adults of less than 95th percentile stature would have difficulty in bracing against the foot rest to support their lower body. The suggested solution was to fit a staged foot rest and provide better padding on the lap bars. These suggestions have been discussed with the manufacturer and HSE are working with them on the details for implementation.

Discussion

The methods for identifying static factors have proved to be effective in investigations of several incidents. Five of the most common components have been analysed to date but more have been identified. The method is still under development and as new rides are produced it is expected that new components could appear. This means that the anthropometric database of containment components will need to be updated to include these.

The methods used for assessing dynamic factors have, so far, been mostly subjective. This makes it difficult to compare one ride with another, even if they are of the same type. However, video recordings have been used successfully in incident investigations, as outlined above. Work is being carried out in HSE to assess objective methods of acceleration measurement and analysis methods which interpret the results in terms of their effects on the safe containment of passengers.

The work carried out to date on the psychology of passenger containment has proved useful in determining predictable misuse of the containment system. It has also helped to identify common passenger behaviour patterns on particular types of ride. However, more work is required in this area to build on the knowledge gained so far. Children's rides are of particular concern as, in many cases, little attention is paid to containment as the rides are less violent in motion. However, children's behaviour is often less predictable than adults and they may have less appreciation of the consequences of their actions.

Conclusions

Considering containment systems rather than just passenger restraints gives a better understanding of what is required for the safe containment of passengers. Identification of the three factors influencing the interaction of passengers with the containment system has helped to produce a structured method for ergonomics assessment of such systems. Use of this methodology on actual incidents has shown that although it does not provide all the answers, it is very efficient in identifying the problems.

HSE inspectors have been given training in the use of this structured method for the identification of problems and there are plans to produce training material for wider consumption.

References

Jackson, J.A. 1993, Passenger containment on amusement devices. In E.J. Lovesey (ed.), *Contemporary ergonomics 1993,* (Taylor & Francis, London) 184-189.

Pheasant, S. 1990, Anthropometrics an introduction, BSI PP7310.

A MULTI-LEVEL APPROACH TO ERGONOMIC AND RELATED ISSUES IN THE B.C. HEALTHCARE INDUSTRY

Julia M. Rylands, Ian Pike, Lesley Allan, Mieke Koehoorn,
Jan Mitchell and Dennis Davison

Employee Health and Safety Services, Healthcare Benefit Trust,
Suite 1200, 1333 West Broadway,
Vancouver, B.C. Canada, V6H 4C1

Problems resulting from poor ergonomics in healthcare facilities are well
known and well documented. These problems are generally multifaceted,
ranging from back injuries due to patient handling, to high absenteeism
resulting from physical and emotional stress due to increased workloads.
We have developed a program that includes a number of strategies to assist
the industry in taking both a proactive and multifaceted approach to
creating a "safe, healthy and effective workplace". The initiatives of the
healthcare facilities who adopt one or all of these strategies, needs to be
both sustained and monitored in terms of their effectiveness. Our program
includes the development and validation of a number of metrics by which
this effectiveness can be judged in the coming years.

Introduction

Absenteeism and injuries statistics in the B.C. healthcare industry have surpassed
many other industries, including heavy industry such as forestry. This is not a new
situation and is the case in many other jurisdictions. Efforts have previously focused on
reducing musculoskeletal injuries (MSI), specifically back problems in nurses. Back
training programs proliferate and much effort goes to reducing manual handling both in
nursing and other areas. It is recognised, however, that training alone will not sustain
reductions in injuries (Stubbs, Buckle, Hudson, Rivers and Worringham, 1983). Some
ergonomic interventions have concentrated on improving the physical and biomechanical
aspects of the job (Garg and Owen, 1992). Others have proposed ergonomic programs
of a more participatory nature (McAtamney and Corlett, 1992).

The approach to absenteeism has dealt more with role conflict, job stress (Burton,
1986; Toivanen, Helin, Hanninen, 1993)) and patient load (Bourbonnais, Vinet, Vezina,
Gingras, 1992). Interventions have been more organisational in emphasis. For example,
quality circles, introducing EAP services, relaxation therapy and scheduling
modifications.

These various approaches have resulted in some impact on reducing the problem. However, it is recognised that dealing with all levels of a problem will bring the greatest returns (Hendrick, 1994). Our program is built on a framework as shown in Figure 1. The strategies discussed below were developed to assist facilities recognise the different facets of their problem, give them ideas about what they should be striving for, and guidance as to how to get there.

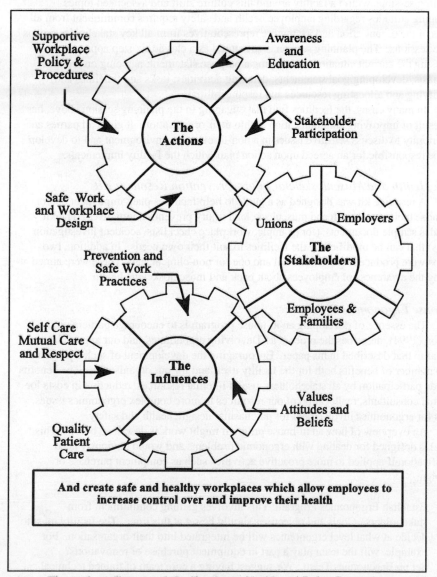

Figure 1. A Framework for Employee Health and Safety Promotion

Strategies

Organisational Culture Change

The Healthcare industry in B.C. have a new mandate to refocus on patient care with the community actively involved in deciding where resources are allocated (Ministry of Health, 1994). The concept of improving patient care by caring for the employee is not commonly accepted, especially when required resources are scarce. We facilitate planning sessions within a facility around this culture shift and associated topics. Changing attitudes regarding employee health and safety requires commitment from all levels of the organisation and we require representatives from all key stakeholder groups at the sessions. The planning session is structured as a classic six step approach: assessing the current situation; developing a purpose statement; agreeing on areas of emphasis; developing goal statements; detailing activities; tasks and timelines and, identifying and allocating resources and responsibilities.

In many cases, the facilities find that engaging in the planning session above, has the result of improving communication within their organisation. It gives all parties an opportunity to discuss sensitive issues in a non-threatening environment and to develop and be responsible for an agreed upon action plan which the facility implements.

Back Health and Musculoskeletal Injury Prevention Resource Kit

A resource kit was designed as a guide to help facilities plan, implement and/or revitalise their back health and musculoskeletal injury prevention initiatives. The kit includes sample documents, (for example, workplace checklists, accident investigation forms) that can be modified by the facilities to suit their own needs. In addition, two videos were produced, one for clinical and one for non-clinical staff. These were aimed at raising the awareness of employees about back and musculoskeletal health.

In-house Ergonomics Program

The essence of developing an in-house program is to encourage participation. Wilson (1994), describes the process as 'Devolving ergonomics' and our approach is similar to that described in his paper. Encouraging the development of such a program has a number of benefits both for the facility itself and for our organisation. The benefits include participation by all stakeholders, use of internal resources, reduction in costs for external consultants, reallocation of our resources to more complex ergonomics issues, focus for ergonomics involvement for purchasing, design, health and safety.

An overview of how an in-house program might work is described below. This model is designed for dealing with ergonomic problems and would obviously need modification if applied to more proactive activities such as equipment purchase or redesign.

1.   Establish Ergonomics Program: This involves gaining commitment from stakeholders. Goals and objectives should be set at this time. The facility must to decide at what level ergonomics will be integrated into their organisation. For example, will the team play a part in equipment purchase or renovations?
2.   Set up Ergonomics Team : We suggest having a core team of trained technical personnel (see *Training of in-house technical personnel*). These personnel would be responsible for undertaking the detailed problem assessment as well as assisting with other steps in the process. The core personnel should be supplemented from

other areas and Table 1 lists appropriate areas with examples of reasons for inclusion. We suggest that key characteristics for team members are enthusiasm, buy-on to the objectives of the team, good communication and committee skills.

Table 1. Ergonomics Program Team

Area Represented	Reason
Administration/Human resources	Long range planning, finance/budget, Collective agreement
Union	Collective Agreement
Health and Safety	Occupational Health and Safety committee liaison
Employee representative from target area	Users perspective, coordinate users and input
Supervisor from target area	Supervisor and department's perspective
Technical Core	
Physiotherapist or Occupational therapist	Knowledge of biomechanics and musculoskeletal problems
Engineer/maintenance	Solution suggestions, feasibility assessment of solutions

3. Problem Identification: Identifying a problem involves examining sick leave and injury data as well as using workplace checklists, worker reporting systems and occupational health and safety audits. Often choices have to be made between the severity of a problem and the number of people affected. Priorities as to which order each area in the facility is to be assessed should be communicated to the staff, along with appropriate justification.
4. Problem Analysis: This should start with user group input to identify problems, and build up credibility and trust. Facilitated focus groups can be used. Problems can then be assessed at a technical level using appropriate tools (task analysis, process flow, postural analysis)
5. Selection of Solutions: A second focus group could be run with the user group to brainstorm solutions that can then be further investigated. Selection of solutions should involve a feasibility assessment by both technical resources and the users.
6. Implementation: This involves setting up activities, tasks and timelines to assist in meeting the objective of the team. Implementation also includes preparing a proposal for management and communicating the change process to the users.
7. Evaluation: The evaluation should be on both the process and the outcome. Evaluation metrics should tie in with those used in the Problem Identification step. A technical assessment should also be completed to ensure that the original problem has been solved and no new ones created. The cycle can then start again either at Step 2 if the team needs to change or at Step 3.

Training of in-house technical personnel

A three day workshop has been designed to train in-house technical personnel. The course is practical in focus and includes 33% hands-on plus 44% discussion and group work. The workshop comprises the following modules: Introduction to Ergonomics; Problems in Healthcare Industry; Workers Compensation Board of British Columbia Draft Ergonomics Regulations; Manual Materials Handling; Repetitive Strain Injury; Workplace Design and Anthropometry; Information Processing; Organisational

Factors; Environmental Ergonomics; Technical Assessments Methods; Equipment Purchase; Ergonomic Program Planning Cycle and, Healthcare Video Case Studies.

Between April and December 1994, we had delivered the workshop to over 180 participants. Many of these have now instigated their in-house program and have started interventions ranging from office ergonomics assessments to bathroom and kitchen redesign.

Program Maintenance and Effectiveness

The maintenance and success of the program will be dependant on the stakeholders within the facilities. Our task is to assist them in achieving success by predicting and accommodating their needs. Currently we provide networking between facilities which has not existed before to any great extent. This involves a newsletter as well as a referral service. For example, if one hospital want to buy a new lift then we will put them in touch with similar facilities that have recently evaluated and purchased lifts. This information exchange is starting to work well between our ergonomics workshop participants who are exchanging information about what strategies work and what do not. We are planning to extend this network by enhancing our facilities with information on their interventions, equipment evaluations, design issues etc. Our workshop participants also provide ongoing training and education for their own people. One of their tasks is to increase awareness of ergonomics to gain commitment. Some have run sessions for all sectors of their facilities and others are concentrating on gaining commitment from management. Where facilities have problems in getting started, we will assist either by talking to the relevant people or by putting them in touch with facilities with 'success stories'.

When an ergonomics team is undergoing their first problem assessment, we will provide on-site assistance to them if and where necessary. In many cases to date, our assistance has been sought where a team is tackling the redesign of an area. This has given us the opportunity to help them with different analysis techniques that they can use in the future. Redesign issues are becoming more and more prevalent as the focus of healthcare changes from acute to longterm and the current hospital designs do not match the needs. We are also responding to requests to work with architects and planners. However, in the longer term we need to improve space planning guidelines and ensure that in-house ergonomics teams are involved throughout any planning and design initiatives.

The effectiveness of our program cannot be judged appropriately unless there are well developed metrics and an effective way of collecting and analysing the data. Facilities vary in their level of computerisation. Human resource data is not necessarily integrated with any other data. In many facilities the accident and injury data is collected only for WCB purposes. The data is scant and with little detail and often is not in a database. We are in the planning stages of developing an integrated health and safety data collection and management system which will be given to the industry. It will enable facilities and ourselves to track, for example, health and safety related data, intervention activity and outcomes. During this development we will be assisting the industry in the development and validation of suitable outcome metrics.

The success of our approach will be judged by whether in the short term we have an impact on absenteeism, injuries and in the longer term, whether the employees have a better quality of life both at work and at home. We are striving towards a modification of knowledge, attitudes and beliefs which should lead to modified behaviour whether this be in lifestyle, proactive approach to the workplace, or safer work styles.

Summary

Our approach to dealing with high absenteeism and injury rate in the B.C. healthcare industry is to assist the facilities in recognising that the problem is at both an organisational and operational level. We have developed some strategies to assist them in dealing with the problem and assessing the success of their efforts. Our strategies will continue to develop in response to the needs of the industry and the problems they face. In our framework, we do not identify ergonomics as a discrete component. Ergonomics, health promotion, industrial hygiene, and health education are all disciplines that are brought to bear in a systematic and integrated approach to the problem of creating safe, healthy and effective workplaces in the B.C. healthcare industry.

References

Burton, G.E., 1986, Quality circles in a hospital environment. *Hospital Topics*, **64**, 11-7.
Garg, A. and Owen, B., 1992, Reducing back stress to nursing personnel: an ergonomic intervention in a nursing home. *Ergonomics*, **35**, 1353-1375.
Hendrick, H., 1994, Future directions in Macroergonomics. In *Proceedings of the 12th Triennial Congress of the International Ergonomics Association*, (HFAC/ACE) **1**, 41-43
McAtamney, L. and Corlett, E.N., 1992, Ergonomic workplace assessment in a health care context, *Ergonomics*, **35**, 965-978.
Ministry of Health, 1993, New Directions for a healthy British Columbia.
Stubbs, D.A., Buckle, P.W., Hudson, M.P., Rivers, P.M. and Worringham, C.J. 1983, Back pain in the nursing profession, II. The effectiveness of training, *Ergonomics*, **26**, 767-779.
Toivanen, H., Helin, P. and Hanninen, O. 1993, Impact of regular relaxation training and psychosocial working factors on neck-shoulder tension and absenteeism in hospital cleaners. *Journal of Occupational Medicine*, **35**, 1123-1130.
Wilson, J.R. 1994, Devolving ergonomics: the key to ergonomics management programs, *Ergonomics*, **37**, 579-594.
Workers Compensation Board of British Columbia, 1993, Draft Ergonomics Regulations and Codes of Practice.

IDENTIFYING RISK FACTORS FOR THE DEVELOPMENT OF WORK RELATED UPPER LIMB DISORDERS

Glyn Smyth and Roger Haslam

Department of Human Sciences
Loughborough University of Technology
Loughborough
Leicestershire
LE11 3TU

The aim of this study was to measure the effectiveness of the 1994 HSE checklist in assessing risks associated with work-related upper limb disorders (WRULD). Analysis of five workstations, in two factories, provided subjective and objective data as to the presence of unacceptable WRULD risk factors. A laboratory based study, using five ergonomists and five non-experts from industry, compared the relative effectiveness of HSE, TUC and RULA methods for identifying those unacceptable risk factors present.

The HSE checklist was the most effective method in identifying the WRULD risk factors of repetition, force and extremes of environmental temperature. The HSE checklist was the least effective method for identifying the other risk factors associated with the development of WRULD and the results did not correlate with the operators' subjective discomfort.

Introduction

There is an increasing use of checklists for assessing risk factors associated with the development of work-related upper limb disorders (WRULD). The Health and Safety Executive (HSE) published a booklet in April 1994 titled 'Upper Limb Disorders: assessing the risk', incorporating a checklist to help employers decide if their employees are at risk.

Checklists are generally designed as sensitive, rapid screening tools, to identify jobs with potentially harmful ergonomic risk factors. As a screening tool checklists are used as the basis for deciding whether further, more detailed assessment is required. Checklists should therefore be highly sensitive, more likely to misclassify an 'acceptable' job as a 'problem' job (a false positive), than to misclassify a 'problem' as 'acceptable' (a false negative).

This study sought to measure the effectiveness of the 1994 HSE checklist in assessing the areas of risk at five workstations, where the operators were involved in highly repetitive work. Ten male operators were analysed, each workstation having two operators. Workstations one, two, three and four were at a tool foundry, and workstation five was at an electromagnetic meter manufacturing company. Workstations one and two involved linnishing/polishing forged metal tools against a static belt grinder. Workstation three involved painting tools mounted on a conveyor, using a hand held paint gun. Workstation four involved forging heated tool ends by placing them in semi-automatic forge (foot controlled) using tongs. Workstation five involved the joining together of small electromagnetic components, firstly by hand and then using a press.

To assess the HSE checklist, it was necessary to have specific criteria against which to judge its effectiveness in identifying areas of risk. Therefore the research involved:

- A literature search to investigate what current research defines as potential risk factors, and criteria to determine when each risk factor becomes unacceptable.
- Application of these criteria to the workstations analysed, thereby determining unacceptable risk factors and assessing the HSE checklist's ability to identify them.
- The collection of subjective data from the operators involved, to identify the extent of WRULD present and subsequently whether this correlated with the HSE checklist findings.
- The use of two other common risk assessment methods, to assess their sensitivity in relation to the HSE checklist, in identifying areas of unacceptable risk.

Causative factors and criteria to determine acceptability

The complexity of musculoskeletal disorders and their causes is recognised in the literature, where not only the physical characteristics but psycho-social and organisational factors are considered. Factors which have been shown to have a causal link with WRULD have been determined, the literature specifying criteria that would allow one to imply an unacceptable level of risk for each factor. The following risk factors have been extracted from the literature:

Repetition
The literature suggests that a cycle time of less than 30 seconds, or if more than 50% of the cycle time is spent in the same fundamental position/action, is unacceptable.

Force
When using a power grip, forces above 4 kg are unacceptable. When using a pinch grip, an unacceptable force is above 1 kg.

Posture
Table 1 outlines specific 'involuntary' joint postures, i.e. postures due to workplace layout and not personal choice, that have been associated with WRULD development.

Physical Risk Factors
Static posture, local mechanical stresses, vibration, environmental extremes of hot and cold and poor lighting have all been linked to musculoskeletal injuries and therefore are unacceptable. Individual differences in operator capabilities, experience, anthropometry and task fitness, can contribute to the development of WRULD and need to be assessed.

Table 1. Unacceptable postures associated with WRULD development

shoulder	- flexion/abduction >60°, 20°-60° A/U, extension <0°, lateral rotation >0°, forced medial rotation/adduction
elbow	- flexion >115°, extension <50°, Pronation >30°, supination >60°
wrist	- flexion >30°, Extension >45°, radial deviation >5°, Ulna deviation >15°
trunk	- flexion >60°, 20°-60° A/U, extension >0° A/U, lateral flexion/rotation >20°
	- trunk + neck > 50°
neck	- flexion >20°, extension >0°, lateral flexion >20°, rotation >45°

A/U denotes postures that might be acceptable or unacceptable, depending on the absence or presence of any other physical risk factor or whether the body part is supported or not.

Psycho-social and Work Organisation Factors

There is a positive association between psycho-social stressors, work organisation and musculoskeletal disorders. As there seems to be a multiplicative effect when risk factors are combined, detrimental psycho-social stressors and work organisation are unacceptable.

Subjective Analysis

The aim of the subjective analysis was to obtain subjective data as to the presence and extent of musculoskeletal disorders. The subjective analysis was undertaken using two tools; a subjective discomfort analysis (Corlett and Bishop, 1976) and the Nordic Musculoskeletal Questionnaire (NMQ) (Kuorinka et al, 1987).

Materials

A variant of the postural discomfort scale (Corlett and Bishop, 1976), with additional regions for the elbow, wrist and hand was selected. This method aimed to identify body parts in which discomfort was experienced and the level of that discomfort.

A modified version of NMQ was used, changes were to include questions to indicate whether reports of symptoms were believed to be work-related.

Table 2. Work-related symptoms

	WORK-RELATED PROBLEMS IN THE PAST 7 DAYS	WORK-RELATED PROBLEMS IN THE PAST 12 MONTHS	PREVENTING THE CARRYING OUT OF NORMAL ACTIVITIES	FREQUENCY	SEVERITY
WORKSTATION ONE	HAND AND WRIST	HAND AND WRIST	HAND AND WRIST	WEEKLY	MODERATE
WORKSTATION TWO	HAND AND WRIST	HAND AND WRIST	HAND AND WRIST	DAILY	SEVERE
WORKSTATION THREE	NONE	SHOULDER	NONE	OCCASIONALLY	MILD
WORKSTATION FOUR	NONE	FOREARM	NONE	WEEKLY	MILD
WORKSTATION FIVE	HAND AND WRIST	HAND, WRIST AND SHOULDER	SHOULDER	DAILY	MODERATE

Results

Table 2 shows the reported work-related discomfort for either operator at each workstation, as determined by the NMQ. The results of the subjective discomfort analysis were in agreement with these, but also indicated some lower back discomfort arising from workstations one and five.

Workstation analysis

The workstation analysis was undertaken to determine the presence of unacceptable WRULD risk factors for each workstation against the criteria determined for this study.

Repetition

Repetition was calculated from the number of pieces handled during the working day. This was related to the task analysis to determine the degree of repetition for each piece.

Force Analysis

Force was analysed using either a spring balance, to determine weight/simple actions, or using a dynamometer, where a more detailed method was applicable.

Lighting, Vibration, Thermal Environment, Work Organisation, Individual Differences, Local Mechanical Stresses and Psycho-social Factors

Using this study's criteria the above factors were assessed through direct observation/measurement and in discussion with the employers and the operators.

Postural and Temporal Analysis

Videotaping was used to analyse the operators' movements. Operators were filmed from appropriate angles, with the use of a stop watch superimposed on the video film to allow timing of static postures.

Using a frame by frame play-back system, a second video showing two typical work cycles for each workstation, from each angle, was compiled. All subsequent video analysis was undertaken using this edited videotape.

The video was initially analysed with respect to the task and secondly with respect to posture. From this it was possible to construct a matrix giving potential risk factors for WRULD for each workstation (Table 3). This analysis was repeated by another ergonomist and the results found to be consistent.

Table 3. Acceptability of WRULD risk factors Risk factors

workstations		repetition	force	hand	wrist	forearm	elbow	shoulder	neck	trunk	static posture	local stresses	lighting	vibration	environment	psycho-social	organisational	individual
	1	U	U	U	U	A	A	U	U	A	U	U	A	A	A	A	A	U
	2	U	U	U	U	U	A	U	U	U	A	U	U	A	A	A	U	U
	3	U	A	A	U	A	A	U	A	A	U	A	A	A	A	A	A	U
	4	U	A	A	U	U	A	A	A	A	U	U	A	A	U	A	A	U
	5	U	A	A	U	U	U	U	U	U	A	U	A	A	A	A	U	U

Results
Table 3 shows the risk factors associated with WRULD and whether they considered acceptable (A) or unacceptable (U), using this study's criteria.

Comparative analysis

Five ergonomists (with a formal qualification in ergonomics) and five non-experts (two safety representatives, one foreman, one production manager and one occupational physiotherapist) were asked to analyse each workstation using the HSE, TUC (Trade Union Congress, 1994) and RULA (McAtamney and Corlett, 1992) methods. The expert subjects were asked to analyse the edited video using the checklists and RULA. The non-expert subjects were asked to analyse the edited video using the two checklists and the checklist section of RULA. Further physical information was available to the subjects, e.g. repetition/number of pieces handled per day. This information was only supplied if asked for, i.e. prompted by the checklists or RULA.

As subjects were unable to conduct a 'walk through', certain questions in the checklists and RULA were impossible for the subjects to answer. To deal with this, these issues were discussed with either the production manager or the health and safety representative in each factory and the answers to questions, that required site knowledge, were based on their consensus opinion.

The subjects' responses were measured against Table 3 with respect to the unacceptable WRULD risk factors, such that mention of an unacceptable risk factor was noted. This enabled the identification of unacceptable risk factors detected by each assessment method for each workstation.

Results
The inability of the HSE, TUC and RULA methods to note unacceptable areas is shown in table 4.

Table 4. Number of unacceptable risk factors not identified

	posture	repetition	force	static posture	local stresses	environment	psycho-social	organisational	individual
HSE Checklist	46	0	0	40	17	5	19	1	48
TUC Checklist	47	9	1	6	1	10	0	0	0
RULA	2	2	1	0	12	8	0	0	0

The HSE was the most effective method for identifying unacceptable risk factors for repetition, force and environmental extremes of temperature.

The HSE and TUC checklists did not identify unacceptable risk factors associated with posture as effectively as RULA. This was not necessarily due to RULA employing a postural analysis, as both non-expert and expert results were similar, suggesting that a checklist alone can be equally sensitive as a postural screening tool.

The HSE checklist was least effective in identifying unacceptable static postures when compared to the TUC checklist and RULA.

Unacceptable local mechanical stresses were present in three workstations, the HSE checklist identified this as a problem in less than 50 % of cases.

For individual differences and psycho-social factors, only three subjects identified these factors as unacceptable using the HSE checklist.

Discussion

The greatest subjective discomfort associated with workstation one was in the hand and wrist. All subjects noted unacceptable hand and wrist postures using the HSE checklist. It would appear that the HSE method was effective in detecting potential risk factors, but this may not be the case. Using the HSE checklist only 30% of the subjects noted local mechanical stresses to be unacceptable and it is unknown whether the hand and wrist discomfort was related purely to posture and/or local mechanical stresses. A similar situation was found to be present at workstations two and four, where 70% of subjects identified unacceptable local mechanical stresses, using the HSE checklist.

Workstation one's reported low back discomfort was unlikely to be solely due to poor posture as this was deemed acceptable; it may have been due to prolonged static posture or differences between individual operators. The case for the cause being due to individual differences is strengthened by the fact that workstation two, a virtually identical task, reported no low back discomfort. Of the subjects using the HSE checklist, 30% and 0% reported unacceptable static posture and individual differences respectively, at workstation one. A similar situation with low back discomfort was noted at workstation five.

Workstation five's reported hand and wrist discomfort was noted as an unacceptable wrist posture by 80% of the subjects using the HSE checklist.

The results suggest that the HSE checklist does not consistently identify the risk factors for WRULD.

Checklists are designed as sensitive, rapid screening tools to identify jobs with risk factors for WRULD. Therefore they need to be sensitive enough to identify all the risk factors present and identify them consistently.

It is unknown how different exposure variables can be interpreted as a single estimate of cumulative exposure. Several studies have noted that individual risk factors are not only relevant in themselves but when combined with other factors the effect seems to be multiplicative. Until research has developed a more accurate picture of how the various risk factors interact to cause WRULD, assessment methods should consistently identify all the potential risk factors.

References

Corlett, E. N. & Bishop, R. P. 1976, A Technique for Assessing Postural Discomfort. Ergonomics 19, 2, 175-182.

Kuorinka, I. Jonsson, B. Kilbom, A. Vinterbergh, H. Biering-Sorrendson, F. Andersson, G. & Jorgenson, K. 1987, Standard Nordic Questionnaires for the Analysis of Musculoskeletal Symptoms. Applied Ergonomics 18.3, 233-237.

McAtamney, L. & Corlett, E. 1992, R.U.L.A.- Reducing the Risk Of Work Related ULD - A Guide and Methods. Institute of occupational Ergonomics University of Nottingham Trade Union Congress 1994, TUC Guide to assessing WRULD Risks. College Hill Press, London.

HEALTH AND SAFETY LEGISLATION : IMPLICATIONS FOR JOB DESIGN

Tom Gough

Division of Operational Research and Information Systems
School of Computer Studies
University of Leeds
Leeds, LS2 9JT

Despite the fact that concern for health and safety in employment has been reflected in legislation over many years and most recently in relation to the use of computing equipment, designers of information systems appear to be largely unaware of the issues that need to be addressed in the information systems development process if the requirements imposed by such legislation are to be properly satisfied. There is the impression, reflected in the information systems design methodologies that health and safety is someone else's responsibility. This paper explores the potential implications for information systems design of the requirements initially set out in Article 118A of the Treaty of Rome, with particular reference to job design. It will also seek to draw attention to the obligations of designers, as employees, and argue that designers must ensure that job design is an integral part of information systems development.

Introduction

The importance of health and safety at work has been a subject of concern for generations. In fact Hippocrates was concerned about lead toxicity in the mining industry in the fifth century BC! Watson (1984) suggests "One of our highest priorities in public health is the protection of millions of Americans during their working lives". This concern for health and safety in employment has been reflected in legislation directed at that concern and reflecting the priorities at the time. A brief historical review of UK legislation will be found in Equal Opportunities Commission (1978). With the rapid growth in the use of interactive systems and the subsequent attention given to the human-computer interface there has been an increase in interest in some aspects of health and safety, especially those associated with VDUs - see Willcocks and Mason (1987) for a reasonably succinct summary and the International Labour Office (1989) for a more detailed review.

Article 118A of the Treaty of Rome resulted in a set of directives (agreed on 29 May 1990) aimed at bringing about improvements in the working environment for the health and safety of workers and at ensuring that all member states have comprehensive and comparable health and safety legislation. In the UK the directives were implemented via new regulations and codes of practice issued under the Health and Safety at Work etc Act of 1974 by the Health and Safety Commission and the Health and Safety Executive. In the advice issued reference is also made to British Standard (BS 7179 : 1990) which provides specific guidance on some of the matters covered by the directives.

However, whilst most systems designers would find it hard to disagree with the suggestion that one aim of systems design should be to produce systems which do not put the health and safety of the users of such systems at risk, it is not clear how the achievement of this aim is reflected in information systems design theory or practice.

Information Systems Design

Little attempt seems to have been made generally to incorporate consideration of health and safety into the information systems design process. Discussion tends to be fragmentary or even non-existent. The publication of the directives resulted in some limited additional interest in the information systems design community, with attention focussed almost entirely on the workstation and, in particular, the VDU. The implementation of the resultant regulations in the UK has seen little increase in the discussion of the design implications and little change in the tenor of what discussion there has been. The focus remains firmly fixed on the 'hardware'. This is perhaps to be expected since the impression in the information systems literature is that health and safety is not the responsibility of the information systems designer.

This criticism of insufficient attention to health and safety is valid across the range of information systems development methodologies. It is ignored almost completely in the traditional systems analysis approach (see, London,1976, Bingham and Davies, 1978, Kilgannon, 1980, for example). The structured approach directs attention to the structure of data and programs (see, Yourdon and Constantine, 1978, Gane and Sarson 1979, for example) or even to the whole analysis and design process as in Cutts (1991) but the structure of the workplace (and the job) is largely ignored. A shift in emphasis to a more people-centred approach only leads to a marginal improvement (Mumford and Weir, 1979), if any (Norman and Draper, 1986). The centrality of 'human activity' to Soft Systems Methodology does not seem to result in any attempt to address health and safety explicitly, in either its original form as set out by Wilson (1990) or its contemporary form as described by Checkland and Scholes (1990). The increasing interest in human-computer interaction has probably narrowed rather than widened the health and safety focus, for example, in Shneiderman (1992) the emphasis is on performance, not health and safety, there is a single paragraph in Booth (1989) and no discussion in Johnson (1992), in Lindgaard (1994) or in Preece (1994). Health and safety is not regarded as a critical issue by Boland and Hirschheim (1987) nor included in human factors by Christie (1985). It is not evaluated in the wide-ranging review of information systems development by Avison and Fitzgerald (1988).

Discussions of information systems design also contain little discussion of health and safety as an issue in the practice of the design and implementation of information systems, apart from some limited discussion of whether or not there are risks associated with the use of VDUs.

In brief then health and safety is not seen as a design issue by information systems designers, whether theorists or practitioners.

Article 118A

As noted earlier, Article 118A was introduced to bring about improvements in the working environment and ensure equality of health and safety legislation in member states of the European Union (then Community). The proposals approved in May 1990 provide the basis of the subsequent legislative regulations in the UK and elsewhere. The proposals comprised a framework directive and five 'daughter' directives, specifying minimum health and safety requirements for : the workplace; the use of machinery, equipment and installations; the use of personal protective equipment; visual display units; and the handling of heavy loads. The framework directive sets out the essential responsibilities of employers and employees. In the UK, the directives were implemented through instruments provided by the Health and Safety at Work etc Act of 1974 and the new approved Codes of Practice came into effect on the 1st January 1993.

Particular attention since the publication of the directives has been paid to the VDU directive and subsequently, to its implementation via the Health and Safety (Display Screen Equipment) Regulations 1992 (Health and Safety Executive, 1992). The directive and the regulations require employers : to analyse display screen workstations to evaluate health and safety conditions, taking appropriate measures to remedy any risks found; to ensure workstations entering service after 31 December 1992 meet the requirements contained in the annex to the directive, which sets standards for display screens, keyboards, furniture, lighting, the working environment, task design and software; to ensure that workstations in service before 31 December 1992 are adapted to comply with the requirements of the annex by 31 December 1996; to plan activities so that daily work on a display screen is periodically interrupted by breaks or changes of activity; to give workers an entitlement to an eye and eyesight test before starting display screen work, at regular intervals thereafter, and if they experience visual difficulties.

As already noted, discussion has tended to focus almost exclusively on the 'hardware' aspects of the legislation and particularly the VDU. One consequence noted by Damodoran (1994) is that the display screen regulations "have sparked a service industry of growing proportions". A second consequence is that the references to task design and software have been largely overlooked, as perhaps has the question of who is responsible for appropriate task design.

Job Design

Regulation 4 of the Display Screen Regulations (Health and Safety Executive, 1992) covers the daily work routine of users where discussion of the nature and timing of breaks includes an oblique reference to job design. Further reference to elements of job design appear in Annex A (after a much longer discussion of the hardware/physical issues).

Unfortunately, job design is paid little more attention than health and safety in most information systems design theory and practice. Job design, however, is a component of socio-technical approaches to information systems design, but as, Mumford and Sutton (1991) note, socio-technical design in the UK" is conspicuous more by its absence than its presence" despite the fact that the UK was the birthplace of the approach and the substantial evidence that adoption of socio-technical principles bring productivity and other benefits. Otherwise the questions posed, for example, in an early paper on job design by Davis, Canter and Hoffman (1955), on the place of job design, on criteria for choosing designs, on responsibility for job design and on the effectiveness of job designs are simply not addressed by most information systems design methodologies and rarely appear in reports of information systems design practice.

A set of job design principles are set out in Mumford (1993) as part of the ETHICS method. The six principles provide a succinct summary of the main issues to be addressed in designing jobs: a good fit with the needs of the person doing the job, work variety and different skills; opportunity to use judgement and make decisions; opportunity to do a complete job; scope for learning; a worthwhile job.

In Regulation 4 there is a paragraph suggesting that wherever practicable users should be allowed some discretion as to how they carry out tasks and some control over the nature and pace of the work. Annex A points out that good design of the task can be as important as making correct choices in relation to the physical aspects of the work environment. It expands on this with references to designing jobs in a way that offers users variety, opportunities for learning and appropriate feedback. There is a later reference to allowing users "to participate in the planning, design and implementation of work tasks whenever possible". Further job design elements are to be found in the discussion on the design, selection and modification of software.

The degree of correlation between the content of the display screen regulations and the job design principles outlined suggests that meeting the legislative requirements would also produce better systems and increase productivity.

There remains the question of responsibility for job design. As Willcocks and Mason (1987) note, in discussing the reduction in job quality often associated with the introduction of computer systems, "Such job degradation may arise almost accidentally, as a result of the failure to operate any conscious job redesign policy". Job design or redesign appears to be no one's responsibility and its general absence from the information systems design agenda has already been noted. The Management of Health and Safety at Work Regulations 1992 (Health and Safety Commission, 1992) make clear that primary responsibility for health and safety in all its aspects rests with employers and hence overall responsibility for job design is theirs also. However, the Regulations also set out the personal obligations of employees in relation to the health and safety of themselves and others in the workplace.

It may be difficult for information systems designers to argue in future that it is not their responsibility to address the job design issues raised in the design, development and implementation of new information systems and the management of the information systems function will need to ensure that the approaches adopted for information systems development facilitate the effective addressing of these issues.

Conclusion

Despite the long-standing concern about health and safety in the workplace generally and the particular more recent concerns about VDU safety there is little sign of serious attention being paid to health and safety in information systems design theory and practice. Furthermore, job design appears on few information systems design agendas.

Recent legislation provides information systems designers with an opportunity to justify the initial additional design costs of turning good intentions on improving information systems design into good quality operational information systems, addressing health and safety requirements effectively, to the benefit of their employers and their fellow employees, as well as satisfying their own and their employers' legal obligations.

References

Avison, D. E. and Fitzgerald G. 1988, *Information systems development - methodology, techniques and tools*, (Blackwell Scientific).

Bingham, J. E. and Davies G. W. P. 1978, *A handbook of systems analysis*, 2nd edn (Macmillan).

Boland, R. J. and Hirschheim, R. A. (eds.) 1987, *Critical issues in information systems research*, (Wiley).

Booth, P. 1989, *An introduction to human-computer interaction*, (Lawrence Erlbaum Associates).

Checkland, P. and Scholes, J. 1990, *Soft systems methodology in action*, (Wiley).

Christie, B (ed.) 1985, *Human factors of information technology in the office*, (Wiley).

Cutts, G. 1991, *Structured systems analysis and design methodology*, 2nd edn, (Blackwell Scientific).

Damodoran, L. 1994, De-bunking some regulations, *Computer Bulletin*, **June**, 8 - 9.

Davies, L. E., Canter, R. R. and Hoffman, J. 1995, Current job design criteria, *Journal of Industrial Engineering*, **vol. 6, no. 2**, 5 - 11.

Equal Opportunities Commission 1979, *Health and safety legislation - should we distinguish between men and women?*

Gane, C. and Sarson, T. 1979, *Structured systems analysis : tools and techniques*, (Prentice Hall).

Health and Safety Commission 1992, *Management of health and safety at work - approved code of practice*, (HMSO).

Health and Safety Executive 1992, *Display screen equipment work-guidance on regulations*, (HMSO).

International Labour Office 1989, *Working with visual display units*, Occupational Safety and Health Series No. 61.

Johnson, P. 1992, *Human computer interaction - psychology, task analysis and software engineering*, (McGraw-Hill).

Kilgannon, P. 1980, *Business data processing and systems analysis*, (Edward Arnold).

Lindgaard, G. 1994, *Usability testing and systems evaluation - a guide for designing useful computer systems*, (Chapman and Hall).

London, K. 1976, *The people side of systems*, (McGraw-Hill).

Mumford, E. 1993, *Designing human systems for health care - the ETHICS method*, (4C Corporation).

Mumford, E. and Sutton, D. 1991, Designing organisational harmony and overcoming the failings of current IT methodologies, *Computer Bulletin*, **August**, 12 - 14.

Mumford, E. and Weir, M. 1979, *Computer systems in work design - the ETHICS method*, (Associated Business Books).

Norman, D. A. and Draper, S. W. (eds.) 1986 *User centred systems design - new perspectives on human-computer interaction*, (Lawrence Erlbaum Associates).

Preece, J. 1994, *Human-computer interaction*, (Addison-Wesley).

Shneiderman, B. 1992, *Designing the user interface-strategies for effective human-computer interaction*, 2nd edn, (Addison-Wesley).

Watson, W. C. 1984, Public health and the workplace. In B. G. F. Cohen (ed.), *Human aspects in office automation*, (Elsevier).

Willcocks, L. and Mason D. 1987, *Computerising work - people systems design and workplace relations*, (Paradigm).

EXAMINING THE HUMAN FACTOR IN SAFETY MANAGEMENT IN THE MARINE INDUSTRY

Jason Leonard & **Jane Rajan**

Total Oil Great Britain Ltd   *ergonomiQ*
Total House   *11 Wendover Close*
4 Lancer Square   *St. Albans*
London W8 4EW   *Herts AL4 9JW*

The introduction of the International Code of the Safe Operation of Ships and for Pollution Prevention (the International Safety Management (ISM) Code) by the International Maritime Organisation (IMO) (1994) has required the marine industry to look at Safety Management Systems (SMS). This paper presents a study which looked at the specific human factors problems faced by the marine industry and developed a framework which could be used as the basis to examine the human factor when establishing a Safety Management System specific to the marine industry.

Introduction

The contribution of poor safety management to major disasters has been well documented. Reason (1990) quotes a series of accidents which occurred during the 1980s where management deficiencies were implicated as a major causal factor including Bhopal , the Challenger shuttle, Chernobyl and the King's Cross Underground fire. The importance of safety management has been stressed in the official enquiries, for example from the Clapham Junction Rail crash Anthony Hidden QC stated: *"It is not enough to talk in terms of 'absolute safety' and of 'zero accidents'. There must also be proper organisation and management to ensure that action lines up with words"* (Hidden 1989). From the inquiry into the Piper Alpha Disaster Lord Cullen stated: *"I am convinced from the evidence...that the quality of safety management...is fundamental to off-shore safety. No amount of detailed regulation for safety improvement could make up for deficiencies in the way that safety is managed"* (Cullen 1990).

Recent marine disasters such as the Herald of Free Enterprise capsize in March 1987 and the Scandinavian Star fire in April 1990 highlighted the important role that Safety Management Systems (SMS) have to play in improving safety in the marine industry. These accidents have been the stimulus which have led to the introduction of the International Code of the Safe Operation of Ships and for Pollution Prevention (the International Safety Management (ISM) Code) by the International Maritime

Organisation (IMO) (Bengston 1993). The cost of ineffective safety management is high. Eves (1994) talks of astonishment when collating the information for the HSE publication "The Cost of Accidents at Work" with one company stating that accidents and other associated losses had cost the company 37% of their profits. In the marine industry the cost of insurance has increased from less than 10% to over 45% of total operating costs over the last 10 years (Anderson 1993). The Chevron Shipping Company have calculated that every personal accident in its fleet costs the company $14,000 (Parker 1990). The European ferry business, for example, has come under increased commercial pressure with the opening of the Channel Tunnel at the same time that accidents such as the capsize of the Estonia killing over 800 people in September 1994 has brought safety of the industry into the public eye. Both of these factors have emphasised the priority that should be placed on safety management.

Systems for safety management have been widely researched and applied within the context of other safety critical industries such as offshore and railways. The development of these methods and systems is often based on models and theories of accident causation and the associated human contribution to failure. However the marine industry has unique attributes which prevent the direct transfer and application of these systems. This paper outlines a project to examine safety management issues in the marine industry, and factors which contribute to the effective development and use of a Safety Management System. The result is a classification framework which identifies the relevant human factors issues contributing to accidents in the marine industry. This framework can be used to systematically identify the human factors issues which need inclusion in a marine Safety Management System.

Approach

A four step approach aimed to combine different sources of data concerning human failures in a marine context, to give a structured and comprehensive framework of the human factors issues which need to be identified and defended against to improve safety. First, the literature review examined work done on safety management and identified human factors issues which have been identified as important to look at as part of a SMS. Secondly, structured interviews with domain experts were used to collect information for the task analysis, find out about the marine context in which the legislation is to be applied, and to discover experiences gained in setting up a SMS if this had been done at all. The third part was the task analysis which aided in the identification of areas where possible human failures could occur. The fourth part was an analysis of accident data to identify human factors issues which have contributed to marine accidents. A standard datasheet allowed the systematic collection of data in a format which facilitated comparison between accidents. Following the literature search it was decided that the analysis would attempt to establish the underlying causes as well as the direct causes of an accident, to reflect current ideas on accident causation such as those of Reason (1990). 10 accidents were analysed and the datasheets refined on the basis of the type, format and content of data examined. An example of the datasheet format is shown below (Table 1).

Table 1. Marine Accident Reporting Sheet

Context: Name of Ship: Ship Details Type: Age (years): Weather: Crew involved:	Date of Accident: Weight: Company: Flag of registration: Cargo:
Location of Accident/ Stage of voyage (including time):	
Direct Cause of accident/problem/trigger mechanism: (free text)	
Underlying /latent failures: (classified according to Table 2)	
Any other information:	
Information Source:	

Results:

Table 2 shows a summary of the framework developed. This was developed by combining the results from the different analyses described above. The framework outlines human factors issues to be addressed when looking at safety management in a marine context.

Table 2. Summary of Classification Framework

1.COMMUNICATION Open Two-way Communication Safety Information Incident Reporting Maintenance	2.ORGANISATIONAL PRESSURES Operation Overload and Fatigue Selection Maintenance
3.PROCEDURES AND STANDARDS Organisation Design Procedures Incident Reporting Emergency and Safety Deck Engineering	4.TRAINING Team training Emergency Training Organisation of Training Training for All Personnel Specific Training
5.RESOURCES Economic Equipment Safety Equipment Maintenance	6.PHYSICAL WORK ENVIRONMENT Physical Environment Equipment Design
7. ALLOCATION OF RESPONSIBILITIES Organisation Change	

Discussion

Part of the study was a review of SMS in other industries together with their approaches to the development of a good safety culture. This was an important element because safety culture will influence the acceptance and compliance with the SMS set up (Whittingham and Hollywell 1994). It became apparent from this study that safety culture within the marine industry is adversely affecting the development of SMS. One influencing factor is that of enforcement. The ISM was produced by the International Maritime Organisation but it requires member nations to enforce the code on an individual basis. There is a great variation in the enthusiasm with which this is done. Because much of the marine legislation is enforced in this way it has led to the increased registration of vessels with so called "flags of convenience". This is in contrast with, for example the Offshore Installations (Safety Case) Regulations 1992 which are policed by the country which produced them. This is just one factor which will influence the enthusiasm with which the ISM code is taken up.

Bird (1985) describes the classic iceberg model where accidents are merely the tip and "near misses" occur in much greater numbers. Whilst the present study only examined accidents to investigate human failures a further study could make use of the Nautical Institute's near miss reporting system called MARS (Marine Accident Reporting Scheme) which is a regular item reported by Captain R. Beedal in the Institute's publication "Seaways". This would provide a more comprehensive database of the types of accidents occurring in the marine industry on which to base analysis.

The database which was to be the basis of the original study did not have sufficient detail for an in-depth analysis to gain an understanding of the underlying human factors involvement, so various other sources had to be used. It was also difficult to obtain more than one source of data for all but the largest of disasters. This made it difficult to cross-check the facts and build a more complete picture of the events. As a result, the in-depth analysis using different sources was limited to 31 accidents, with an additional 90 accidents analysed from one source only. Collection of more data would allow the analysis of frequencies of the types of human factors contribution to accidents which would be useful when deciding on intervention strategies. A further limitation of this study was that the task analysis was carried out on the basis of verbal reports without an observational study. A comprehensive task analysis would be a study of its own requiring more time than that available for this study. The interviews with marine experts were useful to gain an insight into the industry and the varying attitudes towards safety which will influence the uptake of the ISM code. Further study is also necessary to test the robustness of the framework which has been developed. If the framework developed proves sufficiently robust then it could provide the basis for the development of a safety management audit specifically for the marine industry.

References

Anderson, P. 1993, Facing rising cargo claims with a new approach to loss prevention. In *Nautical Institute Project '93 Accident and Loss Prevention at Sea International Conference and Workshops November 1993*, (Nautical Institute, London)

Bengston, S. 1993, Safety and quality management in international passenger ship operation. In *Cruise and Ferry 93 International Conference and Exhibition, Olympia, London, 11-13 May 1993, Vol 1*, (Business Meetings Ltd, Rickmansworth, Herts)

Bird, F. and Germain, G. 1985, *Practical Loss Control Leadership* (International Loss Control Institute, USA)

Cullen, W.D. 1991, *The public inquiry into the Piper Alpha disaster*, 2nd edn (HMSO, London

Eves, D. 1994, Safety management:are we learning the lessons? *Process Safety Progress*, **13** (American Institute of Chemical Engineers)

Hidden, A. 1989, *Investigation into the Clapham Junction Railway Accident, Department of Transport* (HMSO, London)

Parker, C.J. 1990, Why safety must move higher up the agenda, *Seaways*, August, 2-6.

Reason, J.T. 1990, *Human Error* (Cambridge University Press, Cambridge)

Whittingham, R.B. and Hollywell, P.D. 1994, Ensuring you take credit for an effective safety management system. In *Incorporating the Human Factor into Offshore Safety Cases, Conference 1/3/94, Aberdeen, Scotland.*

THE PSYCHOPHYSICAL APPROACH IN UPPER EXTREMITIES WORK

Jeffrey E. Fernandez[1], Tycho K. Fredericks[1], and Robert J. Marley[2]

[1]Industrial & Manufacturing Engineering Department
Wichita State University, Wichita, Kansas 67260 (USA)

[2]Industrial & Management Engineering Department
Montana State University, Bozeman, Montana 59717 (USA)

This paper represents the latest in a series of reports which have advocated the use of the psychophysical method of adjustment to derive limits for work involving upper extremity stressors for the purpose of reducing the risk of musculoskeletal disorders. The paper summarizes task estimates of the maximum acceptable frequencies (MAF) performed under varying force, posture, and duration requirements. Results of studies recommending proper adjustment period as well as the reliability of MAF estimates are provided. The authors recommend that the method of adjustment be used in order to establish realistic and reasonable task design guidelines in the absence of well-defined and acceptable biomechanical or physiological models.

Background

Musculoskeletal disorders (MSDs) have seen a dramatic increase in industrial incidence in recent years (US Labor Dept., 1993). MSDs, which may be referred to in the literature by a variety of terms such as "cumulative trauma disorders," "repetitive strain injuries," "repetitive motion disorders," etc., are often regarded as physical conditions resulting from the summation of micro-trauma exposure (Kroemer, 1989). Repetitious and forceful tasks have been related to the development of upper-extremity tendon and nerve entrapment disorders (Stock, 1991). Several other work factors that have consistently been linked to these and other MSDs include awkward posture, vibration, cold temperature, mechanical stress, and lack of adequate recovery time. The control of work-related MSDs has been traditionally sought through the application of ergonomic principles and proactive medical management. But a dilemma faced by ergonomists is that there are currently no definitive data concerning what constitutes excessive quantities of exposure to upper-extremity MSD risk factors based upon objectively-defined biomechanical or physiological criteria (Kroemer, 1989). A "conceptual" model for establishing quantitative guidelines for jobs with multiple risk factors for carpal tunnel syndrome and other MSDs was proposed by Tanaka and McGlothlin (1993), though there is currently no data available upon which to base the model. Kilbom (1994a, 1994b) recently reviewed numerous published and unpublished studies on the subject which used a variety of criteria. While several of the studies reviewed provide guidelines for a specific risk factor under precise conditions, she ultimately concluded that the interaction between risk factors is mostly unknown. This is particularly noteworthy since most industrial work activities involve multiple risk factors of varying degrees. The studies reported in this paper utilize the psychophysical approach in the absence of comprehensive biomechanical and/or physiological models, since it allows for the integration of multiple risk factors.

The Psychophysical Approach

The study of psychophysics can be referred to as one of the original "schools" of psychology originating with Fechner's work in the mid-1800's when the relationship between human sensations and physical stimuli were considered (Stevens, 1975). Today, the "psychophysical approach" actually refers to one of several separate and distinct methodologies which are generally designed to either determine a threshold (or tolerance), or a subjective assessment of stimulus intensity. The methodologies do have in common a foundation in what is known as the "Stevens Power Law" which describes the magnitude of all human sensation as a non-linear function related to the intensity of a physical stimulus (Stevens, 1975). Gescheider (1984) outlined the four basic methods and associated protocols in the psychophysical approach. The "method of constant stimuli" seeks to determine absolute thresholds, difference thresholds, or points of subjective equality between stimuli. This method receives its name due to the constant number discrete stimulus intensity levels presented to the subject along equidistant scale intervals. The "method of limits" also seeks to determine thresholds or points of subjective equality. However, in this method, the stimulus is presented to the subject in continuous, ascending or descending levels of intensity. Such a protocol is often used to establish hearing thresholds. The "method of ratio scaling" seeks to derive a relation between subjective assessment and a numeric scale. There are four common protocols in ratio scaling: 1) ratio production; 2) ratio estimation; 3) magnitude production; and 4) magnitude estimation. The production protocols require the subject to produce or recreate a stimulus intensity level based upon a numeric or graphical scale presented with the stimulus continuum. The estimation protocols, on the other hand, require the subject to provide a numeric value which would best represent a particular intensity level. Arguably the best known application of a estimation protocol is the Rating of Perceived Exertion (RPE) scale (Borg, 1970) and its many derivatives.

While also seeking to establish a threshold, the "method of adjustment" is fundamentally different from the previous methods in that the subject, rather than the experimenter, has direct control over the stimulus intensity. The protocol asks the subject to match a predetermined level of intensity or other criteria. This method has been used successfully in manual material handling studies to establish acceptable weights of lift under a variety of task factors (i.e., Snook and Ciriello, 1991). The authors of this paper contend that ratio scaling is valid for making inter- and intra-task comparisons or attribute evaluations, but that the method of adjustment is the most appropriate method for establishing specific guidelines and parameters for the work designer, assuming a relevant criteria is stipulated, since the subject must adjust to an acceptable workload given the total task demand and environment. Thus, the remainder of this paper summarizes a series of experiments by the authors and their colleagues which utilized the psychophysical method of adjustment in analyzing jobs with multiple upper-extremity MSD risk factors.

Psychophysics in Hand-Wrist Intensive Work

Marley (1990) adapted a psychophysical adjustment protocol for use in establishing maximum acceptable frequency (MAF) for a sheet metal drilling activity which required various wrist postures. This protocol has been described extensively and results have been found to be reliable (Fernandez, Dahalan, Klein, and Marley, 1993; Kim, Marley, Fernandez, and Klein, 1994; Marley and Fernandez, (in press)). Kim and Fernandez (1993), Viswanath and Fernandez (1992), and Davis and Fernandez (in press) have investigated gender effects and other postural combinations. The results of these studies have demonstrated that wrist deviation significantly reduces MAF, males generate significantly greater MAF values than females, and increases in applied force significantly decreases MAF.

Willis (1994) investigated the repeatability and reliability of the psychophysical approach as it pertains to upper extremity drilling work. His results showed that the MAF selected after an initial 20-minute bout did not differ significantly from the 4th consecutive bout (over a 2-hour period with frequency randomly set at the beginning of each bout). In addition he found no significant circadian effect, but there was up to 12% variability in MAF between periods of 7 to 10 days in which no drilling activities were performed.

The 20-minute adjustment period commonly used in many of the psychophysical studies has been generally accepted as representative of a normal 8-hour working day. Recently, the appropriateness of this assumption as related to upper extremity experiments was investigated (Muppasani, 1994). Results indicated that the 20-minute adjustment period consistently overestimates MAF for a normal 8-hour working day. An extended adjustment period of one hour appears to be a better estimation of a persons' MAF. It was also determined that the overestimation between the 20-minute MAF values and the 1-hour MAF values can range from 39% to 48% (Figure 1). This difference was more pronounced in the deviated wrist postures. These results can be compared with psychophysical studies over extended periods for lifting tasks where the acceptable weight was overestimated in a 20 or 30 minute adjustment period (Mital, 1983; Fernandez, 1986). Table 1 summarizes the MAF limits developed for a drilling task that have been obtained in studies conducted by Fernandez and associates.

In addition to these studies, researchers have conducted psychophysically adjusted frequency experiments in grasping and pinching tasks (Dahalan and Fernandez, 1993; Klein, 1994). Dahalan and Fernandez (1993) determined MAF for a simulated gripping task at different gripping forces and gripping durations using the psychophysical protocol. Results showed that MAF was significantly reduced as gripping force and gripping duration increased. Similarly, Klein (1994) conducted an experiment utilizing the psychophysical approach to establish the MAF for a pinching task. Results indicated MAF significantly decreased with a deviation in wrist posture. Three levels of pinch force (15%, 30%, and 50% of the maximum voluntary contraction (MVC)) along with 3 levels of duration (1, 3, and 7 seconds)

Table 1. MAF per minute at various wrist postures.

Wrist Posture	Degree Deviation	Males			Females		
		Mean	STD	N	Mean	STD	N
Neutral	0	16.57	3.47	33	12.39	1.22	39
Flexion	10	13.42	2.33	15	10.61	0.80	27
	20	12.64	0.42	27	9.48	1.02	27
	25	- - -	- - -	- - -	9.40	2.63	12
	30	10.68	2.00	15	- - -	- - -	- - -
	40	8.97	0.16	27	- - -	- - -	- - -
	50	- - -	- - -	- - -	8.22	3.22	12
Extension	20	- - -	- - -	- - -	11.79	2.74	12
	40	- - -	- - -	- - -	11.29	1.89	12
Ulnar Deviation	15	- - -	- - -	- - -	11.30	2.36	12
	20	- - -	- - -	- - -	12.52	2.83	12
	30	- - -	- - -	- - -	10.40	2.72	12
	40	- - -	- - -	- - -	13.29	3.23	12
Radial Deviation	10	- - -	- - -	- - -	12.06	2.63	12
	20	- - -	- - -	- - -	11.38	2.91	12

[Adapted from Marley and Fernandez (in press); Davis and Fernandez (in press); Willis (1994); Kim and Fernandez (1993); Viswanath and Fernandez (1992); Muppasani, 1994.]

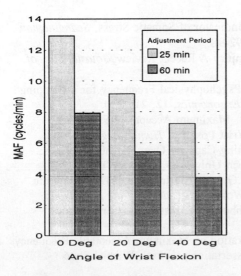

Figure 1. Wrist Angle vs. MAF

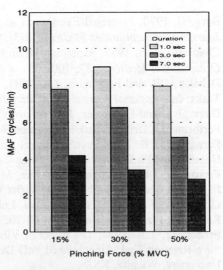

Figure 2. Pinch Force vs. MAF

were investigated. Results showed that MAF significantly decreased as pinch force and duration increased (Figure 2). In both studies, the MAF values derived were supported by physiological data.

It has also been hypothesized that industrial tasks which involve hand tools that vibrate may contribute significantly to the development or exacerbation of MSD (Chatterjee, 1987). A study presently underway (Fredericks, 1994) will attempt to determine the relationship between wrist posture (neutral, 1/3, 2/3 flexion 1/3, 2/3 ulnar deviation) and push force (8, 12 pounds) on MAF for an intermittent riveting task. This particular study will also be measuring vibration levels transmitted to the hand-arm system during a typical psychophysical bout.

Concluding Remarks

This paper has summarized several studies which applied the psychophysical method of adjustment to the analysis of several upper-extremity tasks in order to estimate the MAF performed under varying force, posture and duration requirements. Analysis of other MSD risk factors, such as vibration, are on-going using this methodology.

As ergonomists seek to eliminate or reduce the presence of MSD risk factors, the combined effect of multiple risk factors must also be considered. In the current absence of broad-based biomechanical or physiological models to accurately relate MSD risk, it is proposed that the psychophysical method of adjustment may be utilized to establish reasonable and realistic task parameters. As opposed to other psychophysical methods, the method of adjustment can elicit quantifiable guidelines for the industrial ergonomist. Future research include epidemiological investigations to examine the relationship between psychophysically acceptable exposure levels and morbidity rates.

References

Borg, G. 1970, Perceived Exertion as an Indicator of Somatic Stress, *Scandinavian Journal of Rehabilitation Medicine*, 2(3), 92-98.

Chatterjee, D.S. 1987, Repetition Strain Injury-A Recent Review, *Journal Soc. of Occupational Medicine*, **37**, 100-105.

Dahalan, J.B. and Fernandez, J.E. 1993, Psychophysical Frequency for a Gripping Tasks, *International Journal of Industrial Ergonomics*, **12**, 219-230.

Davis, P.J. and Fernandez, J.E. (in press), Maximum Acceptable Frequencies Performing a Drilling Task in Different Wrist Postures, *Human Ergology*.

Fernandez, J.E. 1986, Psychophysical Lifting Capacity Over Extended Periods, Unpublished Ph.D. Dissertation, Texas Tech University, Lubbock TX.

Fernandez, J.E., Dahalan, J.B., Klein, M.G., and Marley, R.J. 1993, Using the Psychophysical Approach in Hand-Wrist Work, *Proceedings of the MM Ayoub Occupational Ergonomics Symposium*, Lubbock TX: Institute for Ergonomics Research, Texas Tech University, 63-70.

Fredericks, T.K. 1994, The Effect of Vibration on Maximum Acceptable Frequency for a Riveting Task, Unpublished PhD Dissertation proposal, Wichita State University, Wichita, KS.

Gescheider, G.A. 1984, *Psychophysics: Method, Theory, and Application*, Hilldale, NJ: Lawrence Erlbaum.

Kilbom, A. 1994a, Repetitive Work of the Upper Extremity: Part I-Guidelines for the Practitioner, *International Journal of Industrial Ergonomics*, **14**, 51-57.

Kilbom, A. 1994b, Repetitive Work of the Upper Extremity: Part II-The Scientific Basis (Knowledge Base) for the Guide, *International Journal of Industrial Ergonomics*, **14**, 59-86.

Kim, C.H., Marley, R.J., Fernandez, J.E., and Klein, M.G. 1994, Acceptable Work Limits for the Upper Extremities with the Psychophysical Approach, *Proceedings of the 3rd Pan-Pacific Conference on Occupational Ergonomics*, 312-316.

Kim, C.H. and Fernandez, J.E. 1993, Psychophysical Frequency for a Drilling Task, *International Journal of Industrial Ergonomics*, **12**, 209-218.

Klein, M.G. 1994, Psychophysically Determined Frequency for a Pinching Task, Unpublished PhD Dissertation Proposal, Wichita State University, Wichita, KS.

Kroemer, K.H.E. 1989, Cumulative Trauma Disorders: Their Recognition and Ergonomic Measures to Avoid Them. Applied Ergonomics, **20**(4), 274-280.

Marley, R.J. 1990, Psychophysical Frequency at Different Wrist Postures of Females for a Drilling Task, Unpublished PhD Dissertation, Wichita State University, Wichita, KS.

Marley, R.J. and Fernandez, J.E. (in press), Psychophysical Frequency and Sustained Exertion at Varying Wrist Posture for a Drilling Task, *Ergonomics*.

Mital, A. 1983, The Psychophysical Approach in Manual Lifting a Verification Study, *Human Factors*, **15**(5), 485-491.

Muppasani, A.K. 1994, The Effect of Adjustment Time on Maximum Acceptable Frequency for a Drilling Task, Unpublished Masters Thesis, Wichita State University, Wichita, KS.

Stevens, S.S. 1975, *Psychophysics: Introduction to its Perceptual, Neural, and Social Perspectives*, New York: Wiley & Sons.

Snook, S.H. and Ciriello, V.M. 1991, The Design of Manual Handling Tasks: Revised Tables of Maximum Acceptable Weights and Forces, *Ergonomics*, **34**(9), 1197-1213.

Stock, S. 1991, Workplace Ergonomic Factors and the Development of Musculoskeletal Disorders of the Neck and Upper Limbs: a Meta-Analysis. *American Journal of Industrial Medicine*, **19**, 87-107.

Tanaka, S. and McGlothlin, J.D. 1993, A Conceptual Model for Prevention of Work-Related Carpal Tunnel Syndrome (CTS), *International Journal of Industrial Ergonomics*, **11**, 181-193.

U.S. Labor Department, 1993, Occupational Injuries and Illnesses in the United States by Industry, BLS No. 2424, Washington, DC.

Viswanath, V. and Fernandez, J.E. 1992, MAF for Males Performing Drilling Tasks, *Proceedings of the 36th Annual Human Factors Society Meeting*, 692-696.

Willis, M.L. 1994, A Verification study of the Psychophysical Method for Upper Extremity Work, Unpublished MS Thesis, Montana State University, Bozeman, MT.

THE UTILITY OF THE POTENTIAL HUMAN ERROR AUDIT IN AN ERGONOMIC EVALUATION OF PUBLIC PLAYGROUND SAFETY

Sarah Minister and Rachel Benedyk

Ergonomics and HCI Unit
University College London,
26, Bedford Way,
London WC1H OAP.

The Potential Human Error Audit (PHEA) for assessing and reducing risk was applied to public playgrounds, having previously only been applied in an industrial setting. Ergonomic assessments of safety were carried out and relevant parts of the PHEA method were applied to the results. Standards of play environments and equipment provided for play were found to vary greatly. It was found that children regularly displayed hazardous behaviour and that their perceptions of the risk involved in such acts tended to be lower than that of adults. Since most of the controlling measures used to prevent accidents in industry do not apply in this environment, ensuring safety of children in playgrounds must primarily depend upon designing safety into the equipment and maintaining it properly. It was shown that the PHEA method could be potentially useful in the area of playground safety.

The Potential Human Error Audit (PHEA) is an ergonomic method of risk analysis and reduction that was devised by Simpson at British Coal (Simpson, 1992, 1993). Until now the PHEA has been used in an industrial context, where Simpson reports a good rate of accident reduction through its use. This study was set up to evaluate the utility of the PHEA in a consumer product environment, where identifying sources of error could help improve safety of use. The environment chosen was the public children's playground with its associated equipment. Many accidents occur in playgrounds when using such equipment (e.g.. 54,000 playground accidents were treated in hospitals in Britain in 1989 - source Department of Trade and Industry, 1990). Due in part to the introduction of recent standards and to increased awareness and responsibility on the part of manufacturers and providers of playground facilities, the number of deaths and serious accidents has been reduced. However, accidents in playgrounds still account for approximately 6% of child injuries needing hospital treatment (Safety in Playgrounds Action Group- SPAG).

A PHEA involves the following steps (adapted from Simpson 1993):
i) Looking at the system and breaking it into suitable areas, such as task elements or geographical areas; then carrying out a classic ergonomic appraisal of each area to identify as many safety-related potential human errors as possible.

ii) Classifying these errors depending on error type and examining the error precursors and potential latent failures in the system

iii) Seeking specific, agreed solutions to these latent failures from general approaches suggested by the error classification and identifying approaches to remove the influence of latent failures.

In a playground setting, this would involve the following steps:

i) Breaking the system down into areas, namely the separate pieces of play equipment, the environment around them and the behaviour of the children and carrying out an ergonomic appraisal of each area.

ii) Classifying the possible errors and identifying error precursors and potential latent failures (e.g.. inadequate training of playground installers could mean that swings were installed too close to the ground which may cause a child to trip and be hit by a moving swing)

iii) Suggesting solutions and identifying approaches to remove latent failures (e.g.. improved training and instructions for installation)

The ergonomic appraisal of playground equipment and its environment involves assessing each item and its environment for safety, in terms of design and usability, maintenance, installation, siting and the presence of controlling measures. When thinking about the design of equipment, it is important to consider not just whether it is designed safely, but also whether it is designed with the correct purposes in mind. The appraisal of the children's behaviour involves looking at the violations that children commit, how frequently they are committed and how hazardous they are. It also looks at how hazardous the children perceive them to be.

In the case of playgrounds, possible contributing factors to accidents resulting in personal injury are:

- Poorly designed or installed equipment (e.g.. entrapment points)
- Hazards inherent in the environment (e.g.. access to main road nearby)
- Poorly maintained equipment (e.g.. broken steps up to equipment)
- A lack of safety controls (e.g.. no run-off on a slide)
- Inappropriate use of equipment (e.g.. standing up on a swing)
- Lack of training of the users (typical in a consumer environment)
- Lack of supervision of the users (difficult due to nature of users- i.e. children).

The design of equipment is an important factor in preventing playground accidents. King and Ball (1989) say that a significant proportion of accidents associated with playground equipment can be traced to factors such as poor equipment design, layout, installation and inadequate maintenance. Recent standards have been introduced, including the British Standard concerned with Play Equipment Intended for Permanent Installation Outdoors which is BS 5696 (1979). The European Committee for Standardisation, CEN, is currently working on a European Standard which will replace this. The National Children's Play and Recreation Unit (1992) cites poorly designed equipment, not complying with standards, as an important factor contributing to accidents. Some examples of design guidelines and controls against risk are that all rungs and bars, not intended for climbing on, should be vertical, rather than sloping or horizontal, so children are not encouraged to climb on them, and that roundabouts should have a braking mechanism to prevent them going too fast.

The environment around the equipment is very important in terms of avoiding hazards. Inadequate space between equipment can cause accidents in an otherwise safe playground. The Minimum Use Zone or MUZ is defined as the minimum area for each item or group of items that includes the space occupied by the item, the operating area taken by the children on the item and the circulation area round the item (Heseltine, 1988). MUZs should be left clear of obstruction and the MUZs of items in the playground should not overlap. In addition, equipment intended for use by younger children should be physically separated from that intended for older children, since they play in different ways and older children can, even inadvertently, cause young children to have accidents. The majority of playgrounds today have some form of impact absorbent surface underneath most of the equipment, which lessens the severity of the consequence of a fall from a piece of equipment.

The behaviour of children as they play on playground equipment is a major factor related to safety. Children's play involves experimentation, exploration and discovery as they develop awareness of the world around them. In playing on playground equipment, they frequently violate the 'rules' or the accepted use of the equipment. In addition, if the play facilities provided for children fail to provide challenge, excitement and interest and are of poor play value, children will quickly become bored with them. In that case they may either play instead in unsuitable or unsafe places, such as roads, railway lines, building sites and electricity sub-stations, or they may use the equipment in ways for which it was not designed. Such uses include vandalism and play dangerous to themselves and other children. Little, if any formal instruction takes place and few playgrounds have formal supervision. If children are inadequately trained or supervised, accidents frequently result. This is particularly the case for children below the age of 8 whose motor skills are not fully developed and who are less likely than older children to comprehend the danger of situations if they do misuse equipment.

Other factors affecting safety are the installation and maintenance of the equipment, which includes problems caused by wear and tear and also by vandalism.

Data collection for the first stage of this study took place in two types of environment. A self-report questionnaire was administered to children aged 7 to 11 in a junior school, which explored children's risk perceptions relating to their use of playgrounds and their experience of accidents relating to playground equipment. Children were asked to rate how dangerous they felt certain types of equipment were and how dangerous they perceived certain inappropriate activities to be. They were also asked to rate the frequency with which they carried out each of these activities, in line with methods used in the PHEA. Observational and interview techniques were then used at 24 public playgrounds in London. These included using a scoring system (out of 10) to allocate a value (a 'safety value') to each piece of equipment to indicate how risky each was, taking into account its design, installation, maintenance, siting and the presence of controlling measures. A second scoring system (out of 29) was used to rate each playground on the basis of the quality of play experiences it provided (a 'play value' score), which was also seen as contributing to a playground's safety, since children would rapidly become bored with a playground providing low play value and would be likely to misuse the equipment in order to gain excitement from it. Children's use and abuse of equipment was also recorded.

Standards of play equipment and play environments were found to vary greatly. The mean of the safety value for each playground's equipment was calculated and over the

24 playgrounds, these means ranged from 6.3 to 9.9 out of 10. The play value scores ranged from 9 to 24 out of 29. Converting these scores to percentages and combining the safety value and the play value score for each playground yielded an overview of its safety, although it is still vital to consider each component's safety when considering possible sources of hazard, since it only takes one damaged or badly-designed piece of equipment to cause a serious accident. Over the 24 playgrounds, the combined safety scores ranged from 53% to 89%.

From the self-report questionnaires and observation in playgrounds, it was seen that many children regularly display hazardous behaviour, such as climbing up the slope of a slide or jumping off a moving roundabout. Standing up on a swing was the most common misuse of equipment, with 44% of children who answered the questionnaire stating that they did this frequently. It was also the most common type of hazardous behaviour noted in the observational section of this study.

Children's perceptions of risk were considered in terms of pieces of equipment. Climbing frames were perceived to be the most dangerous equipment, with 66% of children thinking they were either fairly or very dangerous. Slides were seen as the safest, with 63% of children thinking they were either fairly or very safe. Judgement tended to be split over the whole range for other types of equipment.

One part of the PHEA method involves calculating the risk of a particular behaviour by taking into account the relative hazard connected with the behaviour and the likelihood of it occurring. In this case, absolute values for risk and likelihood were impossible to calculate, so relative values were given and the behaviours ranked according to relative risk and relative likelihood (or frequency of occurrence). The ratings gained from parents and supervisors were used to rank the behaviours in order of danger or risk involved. The children's questionnaire responses were used to rate the frequency of occurrence of each of the behaviours. The ranks were then summed to give an order of the likelihood that each behaviour would actually cause an accident. Based on these calculations, it was found that the behaviour most likely to cause an accident was that of standing up on a swing, followed by that of climbing up the slope of a slide.

It was found that children have a lower perception of risk involved in such behaviours than the adults who supervise them in playgrounds. The NCPRU (1992) states that children's understanding of potentially dangerous situations differs from that of adults and is age-dependent. Children do not have the experience, predictive ability or physical development of adults. It is likely that a combination of experience, increased skill and improved cognitive ability means that children become more aware of risk as they get older, although in this study no significant differences were found between 7 and 11 year-olds' perceptions of risk.

Very few children were found to refrain from behaviours which they believed to be safe. This implies that they would try any such activity as long as they thought it was safe, although all the behaviours described were ones which supervisors felt were dangerous. This is another illustration of the fact that children's inaccurate perceptions of risk can result in inadvertent dangerous activity.

Common latent failures which were identified in playgrounds are shown in Table 1, along with suggested avenues for approaches towards solutions.

Table 1.

Potential Error	Preferred source of action
Failure to mend/replace damaged and vandalised equipment (especially swings and spring equipment)	Management
Failure to mend/replace/repaint old, worn, rusty equipment, especially roundabouts	Management
Faults in layout (especially placing equipment for younger and older children close together)	Design/Training/Management
Low play value provided	Management
Installation faults (especially swings too high or low)	Training/Management
Entrapment points in equipment	Design
Very high equipment	Design
Misuse of equipment by children	Training/Supervision/Design?

Where management is mentioned as a 'preferred source of action', this may include improving procedures for planning, purchasing, installing, inspecting and maintaining equipment. It may also include improving the safety ethos and raising awareness of certain issues, such as the importance of play value to the safety of a playground. Design is the responsibility of the manufacturers, but playground providers can put pressure on them to design safe equipment by only purchasing equipment that is proved to be safe. Safety Standards should help in this area.

There are several possible sources of future action, where misuse of equipment is identified, none of which are easy to carry out. As Simpson (1993) states, violations are the most difficult area in a PHEA to reduce, with possible approaches being in the areas of training, supervision and design. Both improved training and supervision might be seen as costly and not important enough to warrant public finance, so parents need to teach their own children and improve their perceptions of risk involved as well as provide supervision. Heightened public awareness is necessary here. Another approach worth considering is that of design changes; for example, installing barriers in front of swings, although from observation of children using playground equipment in inappropriate ways, it seems that it is virtually impossible to design equipment so that it will not allow some children to misuse it if they want to. It is the feeling of the investigator that designers and manufacturers should continue their efforts to devise more exciting, but safe equipment.

Overall, playgrounds need more resources to be spent on them, in particular, on planning, equipment, maintenance and regular inspections.

Many of the factors which could commonly be optimised to improve the interaction between users and equipment are inaccessible in this particular consumer product environment. Formal supervision is rare and costly and training of users of the equipment (the children) is sparse, informal and impossible to measure. Also, the use of warning signs is of limited value since many of the users are unable to read or interpret them. User selection in this environment is irrelevant, since any child may use the equipment. Nor is it possible to control user behaviour with rewards for good or safe behaviour and sanctions for bad or dangerous behaviour. Even where design is concerned, playgrounds have an additional problem that workplaces do not generally have, and that is the vast diversity of ability, skills and sizes of body parts of the users. So it would seem that approaches usually open to ergonomists for risk reduction are very limited in this environment, and the only measures that can be relied upon are optimisation of design and maintenance of the equipment and the environment. It may be necessary to 'super-optimise' these features to compensate for the fact that ergonomics principles are stretched to their limits.

As the NCPRU (1992) says, challenge and adventure are natural and important aspects of children's play and should not be lost in playground design, but these must be experienced in a safe environment. Children need to be offered adventure without risk. If playground equipment is designed to be entirely safe, there would be no excitement in its use. As Adams (1988) says, in order to maximise excitement (or to satisfy people's need to take risks), one needs to design in 'invisible safety measures' for which people will not compensate. It seems that the safest playground equipment would look dangerous but actually be very safe so that playing on it would feel as if risks were being taken, thus providing the excitement element.

In the consumer product environment tested, the PHEA was seen to be a useful tool. Data collection was relatively straight forward, with the one major disadvantage being that children do not just use one set of equipment with its inherent design faults and damaged parts; instead it is necessary to talk much more generally and use mean values instead of specific ones. This means that conclusions drawn and ideas for approaches to limit risk, are general and not specific, as has been the case when PHEA has been used previously. Identification of error precursors in an industrial setting would involve accident reports, but such accident reports are not readily available in this context. Even in industry, as Simpson (1993) states, accident statistics rarely give much useful information for use in a PHEA. In addition, latent failures belong to particular playgrounds, along with that playground's equipment, environment and state of management, as well as being concerned with the particular children who use that playground. This means that it is only possible to identify the most frequently occurring latent failures and error precursors and so categorise them. This study has, however, shown the potential usefulness of the method in the area of playground safety.

References

Adams, J.G.U., 1988, Risk Homeostasis and the Purpose of Safety Regulation. *Ergonomics*, No. 4, 407-428.
British Standards Institution, 1979, *Play Equipment Intended for Permanent Installation Outdoors*. *BS5696*.
The Department of Trade and Industry, 1990, *Playgrounds: a Summary of Accidents in Public Playgrounds*. DTI.
King, K. and Ball, D., 1989, *A Holistic Approach to Accident Prevention in Children's Playgrounds*. LSS, London.
Heseltine, P., 1988, *Playing Safe- a Checklist for Assessing Children's Playgrounds*. NPFA and RICA.
The National Children's Play and Recreation Unit for the DES (NCPRU), 1992, *Playground Safety Guidelines*. Crown/ NCPRU.
Safety in Playgrounds Action Group (SPAG), Date unknown, *Action Sheet*.
Simpson, G.C., 1992, Human Error in Accidents: Lessons from recent disasters. *Proceedings of the Institution of Mining Engineers International Symposium*.
Simpson, G.C., 1993, Promoting Safety Improvements via Potential Human Error Audits. *Proceedings of the 25th Conference of Safety in Mines Research Institutes*. Pretoria, S.A.

WORK RELATED UPPER LIMB DISORDERS IN A METER MANUFACTURING COMPANY: REDUCING THE RISKS

Tim Bentley and Roger Haslam

Department of Human Sciences,
Loughborough University of Technology
Loughborough, Leicestershire.
LE11 3TU

Work related upper limb disorders (WRULD) are particularly prevalent in industries where repetitive hand intensive work is carried out. In the meter manufacturing company where this study was conducted, WRULD had been identified as being particularly problematic on two workstations. One of the operations required repetitive lever pushing, knob twisting, and tweezer use, while the second required repetitive use of a pull down power screw. Application of ergonomics investigation methods revealed problems of posture, force, and mechanical stress. Remedial measures included redesign of handles to improve wrist posture and hand grip, and repositioning of handles and components. A participative methodology was employed, utilising expertise within the company, and reducing resistance to change. Recommendations were also provided concerning the organisation of work.

Introduction

Work related upper limb disorders (WRULD), also known as repetitive strain injuries, cumulative trauma disorders, and a large number of other names, is an umbrella term for syndromes which affect muscles, tendons, nerves and blood vessels. These disorders frequently occur in the forearm, wrist, and hand area of the upper limb, examples being tenosynovitis and carpal tunnel syndrome. Disorders are also common in the neck and shoulder area, such as thoracic outlet syndrome, and shoulder cuff tendinitis.

WRULD have been reported with increasing frequency during the last two decades, with an associated increase in concern from industry. Thompson (1988), reported the cost of these type of injuries to UK industry for the year 1988 was in the region of one billion pounds.

The present investigation arose as a result of the high level of complaints of neck, shoulder, arm, wrist, and hand discomfort from workstation operators in a meter manufacturing plant. Two workstations were identified as being particularly problematic in terms of operators suffering from upper limb disorders and taking sick leave as a consequence. Both jobs investigated were hand intensive, requiring repetitive wrist and hand movements. Operators spent an average of about seven hours per day carrying out the same short cycle tasks, with rotation to other jobs being very rare. For the Watch Lathe operation, the operator was required to stake a pin into a mount by first positioning the pin into the mount and then driving it in with a series of pushes of a lever with the right hand. The stake was then checked by pulling with tweezers, and finally trued by observing a monitor whilst turning a knob with the left hand. The Divert Screw and Spring assembly operation required the operator to position several components together at the base of the power screw, and then to pull down the power screw with the right hand, this process screwing the spring and screw to the brake magnet.

Methods

The two workstations were investigated for WRULD risk factors by use of the following methods

Task description and temporal analysis

A task description was carried out for both operations, to identify the main task elements and their durations. This was achieved by use of video recordings of a small number of operators working at each workstation. As well as providing the researcher with a thorough description of the task, this process also provided a framework for the posture and force analysis.

Posture and force analysis

The work elements comprising the two tasks were recorded on an upper extremity postural recording sheet (Armstrong et al, 1982). For each of these elements the postural recording sheet allowed the researcher to note the position of the shoulders, elbows, wrists, and hands. This method allows the position of the joints to be analysed for each degree of freedom: for the shoulder three degrees, and two degrees each for the elbow and wrist. Ranges of angles were used to estimate the position within a specified range. The position of the hand was categorised by determining the dominant grip type and position of the fingers. Significant forces were also identified. This was achieved by observation of the operator carrying out the task, and by asking the operators about the level of force used.

Body Part Discomfort Analysis

A body part discomfort analysis was conducted to investigate the pain and discomfort experienced by operators, the area of the body where discomfort was experienced, and the development of discomfort throughout the working day. A body map technique was used, as described by McAtamney and Corlett (1982). Operators were shown a body map diagram at five regular intervals throughout the working day and asked to indicate areas of the body where discomfort was being experienced.

Interviews with operators

Operators were interviewed by the researcher at a quiet location away from the area where they worked. Interviews were semi-structured, with questions designed to provide useful information on the operators' opinions of the major problems with the workstations, how they considered the workstations could best be improved, and issues concerning the psycho-social environment and the organisation of work. Particular effort was made to take account of the workers' views. As well as providing useful ideas for redesign of the workstations, operator participation in the planning of redesign is important for increasing 'ownership' of any changes made, and reducing resistance to change.

Physical Workplace Assessment
        Measurements were taken of the important workstation dimensions. These included the height of handles, distance of components from operators, and bench height. These dimensions were then compared with anthropometric data.

Results for the Watch Lathe operation

Posture and Force Analysis
        The following major postural problems were identified: supination of the right forearm, and ulnar deviation of the right wrist whilst pushing the lever when staking the pin; supination and pronation of the left forearm, and ulnar and radial deviation of the left wrist whilst turning the knob when truing the pin. Significant forces were identified for the following task elements: lever pushing with the right hand; knob turning with the left hand; tweezer gripping and pulling with the right hand (see figure 1 for photograph of workstation).

Body Part Discomfort Analysis
        Discomfort was reported to the right palm as a result of pressure from gripping the lever. Discomfort was also reported from gripping the tweezers with the right hand (tweezers were never put down and subsequently caused local mechanical stress to the palm and fingers during lever pushing). Discomfort was also reported in the neck, shoulder, forearm and wrist. Discomfort to these areas tended to increase throughout the working day.

Summary of interviews with operators
        Interviews with operators revealed problems with the design and layout of the workstations. Handles were difficult to grip and caused discomfort due to their shape, texture, and size. There was insufficient padding on seats and too little leg room. Operators suggested a number of redesign ideas which were subsequently incorporated into the redesign measures. Operators complained of boredom and expressed a desire to rotate to other jobs more often. The piece-rate payment system led to anxiety among operators, with fear of not meeting their production targets causing them to work excessively fast for long periods in order to make up time. Greatest levels of job satisfaction were experienced when operators helped to train new employees.

Results for the Divert Screw and Spring Assembly operation

Posture and Force Analysis
        The following postural problems were identified: abduction of the right shoulder whilst reaching for components, and abduction of the left shoulder whilst placing assembled components into the machine tool; radial deviation and extension of the right wrist whilst pulling the power screw handle downwards. Significant force was identified for the pulling down of the screw handle against the tension of its spring (see figure 3 for photograph of workstation).

Body Part Discomfort Analysis

    All operators reported discomfort to the palm of the right hand, due to local mechanical stress caused by the ball shaped handle of the power screw. This discomfort was reported throughout the day. Discomfort to the neck and shoulders was reported by operators increasingly through the day.

Summary of interviews with operators

    Operators identified a number of problems with the workstations, with particular concern centring on the discomfort experienced to the neck shoulder and hands when operating the pull-down power screw. Operators were able to suggest redesign ideas for the shape and position of the handle of the power screw. Boredom was a major problem at these workstations, with the task cycle being just 8 seconds. Job satisfaction was at its highest when operators were training new employees or dealing with problems with their workstations or those of others.

Workstation Redesign Recommendations

    Recommendations for redesign of workstations were produced in order to remove or reduce WRULD risk factors identified. A participative methodology (Eason, 1990), was employed, with input to redesign measures provided by the researcher, operators, setters, production engineers, and the Health and Safety Engineer. It was anticipated that this approach would produce optimum redesign solutions by utilising the expertise within the company, and serve to increase ownership of any changes implemented, reducing resistance to change.

Redesign of Watch Lathe workstation

    The Watch Lathe lever was redesigned as follows :
i) an angled handle replaced the straight handle previously used (see figure 1), this allowed operators to keep straighter wrists when pushing the lever
ii) the handle was made adjustable in position, such that the operator could move it towards or away from her body

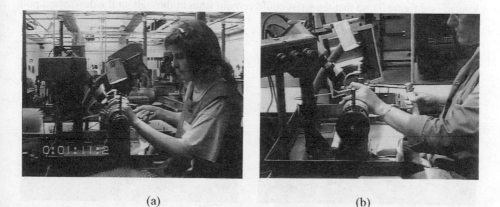

(a)   (b)

Figure 1. Watch Lathe workstation: the original lever (a) and redesigned lever (b).

iii) the handle grip revolved so that when the lever was pushed from right to left the wrist maintained a straight position, rather than having to twist with the lateral movement of the handle
iv) a comfortable soft rubber handle grip was provided
v) the dimensions for handle length and diameter were determined by the operators' grip preferences.

The Watch Lathe knob was redesigned as follows :
i) the small knob which required the operator to grip and turn was replaced by a disc of greater diameter, which was operated by rolling the hand across the top of the disc (see figure 3). The redesigned knob or disc had a soft rubber surface, providing comfortable contact and reducing slippage while providing grip.

(a)   (b)
Figure 2. The Watch Lathe workstation: the original knob (a) and redesigned disc (b)

Redesign of Divert Screw and Spring assembly workstations
The Divert screw and spring assemble workstation was redesigned as follows :
i) the existing ball handle was replaced with a cylindrical handle which was suspended and hinged (see figure 3), so that the wrist could remain straight throughout the downward pull

(a)   (b)
Figure 3. Divert Screw and Spring workstation: the original handle (a) and the redesigned handle (b).

ii) the handle was cranked to the right to the position considered most comfortable by the operators

iii) handle dimensions of length and diameter were determined on the bases of operators' preferences of grip

iv) the handle was covered in soft rubber for a comfortable grip.

Other redesign measures for the Divert Screw and Spring workstation included :
i) replacement of the spring from the power screw with weights to return the power screw to its high position following each pull

ii) repositioning of the components trays in front of, and angled towards operators.

Reanalysis of the workstations following implementation of redesign prototypes

Reanalysis showed all the major non-natural postures were either removed or greatly reduced for both workstations. This was verified by undertaking a postural analysis. The force required to pull down the power screw handle was greatly reduced by removal of the spring, while operator interviews revealed that less force was required to push the Watch Lathe handle with the right hand, and roll the knob with the left hand. Interviews suggested a general satisfaction in the redesign efforts.

Recommendations concerning the organisation of work

Attention was drawn to the need to ensure all employees received training in the use of the redesigned workstations. It was suggested that production engineers and purchasers of workstation equipment and tools should understand principles of workstation and tool design. Operators should be educated to spot WRULD symptoms, enabling them to be detected at an early stage. Problems with the job design and the company's payment system were highlighted, with movements towards cell-based work and payment encouraged. A system of active microbreaks was also recommended (Delgardo et al, 1991), with operators stopping work for short breaks whenever discomfort is experienced and performing prescribed stretching exercises.

References

Armstrong, T., Foulke, J.A., Joseph, B.S. and Goldstein, S.A., 1982, Investigation of cumulative trauma disorders in a poultry processing plant, American Hygiene Association Journal, 43, 103-116.
Delgardo, E., Bustos, T., and Genaidy, A., 1991, Active microbreaks effects on musculoskeletal discomfort perceived in meatpacking plants. In W. Karwowski and J. Yates, (eds), *Advances in Industrial Ergonomics and Safety 111*, (Taylor and Francis, London).
Eason, K., 1990, New systems implementation. In Wilson, J. and Corlett, E, (eds), *Evaluation of human work*, (Taylor and Francis, London).
McAtamney, L. and Corlett, E., 1992, Reducing the risk of work related upper limb disorders: A guide and methods, Institute for Occupational Ergonomics, University of Nottingham.
Thompson, D., 1989, Identification of Causes and Prevention of work related upper limb musculoskeletal disorders, In Megaw, E. (ed), *Contemporary Ergonomics 1989*, (Taylor and Francis, London), 394-400.

DISTRIBUTED DECISION MAKING IN NUCLEAR EMERGENCY MANAGEMENT

Jane Carthey

Industrial Ergonomics Group,
Department of Manufacturing and Mechanical
Engineering,
The University of Birmingham,
Edgbaston, Birmingham, B25 2TT.

This paper outlines a methodology which is being used to analyse nuclear emergency management. The methodology uses the observation of emergency exercises, semi-structured interviews with Nuclear Power Plant (NPP) emergency management personnel, and critical incident technique interviews. This paper mainly focuses on the observational data collection and analysis stage. The study aims to gain an understanding of the communications, decision making and roles and responsibilities of operators in nuclear emergency management. It is concluded that observational methods are a good way to elicit emergency management information. Also, traditional task analysis methods, specifically link analysis and operational sequence diagrams (OSD's) are useful for representing and summarising communication data.

Introduction

Nuclear emergency management involves the organisation of personnel and the allocation of resources to control a nuclear accident or incident. Communication, decision making and performing physical actions at appropriate times are central to efficient nuclear emergency management.

The United States National Research Council (1990) has defined Distributed Decision Making (DDM) as, "...the task faced by organisations in which the information and responsibility for decision making is distributed among individuals within the organisation, who are often distributed geographically."

Nuclear emergency management is a DDM context in which information must be communicated between geographically separate units for the successful control of accident conditions. This information includes regular updates on radiological hazards on and off-site, current plant status, missing personnel/casualties, available resources (equipment and human), reactor damage and progress with effecting repairs. The efficient dissemination of this information is necessary to achieve the multiple objectives of nuclear emergency management. These objectives include;

- returning the plant to a safe state
- consideration of the general public who live in close proximity to the site
- repair of damage to the reactor
- identifying and locating on-site casualties.

liaison with off-site organisations including, the local emergency services (fire, police and ambulance), representatives from the Health and Safety Executive (HSE) and the Ministry of Agriculture, Fisheries and Food (MAFF).

Nuclear emergency management personnel have defined roles and responsibilities which relate to the above objectives. These roles and responsibilities are formally specified in the Emergency Plan- a document which details the company's accident management strategy. By defining the roles and responsibilities of personnel the Emergency Plan makes explicit the formal communication links between emergency management units, who should use these links and who has decision making authority for specific emergency management tasks. The Emergency Plan also outlines the conditions for declaring and cancelling a nuclear emergency, the stages of a nuclear accident, how mustering and roll call should be organised and how the coordination with external organisations should be arranged. Further information on nuclear emergency planning can be found in, 'Arrangements for Responding to Nuclear Emergencies' (HSE, 1990).

Research Needs For Distributed Decision Making

There are problems associated with the study of DDM systems. From an academic perspective there is no widely accepted model of DDM processes. Two practical problems also exist: Studying DDM systems is resource intensive and the complexity of DDM systems makes it difficult to determine which areas of the system the research should address. Nuclear emergency management poses the further problem that accidents and incidents are infrequent events. This means that the DDM system which has been devised for their control is used infrequently under real-life conditions. Thus, it is not easy to determine what effects real-life incidents could have on the system. However, nuclear emergency management systems are supported through regular training exercises. Given the fact that access to observe nuclear emergency management is unlikely to be granted in real incidents, the training exercises allow research into simulations with high environmental and psychological fidelity.

Aims of the Study

This study aims to gain an insight into communication, decision making and the roles and responsibilities of operators in nuclear emergency management. This is achieved via:

(i)   Observation of emergency exercises at five NPP's in the United Kingdom.
(ii)   Semi-structured interviews with emergency management personnel.
(iii)   Critical incident technique interviews with emergency management personnel who have been involved in minor site incidents.
(iv)   Expert judgement interviews which assess the timeliness of decision making and actions during nuclear emergency exercises.
(v)   Analysis of company documentation on emergency management.

This paper focuses on the first part of the study- the observational data collection and analysis methodology.

Observational Data Collection and Analysis Methodology

Emergency exercises involve setting up the formally specified nuclear emergency management system described in the Emergency Plan. Emergency exercises allow NPP personnel the opportunity to practice their emergency management roles under simulated accident conditions.

The nuclear emergency management systems investigated in this study comprise a number of on-site and off-site units. The methodology described here has been used to analyse data gathered from on-site emergency exercises. The on-site units include the Main Control Room (MCR), Emergency Control Centre (ECC), the Access Control Point (ACP) and the First Aid Centre (FAC). In each of these units there are several personnel performing a variety of emergency management tasks.

Pilot studies in two emergency exercises were used to determine the primary areas and roles of importance. From these studies it was decided that the observations should focus on the MCR, the ECC and the ACP, which were found to be the primary areas of communication and decision making activity during on-site accident management.

Trained observers took verbatim notes of the communications that took place between personnel in these three locations. The observers were given a primary focus of attention which was either the Emergency Controller (EC), the Shift Charge Engineer (SCE) or the Access Control Engineer (ACE), depending on whether they were in the ECC, MCR or ACP respectively. The observers were also instructed to record as much of the communication between other personnel as possible, for example, between the Emergency Reactor Physicist and the Emergency Health Physicist in the ECC.

Following the exercise the observer notes were transcribed by the researcher. A quality assurance phase was used to ensure that the researchers interpretation of the observers notes was correct. The observers notes were then compiled into an incident chronology. The incident chronology showed the communications within and between the MCR, ACP and ECC throughout the duration of the exercise.

The incident chronology data was then presented in Temporal, Partitioned Operational Sequence Diagrams (TPOSD's) (Bateman,1979; Kirwan and Ainsworth, 1992). These diagrams showed the following:

(i)   Who communicated with who (both within and between the three locations).
(ii)   The time (in minutes) at which communications occurred.
(iii)   The direction of the communication (i.e. who were the sender(s) and receiver(s) of messages).
(iv)   When decisions were made during the exercise, including who made decisions and who provided information relevant to them.
(v)   Communication and decision making failures, role confusions and action failures.
(vi)   Central aspects of the scenario, for example, loss of grid.
(vii)   The medium of communication, for example, phone, face to face, fax, tannoy.

In the next stage of analysis a micro and macro communications analysis of the TPOSD's was carried out.

Micro Level Communication Analysis

The micro-level analysis involved analysing individual units of communication. In the present study two types of micro-level analysis were carried out; a semantic analysis (i.e. meaning) and a structural analysis (who communicated with who).

For the semantic analysis a set of communication categories relevant to nuclear emergency management were devised. These categories were based on previous research and preliminary observations of the raw data. These are shown in Table 1.

The information presented in the TPOSD's was classified using the communication categories shown in Table 1. The level of categorisation was communication topics. Thus everytime there was a change in the topic of conversation the new message or phrase was categorised. The communication categories were then represented on the TPOSD's.

The semantic analysis gave insights into emergency management roles based on the types of information communicated to and from a person. These insights included who had access to different types of emergency management information (for example, radiological versus plant status information) and to which other personnel this information was disseminated. The semantic analysis also showed where authority exists in the emergency management system, by identifying who issues instructions or sanctions decisions and actions.

Table 1; The Communication Categories Used To Classify Emergency Management Communication

COMMUNICATION CATEGORY
1. STATE INTENTIONS
2. INSTRUCTIONS
3. STATE POSITION
4. AGREEMENT
5. ROLE CLARIFICATION
6. DISCUSS FUTURE ACTIONS
7. CONFIRMATION OF INFORMATION.
8. ADVOCACY
9. SANCTIONING
10. QUESTIONS
11. PROGNOSIS
12. INFORMATION UPDATES;
- Radiological Information Update
- Plant Status Update
- Equipment Information Update
- Casualty Status Update
- Team Status Update
- Damage/Repair Information Update
- Fire Status Update
- Emergency Services Information Update.
13. INFORMATION REQUESTS;
- Radiological Information Request
- Plant Status Request
- Equipment Information Request
- Casualty Status Request
- Team Status Request
- Damage/Repair Information Request
- Fire Status Request
- Emergency Services Information Request
- Resources Request

The structural analysis of emergency management communication identified who communicated with who, how frequently communication took place and whether communication was two-way or one-way. Two-way communication occured where two personnel both initiated communications to each other. One-way communication occured when one communicative partner always adopted the role of sender and the second person always received messages but never sent them. The results of this analysis were summarised in a link analysis format (see Chapanis, 1959).

The findings from the micro-communication analysis will be compared to the Emergency Plan. One can postulate that the Emergency Plan provides a normative model of nuclear emergency management whereas the TPOSD's show the actual process. The comparison of the TPOSD micro-analysis findings with the Emergency Plan can show how simulated accident management relates to the formally specified system (in the Emergency Plan). This will show where the formally specified roles, responsibilities and communication links match those seen in exercises. Conversely it will also identify where informal communication links are used, and where role boundaries are crossed.

Macro-Level Communications Analysis

Macro level analysis is a broader form of communication analysis which traces the journey of different types of information through a distributed system. The TPOSD's provide an ideal format for carrying out this type of communications analysis. In the present research context macro-level analysis can be achieved by tracing the flow of pieces of information through the temporal, partitioned OSD's. This shows;

(i)   Where information relating to different parts of the emergency management process originates from, i.e. the information source.
(ii)   Who receives information relevant to different emergency management tasks.
(iii)   The types of decisions that different types of emergency management information are used for.
(iv)   How emergency management responsibilities are disseminated through the system using instructions.
(v)   Where communication and decision making failures occur. The causes of these failures can also be identified, for example, competing tasks and interpolated communications which interfere with actions being carried out.

To summarise, the macro-communication analysis allows a communication based analysis of the central aspects of DDM. This communication based analysis identifies where DDM failures occur, what types of failures occur and why they occur.

Interviews With Nuclear Emergency Management Personnel

Three types of interviews are also being conducted in this study. Firstly, semi-structured interviews with emergency management personnel are being carried out to examine subjective perceptions of roles and responsibilities, past experiences in emergency exercises and emergency management training. Secondly, critical incident technique interviews are being used to gain insights into real life accident management, including the effects of stress on emergency management performance. These interviews are designed to address the limitation that the stressful conditions of real life accident management are not seen in simulated accident conditions. Finally, expert judgement interviews will be carried out which aim to assess whether actions and decisions taken during the exercises were timely or not. These interviews will use company personnel with training and emergency management expertise to assess TPOSD data on the timing of decisions, for example, evacuation of off-site areas and the use of plant safety systems.

Taken together, the information gathered from the various types of interviews will provide supporting evidence for the findings of the TPOSD analysis. The interview data will also highlight further types of DDM failures which occur in nuclear emergency management. In particular the effects of stress on performance and subjective perceptions of the good and bad aspects of the current emergency management system will be shown. Finally, the interviews will provide support for the good performance strategies identified through the emergency exercise observations. These can then be incorporated into current training practices.

Conclusion

This paper has suggested that observation is a useful method for collecting information on DDM systems, specifically nuclear emergency management. It has also shown that link analysis and operational sequence diagrams are useful for representing emergency management communication data. The methodology used here may be extrapolated and used in other DDM contexts, for example, military systems and air traffic control. However, it is important to find supporting evidence for this data, for example, via the types of interviews which are used in this study.

References

Bateman, R.P. (1979), Design Evaluation of an Interface Between Passengers and an Automated Ground Transportation System. In. *Proceedimngs of the Human Factors Society 23rd Annual Meeting, Boston, Massachusetts, October 29-November 1.* The Human Factors Society. 119-123.

Chapanis, A., (1959), *Research Techniques in Human Engineering.* Baltimore: John Hpokins.

Fischhoff, B. and Johnson, S. (1990), The Possibility of Distributed Decision Making. In, *Distributed Decision making: Report of a Workshop.* The United States National Research Council, Washington D.C. National Academic Press.

Health and Safety Executive (1990), *Arrangements for Responding to Nuclear Emergencies,* HMSO, England.

Kirwan, B. and Ainsworth, L. (1992), *A Guide to Task Analysis.* (Taylor and Francis, London).

ASSESSING WORK RELATED UPPER LIMB DISORDERS IN A BRICK MAKING FACTORY

Gurpreet Basra   **Joanne O. Crawford**

Industrial Ergonomics Group   *Industrial Ergonomics Group*
School of Manufacturing and   *School of Manufacturing and*
Mechanical Engineering   *Mechanical Engineering*
The University of Birmingham   *The University of Birmingham*
Edgbaston, Birmingham   *Edgbaston, Birmingham*

The management, at a brick-making factory, commissioned this study to carry out a detailed examination of the workplace, the workforce and work related upper limb disorders, as staff had been suffering from a variety of upper limb problems. The main objective was to find out why some staff were experiencing problems, whilst others, doing the same job, were not. This was done by carrying out a matched case-control study of the workforce. Data collection methods included a questionnaire survey and video-taping employees doing their everyday tasks. The analysis of the video material was broken onto three fundamental components, methods-time measurement, posture analysis using OWAS and a risk factor analysis. As a result of the investigation a number of work and workplace changes were recommended to alleviate the present situation.

ntroduction

Occupational activities associated with the onset of upper limb disorders (ULD's) arise rom ordinary movements that may include gripping, twisting, reaching and moving, for xample. What makes them hazardous is the prolonged repetition, often in a forceful and wkward manner, without sufficient rest or recovery (Ayoub, 1990; Genaidy et al, 1990; oseph and Dagg, 1987; Parker and Imbus, 1992; Putz-Anderson, 1988). The nature of our vorking environments has also bought about a significant increase in cases of ULD's. An geing working population; machinery used by women designed for men; increased global ompetition and production levels; the introduction of production lines; consumer demands and 1e rapid increase of office automation, all have diverse affects on the employee. The role of 1e ergonomist is therefore to assess how these changes have affected the workforce and :commend preventative actions.

A number of studies have been carried out to investigate the prevalence of cumulative ·auma disorders (CTD's) in a number of different industries. For example Armstrong et al, 982 investigate cumulative trauma disorders in a poultry processing plant; Dimberg, 1987,)oks at the prevalence and causation of tennis elbow in a population of workers in an engineering industry and Buckle and Baty, 1986, study the ergonomics aspects of tenosynovitis and carpel tunnel syndrome in production line workers. In the literature reviewed two papers discussing musculo-skeletal problems in brick-related industries were found. Ferreira and Tracy, 1991, investigated the affect of handling techniques on the occurrence of ULD's and Malchaire and Rezt-Kallah, 1991, looked at the workload of bricklayers in the steel industry.

In the present study management at the company were concerned about the increasing levels of ULD's amongst their employees, the main objective of the study was therefore to identify causal factors and recommend methods to eliminate or reduce the effects of these stressors. Most of the 131 employees work on a shift-work pattern. There are three shifts; morning, afternoon and night, which are rotated on a weekly basis. Shift workers are allowed one 15 minute break during their working day. The small number of employees who work days are allowed one 30 minute break. All employees work on a piece work or piece work related pay structure.

Methodology

The investigation was undertaken in two main stages. The first stage, the questionnaire survey, was intended to obtain general information about all employees and used as a means of selecting subjects for the next stage of the investigation, the video recording. For simplicity the methodology will be divided into three parts;
- questionnaire, design and development
- video recording and analysis
- environmental survey

1. Questionnaire

A questionnaire survey proved to be the most obvious means of collecting the necessary data for the preliminary stage of the investigation. The questionnaire used in the survey, was based primarily on the questionnaire concerning musculo-skeletal troubles in different body regions, designed by a project group in the Nordic Council of Ministers (the questionnaire is described by Kuorinka et al 1987). Questions relating to employee background were added and the Nordic questionnaire modified slightly. The final questionnaire consisted of three main sections;

Section A - asked a series of general questions

Section B - looked at areas of the body in which pain or discomfort may have been experienced

Section C - asked more detailed questions about the pain and discomfort experienced and how this affected work and leisure activities.

To ensure each and every employee received a questionnaire, management issued them with employee payslips. After analysis of the questionnaires a group of subjects were chosen to be filmed. A number of cases and controls needed to be identified, and they were eventually selected by age matching.

2. Video recording and analysis.

Each subject was filmed, doing their everyday job, for about 5 minutes. The final number of subjects filmed totalled fourteen, nine cases and five controls. The analysis of the video material involved the investigation of three components:-

a). Work methods analysis - All tasks are made up of a number of steps or elements (Barnes, 1980), which can be described as fundamental movements or acts, reaching, grasping and positioning for example. Task elements of the repetitive jobs were identified and the time taken to complete each element was measured. (For non-repetitive tasks, this kind of analysis was not possible). This widely used motion classification system is called Methods-Time Measurement (MTM).

b). Posture analysis - An analysis of the posture at each element was carried out, using the Ovako Working Posture Analysis System (OWAS), Karhu, 1977. A percentage of task time spent in a stressful posture was evaluated and an overall classification was made accordingly. Classification categories have been developed as follows,

Class 1 - normal postures which do not need any special attention except in some special cases
Class 2 - posture must be considered during the next regular check of working methods
Class 3 - postures need consideration in the near future
Class 4 - postures need immediate consideration

c). Risk factor analysis - A risk factor is defined as an attribute or exposure that increases the probability of disease or disorder (Last, 1983). The analysis was done by, once again observing the video material. Hand arm postures were noted, adequacy of workspace design and any other observations were recorded. The effort required to carry out the task was evaluated using table 1. This table attempts to summarise four factors which will result in the employee working with an elevated effort level, and is based on Huppes, 1993. The table has not been validated, it has been designed to give a relative impression of the task requirements based on the information obtained from the video and from supervisors.

Table 1. Effort levels

	Light	Medium	Heavy
Distance	<1m	1 - 3m	>3m
Weight	<6 kilos	6 - 8kilos	>8kilos
Frequency	<30%	30 - 50%	>50%
Height from floor	Waist height	Pallet height	Floor

3. Environmental survey.

A very crude survey of the environment was carried out. This involved measuring lighting levels, noise levels, temperature and relative humidity at each work station. These levels were compared with the generalised optimum levels taken from Pheasant, 1987.

Lighting - for rough or coarse benchwork; assembly work; inspection tasks, recommended level is set at 540lx with a minimum set at 325lx.

Noise - For workshops an approximate maxim for auditory comfort is 65dBA

Temperature - 22°C, most comfortable year round temperature for sedentary people.

Relative humidity - 40 - 50% is the ideal.

Results

Of the 131 questionnaires administered, only 39 were returned, and of these only 23 were willing to be filmed. The following data was collated from all 39 questionnaires returned, this material is purely for diagrammatic purposes. Table 2 gives an overview of the age, height and weight of the respondents.

Table 2. Overview of respondents

	Mean	Std. Dev.	Max.	Min.
Age	40.97	11.38	62	21
Height	5.07	0.177	6.04	4.10
Weight	12.06	1.71	16.3	9.0

The nature of the investigation required the identification of cases, those experiencing problems and controls, those not experiencing problems. 31% of the respondents were classified as cases and 69% as controls. Of the cases, it was found that 26% had changed jobs at some point due to the discomfort they were experiencing.

The prevalence of upper limb problems did not seem to increase with the increasing number of years on the job. Workers employed with the company for over 20 years displayed the same number of complaints as workers employed with the company for less than a year. The highest occurrence of upper limb discomfort was amongst those employees employed between 5-20 years.

In order to obtain a clearer picture of each subjects exposure to CTD stressors, questions were asked about their leisure activities, as the affects of non-occupational factors on the occurrence of ULD's has been recognised. All but four of the "cases" took part in one or a combination of activities which could have diverse effects in the development of ULD's, namely gardening, DIY, sports and the playing of a musical instrument.

Each subject was analysed, from the video, in the manner set out in the methodology. The MTM illustrated the short task cycle time of all the jobs studied. The posture analysis revealed that nearly all subjects, controls and cases, spent more than 50% of their task time in a detrimental posture, with some spending upto 90%. The risk factor analysis and the direct case control comparisons, disclosed several areas for improvement. The most obvious problem observed was the constant need for bending and twisting, this was also highlighted in the posture analysis. Due to the shape of the bricks, workers were having to adopt poor hand arm postures. Pulp grasps with wrists in extension, forceful wrist movements when using tools and vibrations from machinery, were the three most common observations. Unequal loading and the absence of protective clothing in the form of gloves was also noted. All subjects worked at medium to high levels of effort.

Generally speaking, the environmental conditions were not grossly over the optimum levels. The noise level was exceeded at every work station, similarly temperature and relative humidity were on the whole higher than recommended levels. Lighting levels exceeded minimum requirements.

Discussion

The study was not comprehensive or exhaustive by any means. Problems arose in the first stage of the investigation, the questionnaire survey. Approximately 30% of the entire workforce completed the questionnaires and this caused a number of problems. Firstly, there was an insufficient pool of employees from which to select the subjects. An ideal situation would have allowed, for each task type, a case and a control, who could be roughly matched also by age and experience. Due to the poor response rate selection of subjects could only feasibly be done by age matching. Secondly, all the different jobs within the plant could not be represented, consequently there was an inadequate identification of all employee problems and workplace problem areas. The poor response rate could be attributed to employee fear of the management's response. The questionnaire was issued by the management, via employee payslips. This could have given the impression that the management were directly conducting the investigation. Employees felt that if they complained about their work, management would either replace them or transfer them to lower paid jobs. Similarly, although complete confidentiality was ensured employees were concerned about being singled out, rather than risk this they simply refused to complete the questionnaires.

Analysing the film of each subject, also posed a number of problems. Due to the initial poor response from the questionnaire survey, selecting appropriate cases and controls proved to be a difficult task. The actual jobs of the two groups did not generally correlate and it was evident, even before the analysis stage, that it would not be possible to make direct comparisons. The cases generally did high repetitive tasks, the controls generally did maintenance work which meant their jobs changed from day to day depending on what was required of them.

A number of general deficiencies were highlighted, recommendations to overcome and reduce these shortcomings will now be discussed.

1. A variety of different handling techniques were observed. A number of employees were seen using pulp pinch type grips with their wrists in extension, causing a lot of stress on the hand. Company wide training on handling techniques and the affects of bad handling methods, with refresher courses when appropriate, may standardise handling techniques and possibly reduce the number of problems.

2. Again by direct observation, it was found that most of the stressful postures of the fingers, wrists and arms were caused by the need to place loads in precise locations. The Manual Handling Directive suggests that to reduce injury, one must first put down the load, then adjust into position. The Manual Handling Directive must be implemented to ensure the company complies to its recommendations.

3. Employees should be encouraged to do hand care exercises. By doing these exercises several times a day, hands will be given a break from repetitive work and will help keep them flexible. The exercises should be done every hour or when hands start to get tired. Appropriately placed posters throughout the plant would serve as a constant reminder.

4. The major work station design flaw, was the constant need for bending and twisting. It is apparent that in some sections of the plant, layout redesign could not be easily and cost effectively introduced. A few modifications would however, alleviate some of the problems. In the packing department, all the pallets are placed on the floor, loading and unloading is done at this height. By simple introduction of tables, thus bringing the work to waist height, most of the bad postures will be eradicated. By placing these tables at the and of a conveyor, where appropriate, the need for twisting will also be eradicated.

5. At present all employees are allowed only one break. It has been found that taking breaks actually improves performance, so increasing the number of breaks is an option that needs consideration. There would be no need to shut down the machinery, should the number of breaks be increased, if break times were staggered.

6. Piece-work pay has been linked with the occurrence of upper limb disorders, as people tend to work through breaks and exceed their physical ability to receive an incentive. The "piece work" wage structure should be replaced by a structure mutually agreed by both management and workforce.

Conclusions

The study has identified the occurrence of upper limb disorders, in this particular factory. A number of possible causes have been highlighted and recommendations to overcome these problems suggested. Further work may involve compulsory screening of all employees, medical examinations, a more detailed survey of the environment and a more in depth study of the work organisation, in terms of the task layout, sequence of operations and the equipment used.

References

Armstrong, TJ; Foulke, JA; Bradley, SJ and Goldstein, SA, 1982. Investigation of Cumulative Trauma Disorders in a Poultry Processing Plant. *Amer. Indust. Hygiene Ass. J.*, **43**,103-116

Ayoub, MA, 1990. Ergonomics Deficiencies I: Pain at Work. *Journal of Occupational Medicine*, **32/1**, 52-57

Barnes, RM, 1980. *Motion and Time Study. Design and Measurement of Work.* 7th Ed. John Wiley, New York.

Buckle, PW and Baty, D, 1986. Ergonomic aspects of tenosynovitis and carpel tunnel syndrome in production line workers. In: *Contemporary Ergonomics* 1986, edited by DJ Oborne. London: Taylor and Francis, 237-241

Crawford, J, 1994. *Workplace and Upper Limb Disorder Assessment.* Unpublished report: University of Birmingham

Dimberg, L, 1987. The prevalence and causation of tennis elbow (lateral humeral epicondylitis) in a population of workers in an engineering industry. *Ergonomics*, **30**, no. 3, 573-580

Ferreira, DP and Tracy, MF, 1991. Musculoskeletal Disorders in a Brick Company. In: *Contemporary Ergonomics.* Edited by EJ Lovesey. Taylor and Francis, London. 475-480

Genaidy, AM; Bafna, KM; Delgado, E; Mhidze, A, 1990. An Ergonomic Study for the Control of Upper Extremity Cumulative Trauma Disorders in a Manufacturing Industry. In: *Advances in Industrial Ergonomics and Safety II,* Edited by B Das. Taylor and Francis, London. 245-250

Huppes, G; Maas, K, 1993. Repetitive work and RSI: Recognising work situations at risk and instrument for diagnosis. In: *The Ergonomics of Manual Work.* Edited by Marras, WS, Karwowski, W, Smith, JL and Pacholski, L, 319-322

Joseph, BS and Dagg, J, 1987. A System Designed to Measure Some of the Occupational Risk Factors Associated with Upper Extremity Cumulative Trauma Disorders. In: *Trends in Ergonomics Human Factors IV, Proceedings of the Annual International Ergonomics and Safety Conference.* Held in Miami, Florida, USA, 9-12 June 1987. Edited by Asfour. North-Holland, Amsterdam. 1013-1020

Karhu, O; Kansi, P; Kuorinka, I, 1977. Correcting Work Postures in Industry: A Practical Method for Analysis. *Applied Ergonomics*, **8**, 199-206

Kuorinka, I; Jonsson, B; Kilbom, A, 1987. Standardised Nordic Questionnaire for the Analysis of Musculoskeletal Symptoms. *Applied Ergonomics*, **18**, 233-237

Last, JM, 1983. *Dictionary of Epidemiology.* Oxford University Press, New York.

Malchaire, JB and Rezt-Kallah, B, 1991. Evaluation of the Physical Workload of Bricklayers in the Steel Industry. *Scandinavian Journal of Work, Environment and Health.* **17/2**, 110-116

Parker and Imbus, 1992. *Cumulative Trauma Disorders.* Current Issues and Ergonomic Solutions: A Systems Approach. Lewis Publishers. Boca Raton

Pheasant, S, 1987. *Ergonomics - Standards and Guidelines for Designers.* BSI, Milton Keynes

Putz-Anderson, V, 1988. *Cumulative Trauma Disorders: A Manual for Musculoskeletal Disease of the Upper Limbs.* Taylor and Francis, London

Human Performance

ELECTROPHYSIOLOGICAL CORRELATES OF SLEEP LOSS

*Derya Sürekli - Dr. J. Empson

**Hüseyin Hüsnü Paşa Sok.
9/9 Feneryolu / ISTANBUL / TURKEY*

One electroencopholographic measure was used in order to assess work performance under sleep deprivation. In this experiment, performance was assessed after sleep deprivation by using Contingent Negative Variation. Contrary to the "Lapse hypothesis" for sleep loss effects both fast and slow responses were increased by sleep loss, rather than slow responses only. CNV amplitudes (mean, late, peak) were unaffected by sleep loss. Seperate analyses of the potentials associated with fast and slow responses showed no differences, and similarly no effects of sleep loss on middle and late CNV amplilute. Early, frontal CNV amplitudes were increased by sleep loss, both overall and uniformly for fast and slow response sweeps. P2 amplitude to the warning stimulus was significantly reduced by sleep loss, a finding not previously reported. It is hoped that further research will confirm this interesting finding.

Introduction

Sleep Loss and CNV

Contingent Negative Variation (CNV) is an event-related brain potential that appears between the occurence of two successive stimuli, such as in constant-foreperiod reaction time where one stimulus is a preparatory stimulus for response to the other (Walter et al 1964). Electronegativity progressively develops between S1 and S2. These effects are generally emphasised if the subject is instructed to perform an operant response to indicative stimuli which become imperative. Hence CNV development appeared to be a function of the probability of occurence of the second stimulus (Irwin et al 1966).

The reduction in CNV amplitude with sleep loss, observed by Naitoh et al (1971) could be attributed to a general decrease in negativity associated by reduced arousal. On the other hand, following the lapse hypothesis, one might predict that CNV was reduced because of the inclusion of trials in which the subjects were not paying attention to the warning stimulus, during lapses of attention. If the latter explanation was correct, the CNV's associated with fast reactions should be the same after sleep loss as when rested, while they should be reduced in amplitude when preceding slow reactions (when the subject was essentially unprepared, having failed to pay attention to the warning stimulus). Less specificaly one could argue that despite having noticed the warning stimulus on all trials, slow responses should be associated with reduced CNV in that both indicate a transient lowering arousal or activation.

Hypotheses that will be tested by this experiment

1. Naitoh et al (1971) and Gauthier and Gottesman (1983) found a reduction in CNV amplitude after periods of sleep loss of 24 and 48 hours respectively. It is expected to replicate this effect in this study.

2. The lapse hypotheses would appear to predict that any effects of sleep loss are transitory or phasic a) Fast trials should therefore be just as fast after sleep loss as when rested, while slow trials should become slower b) It should also be predicted that the effects of sleep loss on the CNV should be confined to trials when subjects were performing at their worst, and that CNV preceeding normally fast responses should be unaffected by sleep loss.

3. P2 amplitude is routinely assessed, although it is unclear what effect sleep loss should have on this component.

Method

Subjects

7 students from the psychology department of the University of Hull were the subjects of this study. Subjects came to the psychology department at 10 pm. They were sleep deprived in a room in which they spent time reading, watching tv or video, playing the cad and computer games.

EEG Recording

EEG recordings were taken from silver-silver chloride electrodes at FZ, CZ and PZ referred to the right mastoid EOG was recorded from above below the subject's right eye and forehead earth. 16-Channel Siemens Elema-Shonander electroencopholograph was used. Digital sampling rate was 150 Hz. The reaction time was measured by BBC computer which was programmed in BBC basic. Warning stimulus (WS) was auditory (85 dB) last 40 ms. The imperative stimulus (IS) was a visual display of the words "PUSH THE BUTTON". The interstimulus interval (ISI) between these two stimulus was 1.5 secs. EEG was filtered and amplified at a time constant which is 1.2, and EEG data was avaraged using an averaging programme. Signals coming from EEG fed into an Anolague Digital Converter (ADC) where the anolague signals were converted into numbers. Each sweep comprised four arrays of 307 data points (3 EEG, 1 EOG). The first 7 points were sampled before S1 and acted as a pre-warning stimulus baseline from which CNV measure would be calculated.

The CNV peak amplitude was calculated as the difference between the mean basaline value and the maximal value at 1500 msec after the warning stimulus CNV mean amplitude was calculated as the difference between the mean baseline value and the average value of all points between S1 and S2. Early CNV amplitude was calculated as the difference between the mean baseline value and the mean value between 475 msec S1 (75th data point) and 750 msec (112 data point). Late CNV was calculated as the difference between the mean baseline value and the mean value between 1250 msec after S1 (187th data point) and 1500 msec (225th data point). P2 amplitude was calculated as the difference between mean baseline value and the mean value between 150 ms and 300 ms (30th and 60th data point).

Results

We organized our data (Reaction times, CNV amplitude measures and P2 amplitude) to analyse into two section. In the first part CNV data was analysed into two conditions (SD\Rested). In the second section, the data associated with 8 fastest and 8 slowest responses were analysed after normal sleep and sleep deprivation.

In the first part of analysis, there was no statistically significant difference between two conditions for reaction times. The only significant difference on CNV measures was found for Early CNV amplitude. Early CNV amplitude was increased, especially on the frontal areas ($p<0.0001$). As far as P2 measures is concerned in the first part of analysis, sleep deprivation decreased P2 amplitude significantly ($p<0.05$).

According to the second part of analysis, Table below shows mean reaction times for the 8 fastest and slowest responses associated with EEG recordings after normal sleep and sleep deprivation.

Table 1. Reaction Times

	F	df	P
Fast and slow reactions	76.78	1,6	0.005
Rts after SD us NS	8.85	1,6	0.05
Interaction	0	1,6	NS

ANOVA showed that reaction times are significantly lower after sleep loss than normal sleep ($F=8.85$, $p<0.05$). Not surprisingly, the difference between fast and slow responses was found to be statistically significant ($F=76.78$, $p<0.0005$). Contrary to what the lapse hypothesis would predict, there was no sleep x speed interaction at all ($F=0.000$). It was found that both fast and slow responses are affected in the same way. Figure below shows main effect of sleep loss while there is no interaction effect at all.

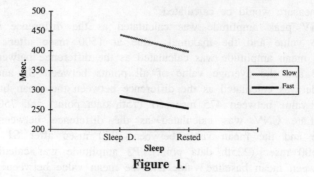

Figure 1.

Table 2. EARLY CNV AMPLITUDE (between 500 msec and 750 msec) associated with fastest and slowest responses.

	F	df	P
Scalp	11.27	3,72	<0.001
Speed	.02	1,72	NS
Sleep	4.27	1,72	<0.04
Scalp x speed	0.001	3,72	NS
Scalp x sleep	.69	3,72	NS
Speed x sleep	.02	1,72	NS
Scalp x speed x sleep	.12	3,72	NS

Unexpectedly, it was found that after sleep deprivation CNV amplitude significantly increased compared to during the rested periods (p<0.04). There were large differences between different scalp locations, frontal early CNV. ANOVA indicated no interaction effect at all either between scalp location and conditions or involving fast versus the slow responses.

Table 3. P2 AMPLITUDE (150 msec and 300 msec) associated with fastest and slowest responses.

	F	df	P
Scalp	1.49	2,72	NS
Speed	.41	1,72	NS
Sleep	9.25	1,72	<.03
Scalp x speed	.04	2,72	NS
Scalp x sleep	.48	2,72	NS
Sleep x speed	.29	1,72	NS
Scalp x sleep x speed	.004	2,72	NS

P2 amplitude decreased significantly after 24 hours sleep deprivation. While there was no interaction effect at all with any other variables.

Discussion

According to the result of the first group of analyses, the CNV (mean, peak and late) amplitudes were not reduced after 24 hours sleep deprivation. In fact, unlike Naitoh's (1971) and Gauthier and Gottesman (1983) studies, after sleep deprivation, early CNV amplitude was increased especially on the frontal areas. According to the result of the second group of analyses, both fast and slow responses are slowed down by sleep loss. As far as slow potential data is concerned, CNV amplitudes (When fast and slow responses are considered together) are not reduced by sleep loss, though this effect is non-significant. However, it was found that early CNV amplitude associated with fast and slow responses was higher on frontal areas compared to parietal and central areas (p<0.001). Secondly, after sleep deprivation early CNV amplitude significantly increased compared to that of rested periods (p<0.04). In the case of P2 amplitude, it was found that after 24 hours sleep deprivation, P2 amplitude significantly decreased.

As far as lapse theory is concerned a subject's best performance is impaired following sleep loss as his worst performance. Williams, Goodnow and Lubin (1959) stated that when RT measures increase with sleep loss, the effect is wholly due to the fact that slow RTs get slower whilst the fast RTs remain unchanged. In the present study, fast response impairment after sleep loss was at least as strong as slow response impairment after sleep deprivation. This finding reinforced the position put forward by Kjellberg (1977) that lapses are not abrupt and discrete periods of lower arousal, but that there is a fluctuation of arousal level, which at a critical point we categorize as "performance lapse". The present experimental results suggest that arousal level is lower following sleep loss by looking behavioral data. Although it is not as lapse theory would suggest, at predeprivation levels sudden, periodic troughs are associated with the slow responses which occur during lapses. On the basis of the present study, it is reasonable to suggest that arousal level fluctuates for behavioral performance around a lower level than that of the non-sleep deprived.

As far as slow potential data is concerned, CNV's were not reduced after sleep deprivation. According to Hillyard (1973), it is an anticipation of stimuli that demand a motor response or command the subject's attention or interest. So sleep loss is the factor which decreases the subject's attention or alertness as lapses increase then CNV magnitude is also reduced. However, this simple causal relation is not enough to explain the effects of sleep loss on CNV amplitude as the result of this study indicates. Lapse hypothesis deals only with one dimension of the effects of sleep loss which is dearousal. Obviously pure effects of sleep loss which is dearousal. However, some situational or experimental conditions increase arousal level by putting more effort into keeping the performance level up against the lowering effect of sleep deprivation (Malmo, 1960, Corcoran, 1962, Wilkinson, 1962). In this regard, CNV may reflect a conscious effort to increase the level of performance as well as other motivational and emotional aspects of situation.

The findings about P2 amplitude have not been reported before. Previous workers like Gauthier and Gottesman did not look at the sleep loss effect on the P2 component. Present finding about the decreasing effect of sleep loss on P2 amplitude suggests that there should be further research on P2 amplitude after sleep deprivation.

References

Corcoran, D.W.J. 1962, *Noise and loss of sleep.* The Quarterly Journal of Experimental Psychology, **14**, 178.

Gauthier, P and Gottesman, C. 1983, *Etude de la variation contingente negative et de l'monte post imperative on presence d'infere,* Electroencepholography and clinical Neurophysiology, **40**, 143-158.

Hillyard, S. 1975, *Relationship Between the Contingent Negative Variation (CNV) and Reaction Time,* Physiology and Behavior, **4**, 351-357.

Irwin, D, Knott, J.R, Dale, W, McAdam, R.C. 1966, National Determinants of the "Contingent Negative Variation", Electroencepholography and Clinical Neurophysiology, **21**, 538-543.

Kjellberg, A. 1977, *Sleep Deprivation* and some Aspects of Performance. Waking and sleeping, **1**, 39.

Malmo, R.B and Surwillo, W.W. 1960, *Sleep Deprivation: Changes in Performance and Physiological incidents of activation,* Psychological Monographs, **74**, 1-23.

Naitoh, P, Jhonson, L.C and Lubin, A. 1971, *Modification of Surface Negative Slow Potential (CNV) in the Human BRain after loss of sleep,* Electroencepholography and Clinical Neurophysiology, **30**, 17-32.

Walter, G, Cooper, R, Aldridge, V.J, McCallum, W.C. and Winter, A.L. 1964, Contingent Negative Variation: An Electric sign of motor association and expectancy in the human brain, Nature, **203**, 380-384.

Wilkinson, R.T. 1962, Muscle Tensio during mental workload under Sleep Deprivation, Journal of experimental Psychology, **64**, 565-571.

THE EFFECT OF SLEEP LOSS ON ADAPTIVE CONTROL

Nik Chmiel

Department of Psychology
University of Sheffield
Sheffield S10 2TN
UK

Adaptive control is concerned with continual adjustment to meet changing task demands, and has been supposed to involve two levels of mental control. Broadbent (1971) argued that sleep loss affected the lower level, making it more inefficient, but that upper level activity could compensate for this before it became fatigued, whereupon performance would suffer. In contrast to the perceptual-motor tasks on which this view was based, the task reported here made demands on cognitive compensatory processes. The effects of sleep loss on this task suggested that the upper level of control should not be thought of as subject to a continuum of activity, but should be considered as changing in efficiency along a dimension like information processing rate. Cognitive compensatory processes appeared to retain their function under conditions of sleep loss and fatigue, but took longer to operate.

INTRODUCTION

Adaptive control describes the general situation where a person is attempting to control some process in order to meet specified output targets which are continually changing. An example from perceptual-motor behaviour is the adjustment pilots need to make to aircraft controls to effect a safe landing in light of feedback from instruments and the environment. A more cognitive example is the control of electricity production where operators meet changing demand by altering system parameters via computers. The questions raised in this paper are: what are the effects of sleep loss on this kind of process, and how should they be understood?

Broadbent (1971) outlined a framework for approaching these questions by considering the effects of stressors on continuous perceptual-motor performance. In the serial reaction task studied, subjects sat before a board on which there were five lights, arranged in a pentagon, each adjacent to an associated touch pad. One of the lights could come on at any time, whereupon the subject was required to touch the associated pad with a stylus. Touching the pad turned the light off. In the unpaced version of the task another light would come on immediately. The speed of reaction, the errors made, and the number of reponses which took longer than 1.5 seconds (gaps), could be measured. The effects of fatigue could be explored by prolonging the work period, usually to half an hour, and other stressors could be introduced as required. Under normal conditions it was found that the average rate of work remained constant over half an hour, however the number of gaps increased toward the end of the work period. Broadbent summarised the effects of certain stressors on this task by observing that heat produced errors immediately, noise produced errors and sleeplessness produced slow reactions, but not until the work period had proceeded for some time.

In light of this kind of data Broadbent (1971) suggested a lower level for whose activity there was an optimum, and an upper level whose increasing activity reduced the effects of sub or super optimal activity in the lower level. He suggested that the lower level was concerned with the execution of well-established decision processes, and the higher level was concerned with monitoring and altering the parameters of the lower level in order to maintain constant performance. He extended his account, concerning the way in which these levels operated, to propose that a true multi-level explanation entailed that the two levels operated simultaneously, that they were different mechanisms, that they operated over different time-spans, and that the upper could alter the lower, and not merely start or stop its activity (Broadbent, 1977).

A key aspect of the levels of control framework is that the cognitive system can adjust itself, within limits, to maintain performance in the face of stressors. There is an strong emphasis, therefore, on compensatory processes and mechanisms, and self regulation of the cognitive system. In this vein Broadbent (1971) argued that sleeplessness affected the lower level, and fatigue affected the upper. Thus sleep loss would not produce an effect on serial responding immediately because the inefficiency of the lower level, relative to a person who had slept, would be corrected by the upper level, until that level became less active through fatigue.

SLEEP LOSS, FATIGUE AND ADAPTIVE CONTROL

Serial reaction requires continual adjustment to changes in the environment, which demand a response. A cognitive task which shares this feature was outlined by Broadbent (1977) in the service of investigating levels of control and is the one reported on here.

The general task that Broadbent described was a simulation of a city transport system. Subjects had to achieve target ranges of the number of people travelling on buses (load per hundred buses) and the number of car-parking spaces available in the city. They could do so by adjusting the time interval between buses and the fee charged for car-parking. The input variables of Interval between buses and Fee charged were related to the output variables of Load and Spaces by two simultaneous equations which subjects were not told about. The equations were $L = 220t + 80f$, and $S = 4.5f - 2t$, where t & f are the bus interval and parking fee respectively, and L & S are the bus load and parking spaces. Broadbent argued that if, after each decision about Load and Spaces, the targets changed such that they increased each time (constituting a ramp input), then the behaviour demonstrated by the subject could only be characterised by a cybernetic mechanism at least as complicated as an 'adaptive controller'. The crucial aspect of such a mechanism, from the levels perspective, is that it has two levels, and adjusting to varying targets requires the upper level of control (see Chmiel, Totterdell & Folkard; in press).

In the study reported here the task was presented in a graphical form which consisted of two graphs showing the targets to be achieved (these followed a sine wave function, and targets on each graph were in phase), and the result of the last decision about interval and fee on load and spaces respectively. In both cases the X axis showed the decision numbers the subject had completed. The Y axis was marked with an upper and lower limit for load per hundred buses, and car-parking spaces respectively. A trial consisted of 10 decisions (about interval and fee) whereupon a new trial was initiated with fresh starting values for the target sine wave functions. Subjects did not have a record of their past decisions and their consequences. This step was taken to ensure that this kind of information did not act as a memory aid during each trial, thus helping to ensure the adaptive part of the task was preserved.

Two major measures collected from this task are: 'performance level', that is the number of decisions about bus interval and fee which entail the two targets being achieved simultaneously; and 'work rate', that is the overall number of decisions achieved in a work period, regardless of outcome. Subjects received a bonus which they were told was related to their performance on the work task. Performance level, rather than work rate, was emphasised.

The transport task formed part of a much larger set of observations, not connected with immediate concerns about levels of control, but designed to investigate body rhythms and performance. The study is taking place over a long period. The time span is required because subjects are placed in isolation for 17 days at a time, and only three people can occupy the isolation unit at any one time. Physiological, other cognitive performance data, subjective ratings of alertness, and questionnaire data are collected during this period. The data from the adaptive control task are presented for the six subjects run to date. These were female members of Manchester University who were paid to participate. Their ages ranged from 18 to 20 years. The control task was run on a Research Machines Notebook Micro-computer.

Data are presented concerning sleep loss of up to one night. However much early discussion of the effects concern loss of at least one night's sleep, and it was this data which caused Broadbent to argue that sleep loss affected the lower level. As is now well recognised sleep consists of at least Rapid Eye Movement (REM) and Slow Wave (SW) sleep. The majority of SW sleep occurs in the early part of sleep. Thus depriving subjects of only part of a night's sleep has different implications for the amount of REM and SW sleep lost, depending on the amount and timing of partial sleep loss. In the present study the values of partial loss chosen were connected with controlling amounts of SW sleep for reasons not related to the adaptive control task.

The sleep dose days consisted of 100%, 67%, 33% and 0% of estimated slow wave sleep, with subjects staying up to achieve the sleep loss. Thus 0% represents a night without sleep, and 100% represents a normal night's sleep of approximately 8 hours. The other percentage values correspond to approximately 2.75 and 1 hours sleep. The dose days were on days 3, 7, 11, and 15 following entry into the isolation unit. Half the subjects received a dose order of 100, 0, 67, 33 percent, and the other half received an order of 33, 67, 0, 100 percent.

Before entering the isolation unit each subject practised the control task for four one and a half hour sessions. The last practice session was on the evening that subjects entered the isolation unit. There were four work sessions on days 1,2,3,4,6,7,8,10,11,12,14,15,16 of the study. Work sessions lasted one and a half hours and were timed to start at 09.30, 11.30, 14.15 and 16.15.

The adaptive control task was presented in blocks of 10 sets of 10 decisions, with a short pause between each during which data were written to the computer memory. Average performance level during a work period was derived from these blocks by taking the number of decisions in a set of 10 which lead to the target values being achieved, and then averaging over the work session. Thus performance level scores represent so many sucessful decisions out of a maximum of 10. Work rate was calculated as the number of blocks in each session. Each block consisted of 100 decisons concerning the two variables under the control of the subject.

Taking performance measures throughout the day means the measures may be sensitive to the effects of time-of-day on performance, rather than fatigue. Broadbent (1971) did consider time-of-day effects in relation to sleep loss, and argued they could be different. He also made the point that performance tends to be correlated with variation in body temperature. On this ground performance should increase throughout the day, whereas any effect of long work should show the opposite effect. Subsequent studies of performance variations with time-of-day show this view is too general, and that different activities may show different patterns. Thus short-term memory may be better in the morning, whereas speeded response improves in the afternoon (Folkard & Monk, 1983), and different attentional phenomena behave differently according to time of day (Broadbent, Broadbent & Jones, 1989). Thus it would be beneficial to control for time-of-day effects. One way to do so, which we have adopted here, is to compare the performance of interest in each time period with that achieved in the same time period the day before. A further advantage of this kind of analysis is that the influence of potential practice effects are better controlled. This procedure has a disadvantage, however, of potentially obscuring the effects due to work session which could be a function of fatigue, or time-of-day, because these effects may also be present on the control day. Therefore a fuller picture of the pattern of performance data can be gained by also analysing the performance on the sleep dose days alone. The latter is done in the second part of the results. Effects which were significant are reported.

RESULTS

1. Difference scores

Table 1 shows performance level differences between days following sleep dose and the day before for each work session. A positive score indicates a lower performance following sleep dose.

work session		1	2	3	4
SW	0	.47	-.07	.67	.28
sleep	33	-.23	-.20	.02	-.27
dose	67	.13	-.08	.03	.17
(%)	100	-.17	.05	-.20	.05

Table 2 shows work rate differences between days following sleep dose and the day before for each work session. A positive score indicates a lower work rate following sleep dose (1.0=100 decisions).

work session		1	2	3	4
SW	0	1.67	1.67	1.83	2.33
sleep	33	1.17	1.00	1.67	0.00
dose	67	1.17	1.83	0.00	1.00
(%)	100	0.00	.50	-.67	.33

There was a significant effect of dose, $F(3,15)=4.84$, $p=0.015$, indicating that work rate dropped significantly following sleep loss compared to the day before. Inspection of table 2 shows this drop to be greatest after losing a whole night's sleep. The drop is greatest in the last work session of the day following loss of a whole night's sleep, but the interaction between dose and session was not significant, $F(9,45)=1.65$, $p=0.13$.

2. Dose day scores

Table 3 shows performance level for days following sleep dose for each work session.

work session		1	2	3	4
SW	0	5.42	5.75	5.12	5.37
sleep	33	5.32	5.47	5.45	5.57
dose	67	5.68	5.72	5.60	5.80
(%)	100	5.50	5.47	5.53	5.57

Table 4 shows work rate for days following sleep dose for each work session (1.0=100 decisions).

work session		1	2	3	4
SW	0	3.67	3.50	3.67	2.83
sleep	33	3.33	4.50	3.50	4.83
dose	67	4.17	3.83	5.17	4.50
(%)	100	4.17	4.00	4.33	4.50

The interaction between dose and session was significant, F(9,45)=2.14, p=0.045. Inspection of table 4 shows that work rate was slowest during the last work session of the day following loss of a whole night's sleep. However, there is also some variability in the work rate for other times of day and sleep dose.

LEVELS, CONTROL AND SLEEP LOSS

The results can be summarised as showing no performance level decrement, either following sleep loss, or over the work sessions. Work rate did prove sensitive to both sleep loss and work session, suffering following a whole night without sleep when compared to the previous day's rate. When considering the effect on work rate of losing a night's sleep over the course of the following day, rate suffered particularly during the last work session, compared to the first three. Thus the effects of sleep loss were influenced by time on task through the day.

The data from this study point to an interesting observation, namely that work rate drops following sleep loss, but performance level is maintained. The results are from six subjects, and therefore should be treated as preliminary, especially as regards non-significant effects. However it is worth noting that such results are consistent with others showing that performance can be maintained following prolonged work, and are based on a very large number of observations per subject.

How should these data be understood in terms of the idea of levels put forward by Broadbent, and what do they mean for his suggestion that fatigue affects the upper, and sleep loss the lower level only? It would follow that because subjects can maintain their performance level, and this involves the upper level of control, then that level must be capable of achieving the activity needed regardless of sleep loss or fatigue. The conclusion then is that fatigue does not affect the upper level, contrary to the speculation put forward by Broadbent (1971), unless the fatigue manipulation turned out not to be sufficiently strong. The latter could be the case, but appears less likely in light of the effect of work session on work rate. It would follow also that if work rate is a function of lower level efficiency then it is the lower level that is affected by sleep loss, as Broadbent thought, and also by fatigue, counter to his reasoning.

Accepting this line of argument would lose the attractive idea of performance in the face of a stressor being held optimally constant, until a compensatory mechanism is fatigued. This idea has very general currency in accounting for a wide range of observations of the effects of workload and fatigue on performance over time (see, for example, Hockey, 1993). The argument above is based on the premise that the two measures used can be associated with different levels of control. In principle this premise is questionable. It is true that performance level must be at least a function of the upper control level, and this level must be used throughout. However it is also the case that, cybernetically speaking, what the upper level does in the adaptive controller is to take an error signal and adjust the parameter(s) of the lower level in order to try to reduce the error to zero.

The output of the system is therefore the product of the two levels acting together, and hence performance level is surely a function of the lower level state as well. Equally it must be the case that work rate could be a function of the efficiency of the two levels in principle. For example, the time taken by the upper level to effect a reparameterisation of the lower level could lengthen, in principle, following sleep loss or fatigue. Thus we should consider the system from the point of view of the two levels acting in concert. In particular we should take seriously the possibility that subjects are able to trade-off performance level against work rate when the system is influenced by stressors. Data from choice reaction tasks demonstrate that subjects are able to make such a trade-off, and use error feedback from the task to do it (Rabbitt & Maylor, 1991).

The data, and others like them, could then be interpreted by considering that the system as a whole becomes more inefficient following sleep loss. How this inefficiency manifests itself, though, is under the control of the subject, who can respond to demands in the task. Thus performance level could be preserved at the expense of work rate. In terms of levels of control, the upper level is able to function well enough to maintain performance, but needs extra time to overcome the inefficiency in the system. This inefficiency could be anywhere in the system in principle: at the upper or the lower or both levels, or in the rate of processing feedback. If at the upper level we should not think of efficiency in terms of activity, but rather in terms of something like processing rate. If the inefficiency is at the lower level we could still regard it as a general sub- or super-optimal state. However the point we wish to emphasise is that our measures cannot, in principle, be sufficiently diagnostic with respect to levels of control, to draw any firm conclusion. A similar point has been made by Rabbitt & Maylor (1991) in relation to controlling the relationship between speed and error in Choice Reaction tasks. They conclude that although the data are "obviously interesting and important reflections of...changes in state [they] are not at all analytic of changes in the efficiency of processes of control and comparison, of changes in the sensitivity with which feedback is monitored ..." (p281). In this sense we hope we have illustrated that the framework concerning levels proposed by Broadbent can only be applied to our data post hoc. Its value is as an interpretive framework which emphasises adaptive and compensatory processes, and as such it has also recently been applied to understanding the effects of neurotransmitters (Robbins & Everitt, 1987), and workload and effort (Hockey, 1993).

REFERENCES

Broadbent, D.E. (1971). Decision and Stress. London: Academic Press.
Broadbent, D.E. (1977). Levels, Hierarchies, and the locus of control. Quarterly Journal of Experimental Psychology, 29, 181-201.
Broadbent, D.E., Broadbent, M.H.P., & Jones, J.L. (1989). Time of day as an instrument for the analysis of attention. European Journal of Cognitive Psychology, 1, 69-94.
Chmiel, N., Totterdell, P., & Folkard, S. (in press). On adaptive control, sleep loss, and fatigue. Applied Cognitive Psychology.
Folkard, S., & Monk, T.H. (1983). Chronopsychology: circadian rhythms and human performance. In Physiological correlates of human behaviour. London: Academic Press.
Hockey, G.R.J. (1993). Cognitive-energetical control mechanisms in the management of work demands and psychological health. In A. Baddeley & L. Weiskrantz (Eds.) Attention: selection, awareness, and control. A tribute to Donald Broadbent. Oxford: O.U.P.
Rabbitt, P.M.A., & Maylor, E.A. (1991). Investigating models of human performance. British Journal of Psychology, 82, 259-290.
Robbins, T.W. & Everitt, B.J. (1987). Psychopharmacological studies of arousal and attention. In S.M. Stahl, S.D. Iversen, & E.C. Goodman (Eds.). Cognitive Neurochemistry. Oxford: O.U.P.

HUMAN RELIABILITY IN SAFETY-CRITICAL SYSTEMS

Megan R. Brown and *Paul D. Hollywell

*Ergonomics Unit,
University College London,
26 Bedford Way, London WC1H 0AP.*

** EWI Engineers & Consultants,
Electrowatt House, North Street,
Horsham, West Sussex RH12 1RF.*

This paper proposes that there are parallels between designing the
hardware and human aspects of a Safety-Critical System (SCS). The
aims, when designing both these elements are to produce highly reliable
equipment and to reduce operational errors. It is suggested that high
levels of hardware and human reliability depend on three important design
factors: quality, redundancy and diversity. Consideration of all three
factors is required if high reliability is to be achieved through the
appropriate design and operation of a SCS. This paper presents guidance
on achieving human reliability in SCSs. It also describes an engineering-
based approach to support designers and ergonomists alike in the
realisation of such designs.

Introduction

Recently there has been a growing interest in Safety-Critical Systems (SCSs) within
a wide range of industries. These areas include the nuclear, aerospace, oil and gas,
chemical, transport and medical sectors. A SCS often consists of two main elements:
hardware (and software if a computer is involved) and human activity associated with
the operation of the hardware. Research into SCSs, particularly the issue of software
reliability, has drawn a great deal of attention over the past few years. Also, it has
become clear from major accidents that the human element is vitally important when
considering the reliability of SCSs. The human element should always be fully
considered during the design process if human-related system failures are to be
avoided.

This paper proposes that there are useful parallels between designing the hardware
and human aspects of a SCS. These parallels can support designers and ergonomists
alike when attempting to design and operate SCSs that can achieve high reliability.

499

Designing High-Reliability Equipment

Over the years, when attempting to design high-reliability equipment, there has been a shift away from relying on single components always performing satisfactorily to achieve an acceptable level of reliability. More recently the design of high-reliability equipment for inclusion in SCSs has involved the combination of three distinct approaches: the design factors quality, redundancy and diversity. These factors are shown schematically in Figure 1, below, and are considered the three primary dimensions of the general design-for-reliability problem.

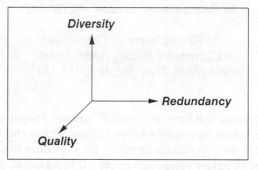

Figure 1. Dimensions of Reliability

Quality in Equipment Design

The quality approach to equipment design takes special care in selecting the most suitable components. It invests in quality control, quality assurance and the de-rating of components (ensuring that components are used at levels set below their peak loading). Thus, the quality approach focuses on improving quality, thus maximising the reliability that is achievable for each component.

Redundancy in Equipment Design

The redundancy approach to equipment design uses more than one component to perform a function, for which a single component is only strictly necessary. For example, many industrial plants have redundant (or backup) equipment items that are available if there is a failure with the normal item, e.g. pumps, valves, cooling, power supplies. The addition of redundant components increases total reliability.

The reliability benefits of redundancy, however, can only be fully realised if true independence exists between the redundant components. If the redundant components are not truly independent, the system can be vulnerable to failures caused by common single factors. These types of failures are commonly known as Common Mode Failures (CMFs) or Common Cause Failures (CCFs). CMFs usually occur when redundant components are caused to fail in the same way due to a common internal cause that is inherent in that particular component, e.g. faulty design or maintenance on two or more identical electrical devices. CCFs usually occur when an external cause exerts an influence on redundant components such that those components can no longer be regarded as failing independently, e.g. common power supply to redundant electrical devices. To overcome the potential weaknesses of a system design due to CMFs or CCFs, an additional approach is therefore required to improve reliability.

Diversity in Equipment Design

The <u>diversity</u> approach to equipment design seeks to reduce any potential problems with the system that could result from a CMF or a CCF by introducing some diversity into the design. The addition of diverse components further increases total reliability. An illustration of the inclusion of diversity in a design, is the navigation system used on most flight decks of aeroplanes. These normally comprise two types of compass technologies: a magnetic and a gyroscopic compass. Thus, there are both redundancy and diversity within the compass design. If one type of compass becomes faulty, the other type of compass can be used as an alternative. It is extremely unlikely that the cause of a fault in one compass would occur in the other compass due to the diverse types of technology involved. Thus, the vulnerability to CMFs and CCFs is reduced and the total system reliability is improved.

Reliability Design Process

It is proposed that to maximise the reliability of a SCS, the design process should: (i) employ high-quality components (in terms of their selection and use), (ii) have some redundancy of function, and (iii) have some degree of diversity to overcome the potential reliability problems that can occur with using redundant components. These elements of the design process are shown schematically in Figure 2, below.

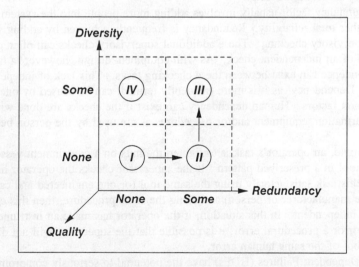

Figure 2. Elements in the Design of High-Reliability Equipment

An initial design might begin without any redundancy or diversity (Element I). As improved reliability is required, so redundancy is introduced (Element II). Then some diversity may be considered to combat the problem of CMFs and CCFs (Element III). It is hard to imagine a situation where some diversity exists without any redundancy (Element IV). The quality design factor may be considered for all elements of the design process. Designs of equipment for SCSs will, therefore, normally involve employing the most effective combination of the above three dimensions of the general design-for-reliability problem.

Designing for Human Reliability

There are strong parallels between designing high-reliability equipment and designing human reliability into the operation of any SCS. Again, the three primary dimensions of reliability shown, in Figure 1, are important if human error is to be reduced. In the case of designing for reliable operation, however, quality, redundancy and diversity have to be translated into human-related terms.

Quality in Human Design

The quality factor focuses on the systematic application of good ergonomics principles and guidelines to reduce human error. Relevant guidance is given in standard ergonomics reference material, such as the HFRG guide to reducing human error (HFRG, 1991). Parallelling equipment design, the quality factor focuses on appropriate selection and training of personnel for the reliable operation of the SCS. Additionally, this approach attempts to encourage reliable operation of a system by applying good human-interface design practices and proper design of tasks, such that personnel are seldom required to carry out tasks under conditions of extremely high workload.

Redundancy in Human Design

The redundancy factor usually involves adding more people into the system to achieve higher total reliability. Redundancy is frequently achieved by adding some form of supervisory checking. These additional supervisory checks can often create the illusion of an independent check. As with equipment design, however, a lack of true independence can exist between these checking tasks. This lack of independence, or 'Human Dependency' as it is more frequently termed, can be caused by one or more common factors. Human dependency can exist if the checks are done with similar information, equipment and/or procedures as are used by the person being checked.

For example, an operator's task is to tighten the nuts on a containment vessel using a torquing tool in a prescribed pattern. If the supervisor checks the operator has completed this task satisfactorily using the same tool (or one engineered and calibrated by the same manufacturer or person) and using the same procedure, then this check will not be independent. In this situation, if the operator has made an instrument reading error or a procedural error, it is possible that the supervisor will not discover it, falling foul of the same human error.

Human Dependent Failures (HDFs) have the potential to seriously compromise the assumed independence of tasks designed to be redundant. This can significantly reduce the reliability of SCSs. There, therefore, needs to be a full and proper appreciation of the potential impact of HDFs on SCSs in order for an assessment of reliability of the whole system (hardware and human) to be a realistic one. Fortunately, the subject of HDFs is becoming more widely understood (Hollywell, 1994) and improved techniques for including them within risk assessments are continually being developed (Brown, 1994).

As with equipment design, to overcome the problem of HDFs some form of diversity must be introduced between related checking tasks. Strictly speaking, in human systems the introduction of redundancy (i.e. adding more people) will introduce some degree of diversity. This is because no two people function in exactly the same way. This diversity is extremely limited, however, when compared to the strength of some of the human error-producing mechanisms that could be involved in the related tasks. Significant diversity can only be accomplished by ensuring that major differences exist between the related tasks.

Diversity in Human Design

The diversity factor, normally concerns itself with the design of the tasks to be carried out by the personnel operating the system. For the reasons given in the previous section, this is the only way in which significant diversity can be introduced into the design of a SCS.

Let us continue the example given in the previous section. If the supervisor employs a different procedure to check the tightness of the nuts, and uses a different torquing tool than the one used by the operator, then the checking task will be independent of the original operational task. As a result, it is more probable that the supervisor will identify any errors made by the system operator. Therefore, the introduction of diversity between system-related tasks and activities leads to an increase in total system reliability.

In Hollywell (1994), the author suggests a wide range of strategies for the inclusion of diversity into human engineering design to achieve human reliability. It is important to note that there can be some practical problems in applying diversity-of-design strategies. Potential conflicts can exist between maintaining quality while attempting to obtain diversity. Some diversity-of-design solutions tend to go against current standardisation practices, designed to enhance quality. The applicability of such diversity-of-design measures should, therefore, be assessed in detail within its context of use.

In designing for reliable operation, unlike designing equipment, it is possible to introduce diversity into a system design without there being any preexisting redundancy present. This is because it is possible for one person to carry out a single, important function or task in more than one way. For example, the combined use of row and column checks when working with tables of numbers can often lead a person to identifying their own simple arithmetic slips. Another example, is that a person can be trained and instructed to use double-entry bookkeeping to keep a set of accounts. The inbuilt diversity within this method of keeping a set of accounts is claimed to increase a person's ability to detect and correct their own accounting errors.

Human Reliability Design Process

It is proposed that to reduce human error and maximise total system reliability, designing for human reliability should follow a similar process as that used for equipment design. The human design process should: (i) apply sound ergonomics principles and guidelines, (ii) consider adding supervisory checking procedures to improve task reliability and (iii) ensure that significant diversity exists between related tasks, particularly checking tasks, to minimise human dependency. Alternatively, consider introducing some diversity within the original task, thus improving human reliability without the additional involvement of another person. All these elements of designing for human reliability are shown schematically in Figure 3, below.

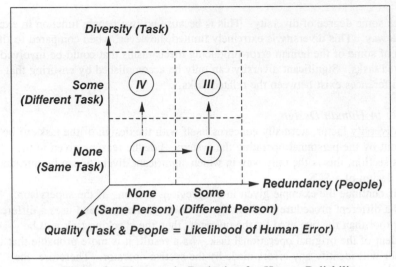

Figure 3. Elements in Designing for Human Reliability

An initial design might begin with one person carrying out a single task (Element I). As a related task, such as a supervisor's check, is introduced to improve human reliability, this provides some redundancy (Element II). To overcome human dependency problems that may accompany the related task, the related task is designed in a way that provides some diversity (Element III). Alternatively, improved human reliability could be achieved by enhancing the error-detecting and correcting abilities of the person carrying out the original task by getting the same person to carry out the same task in a different way (Element IV). The quality of the ergonomics can be considered extremely relevant to reducing human error associated with all elements of designing for human reliability. Maximising human reliability for the operation of a SCS needs, therefore, to take proper account of all the above design elements.

Conclusion

An effective SCS design will need to involve the 'optimal' solution of the general design-for-reliability problem, described in this paper, for both the hardware and human aspects of the proposed SCS. Systematic consideration of the quality, redundancy and diversity design factors is necessary. A practical method of achieving this solution is by using proven quantitative techniques, such as Quantitative Risk Assessment (QRA) supported by Human Reliability Analysis (HRA).

References
Brown, M.R. 1994, A Qualitative Approach to the Incorporation of Human Dependent Failures into Human Reliability Assessment, MSc (Ergonomics) Project Report, Faculty of Science, University of London.
HFRG 1991, The Guide to Reducing Human Error in Process Operations, (SRDA).
Hollywell, P.D. 1994, Incorporating Human Dependent Failures in Risk Assessment to Improve the Estimation of Actual Risk. *Proceedings of Risk Assessment and Risk Reduction Conference, Aston University, 22 March.*

SOME EFFECTS OF ANTICHOLINERGIC DRUGS ON PERFORMANCE

Sona J Toulmin and Anthony Wetherell

Chemical and Biological Defence Establishment
Porton Down
Salisbury
Wiltshire SP4 0JQ

People rarely perform at 100% all the time, and one of the reasons is that they take drugs. Most people will take drugs at some time, some people might have to take them all the time, and all drugs have side effects which may impair performance directly, or indirectly by affecting mood.
Thousands of drugs are available, and this paper considers anticholinergic drugs such as atropine and hyoscine, which are widely taken for a variety of conditions including asthma, motion sickness, organophosphorus poisoning, and cases involving smooth muscle spasm such as dysmenorrhoea, intestinal colic, constipation, and ulcers. Anticholinergics impair memory, numerical and verbal reasoning, reaction time, eye-hand coordination, attention, and driving behaviour, and their use should be taken into account by ergonomists and designers.

Introduction

It has been said that ergonomics began by designing for super-people, then for normal people, and more recently for people whose performance is impaired. Performance may be impaired by many factors, eg physical disability, fatigue, and illness, but one that ergonomics has so far tended to ignore is drugs. Almost everyone will take drugs at some time, some people may have to take them all the time, and all drugs have side effects which can impair performance directly, or indirectly by affecting mood or other subjective state. Thus, people who are taking drugs may not be able to cope with equipment or procedures that have been designed with healthy people in mind.

There are far too many drugs to cover in one paper, and we have concentrated here on anticholinergics, sometimes called antimuscarinic, or belladonna drugs, a class which includes atropine, named after Atropos, the oldest of the three Fates, who cuts the thread of life, and hyoscine, also known as scopolamine, the "truth serum" of spy stories.

These drugs are widely distributed in nature, especially in Solanaceous plants. Atropine is found in *Atropa belladonna*, or deadly nightshade, and in *Datura stramonium*, also known as Jamestown or Jimson weed, stinkweed, thornapple, and devil's apple. Hyoscine is found chiefly in *Scopola carniolica*, and in the shrub *Hyoscyamus niger*, or henbane.

Anticholinergic drugs have been known and used by physicians for centuries: preparations of belladonna were known to the ancient Hindus, and extracts of deadly nightshade were a favourite among poisoners of the Middle Ages. Belladonna, "beautiful lady", is named after ladies who used to drop tinctures of deadly nightshade into their eyes to dilate their pupils and make them more alluring.

Chemically, the drugs are organic esters formed from tropic acid and complex organic bases, usually tropine (tropanol) or scopine. Pharmacologically, anticholinergics block the action of the neurotransmitter acetylcholine at muscarinic receptor sites. Thus, low doses will depress salivary, bronchial and sweat secretion; higher doses will dilate the pupil, inhibit accommodation of the eye, and increase heart rate; even higher doses will inhibit micturition and decrease the tone and motility of the gut; still higher doses will inhibit gastric secretion. The drugs also affect the central nervous system, generally causing stimulation in low doses, depression in higher doses, and hallucinations in very high doses.

Anticholinergics are used clinically a) as pre-anaesthetic medications, b) to treat Parkinson's disease, c) as adjuncts in the treatment of myasthenia gravis, d) as antispasmodics for various conditions involving smooth muscle spasm, eg dysmenorrhoea, biliary, intestinal and renal colic, constipation, gastric and duodenal ulcers (reduce gastric motility and acid secretion), e) as constituents of anti-asthmatic preparations (dilate bronchioles and reduce secretions), f) as mydriatics (increase pupil size) and cycloplegics (relax ciliary muscle) to aid eye examinations and to treat acute iritis and keratitis, g) to treat motion sickness, and h) to treat and pretreat organophosphorus poisoning (block acetylcholine actions).

Atropine and hyoscine generally have similar actions, but differ slightly: hyoscine has a stronger action on the iris, ciliary body, and salivary, bronchial and sweat glands; atropine is the more potent on heart, intestine and bronchial muscle, has a more prolonged action, and does not depress the central nervous system as much as hyoscine in normal clinical doses.

The exact scale of use of anticholinergics is not known, but in the year ending March 1994, 232000 prescriptions were written for one dysmenorrhoea preparation alone, 3000 prescriptions were written for other products containing belladonna, 13000 for products containing atropine sulphate, 20000 for products containing hyoscine butylbromide, and 16000 for products containing another anticholinergic, propantheline bromide. Preparations containing anticholinergics should come with instructions to avoid driving or operating machinery, but while it would be good to think that people would comply, it is certain that some people would not. Even those who do comply might misjudge the time of onset, duration, type, or severity of effects, or might misinterpret the term "machinery".

Hyoscine

Hyoscine has been widely studied, mostly in terms of memory, and several studies have found that the drug impairs memory in subcutaneous (sc), intramuscular (im), intravenous (iv) and oral doses from 0.35 to 1mg (eg Colquhoun 1962; Hrbek et al 1970; Dundee & Pandit 1972; Drachman & Leavitt 1974; Poulton & Edwards 1974; Ghoneim

& Mewaldt 1975; 1977; Crow et al 1976; Frumin et al 1976; Drachman 1977; Petersen 1977; 1979; Sitaram et al 1978; Liljequist & Matilla 1979; Mewaldt & Ghoneim 1979; Caine et al 1981; Frith et al 1984; Rusted & Warburton 1988; Wesnes & Simpson 1988; Brazell et al 1989). Hyoscine particularly affects memory for new information (Ostfeld & Aruguete 1962; Safer & Allen 1971; Crow & Grove-White 1973; Crow et al 1976; Ghoneim & Mewaldt 1975; 1977), and impairs memory storage but not recall processes (Doherty & Wetherell 1991).

Several studies have been concerned with the effects of hyoscine on vigilance and attention: impairments have been reported following sc, iv and oral doses ranging from 0.6mg to 1.2mg (eg Colquhoun 1962; Ostfeld & Aruguete 1962; Safer & Allen 1971; Drachman et al 1980; Wesnes & Warburton 1983; 1984; Wesnes & Revell 1984; Parrott 1986; Broks et al 1988; Huff et al 1988; Wesnes et al 1988).

Relatively few studies of other cognitive and psychomotor effects of hyoscine have been carried out. Ketchum et al (1973) reported that 17µg im hyoscine per kg of body weight impairs numerical facility, and Poulton and Edwards (1974) that 1mg oral hyoscine impairs tracking accuracy. Wesnes et al (1988) reported that 0.6mg sc hyoscine impairs choice reaction time, letter cancellation, logical reasoning and rapid information processing (which has a high vigilance component), and Doherty and Wetherell (1990) that 0.6mg im hyoscine impairs numerical, verbal and spatial reasoning, tracking, continuous responding and fine manual dexterity.

Hyoscine can also be administered transdermally to prevent motion sickness. Adhesive patches, usually affixed to the skin over the mastoid process, contain 1.5mg hyoscine, of which about 0.5mg is released over a period of three days. This means of administration is very convenient and efficacious, but can also cause blurred vision, dry mouth and drowsiness (Graybiel et al 1976; 1981; 1982; Price et al 1981; Homick et al 1983; Chandrasekaran 1984; Shattock & Wetherell 1991; Wetherell 1993).

In terms of performance, transdermal hyoscine has been found to impair tapping rate (Hordinsky et al (1982) and mass discrimination (Ross & Schwartz 1984). Parrott & Jones 1985), in a study carried out at sea, reported that transdermal hyoscine increased errors on a letter cancellation test, and that several subjects were unable to complete the tests owing to focussing problems. Parrott (1986) compared transdermal hyoscine with 0.15mg, 0.3mg, 0.6mg and 1.2mg oral hyoscine; he found significant linear oral dose-response effects on continuous attention, continuous performance, alertness and memory storage, and reported that the effects of a transdermal patch were similar to those of the higher oral doses. Shattock and Wetherell (1991) and Wetherell (1993) found that both one and two patches impaired short-term memory and vigilance, and that the effects of two patches were, as expected, worse than those of one patch.

Atropine

In contrast to hyoscine, little experimental research into human performance has been carried out on atropine, but the drug has a long history of use, and there are several clinical reports of its effects. Atropine is similar to hyoscine in that it causes dry mouth, blurred vision, and drowsiness.

In terms of performance, Cullumbine et al (1959) reported dizziness, drowsiness and difficulty in reading after doses of up to 5mg. Moylan-Jones (1969) reported that three 2mg im injections impaired performance of some routine military tasks including digging, vehicle wheel changing, map reading, taking compass bearings, and arithmetic.

Two mg im atropine impairs numerical and verbal reasoning, reaction time, eye-hand coordination and attention from one to four hours after dosing (Holland et al 1979).

The same dose also impairs short-term memory (Wetherell 1980), and reaction time by slowing perceptual and decision, but not motor, process (Wetherell 1983).

In vehicle driving, 2mg atropine impairs drivers' ability, confidence and willingness in driving through narrow gaps, and affects their ability to maintain distance in car-following (Wetherell 1986).

Conclusions

People who take anticholinergics will experience several side effects that could impair performance directly, or indirectly through such effects as restlessness, irritability, disorientation, giddiness and drowsiness. In terms of performance, hyoscine impairs cognitive functions such as short term memory, numerical, verbal and spatial reasoning, and psychomotor functions such as tracking, continuous responding and fine manual dexterity. Transdermal hyoscine patches (for travel sickness) also impair short term memory and other cognitive functions. Atropine has a similar spectrum of activity, and also affects drivers' gap judgements, following headways, and memory for route directions.

Despite warnings not to drive or operate machinery, people might not comply, either deliberately or unwittingly. They may do so deliberately because they want to make a journey, or do not want to lose money or time off work. They may rationalise that a trip to the shops is not really "driving", that motorcycles are exempt, or that their equipment is not really "machinery". They may do so unwittingly because they are unaware that they are affected, they have misjudged the time of onset, duration, type, or severity of effects, or they might genuinely believe that their equipment is not actually "machinery".

Whatever the reason, performance will be impaired for some or all of the time, and the result will be at best loss of production, and at worst, injury or death to themselves or others. Anticholinergics are but one class of drugs among many, and drugs are but one reason. Thus, significant numbers of people or periods of time are involved, and the problem deserves some attention. Ergonomists are taught not to regard themselves as representative of their target population, so they should not think that their population will be healthy just because they are; they should remember that even ergonomists can fall ill and have to take drugs.

References

Brazell, C. Preston, G.C. Ward, C. Lines, C.R. Traub, M. 1989, The scopolamine model of dementia: chronic transdermal administration, *J Psychopharmacology*, **3**, 76-82.
Broks, P. Preston, G.C. Traub, M. Poppleton, P. Ward, C. Stahl, S.M. 1988, Modelling dementia: effects of scopolamine on memory and attention, *Neuropsychologia*, **26**, 685-700.
Caine, E.D. Weingartner, H. Ludlow, C.L. Cudahy, E.A. Wehry, S. 1981, Qualitative analysis of scopolamine induced amnesia, *Psychopharmacology*, **74**, 74-80.
Chandrasekaran, S.K. 1983, Controlled release of scopolamine for prophylaxis of motion sickness, *Drug Dev & Ind Pharm*, **9**, 627-646.
Colquhoun, W.P. 1962, Effects of hyoscine and meclozine on vigilance and short term memory, *Br J Indus Med* **19**, 287-296.
Crow, T.J. Grove-White, I.G. 1973, An analysis of the learning deficit following hyoscine administration in man, *Br J Psychol* **49**, 322-327.
Crow, T.J. Grove-White, I.G. Ross, D.G. 1976, The specificity of the action of hyoscine on human learning, *Br J Clin Pharmacol*, **2**, 367-368.

Cullumbine, H. McKee, W.H.E. Creasey, N.H. 1959, The effects of atropine sulphate upon healthy male subjects, *Quart J Exp Physiol,* **40**, 309.

Doherty, P.C. Wetherell, A. 1990, Cognitive, psychomotor and subjective effects of 0.6mg hyoscine, Unpublished MOD Report.

Doherty, P.C. Wetherell, A. 1991, Some effects of 0.6mg hyoscine on memory, Unpublished MOD Report.

Dundee, J.W. Pandit, S.K. 1972, Anterograde amnesic effects of pethidine, hyoscine and diazepam in adults, *Br J Pharmacol,* **44**, 140-144.

Drachman, D.A. 1977, Memory and cognitive function in man: does the cholinergic system have a specific role? *Neurology,* **27**, 783-790.

Drachman, D.A. Leavitt, J. 1974, Human memory and the cholinergic system, *Arch Neurol,* **30**, 113-121.

Drachman, D.A. Noffsinger, D. Sahakian, B.J. Kurdziel, S. Fleming, P. 1980, Aging, memory and the cholinergic system: a study of dichotic listening, *Neurobiol Aging,* **1**, 39-43.

Frith, C.D. Richardson, J.T.E. Samuel, M. Crow, T.J. McKenna, P.J. 1984, The effects of intravenous diazepam and hyoscine upon human memory, *Quart J Exp Psychol,* **36A**, 133-144.

Frumin, M.J. Herekar, V.R. Jarvik, M.E. 1976, Amnesic actions of diazepam and scopolamine in man. *Anesthesiology,* **45**, 406-412.

Ghoneim, M.M. Mewaldt, S.P. 1975, Effects of diazepam and scopolamine on storage, retrieval and organisational process in memory, *Psychopharmacologia,* **44**, 257-262.

Ghoneim, M.M. Mewaldt, S.P. 1977, Studies on human memory: the interactions of diazepam, scopolamine and physostigmine, *Psychopharmacology,* **52**, 1-6.

Graybiel, A. Cramer, D.B. Wood, C.D. 1981, Experimental motion sickness: efficacy of transdermal scopolamine plus ephedrine, *Aviat Space Environ Med,* **52**, 337-339.

Graybiel, A. Cramer, D.B. Wood, C.D. 1982, Antimotion sickness efficacy of scopolamine 12 and 72 hours after transdermal administration, *Aviat Space Environ Med,* **53**, 770-772.

Graybiel, A. Knepton, J. Shaw, J. 1976, Prevention of experimental motion sickness by scopolamine absorbed through the skin, *Aviat Space Environ Med,* **47**, 1096-1100.

Holland, P. Kemp, K.H. Wetherell, A. 1978, Some effects of 2mg im atropine and 5mg im diazepam, separately and combined, on human performance, *Br J Clin Pharmacol,* **5**, 367P.

Homick, J.L. Lee Kohl, R. Reschke, M.F. Degionanni, J. Cintron-Trevino, N.M. 1983, Transdermal scopolamine in the prevention of motion sickness: evaluation of the time course of efficacy, *Aviat Space Environ Med,* **54**, 994-1000.

Hordinsky, J.R. Schwartz, E. Beier, J. Martin, J. Aust, G. 1982, Relative efficacy of the proposed space shuttle antimotion sickness medications, *Acta Astronautica,* **9**, 375-383.

Hrbek, J. Komenda, S. Macakova, J. Siroka, A. 1970, Effect of physostigmine on the inhibitory action of scopolamine in man, *Act Nerv Sup,* **12**, 273.

Huff, F.J. Mickel, S.F. Corkin, S. Growdon, J.H. 1988, Cognitive functions affected by scopolamine in Alzheimer's disease and normal aging, *Drug Dev Res,* **12**, 271-278.

Ketchum, J.S. Sidell, F.R. Crowall, E.B. Aghajanian, C.K. Hayes, A.H. 1973, Atropine, scopolamine and ditran: comparative pharmacology and antagonists in man, *Psychopharmacologia,* **28**, 121-145.

Liljequist, R. Mattila, M.J. 1979, Effect of physostigmine and scopolamine on the memory functions of chess players, *Med Biol,* **57**, 402-405.

Mewaldt, S.P. Ghoneim, M.M. 1979, The effects and interactions of scopolamine, physostigmine and methamphetamine on human memory, *Pharmacol Biochem Behav,* **10**, 205-210.

Moylan-Jones, R.J. 1969, The effect of a large dose of atropine upon the performance of routine tasks, *Br J Pharmacol,* **37**, 301.

Ostfeld, A.M. Aruguete, A. 1962, Central nervous system effects of hyoscine in man, *J Pharmacology*, **137**, 133-139.

Parrott, A.C. 1986, The effects of transdermal scopolamine and four dose levels of oral scopolamine (0.15, 0.3, 0.6 and 1.2mg) on psychological performance, *Psychopharmacology*, **89**, 347-354.

Parrott, A.C. Jones, R. 1985, Effects of transdermal scopolamine upon psychological test performance at sea, *Eur J Clin Pharmacol*, **28**, 419-423.

Petersen, R.C. 1977, Scopolamine induced learning failures in man, *Psychopharmacology*, **52**, 283-289.

Petersen, R.C. 1979, Scopolamine state-dependent memory processes in man. *Psychopharmacology*, **64**, 309-314.

Price, N.M. Schmitt, L.G. McGuire, J. Shaw, J.E. Trobough, G. 1981, Transdermal delivery of scopolamine for prevention of motion-induced nausea at sea, *Clin Pharmacol Ther*, **29**, 414-420.

Poulton, E.C. Edwards, R.S. 1974, Interactions, range effects and comparisons between tasks in experiments measuring performance with pairs of stresses: mild heat and 1mg l-hyoscine hydrobromide, *Aerospace Med*, **45**, 735-741.

Ross, H.E. Schwartz, E.O. 1984, Can medication interfere with space research? An example from a mass-discrimination experiment on Spacelab 1, *Proc 2nd European Symposium on Life Sciences Research in Space*, 261-264.

Rusted, J.M. Warburton, D.M. 1988, The effects of scopolamine on working memory in healthy young volunteers, *Psychopharmacology*, **96**, 145-152.

Safer, D.J. Allen, R.P. 1971, The central effects of scopolamine in man, *Biol Psychiat*, **3**, 347-355.

Shattock, J.A. Wetherell, A. 1991, Cognitive, psychomotor and subjective effects of transdermal hyoscine, Unpublished MOD Report.

Sitaram, N. Weingartner, H. Gillin, J.C. 1978, Human serial learning: enhancement with arecholine and choline and impairment with scopolamine, *Science*, **201**, 274-276.

Wesnes, K.A. Revell, A. 1984, The separate and combined effects of scopolamine and nicotine on human information processing, *Psychopharmacology*, **84**, 5-11.

Wesnes, K.A. Simpson, P.M. 1988, Can scopolamine produce a model of the memory deficits seen in aging and dementia? In M.M. Gruneberg, P.E. Morris, R.N. Sykes (eds), *Practical Aspects of Memory: Current Research and Issues, vol 2, Clinical and Educational Implications*, (John Wiley & Sons, Chichester) 236-241.

Wesnes, K.A. Simpson, P.M. Kidd, A. 1988, An investigation of the range of cognitive impairments induced by scopolamine 0.6mg s.c. *Human Psychopharmacol*, **3**, 27-41.

Wesnes, K.A. Warburton, D.M. 1983, Effects of scopolamine on stimulus sensitivity and response bias in a visual vigilance task, *Neuropsychobiol*, **9**, 154-157.

Wesnes, K.A. Warburton, D.M. 1984, Effects of scopolamine and nicotine on human rapid information processing performance, *Psychopharmacology*, **82**, 147-150.

Wetherell, A. 1980, Some effects of atropine on short-term memory, *Br J Clin Pharmacol*, **10**, 627-628.

Wetherell, A. 1983, Atropine and diazepam actions on information processing stages in reaction time performance, *Paper to British Psychological Society Conference, York*.

Wetherell, A. 1986, Effects of atropine on drivers' perceptual-motor and decision-making behaviour, In J.F. O'Hanlon, J.J. de Gier (eds), *Drugs and Driving*, (Taylor and Francis, London) 291-302.

Wetherell, A. 1993, Performance effects of physostigmine and scopolamine as nerve agent pretreatments, *Proc 1993 Medical Defense Bioscience Review, vol 2, US Army Medical Research and Development Command*, 653-656.

Controls

ON VARIABILITY IN HUMAN CHARACTERISTICS

H. Kanis and L.P.A. Steenbekkers

Faculty of Industrial Design Engineering
Jaffalaan 9
2628 BX Delft, the Netherlands

Variability is at heart of human characteristics. This paper focusses on the description of this variability between and within subjects. In addition, the patterning of the variation in human performance is briefly discussed with respect to the design of products.

Introduction

Observation on the basis of measuring is always liable to at least some random variation in the results if a measurement is repeated. The specification of this random variation is generally termed as measurement error (cf. ISO 5725, 1986). However, observed variation can partly be due to fluctuations in the measured phenomenon itself. This may particularly be the case for human characteristics since variability is an inherent aspect between and within all biological systems (cf. Newell and Corcos, 1993). The mergence of fluctuations in the phenomenon measured with measurement error generally complicates the description of the observed variation in statistical terms. This complication is studied by comparing the results of repeated measurements of a body dimension with the results for a human force exertion.

Homoscedasticity

In Figure 1 measurement results are shown concerning the head circumference of children between two and thirteen years of age (Steenbekkers, 1993). The data were gathered in a test-retest with a one day to a one week interval. Vertically the within-child differences of the two measurements are plotted, and horizontally the means of these results per child. In view of the small time-interval between the test and the retest it is taken for granted that the object of measurement, i.e. the head circumference, is constant per child. In order to arrive at the specification of a measurement error on the basis of the observed variation in measurement results, it has to be made clear at first that on average this variation occurs randomly per child.

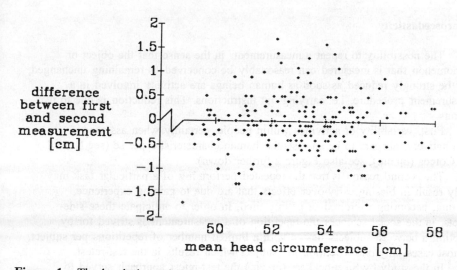

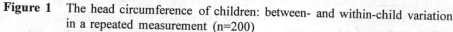

Figure 1 The head circumference of children: between- and within-child variation in a repeated measurement (n=200)

This means that, overall, systematic differences between the results of the first and of the second measurement should be absent. Basically this can be checked by a t-test. However, a t-test on the averages of two consecutive measurement series is blurred if the between-subjects differences are relatively large, as to some extent is apparently the case for the data in Figure 1 (cf. Altman and Bland, 1983). Then a more sensible approach is to eliminate the between-subjects differences by testing the average of all nominal differences of the two results per child against zero. In the application of ANOVA the equivalent of this approach is a two-way ANOVA with the session and the subjects as the independent variables, without an interaction term, and with fixed effects. In some research areas this approach is reported as a 'two way ANOVA with repeated measures design'.

For the data in Figure 1 no significant sequential deviation is found, $F(2,200)=0.70$, $p=0.41$. This implies that, on average, the sequence of measurement results per child could equally well have been the reverse. This conclusion, however, does not yet mean that within-child differences can be seen as occurring randomly throughout the children, while this condition must hold in order to be able to specify a constant measurement error. For the data in Figure 1, the randomness of the within-child differences can be checked by correlating the absolute within-child differences with the corresponding means per child. This results in a Pearson product moment correlation coefficient of $r=-0.04$, which differs not significantly from 0 ($p=0.62$). In statistical terms this constancy of variation is referred to as homoscedasticity (from the Greek word 'skedasis' which means 'dispersion').

The picture in Figure 1 agrees with the idea that the applied method is the dominant source of variation which is constant throughout the children, without a traceable contribution of the objects that are measured, i.e. the childrens' heads. In this case the type of measurement concerned largely resembles measurement procedures in the area of technical research (see ISO 5725, op.cit.). Essential for these procedures is that in principle a measurement can be repeated numerous times for the same object. It is clear that this generally holds for measuring static body dimensions within certain time limits such as the one in Figure 1.

Heteroscedasticity

The possibility to repeat a measurement, in the sense that the object or phenomenon that is measured can reasonably be conceived as remaining unchanged, may be strongly reduced as soon as human beings are actively involved in a measurement procedure, i.e. carrying out instructions. This reduction is due to two reasons.

First, variability of activities within people, certainly when assessed quantitatively, appears to be an essential human characteristic per se (see Newell and Corcos (op.cit.), see also Figure 2 further down).

The second reason is that the repeated performing of a particular task may easily result in biasing carry-over effects that are due to gaining experience, learning, becoming exercised or getting tired. In order to anticipate these side-effects, in the social sciences the repetition of a measurement is strived for by adopting a between-subjects design with a limited number of repetitions per subject, in most cases only one repetition per subject which results in the test-retest.

In the study by Steenbekkers (op.cit.) the test-retest approach served to establish performances of children in carrying out various tasks including the isometric exertion of a force, gradually built up in a few seconds. In Figure 2 the within-child differences in pulling maximally are plotted against the corresponding means. Here a striking relationship shows up as the within-child variation strongly increases with the level of measurement. The association between the absolute within-child differences and the corresponding means amounts to $r=0.47$ which differs significantly from 0 on a $p<0.0001$ level of significance. In statistical terms this phenomenon is referred to as heteroscedasticity. It is noted that for the data in Figure 2 no sequential deviations were found in the two-way ANOVA referred to above, $F(2,782)=0.13$, $p=0.72$, so on average the variation per child, like for the data

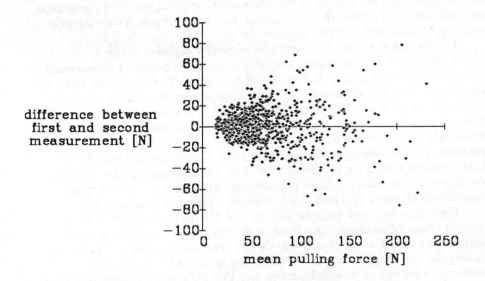

Figure 2   Maximum force exertion by children in pulling: between- and within-child variation in a repeated measurement (n=782)

in Figure 1, can be considered as occurring randomly. From the picture in Figure 2 it is immediately clear that, contrary to the data in Figure 1, the specification of the random variation cannot be a constant.

The specification of random variation

The term 'measurement error'

In Figure 1 the application of a particular measurement procedure can be seen as the predominant source of random variation in measurement results. Although the term 'error' seems debatable, as if the establishing of any 'true value' would be achievable (cf. Lumsden, 1976), following current practice the expression 'measurement error' is yet adopted to address the variation in Figure 1. In Figure 2 the variability in the observed human activity seems to play an important role at least. Thus the term 'error' is avoided here in indicating the observed random variation. Therefore, in this case the variation is referred to as it is, rather than in terms of measurement error.

Homoscedastic variation

The fundamental measure to specify the random variation in the results of a repeated measurement is the estimated standard deviation **s** of the distribution of the measurement results (cf. ISO 5725 (op.cit.)). Instead of **s** also the variation coefficient **s/m** can be used (with **m** as the mean of the observed results), as well as the standard deviation of the distribution of the differences between two single measurements. The latter measure amounts to $s\sqrt{2}$ since standard deviations add up quadratically.

In Figure 1 the expression $s\sqrt{2}$ can be used to compute the measurement error since the within-child differences in the measured head circumference can be conceived as being randomized across the children due to the homoscedasticity of the variation. It is found that $s\sqrt{2}=0.77$cm which gives s=0.41cm as the measurement error. The same outcome is found in two other ways:
- in the two-way ANOVA referred to above as the standard deviation of the residuals;
- according to the formula for the so-called standard error of measurement, $s_{obs}\sqrt{1-r_{12}}$, with r_{12} the correlation between the two series of measurements, i.e. $\{x_{i1}\}$ and $\{x_{i2}\}$, with 'i' denoting the children (i=1 ... 782), and s_{obs} the standard deviation of the distribution of $\{x_{i1}\}$ (or of $\{x_{i2}\}$), that is: accounting for the between-children variation.

The formula for the standard error of measurement is a well-known expression in textbooks on social science research, see e.g. Nunnally, 1978. It is noted that the correlation coefficient r_{12} itself, which features in many Ergonomics/Human Factors publications as criterion for the 'reliability', 'reproducibility' or 'repeatability' of measurement results, is an insignificant measure for addressing measurement error. As can be seen from the expression $s_{obs}\sqrt{1-r_{12}}$ for the standard error of measurement, r_{12} accounts for the relative share of the between-subjects variance in the totally observed variance. Thus a high r_{12} does not necessarily mean that the measurement error would be small, while on the basis of $r_{12}=0$ cannot be concluded that the measurement error is large (Kanis, 1993a).

For reason of completeness it is observed that in case of more than one repetition of a measurement per subject only the two-way ANOVA procedure remains available, since this approach is the only one of the three procedures indicated above which is not restricted to a pairwise comparison or association of observations.

Heteroscedastic variation

Altman and Bland (op.cit.) propose a logarithmic transformation of the data as the most appealing approach that will often be suitable to remove the association. However, for the data in Figure 2 this transformation inverts the heteroscedasticity, from $r=0.47$ to $r=-0.14$ which differs from 0 on a $p=0.0002$ level of significance. It may be questioned though whether a logarithmic transformation is an adequate approach indeed since this procedure actually seems to smooth away unwelcome characteristics of data, rather than to eliminate them.

What may be an effective way to remove heteroscedasticity satisfactorily is by taking the ratio between the within-subject differences and the corresponding means. In this way a homoscedastic variation was attained for force exertions by physically impaired subjects ($n=34$, see Kanis, 1993b). By dividing this 'normalised' random variation by $\sqrt{2}$ (see above) the observed variation could be specified in a variation coefficient that was constant throughout the 34 subjects. For the data in Figure 2 this procedure ends up in an overcompensation since the correlation between on the one side the averages per child and on the other side the absolute values of the within-child differences divided by the corresponding mean amounts to $r=-0.103$ on a $p=0.004$ level of significance.

In order to arrive at a specification of the random variation that is dependent on the level of measurement it seems a reasonable approach to adopt for this dependency the same linear relationship as is found by linearly regressing the absolute within-child differences with the means per child. Then, on the basis of this relationship, the standard deviation of the average of any local within-child differences is found by adjusting the standard deviation of the overall average of the within-child differences with help of the regression coefficient. Next, division by $\sqrt{2}$ specifies the local random variation in measurement results.

This procedure is applied to the data in Figure 2, see Table 1. In this table the first two measures are computed without taking into consideration the heteroscedasticity. The computation on the basis of the adjusted standard deviation of $\{x_{i1}-x_{i2}\}$ shows that the two constant measures to a considerable extent overestimate the variation occurring with children who are relatively weak and,

Table 1. Different measures for the amount of random variation in the results of a test-retest of pulling by children ($n=782$)

Amount of random variation computed on the basis of:		
Standard error of measurement, $s_{obs}\sqrt{1-r_{12}}$ ($r_{12}=0.90$)	Stand. dev. of $\{x_{i1}-x_{i2}\}/\sqrt{2}$	Adjusted stand. dev. of $\{x_{i1}-x_{i2}\}/\sqrt{2}$, for percentiles of $\{(x_{i1}+x_{i2})/2\}$ 5% (23.1N) 50% (54.1N) 95% (144N)
12.2N	12.3N	5.9N 10.7N 24.7N

similarly, underestimate the variation for relatively strong children. It is emphasized that this specification of random variation should be seen as an illustration of a possible way to tackle the variation of heteroscedastic data, rather than a panacea. In particular, the symmetry of the positive and negative within-child differences seems to be crucial. Furthermore, the lineairity of the relationship between the within-child differences and the corresponding means may be questionable. This relationship is known to somewhat level off the wider the range of observations (see Carlton and Newell, 1993). To conclude it is pointed at the irrelevance of $r_{12}=0.90$ as a measure for repeatability (see Table 1), in view of the actually observed variation in Figure 2.

Patterned variability and design

The type of distribution of the variation in human force exertion as given in Figure 2 appears to be the rule rather than the exception, see Newell and Corcos, op.cit., see also Mitka and Baker, 1994. In Ergonomics/Human Factors literature this patterning of human variability is generally ignored, although the relationship between the level of force exertion and its variability was already studied some 100 years ago in psychophysics in view of Weber's law. May intra-individual variation sometimes be difficult to master in statistical terms due to heteroscedasticity, it does comprise relevance for Ergonomics/Human Factors as a research area producing design guidelines. In this respect the issue is that maximum values for force exertion as presented in tables may tell little more than half the story of what is (not) attainable by various users of a product. Here the setting of margins, e.g. in the case of child-resistent packages, should be tuned not only to interindividual variation in human performance but also to the intra-individually occurring variation. As is shown in this paper, this variation may be large.

References

Altman, D.G. and Bland, J.M. 1983, Measurement in medicine: the analysis of method comparison studies, *The Statistician*, 32, 307-317.

Carlton, L.G. and Newell, K.M. 1993, Force Variability and Characteristics of Force Production. In: K.M. Newell and D.M. Corcos (eds.), *Variability and Motor Control*, (Human Kinetics Publishers, Leeds) 15-36.

ISO 5725, 1986, *Precision of test methods*, International Standard Organisation, Geneva.

Kanis, H. 1993a. Reliability in Ergonomics/Human Factors. *Contemporary Ergonomics*, 91-96, Taylor and Francis, London.

Kanis, H. 1993b. Operation of Controls on Consumer Products by Physically Impaired Users, *Human Factors*, 35(2), 305-324.

Lumsden, J. 1976, Test theory, *Annual Review of Psychology*, 251-280.

Mitka, G.A. and Baker, A. 1994, An investigation of the variability in human performance during manual material activities, *Proceedings of the Human Factors Society 38th Annual Meeting*, Santa Monica, CA, USA.

Nunnally, J.C. 1978, *Psychometric Theory*, McGraw-Hill, New York.

Newell, K.M. and Corcos, D.M. 1993, *Variability and Motor Control*, Human Kinetics Publishers, Leeds.

Steenbekkers, L.P.A. 1993, *Child development, design implications and accident prevention*, Delft University Press, Delft.

INTERRUPTION IN HAND CONTROL DURING EXPOSURE TO WHOLE-BODY VERTICAL VIBRATION

G.S. Paddan and M.J. Griffin

Human Factors Research Unit
Institute of Sound and Vibration Research
University of Southampton
Southampton
SO17 1BJ England

An experiment was conducted to investigate the transmission of vertical seat vibration to the hands of seated subjects and to determine displacements of the hands during exposure to whole-body vibration. The subjects, twelve males, sat on a rigid flat seat with their backs in contact with the seat backrest. With random vibration in the frequency range below 15 Hz at a magnitude of 1.0 ms^{-2} r.m.s., measurements were made of triaxial accelerations at the right hands of the subjects. Two postures of the arm were investigated: the whole arm held horizontal and in front of the body (i.e. an elbow angle of 180°), and the arm held nearer to the body with the lower arm horizontal and in front and the upper arm vertical (i.e. an elbow angle of 90°). More motion occurred at the hand when the arm was held with a 180° elbow angle compared with an elbow angle of 90°.

Introduction

The ability of a person to select and operate a switch can be hindered during exposure to whole-body vibration. The switch may be a conventional mechanical switch, a computer keyboard or a touch sensitive screen. Whichever type of switch is used, the interruption and the errors due to the vibration may be related to the relative displacement between the hand and the switch being operated: the smaller the relative displacement, the less the problem.

The present experiment was conducted to determine the transmission of vertical seat vibration to the hands of seated subjects and the effect of arm posture. The displacements at the hands may be used to estimate the sizes of switches required to reduce errors in switch selection during exposure to whole-body vibration.

Equipment and Procedure

The experiment was conducted on an electro-hydraulic vibrator capable of producing vertical displacements of one metre. The subjects sat on a rigid flat seat with the supporting surface 470 mm above a moving footrest. The seat surface was inclined backwards at an angle of 3° to the horizontal. A rigid flat backrest attached to the seat was inclined at an angle of 6° to the vertical and had its lower and upper edges 90 mm and 480 mm respectively, above the seat surface. A thin layer of high stiffness, high friction rubber was glued to both the seat and the backrest surfaces to reduce relative motion between the subject and the seat due to sliding. A loose lap strap was worn by the subjects for safety purposes.

Accelerations of the unsupported hand were measured using an instrumented bar which the subjects held in their right hands. The bar consisted of a wooden rod 28 mm in diameter and 130 mm long with three small translational piezoresistive Entran type EGCS-240B-10D accelerometers mounted on one end. The total mass of the instrumented bar was 105 grams. The three accelerometers were orientated in mutually orthogonal axes to measure motions in the three translational axes of the hand (i.e. fore-and-aft, lateral and vertical). The directional axes measured at the hand were parallel to the equivalent axes of the body of the seated person.

The subjects sat on the seat in an upright comfortable posture with their backs in contact with the seat backrest which extended up to the subject's shoulders (i.e. a 'back-on' posture). Two different postures of the arm were investigated; these resulted from varying the included angle between the upper arm and the forearm. The two included angles were 90° (i.e. the upper arm vertical but not in contact with the body, with the forearm horizontal) and 180° (i.e. both upper arm and forearm held horizontal: subject's arm stretched out to the front). The forearm and the hand pointed in the forward direction for both postures of the arm. The subjects were instructed to maintain the required body position and to avoid voluntary movements of any part of the body during exposure to the vibration. The experimenter ensured that the postures of the arm did not depart significantly from those required.

Twelve healthy male volunteers participated in the study. The physical characteristics of the subjects are shown in Table 1. Eleven subjects were right-handed and one was left-handed.

A vibration data acquisition and analysis system, *HVLab*, developed at the Institute of Sound and Vibration Research in the University of Southampton, was used to control the experiment and analyse the acquired data. A computer-generated random waveform having a nominally flat constant bandwidth acceleration spectrum was used with an acceleration magnitude of 1.0 ms^{-2} r.m.s. at the seat platform. The vibration signal was equalised for the response of the vibrator. The waveform was sampled at 64.5 samples per second and low-pass filtered at 25 Hz before being fed to the vibrator. Signals from four accelerometers (three on the hand-held bar and one on the platform) were passed through signal conditioning amplifiers and then low-pass filtered via anti-aliasing filters (70 dB/octave Butterworth characteristic) set at 15 Hz. These signals were digitised into a computer at a sample rate of 64.5 samples per second. The duration of each vibration exposure was 120 seconds.

Analysis

Transfer functions between vertical seat acceleration and translational hand acceleration were calculated using the 'cross-spectral density function method'. The transfer function, $H(f)$, was determined as the ratio of cross-spectral density of seat and hand acceleration, $G_{sh}(f)$, to the power spectral density of seat acceleration, $G_{ss}(f)$: $H(f) = (G_{sh}(f))/(G_{ss}(f))$. Frequency analysis was carried out with a resolution of 0.126 Hz and 64 degrees of freedom.

Two types of transfer function are presented: the ratio of acceleration at the hand to acceleration at the seat, and the ratio of the displacement at the hand to the acceleration at the seat. The displacements at the hand (in each of the three axes) were calculated in the frequency domain by dividing the acceleration transfer functions by $(2\pi f)^2$ where f is the frequency. This corresponds to double integration of the acceleration time history in the time domain.

Table 1. Physical characteristics of 12 subjects taking part in the experiment.

	Age (years)	Weight (kg)	Height (m)
minimum	23	60	1.68
maximum	47	85	1.89
mean	31.0	72.9	1.80
standard deviation	8.5	7.4	0.065

Results and Discussion

The acceleration transmissibilities between vertical seat vibration and the three translational axes of motion at the hands of all 12 seated subjects are shown in Figure 1. The figure shows data obtained with both postures of the right arm: arm held straight outwards (i.e. 180° elbow angle) and arm held closer to the body (i.e. 90° elbow angle). A large variability in transmissibility is seen between subjects for all axes of motion at the hand. This is particularly the case for fore-and-aft motion with a 90° arm posture: one subject had a transmissibility of approximately zero at about 5 Hz whereas another subject showed a value of about 2.

Figure 2 shows median and interquartile ranges of the acceleration transmissibilities presented in Figure 1. The effect of changing arm posture from 90° to 180° is seen as an increase in motion transmitted to the hands of the seated subjects. There was a significant increase in lateral hand motion for frequencies above 6.4 Hz with the straight arm (Wilcoxon matched-pairs signed ranks test, p<0.01). The peak in the transmissibility for lateral motion shifted from about 4.5 Hz to approximately 7 Hz when the arm was held straight. The transmissibilities to vertical hand acceleration show two peaks: one at about 2 Hz and the second peak at around 5 to 6 Hz. The frequency corresponding to the second peak for vertical motion at the hand increased from about 4.5 Hz to about 6 Hz as the arm angle changed from 90° to 180°.

The data shown in Figure 1 were converted to displacement transmissibilities (i.e. the ratio of displacement at the hand to acceleration at the seat). The corresponding transmissibilities are shown in Figure 3. Median transmissibilities for the data in Figure 3 are presented in Figure 4.

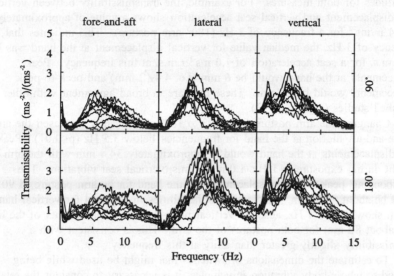

Figure 1. Transmissibilities between vertical seat acceleration and the translational acceleration at the hands of subjects seated in a 'back-on' posture with two arm postures (90° and 180°) (0.126 Hz resolution, 64 degrees of freedom).

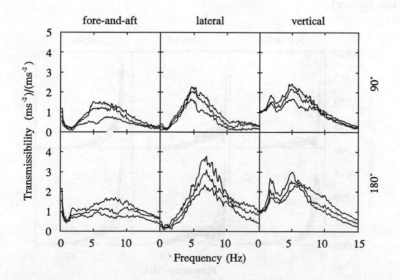

Figure 2. Median and interquartile transmissibilities between vertical seat acceleration and the acceleration at the hands of seated subjects (0.126 Hz resolution, 64 degrees of freedom).

The displacements and accelerations in Figures 3 and 4 apply to r.m.s. magnitudes for both measures. For example, the transmissibility between vertical hand displacement and vertical seat acceleration shows a value of approximately 0.0044 m/ms^{-2} for a frequency of 3 Hz (180° arm posture). This indicates that, for a frequency of 3 Hz, the median value for vertical displacement at the hand was 4.4 mm r.m.s. for a seat acceleration of 1.0 ms^{-2} r.m.s. at this frequency. Peak displacements at the hand would be 6 mm (i.e. 4.4√2 mm) and peak-to-peak displacements would be 12 mm. These data are in broad agreement with other published studies (e.g. So, 1993).

Changing the arm posture from 90° to 180° significantly increased the amount of fore-and-aft motion at the hand for frequencies below 3.5 Hz ($p<0.01$). Peak-to-peak displacements at the hand would be approximately 34.4 mm with the arm held straight during exposure to 1 Hz, 1.0 ms^{-2} r.m.s. vertical seat vibration. The corresponding peak-to-peak displacements at the hand for an arm posture of 90° would be about 14.7 mm. Median transmissibilities in Figure 4 for vertical hand motion show that at 1 Hz, absolute vertical peak-to-peak displacements of the hand were about 80 mm for both postures of the arm. This is consistent with a transmissibility slightly greater than unity at this frequency.

To estimate the dimensions of switches that might be used while being exposed to whole-body vibration in vehicles, it is necessary to consider the relative motion between the hand and the vehicle. The transmissibilities shown in Figures 1 to 4 are derived from the absolute motions of the seat and the hand. Consequently, at low frequencies, they show large values whereas the relative motion between the seat and the hand was low. Calculations show smaller relative vertical displacements at 1 Hz (about 12 mm peak-to-peak for a 90° arm posture and about 62 mm for a 180° arm posture).

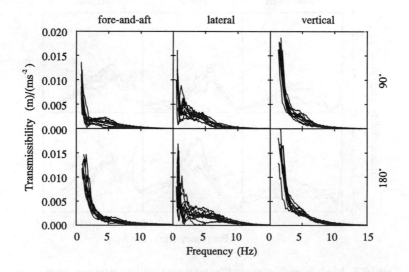

Figure 3. Transmissibilities between vertical seat acceleration and displacements at the hands of the subjects seated in a 'back-on' posture with two arm postures (90° and 180°) (0.126 Hz resolution, 64 degrees of freedom).

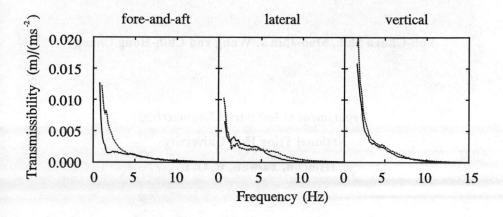

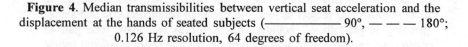

Figure 4. Median transmissibilities between vertical seat acceleration and the displacement at the hands of seated subjects (————— 90°, — — — 180°; 0.126 Hz resolution, 64 degrees of freedom).

Conclusions

Large variability between subjects was seen in the transmission of vertical seat vibration to the hands. At some frequencies, hand motions in the fore-and-aft direction were approximately 2.3 times greater for an arm posture of 180° compared to a 90° arm posture. Extending the arm had only a small effect on hand motions in the lateral and the vertical axes. At frequencies below about 1 Hz the acceleration transmissibility to the hand was greatest in the vertical direction with a transmissibility of approximately unity. Greatest absolute displacements at the hand will therefore occur at low frequencies in the vertical axis. However, relative displacements, which are influenced by a difference in magnitude and any differences in phase at the seat and the hand, were smaller in magnitude.

References

So, R.H.Y. 1993, Aiming by head and finger pointing during exposure to vertical seat vibration: Effects of target position. *United Kingdom Informal Group Meeting on Human Response to Vibration*, Army Personnel Research Establishment, Farnborough, England, 20th to 22nd September 1993.

Acknowledgements

This work has been carried out with the support of the UK Ministry of Defence under an extramural research agreement with the Centre for Human Sciences, DRA Farnborough.

EVALUATING THE FACTORS AFFECTING
MVTE IN VALVE OPERATION

Yuh-Chuan Shih, Mao-Jiun J. Wang and Chih-Hong Chang

Department of Industrial Engineering

National Tsing Hua University

Hsinchu, Taiwan, R. O. C.

Forty subjects (20 males and 20 females) were involved in this experiment. The combined nested-factorial design and block design was employed. In addition to sex and subject factors, the effects of handwheel size and shape, operating height (elbow, shoulder and overhead) and glove (one pair of cotton, two pairs of cotton and rubber) on the maximum volitional torque exertion (MVTE) of operating valve were investigated. The barehanded condition was also involved for comparison. The ANOVA results indicate that all the main effects are significant ($p<0.001$). The gloved MVTE is found to be greater than the barehanded MVTE. The shoulder height has the best MVTE, followed by elbow and overhead heights. As to the size effect, a positive relation between MVTE and size is found. The female MVTE of operating valve handwheel is about 63% of that of male.

Introduction

Valves are commonly used in modern industrial production and daily life. They have become indispensable in almost every facet of modern technology. The valve functions can be categorized into manual operation and automatic operation. The handle of manual valves are either handwheel or lever. The torque reaction force is the main force to operate handwheel valves. The torque exerted by human hands can be classified into three types. They are wrist extension/flexion (E/F) (e.g. loosening or tightening electric connectors), wrist pronation/ supination (P/S) (e.g. operating screwdrivers) and wrist radial/ulnar deviation (R/U) torque reaction forces (Drury 1980). When operating a valve with handwheel, the torque applied to tighten or loosen is the R/U torque exertion.

The handle is an important interface of directly transmitting force generated by human to the workpiece/equipment. About the assessment of handle sizes, the consistent results concluded form many studies indicated that the greater the handle size is, the greater the maximum volitional torque exertion (MVTE) is, regardless of the types of handles (Pheasant and O'Neill 1975, Adams and Peterson 1988, CoChran and Riley

1986). Further for handle shapes, it has been reported to have significant influence on E/F and P/S MVTE (Pheasant and O'Neill 1975, CoChran and Riley 1986, Mital and Channaveeraiah 1988). Furthermore, glove is one of the most popular worker-tool/equipment interfaces and is applied widely in various manual operations. The primary purpose of wearing gloves is to protect hands from injuries, like cuts, bruises, abrasions and touching extreme temperatures. For both lab setting (Riley et al. 1985) and realistic manual operation (Adams and Peterson 1988, Mital et al. 1994), past findings indicated that wearing gloves can increase the E/F and P/S MVTE.

Unfortunately, the evaluation of R/U MVTE is still lacking, especially the interface factors between human and valves. Therefore, there is an obvious need to obtain the related information about the design of manual valve operation. The objective of this study is to determine the influences of worker/handwheel interface factors (such as handwheel shapes and sizes, gloves and operating heights) on R/U MVTE of operating valve handwheel.

Method

Subjects:

Forty student subjects (20 males and 20 females) participated in this study. They were all right-handed and the average age was 21 years old (range = 18-27) for males and 19 for females (range = 18-23). All subjects were in good health and without known hand-arm musculoskeletal disability.

Apparatus and Materials:

A SENSOTEC torque sensor mounted on an adjustable steel frame and a 486 PC with A/D card were used for data collection. Three different types of gloves (single cotton (cotton1), double cotton (cotton2) and rubber) frequently used in the workplace were selected for investigation. Two types of valve handwheel shapes, curved and smooth edges, were employed. Each shape had 5 different sizes, see Figure 1. The detailed illustrations of handwheels are listed in Table 1.

Table 1: The descriptions for handwheel

Shape	Size	Diameter (mm)			Number of curves
		outside	inside	effective**	
Curved	1	58	51	55	8
	2	65	58	62	8
	3	73	65	69	8
	4	104	92	98	10
	5	125	115	120	12
Smooth	1	55	55	55	*
	2	62	62	62	*
	3	75	75	75	*
	4	95	95	95	*
	5	118	118	118	*

**: mean diameter of inside and outside diameters

Experimental Design:

The combined ANOVA of nested-factorial design and block design was applied. The response variable was the R/U MVTE of operating handwheel, and the independent variables were sex, subject, glove, operating height, handwheel size and shape. Subject factor was nested under sex, size factor was nested under shape, and operating height was a blocking factor. Only the subject factor was considered as a random variable. In addition to the three glove conditions, the bared condition was also included for comparison. The operating height was specified at elbow, shoulder and overhead levels and it was adjusted for each subject. A total of 120 treatment combinations (4 bare/gloves × 3 heights × 5 sizes × 2 shapes) was performed by each subject.

Smooth shape   curved shape

Figure 1: The appearance of used handwheels

Experimental procedure:

At the beginning, each subject was instructed about the purpose of this experiment. Five trials were given to make subject familiar with the experimental procedure. The MVTE exerted by each subject was measured in accordance with the procedures outlined in the literature for the measurement of isometric muscle strengths (Caldwell 1974; Chaffin 1975). The operating plane was specified at the transverse plane (parallel to the floor). The reach distance was individually adjusted for the subject to naturally straighten his/her arm to grasp handwheel.

Results and discussions

The ANOVA results depict that all the main effects and first order interactions are statistically significant ($p < 0.001$), see Table 2. The results of Duncan's multiple range test of all subjects, male and female are demonstrated in Table 3. Operating height can change the effective position and motion of hand and arm, and then result in significant change in MVTE. Table 3 shows that shoulder height (5.13 Nm) is the most exertable height followed by elbow height (4.88 Nm) and overhead height (4.77 Nm). Wrist flexion is the main reason to cause the reduction of R/U MVTE at overhead height. Therefore, it is not recommended to install valve in high position.

Table 2 The ANOVA results

Source of Variation	D. F.	M.S.	Error Term	F-Ratio	Significant Level
(S)ex	1	6100.93	Su(S)	40.87	***
(Su)bject(S)	38	149.26	E	281.72	***
(Sh)ape	1	720.24	Su*Sh(S)	242.74	***
(Si)ze(Sh)	8	619.49	Su*Si(S*Sh)	221.88	***
(G)love	3	59.00	Su*G(S)	28.74	***
(H)eight	2	54.28	E	102.45	***
S*Sh	1	45.88	Su*Sh(S)	15.46	***
S*Si(Sh)	8	113.65	Su*Si(S*Sh)	40.71	***
S*G	3	18.56	Su*G(S)	9.04	***
Su*Sh(S)	38	2.97	E	5.60	***
Su*Si(S*Sh)	304	2.79	E	5.27	***
Su*G(S)	114	2.05	E	3.87	***
Sh*G	3	6.40	E	12.09	***
Si*G(Sh)	24	4.86	E	9.17	***
(E)rror	4251	0.53			
Total	4799				

***: significant at 0.1%

As to shape effect, from Table 3, the MVTE of curved-edge handwheel is greater than that of smooth-edge handwheel for both male and female. It is because that curved shape provides favorable finger grasping condition for force exertion, and results in 17% more MVTE than that of the smooth handwheels for both male and female. Wearing gloves also affects R/U MVTE significantly. As it can be seen in Table 3, the gloved MVTEs are greater than the barehanded MVTE (with about from 3% to 9% increase for both males and females). This finding of increased MVTE is consistent with the past findings of glove effect on both P/S MVTE (Mital et al. 1994) and E/F MVTE (Riley et al. 1985, Adams and Peterson 1988). The dominance of rubber and cotton2 gloves is because of the increase of friction between hand/gloves and handwheel, and the reduction of painful feeling.

Further, Table 3 indicates that male MVTE increases as the handwheel size increases, but, female has the greatest MVTE in size 4 (95-98 mm) for both shapes. The reason for the difference is possible that female has smaller hand dimension. Theoretically speaking, the MVTE increases as the leverage increases (Pheasant and O'Neill 1975, Admas and Peterson 1988). This phenomena is also found in the present study under barehanded condition for both male and female (see Figure 2). Figure 2 demonstrates the combined effect of glove and handwheel size. However when under gloved condition, for male, except for cotton2, the greater the size, the greater the MVTE. But the superiority of size 5 in MVTE to size 4 is decreasing as the thickness of gloves is increasing. For female, the positive relation between MVTE and size only occurs under the barehanded condition; the superiority of size 4 to 5 gets larger as gloves get thicker. This is possible due to that gloves can change the effective dimensions of hand (Damon et al. 1966).

It is also found that there is a significant difference between male and female in MVTE. The average MVTE is 6.05 Nm for male and 3.80 Nm for female. The Chinese female's MVTE is about 63% of that of Chinese male. The significant gender effect may be due to the sex differences in physiological capabilities and in pressure-pain threshold. Fransson-Hall and Kilbom (1993) reported that the pressure-pain threshold of female is about 2/3 of male.

Table 3 The results of Duncan's multiple range test

Variable		Rank				
		1	2	3	4	5
Shape	all	curved	smooth			
	male	curved	smooth			
	female	curved	smooth			
Height	all	shoulder	elbow	overhead		
	male	shoulder	elbow	overhead		
	female	shoulder	elbow	overhead		
Size(smooth)	all	5	4	3	2	1
	male	5	4	3	2	1
	female	4	5	3	2	1
Size(curved)	all	5	4	3	2	1
	male	5	4	3	2	1
	female	4	5	3	2	1
Glove	all	rubber	cotton2	cotton1	bared	
	male	rubber	cotton2	cotton1	bared	
	female	cotton2	rubber	cotton1	bared	

Conclusions

From the above discussions, significant effects of glove, operating height, handwheel shape and size on R/U MVTE are found. Wearing gloves is recommended during operating valves for not only protecting hand from injuries, but increase torque exertion. The curved-edge handwheel shape and shoulder operating height are recommended for greater torque exertion. Generally speaking, the greater the handwheel size, the greater the MVTE, but wearing gloves can change the effective grasp span and results in the change of handwheel size having the greatest MVTE.

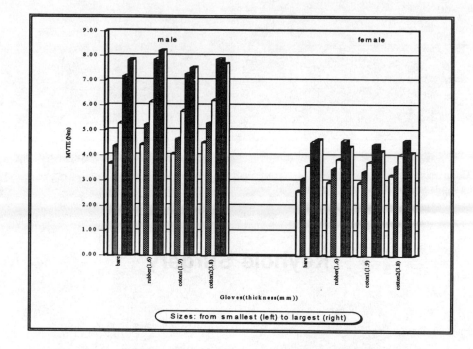

Figure 2: The interaction between glove and size

References

Adams, S. K. and Peterson, P. J., 1988. Maximum voluntary hand grip torque for circular electrical connectors. Human Factors, 30, p733-745.

Caldwell, L. S., 1974. A proposed standard procedure for static muscle strength testing. American Industrial Hygiene Association Journal, 35, 201-206.

Chaffin, D. B., 1975. Ergonomics guide for the assessment of human static strength. American Industrial Hygiene Association Journal, 36, 505-511

CoChran, D. J., and Riley, M. W., 1986. The effects of handle shape and size on exerted forces. Human Factors, 28(3), 253-265.

Damon, A., Stoudt, H. W. and McFarland, R. A., 1966. The human body in equipment design (Harvard university Press, Cambridge, MA)

Drury, C. G. 1980, Handles for manual materials handles. Applied Ergonomics, 11-1, 35-42.

Frasson-Hall, C. and Kilbom, A., 1993. Sensitivity of hand to surface pressure. Applied Ergonomics, 24-3, 181-189.

Mital, A. and Channaveeraiah, C., 1988. Peak volitional torques for wrenches and screwdrivers. International Journal of Industrial Ergonomics, 3, 41-64.

Mital, A., Kuo, T. and Faard, H. F., 1994. A quantitative evaluation of gloves used with non-powered hand tools in routine maintenance tasks. Ergonomics, 37-2, 333-343.

Pheasant, S., and O'Neill, D., 1975. Performance in griping and turning - A study in hand/handle effectiveness. Applied Ergonomics, 6(4), 205-208.

Riley,, M. W., CoChran, D. J. and Schanbacher, C. A., 1985. Force capability differences due to gloves. Ergonomics, 28, 441-447.

Keyhole Surgery

VISUAL DEPTH PERCEPTION IN MINIMALLY INVASIVE SURGERY

Anthony H. Reinhardt-Rutland[1] and Anthony G. Gallagher[2]

Psychology Department, University of Ulster, Newtownabbey, BT37 0QB[1]
School of Psychology, Queens University of Belfast, Belfast, BT7 1NN[2]

Although of potentially immense value, minimally invasive surgery (MIS) has been associated with increased post-operative complication rates. In this paper, it is argued that a problem with MIS resides in the loss of depth information for the surgeon - particularly binocular information - as a result of indirect viewing of the site of operation via a TV monitor. Technological solutions to the problem should be approached with caution; however, training procedures focussing on non-binocular depth information should be useful.

Introduction

Minimally invasive surgery (MIS), often known colloquially as "keyhole" surgery, refers to a surgical technique by which trocars - small-diameter tubes - are inserted into the patient at the site of the operation. Installed in one of the trocars is a computer-chip pencil camera - more commonly known as a laparoscope - which enables the surgical team to view the operation externally on a TV monitor. The trocars also allow control of surgical instruments. Since incisions for the trocars need only be small (10 - 20 mm in diameter), potential gains over conventional surgery reside in post-operative care of the patient regarding pain and infection, which in turn allows financial savings, since hospital stays are shorter. MIS has been applied most notably to cholecystectomy - surgical removal of the gall bladder; other applications include hernia repair and some stomach, prostate and gynaecological procedures (Grace and Bouchier-Hayes, 1990; Leahy, 1989).

However, the use of MIS has not been without problems, with persistent reports of complications following the technique (New York State Department of Health, 1992); in consequence, the UK Department of Health has imposed limitations on the range of operations that may be performed (see The Observer: 23/10/94). While a full account of the difficulties with MIS remains to be determined, one general area of concern must reside in the information-processing difficulties imposed by indirect viewing of the site of the operation and indirect manipulation of surgical instruments. The issue of viewing conditions is the focus of the present paper. In particular, it is argued from a review of appropriate literature that problems of depth perception are inevitable in MIS. Nonetheless, there is reason to believe that appropriate training may overcome some of these problems.

Information for visual depth perception

In normal viewing conditions, as might apply in conventional surgery, there are a number of potential sources of information for depth. These can be labelled *binocular, motion, monocular* and *pictorial*. Binocular information includes retinal disparity: the two retinal images differ because of the separation of the eyes in the head. Closer features of the scene lead to bigger retinal differences than do more distant features. Motion information, arising from the observer's motion, includes motion parallax: features at different distances from the observer move visually across the retina at different rates. The rates are greater for nearer features than for more distant features. Monocular information includes monocular parallax associated with depth-related ocular adjustments, most obviously accommodation - adjustment of lens curvature. Finally, pictorial information includes elements such as edges; these signal discontinuities in depth. Higher-order examples are occlusion - a nearer object may obscure a more distant object - and relative visual size - two objects of similar physical size will have different visual sizes if they are at different distances from the observer. Pictorial information as its name implies is well-known from two-dimensional representations of depth, such as photographs.

Viewing via a TV monitor leads to loss of depth perception. In particular, binocular information is not available; also, monocular information is restricted, particularly since visual motion is unlikely to be correlated with the observer's motion. On first consideration, this might seem unimportant: except in the most restricted conditions, there must be redundancy in depth information, given that each type of information - be it binocular, motion, monocular or pictorial - should convey the same depth in the same viewed scene. However, this assumes that all depth information is equally potent - which is not supported by empirical evidence.

Investigating the different types of information

Simulations
The effects of the different types of information have been investigated using two methods. The first method is to simulate depth in a stimulus that otherwise appears flat: typically, the stimulus is a pattern of densely-spaced dots that appear randomly distributed. Examples are the stereograms developed by Julesz (1971): systematic differences in the dot patterns received by the two eyes elicit a compelling impression of depth. In depth-from-motion simulations, the dots are presented on a VDU screen and move systematically as the observer makes lateral head motions: the head is supported on a moving trolley which affects the motion of the dots on the VDU screen via software (Rogers & Graham, 1979, 1982).

Real depth
One must be concerned with how the evidence from simulations transfers to stimuli truly varying in depth. This leads to the second method of investigation, which is to

arrange that types of information compete. This is particularly convenient with regard to pictorial information, because it is readily manipulated to be non-veridical. An example of such research concerns the perceived orientation-in-depth of trapezoidal surfaces: these convey non-veridical orientations-in-depth, if they are interpreted as rectangular. Estimates of orientation-in-depth were made under static-monocular, moving-monocular and static-binocular viewing conditions (Reinhardt-Rutland, 1990, 1993b). Other studies have concerned the perceived separation-in-depth of two or more objects at different distances from the observer (Hell, 1978; Hell and Freeman, 1977).

Comparing the different types of information

Some general conclusions can be drawn from the research on depth perception. Much of the experimental evidence relates to observers with normal or corrected-to-normal vision; less evidence pertains to those with visual deficits.

Monocular information
While there is evidence that monocular information can effect depth perception, it is generally recognized as weak and unreliable (Foley, 1978). For example, seemingly irrelevant manipulations of experimental design and procedure can affect the pertinent evidence (Reinhardt-Rutland, 1993b). More importantly, when depth information is likely to be monocular alone - as, for example, during static-monocular viewing of highly impoverished displays - there is a strong tendency to perceive the display as flat and equidistant (Foley, 1978; Gogel, 1956; Reinhardt-Rutland, submitted).

Motion information
The evidence regarding motion information is highly equivocal, despite considerable emphasis on such information in "ecological" treatises (Gibson, 1979) and computational algorithms directed at simulating human depth processing (Simpson, 1993): while Rogers and Grahams' (1979, 1982) simulations suggest that motion information is highly effective, all studies entailing conflicting pictorial information suggest the opposite. Reinhardt-Rutland (1993a, submitted) has recently argued that the simulation studies might in fact be explained by pictorial information introduced by motion in the display: in particular, simulations generally introduce edges, which convey discontinuities in depth. The general conclusion at the present stage of research is that motion information *independent* of other information is not effective. Rather, the main value of motion information may be that it enhances the effectiveness of pictorial information; for example, Gibson (1979) describes how motion can elaborate pictorial occlusion in a sequence of *kinetic* occlusion and disocclusion as the observer moves with respect to a pair of objects each at a different distance. Nonetheless, the potential of motion information in enhancing depth perception in MIS is probably limited.

Binocular information
Much of the evidence from both simulations (Julesz, 1971; Rogers and Graham, 1982) and real depth in stimuli (Reinhardt-Rutland, 1990; 1992a) suggests that binocular information is the most important source of depth information at close distances for observers with normal or correctected-to-normal vision. Hence, it should not be

surprizing that MIS can lead to complications: the surgeon cannot accurately perceive where incisions or other manipulations should be performed at the site of operation.

However, a not-insignificant proportion of the general population lacks full stereoscopic vision. It is rarely reported that such individuals are seriously handicapped in tasks requiring accurate depth judgments. Indeed, permanently monocular drivers do not suffer higher casualty-rates than their binocular counterparts; if anything, the reverse is true (Evans, 1991; Hills, 1980). This finding is consistent with an extensive series of studies directed at depth-related tasks and carried out early this century in Germany in relation to the Accident Insurance Law of that country (Hell, 1981). How permanently-monocular individuals adjust has not been fully ascertained. Given the potential redundancy of depth information, permanently-monocular individuals may learn to attend to the types of information available to them. Perhaps also permanently-monocular individuals are more cautious in performing skills requiring good depth perception. This argument is given some support from other roadway evidence regarding a seeming paradox. Young male drivers, who should be the safest in terms of their perceptual capabilities, are the most prone to becoming road casualties; it is assumed that they are impulsive and least attentive to depth information (Evans, 1991). There are compelling reasons for believing that drivers' perception is inadequate in several ways, no matter how good the vision of the individual driver (Reinhardt-Rutland, 1992b). However, visual perception is so phenomenally immediate - "seeing is believing" - that inexperienced drivers may fail to appreciate perceptual problems. Of course, there are major differences in the visual demands of driving and surgery: nonetheless, the driving evidence should not be lightly dismissed.

Pictorial information
Pictorial information is also extremely important and indeed, in competition with binocular information, can in some experimental arrangements be more effective than binocular information for normal observers (Gehringer & Engel, 1986; Stevens & Brookes, 1988).

The significance of pictorial information has often been overlooked in previous research, perhaps because its applicability has seemed to be relatively limited and - more importantly - because it can easily be manipulated to be non-veridical. Nonetheless, pictorial information is often subtle and its effects strongly engrained: recently, Reinhardt-Rutland (1993a, submitted) has shown that the visual length of a triangular surface can be a potent factor in observers determining the horizontal orientation-in-depth of the surface, despite the fact that visual length should in principle only be able to convey a difference in horizontal orientation-in-depth across surfaces.

Implications

It must be recognized that the various research into visual depth perception has entailed stimuli that may not on first consideration seem particularly similar to those encountered by the surgeon. Nonetheless, a number of important points - for example, regarding the importance of binocular information and pictorial information - are based on a diverse range of stimuli and - perhaps as important - a diverse range of experimental procedures; it is reasonable to assume that they form an adequate basis on which research specific to MIS may be generated.

One clear conclusion to be drawn is that the lack of binocular information must pose serious problems for the surgeon with normal or corrected-to-normal vision performing MIS. This suggests that technological development could usefully be directed to introducing binocular information by way of pairs of cameras. A number of issues suggest that this approach should be treated with caution. Firstly, some technique for scaling the stereoscopic views would be required; otherwise, the disparity in the two images is likely to be interpreted in a non-veridical way. Secondly, some unobtrusive technique would need to be developed to convey the two images separately to the two eyes, without causing degradation to other aspects of perception. Finally, the introduction of a further trocar for the second camera moves away from the principle of minimal invasion.

Returning to the current general state of technology, it may well be the case - somewhat paradoxically - that surgeons *lacking* good stereocopic vision may be at an *advantage* in performing MIS, compared with surgeons with normal or corrected-to-normal vision. Drawing inferences from permanently monocular individuals, it is likely that lack of stereoscopic vision can be overcome by relying on other depth information; since there is redundancy in depth information, this is plausible. Another strategy for permanently monocular individuals may be that they perform pertinent tasks more cautiously.

The most obvious type of information to substitute for binocular information is pictorial information, perhaps augmented by motion. This conclusion is reinforced by the relative effectiveness of MIS that is believed to apply in cholecystectomy: in this procedure, the gall bladder is well-separated from surrounding tissue, so that, for example, occlusion can provide powerful pictorial information for the depth of the gall-bladder in relation to the surrounding tissue. In other cases, the pictorial information may be much less obvious; however, drawing from Reinhardt-Rutland (1993a, submitted), it is important to realize how influential pictorial information can be, even when what is conveyed seems almost too subtle to be useful. The other issue is, of course, that pictorial information can easily mislead. Nonetheless, a way to improve performance in MIS may well reside in identifying and enhancing available pictorial information.

In the meantime, the evidence from the experimental study of visual depth perception argues strongly that it is wise to impose restrictions on the use of MIS. Indeed, it is unfortunate that there are not procedures by which radical departures from previous standards are not subjected to wider scrutiny *before* they are applied generally.

References

Evans, L. 1991, *Traffic Safety and the Driver* (Von Nostrand, New York)

Foley, J. M. 1978, Primary depth cues. In R. Held, H. W. Leibowitz and H. L. Teuber (eds.), *Handbook of Sensory Physiology, Volume 8: Perception* (Springer, Berlin) 181-213.

Gehringer, W. L. and Engel, E. 1986, Effect of ecological viewing conditions on the Ames' distorted room illusion, Journal of Experimental Psychology: Human Perception and Performance, **12**, 181-185.

Gibson, J. J. 1979, *The Ecological Approach to Visual Perception* (Houghton-Mifflin, Boston MA)

Gogel, W. C. 1956, The dependency to see objects as equidistant and its inverse relation to lateral separation. Psychological Monographs, **70**, whole number 411.

Grace, P. A. and Bouchier-Hayes, D. J. 1990, Laparoscopic cholecystectomy: the implications for surgical practice and training, Journal of the Irish Coggeges of Physicians and Surgeons, **19**, 180-181.

Hell, W. 1978, Movement parallax: an asymptotic function of amplitude and velocity of head motion, Vision Research, **18**, 629-635.

Hell, W. 1981, Research on monocular depth perception between 1884 and 1914: influence of the German Unfallversicherungsgesetz and of the jurisdiction of the Reichsversicherungsamt, Perception, **10**, 683-694.

Hell, W., and Freeman, R. B. 1977, Detectability of motion as a factor in depth perception by monocular movement parallax, Perception and Psychophysics, **22**, 526-530.

Julesz, B. 1971, *Foundations of Cyclopean Perception* (Chicago University Press, Chicago)

Leahy, P. F. 1989, Technique of laparoscopic appendicectomy. British Journal of Surgeons, **76**, 616.

New York State Department of Health Memorandum, Laparoscopic surgery. Soris 29-20, June 12, 1992.

Reinhardt-Rutland, A. H. 1990, Detecting the orientation of a surface: The rectangularity postulate and primary depth cues, Journal of General Psychology, **117**, 391-401.

Reinhardt-Rutland, A. H. 1992a, Primary depth cues and background pattern in the portrayal of slant, Journal of General Psychology, **119**, 29-35.

Reinhardt-Rutland, A. H. 1992b, Some implications of motion-perception evidence and theory for road accidents, Journal of International Association of Traffic and Safety Sciences, **16**, 9-14.

Reinhardt-Rutland, A. H. 1993a, Judging slant-in-depth of triangular surfaces: A relative size heuristic employed during motion, Irish Journal of Psychology, **14**, 377-378.

Reinhardt-Rutland, A. H. 1993b, Perceiving surface orientation: pictorial information based on rectangularity can be overriden during observer motion, Perception, **22**, 335-341.

Reinhardt-Rutland, A. H. (submitted), Judging the orientation-in-depth of real triangular surfaces, Perception.

Rogers, B. and Graham, M. 1979, Motion parallax as an independent cue for depth perception, Perception, **8**, 125-134.

Rogers, B., and Graham, M. 1982, Similarities between motion parallax and stereopsis in human depth perception, Vision Research, **22**, 261-270.

Simpson, W. A. 1993, Optic flow and depth perception, Spatial Vision, **7**, 35-75.

Stevens, K. A., and Brookes, A. 1988, Integrating stereopsis with monocular interpretations of planar surfaces, Vision Research, **28**, 371-386.

Acknowledgment: We thank Mary McClean for helpful comments regarding this paper.

Late Submissions

COMPARING HEURISTIC, USER-CENTRED AND CHECKLIST-BASED EVALUATION APPROACHES

Mark G. Westwater

AT&T Global Information Solutions
ATMBU, PP&S
Kingsway West
Dundee
DD2 3XX
mark.westwater@dundee.ncr.com
+44 (382) 592516

Graham I. Johnson

AT&T Global Information Solutions
Technology Development
Kingsway West
Dundee
DD2 3XX
graham.johnson@dundee.ncr.com
+44 (382) 592633

This study investigated the effectiveness and efficiency of three usability evaluation and inspection techniques: heuristic (expert) evaluation, checklist-based approach, and empirical user testing using two pen-based Personal Digital Assistants (PDA's) as a case study. The purpose of the study was to provide insight into the major differences, advantages and disadvantages of these three approaches, as this is of interest to human factors practitioners. From the findings, heuristic and checklist evaluation approaches were able to identify many usability problems in a cost-effective manner. Our study demonstrated that user-based evaluations offer much more information on the nature of usability issues, and seem to provide a greater diagnostic value as far as development is concerned

Introduction

It is generally accepted that the software development process is frequently governed by resource constraints, be they financial, time or staffing. A host of "quick and dirty" evaluation techniques (Nielsen, 1992) have evolved which attempt to combat the restrictions placed on the development process. These techniques are collectively known as "walkthrough evaluations" (Nielsen, 1993). The benefits of such techniques are unquestionable and, in recent years, well documented (Karat et al., 1992). Also, the benefits gained from empirical testing are clear, and it has been suggested that these user-based approaches are more effective but less efficient at highlighting usability problems (Karat et al, 1992; Desurvire et al, 1992).

Comparison of usability evaluation techniques has become relatively prolific in recent years but little has been undertaken regarding the appropriateness of the usability checklist approach despite its application in industry (Johnson, 1995). Nielsen and Phillips (1993) concluded that user-based evaluations are still considered the best predictor of field performance (and hence the identification of usability problems) although it was much more expensive than the heuristic evaluation. Karat et al. (1992) found that empirical user testing identified four times as many problems and in many cases was more cost-effective than the walkthrough techniques employed. This is in contrast to Jeffries et al. (1991) who found that walkthrough methods, such as the heuristic evaluation were just as effective and more cost-effective than the empirical testing method at identifying the greatest number of and the most severe usability problems.

Within the Technology Development Group at AT&T GIS, Dundee, a usability evaluation of two hand-held, pen-based Personal Digital Assistants (PDAs), with the twofold purpose of (i) determining design directions for a future interface concept applying similar technology to that of PDAs, and (ii) assessing the relative effectiveness of three usability evaluation techniques, namely an Heuristic (expert) Evaluation (Nielsen, 1992), a usability checklist evaluation (Ravden and Johnson, 1989), and empirical user testing. The following areas were focused upon within the comparison of the evaluation techniques.

- *Usability Problems:* Issues investigated were basic metrics such as the number of usability problems identified and the estimated severity and type of usability problems.
- *Cost-effectiveness:* Regarded as the amount of participant hours required, the amount of human factors involvement and knowledge needed, and the equipment required.
- *Diagnostic Value*: The value of the evaluation data in facilitating redesign and endeavours towards further iterations.

The two PDAs used as the subject matter for this exercise were the AT&T E0440 and the Amstrad PenPad PDA600. They both offer basic functions such as address books, diaries, notebooks, and so forth, and both are operated using pen input. Each has a small screen and relies upon graphical and iconic styles within the user interface.

Heuristic Evaluation

This is a semi-formal inspection or evaluation method in which a number of evaluators derive a list of usability problems guided by accepted "usability heuristics" (Neilsen and Molich, 1990). The method does not prescribe a specific set of methods, rules or procedures to be used and as a result is open to varying degrees of interpretation. This type of evaluation is generally carried out by experts with the interface, either interface designers or human factors practitioners.

User-Centred Evaluation

This approach allows the evaluators to manipulate a number of factors (variables) associated with the interface design and study their effect on various aspects of user performance. User trials, typically within an experimental design, provide information about how the intended or potential end-users will experience the interface, the likely problems associated with, and end-user attitudes towards it. Many factors are considered within user-centred evaluation, with the focus on dependent measures in a controlled environment such as a laboratory.

Checklist-Based Usability Evaluation

The usability checklist applied was that of Ravden and Johnson (1989) which was originally designed as a means of carrying out basic usability evaluations. The checklist consists of nine criterion-driven sections with each section comprehensively broken down into a number of related questions, anchored with simple four point response scales. The checklist also incorporates two general sections, one closed-question section investigating users experiences with the interface and an open-ended section which considers general issues.

The Evaluation Approaches - Procedures

All trials were carried out at AT&T Dundee. Participants were volunteers and from the local organisation.

Heuristic Evaluation

This exercise involved eight participants, four of which are Human Factors practitioners (all novice PDA users) and four PDA expert users (an "expert" was defined as someone who had used such a device, at varying levels, for longer then 6 months). Each participant carried out four basic tasks, two entries into each PDA, an address book entry and a day planner entry in order to compare like with like. Participants were also given ten minutes to explore each of the interfaces.

After task completion, a standard questionnaire incorporating the "usability heuristics" (Nielsen and Molich, 1990) and related items was completed by each of the participants for each PDA.

User-Centred Evaluation

There were eight participants who took part in the trials. All subjects were regular computer users during work and at home (with an average of four to five hours per day). The participants had varying degrees of pen-computing experience ranging from having never used one to being relatively proficient with such devices.

The study was a within subjects design balanced for handedness, sex, level of pen-computing experience, and order of PDA usage. The study consisted of participants undertaking pre-determined tasks on both PDA's over two sessions (two days apart).

Both objective data and subjective responses were recorded. Objective measures were of the form of a number of requests for help, and "slips" and "mistakes" (Norman, 1983) made during the tasks. Subjective data were in the form of a questionnaire, ad hoc verbalization during the tasks, and a semi-structured interview. All trials and interviews were recorded on audio cassette for later transcription and analysis, whilst task performance with the focus upon errors, were observed and noted against the prepared task framework by the experimenter.

Checklist-Based Evaluation

The study enlisted eight participants (four human factors practitioners and four software engineers). The participants carried out the same tasks, receiving the same written instructions, as in the heuristic evaluation with each participant completing two checklists, one for each PDA. Trials, in total, took approximately one hour to complete with evaluators working through the checklist items having experienced the PDAs via the set tasks.

Results

The results of this investigation are grouped into three general categories which attempted to provide useful indications of the ability of the approach to identify usability problems and issues, the costs associated with use of the approach, and the application scope of the approach.

Usability Problems

The total number of usability problems (instances minus repeats) identified was greatest for the checklist-based approach followed by empirical user testing and the heuristic evaluation (87→48→28 respectively). NB. What constituted a usability problem varied across the three approaches.

The variation in the usability problem set (i.e. the range of problems 'found') identified was non significantly greatest for the checklist-based evaluation. As expected there was a positive correlation between the number of usability problems identified and the greatest variability in the identified problem set. Usability issues in the main focused on the system structure (navigation), system feedback, screen organisation etc. Many of these problems were highlighted during the other two evaluation approaches but the "one-off" problems identified were missed during the checklist evaluation.

A majority of the severe usability problems (significant consequence for task completion) were highlighted during the user trials. Both the heuristic and checklist evaluations identified "severe" problems but to a far lesser extent and as previously mentioned pinpointed many "one-off" problems specific to the tasks employed.

Cost-effectiveness

Human factors involvement was, as expected, greatest for the empirical user trials in terms of trial organisation, administration and analysis. However, when considering both the heuristic and the checklist evaluations, human factors involvement was minimal. The checklist evaluation required only the test organisation (2 hours) and a relatively short analysis phase (3.5 hours). Greatest time required for the heuristic evaluation was in terms of organisation which included questionnaire design and task selection, however, this is only with this interpretation of the approach.

Participant hours was also greatest for the empirical user tests (however, this study was over two sessions which effectively doubled the participant hours). The length of time required for the checklist evaluation was due primarily to the length of the checklist. The empirical approach also required the most equipment (lab space availability, and audio recording equipment) whereas only the PDA's were required for the other two techniques.

Human factors knowledge required varied across the three approaches. To continue the trend, the empirical trials incorporated the need for the greatest human factors involvement. This was in terms of questionnaire design, task selection, error identification, statistical analysis and conducting a semi-structured interview.

Diagnostic Value

In terms of the value of data for possible system redesign the empirical user trials delivered the most appropriate data. Usability problems such as lack of system structure and poor task organisation were identified during the user trials. The nature of the usability checklist does not lend itself to identify specific usability problems. However, the checklist highlighted a great number of individual usability issues, although unlike the formal user trials, the context of these problems is absent. Similarly, the heuristic evaluation could provide little context to problems.

Table 1 (below) presents a summary of the key findings for the three evaluation methods used.

Table 1. Summary of Evaluation Approaches

	Heuristic	User-Centred	Checklist-based
Usability Problem Number	28	48	87
Ability to Identify Severe Usability Problems	Medium	High	Medium
Usability Problem Scope	Narrow	Broad	Narrow
Evaluation Planning & Organ.	8 hours	19 hours	2 hours
Evaluation Administration	2.5 hours	16 hours	2 hours
Data Analysis	4 hours	24.5 hours	3.5 hours
Equipment Required	Low	High	Low
Participant Hours	6 hours	16 hours	9.5 hours
HF Knowledge	Medium	High	Low
Diagnostic Value	Medium	High	Low
Appl. within Development	Throughout	Dependent on functionality	Throughout
Flexibility of the approach	Limited	High	Limited

NB. The user-centred evaluation was carried out over two sessions to investigate system learnability. The results reflect the data collated over the two trials.

Discussion

The case study described shows that the heuristic and checklist approaches are relatively cost-effective. The large number of usability problems identified by the checklist approach seems to be a by-product of the method: On closer examination many of these usability problems are actually closely related to one another.

Our overall analysis of the approaches confirms that the user-centred, experimental approach does incur cost of various types although this approach is flexible, successful in identifying major usability issues, and has a relatively high diagnostic value. Illuminating the reasons 'behind' usability problems, and the context and consequences of these problems, not surprisingly is the major strength of this user trial approach.

Our attempts to make comparisons of usability problem numbers, as with other comparative studies, meet with some difficulty in the definition of a 'usability problem'. Also, the degree to which these three examples of the application of the methods meets national standard or typical cases is open to debate. Differentiating the techniques can also prove problematic: For example, the checklist and heuristic evaluation approaches are quite close to one another in terms of basic assumptions (Johnson, 1995).

As far as the development cycle is concerned, the scope of these techniques is apparent. Whilst this study tackled marketed products, there exists real limitations for formal user-based trials early within the development cycle. The human factors knowledge required in the design and conduct of user-based evaluations also seems to differentiate this from the other two methods.

Clearly, this investigation is based upon a limited case study with PDAs. The extent to which we can generalize our experiences from this (type of product and interface) is difficult to judge. It would be interesting to consider products and interfaces in very different categories, as well as other approaches, especially those involving cognitive modeling of the end-user.

Further work

Looking into the relative efficiencies and knowledge requirements of these approaches is relevant to any practitioner who is constantly faced with critical resource issues. Assessing the 'diagnostic value' of these approaches, and their ability to pick up on usability strengths (rather than weaknesses) is also crucially important within development, and ought to be the subject of further work.

Acknowledgment

Our thanks are due to AT&T Global Information Solutions, Technology Development Group, Dundee for support of this work.

References

Desurvire, H., Lawrence, D., and Atwood, M.E. 1991, Empiricism versus judgement: Comparing user interface evaluation methods on a new telephone based interface, *SIGCHI Bulletin,* 23(4), 58-59.

Jeffries, R., Miller, J.R., Wharton, C., and Uyeda, K.M. User interface evaluation in the real world: A comparison of four techniques. *Proceedings of CHI '91 Conference,* New Orleans, LA. 27 April-2 May, (ACM Press, NY.), 119-124.

Johnson, G.I. 1995, The usability checklist approach revisited, To appear in Jordan, P. et al. (Eds.), *Usability Evaluation in Industry,* (Taylor & Francis, London)

Karat, C.M., Campbell, R., and Fiegel, T. 1992, Comparison of empirical testing and walkthrough methods in user interface evaluation. *Proceedings of CHI '92,* Monterey, CA, 3-7 May, (ACM Press, NY.), 397-404.

Nielsen, J. 1992, Finding usability problems through heuristic evaluation. *Proceedings of CHI '92,* Monterey, CA, 3-7 May, (ACM Press, NY.), 373-380.

Nielsen, J. 1994, Usability inspection models. *Proceedings of CHI '94,* Boston, Massachusetts, 24-28 April, (ACM Press, NY.), 8-12.

Nielsen, J., and Molich, R. Heuristic evaluation of user interfaces. *Proceedings of CHI '90,* Seattle, WA, 1-5 April 1990, (ACM Press, NY.), 249-256.

Nielsen, J., and Phillips, V.L. 1993, Estimating the relative usability of two interfaces: Heuristic, formal, and empirical methods compared. *Proceedings of INTERCHI'93,* Amsterdam, the Netherlands, 24-29 April, (ACM Press, NY.), 214-221.

Norman, D.A. 1983, Design rules based on the analyses of human error. *Communications of the ACM,* 26(4), 254-258.

Ravden, S.J., and Johnson. G.I. 1989, *Evaluating Usability of Human-Computer Interfaces: A Practical Method,* (Chichester: Ellis Horwood)

EVALUATION OF STANDING AIDS FOR THE WORKPLACE

R.S.Whistance and R.S.Bridger

*Department of Biomedical Engineering,
University of Cape Town Medical School,
Observatory, Cape 7925, South Africa*

Providing some degree of support to standing workers while still enabling them to be upright and able to work may help to prevent or alleviate some of the problems related to prolonged constrained standing. This study examines the use of three standing aids to determine whether subjects with access to these aids experience significantly less discomfort than a control group. The results show that those subjects using the aids were no more comfortable than the control group although interesting patterns of interaction with the aids emerged.

Introduction

Surveys in factories (Schierhout et al,1993) have shown that constrained standing is a common working posture in many industries. Relatively fixed standing postures adopted throughout the working day over a period of years or even a lifetime can lead to serious long-term disorders like deterioration of the intervertebral discs, varicose veins and thromboses of the leg. (Grandjean,1980)

Workspace design conventionally caters for a limited range of working postures, mainly sitting or standing. However, according to Hewes (1957), the ordinary upright stance is only one of approximately 1,000 different static postures which human beings can assume and maintain comfortably for some time. He suggests that we should explore the possible usefulness of alternate postures when planning workspaces.

In this study, standing aids were used to test the hypothesis that enabling the subject to have some degree of support while remaining in an upright position would decrease discomfort during prolonged constrained standing.

Method

Apparatus

Three standing aids were used in this study:
- a two-meter long 30mm diameter wooden pole with a tennis racquet grip.
- a padded kneerest of adjustable height. The padded area was 500mm by 250mm and was covered in vinyl.
- a footrail with four rails which were 500mm wide and 120mm, 240mm, 360mm and 480mm from the floor.

Subjects

15 healthy male and 26 healthy female subjects participated in the study. The mean age for the males was 22,1 years, their mean weight was 72,9 kg and their mean height was 175cm. The mean age for the females was 21,1 years, their mean weight was 61,4 kg and their mean height was 160,3 cm. Of the 41 subjects, 36 were right handed (87,8%) and 5 were left handed (12,2%).

Procedure

Subjects were randomly assigned to one of four groups:
- used the footrail
- used the kneerest
- used the stick
- had no standing aid.

The subjects stood barefoot or in socks for 240 minutes each with six 3-minute breaks (after each 27 minutes of standing)(Rys & Konz,1989). During the 27 minute session they were confined to a carpeted area 500mm by 500mm and during the 3 minute break they could walk about or rest comfortably. The kneerest and the footrail were constructed so as to fit directly in front of the 500mm by 500mm square. Subjects were videotaped while they watched popular films on video.

At the end of each 27 minute session, subjects were shown a body diagram and asked to point out up to 3 zones where they had experienced discomfort and/or pain. They were also asked to rate their overall comfort on a scale of 1 to 7, where 1 represents the total absence of discomfort and 7 represents acute discomfort or pain.

Observation and data capture

Each videotape was played back and for each session, the number of times the subject made contact with the standing aid, the duration (seconds) of each contact and, in the case of the footrail or kneerest, whether the right or left leg was lifted were recorded.

Results

Interaction with standing aid

Table 1 shows the total contact time overall for all subjects for each standing aid as well as the percentage of the total time available (129600 seconds).

Table 1. Total contact time overall (seconds)

Footrail	17371	(13,4%)
Kneerest	34140	(26,3%)
Stick	12122	(9,4%)

There was significantly more contact overall made with the kneerest than with the other two standing aids. (Anova: F-statistic - 19,4; significance level < 0,01) There was also a significantly greater number of contacts made with the kneerest than with the other two aids.(Anova: F-statistic - 32; significance level < 0,01) For all subjects, 260 contacts were made with the kneerest, 187 with the footrail and 18 with the stick.

Number of contacts with the stick were considerably lower than the other two aids. However, the different standing aids had a significant effect on mean duration. (Anova: F-statistic - 6,6; significance level <0,01) The mean duration for the stick was 673,4 seconds, for the kneerest 130,2 seconds and for the footrail 92,5 seconds. Contacts with the kneerest were therefore not only more frequent than contacts with the footrail but also of longer duration. Contacts with the stick were very infrequent but considerably longer in duration.

Total duration of contacts varied significantly over the eight periods. (Anova: F-statistic - 5,7; significance level < 0,01). **Figure 1** shows how, for all standing aids, total contact time increased the longer the subjects stood.

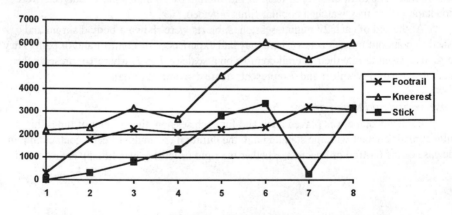

Figure 1. Total contact time (seconds) over the eight 27 minute periods.

As mean contact duration did not vary significantly over periods, increased total contact time is explained by the significant variation of number of contacts over the eight 27-minute periods. (Anova: F statistic 3,2 ; significance level <0,05). **Figure 2** shows how number of contacts with the footrail and kneerest increased over time, particularly in the last two periods. Contacts with the stick remained fairly constant over time.

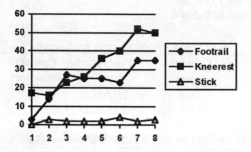

Figure 2. Number of contacts with standing aids over the eight periods.

Comfort Rating

Comfort rating differed significantly between the four groups. (Anova: F-statistic - 11,12, significance level < 0,05). **Table 2** shows the mean ratings.

Table 2. Mean Comfort Ratings

Group	Mean Rating
Footrail	3,50
Kneerest	4,18
Stick	3,66
Control	3,80

There was a significant difference between groups.(Anova test statistic - 11,13; significance level < 0,05) The group using the kneerest reported greater discomfort than the other three groups.

Body Zones

Discomfort was predominantly reported in the lower legs followed closely by the lower back. **Table 3** shows the percentages for legs and lower back for the four groups.

Table 3. Percentages for legs and lower back.

	Footrail	Kneerest	Stick	Control
Legs	43,4%	37,2%	47,7%	43,7%
Lower back	24,8%	36,6%	36,7%	29,8%

On average, legs and lower back accounted for 75% of the body zones indicated while neck, shoulders and thighs accounted for 18,2%.

Preferred Rungs of the Footrail
Results are shown in **Table 4.**

Table 4 - Footrail rung preference.

Rung	Contact (secs)	Percent. total	Contact (number)	Percent. total
1	3578	20,6%	44	23,5%
2	4742	27,4%	56	29,9%
3	5839	33,7%	51	27,3%
4	3170	18,3%	36	19,3%

Subjects moved their feet more frequently onto Rung 2 although they kept their feet longest on Rung 3.

Left and Right side preferences
There was not a significant preference for one side. For the footrail, subjects used their right feet for 58,8% of the contacts and their left feet for 41,2%. For the kneerest subjects used their right legs for 52,7% of the contacts and their left legs for 47,3%.

DISCUSSION

Contacts with the standing aids increased over time and this was accounted for by the increased number of contacts and not by longer duration of individual contacts. The kneerest had the longest contact time overall. Subjects lifted their legs up to the rest 40% more than subjects lifted their feet onto the footrail.

However, the group using the kneerest reported the greatest discomfort. The results clearly show that use of the standing aids did not decrease discomfort. The aids seem to have served more as a means of distraction over the long session which would explain why number of contacts increased over time. It is possible that the subjects felt the need to move more over time due to ischaemia or venous pooling in the legs and feet. The kneerest was padded and it is also possible that this, although apparently offering no significant relief, invited more contacts than the footrail.

The results suggest that rung 3 of the footrail (at 360mm above the ground) was more comfortable than the other rungs as subjects used it the longest. However, rung 2 was used more frequently and it is possible that this was because at 240mm from the ground it was nearer and more accessible.

The subjects showed a marked reluctance to use the stick and it is possible that they did not know how to use it the way a Nilotic herdsman might, for example. Ignorance of how any of the three aids might best be used to alleviate discomfort may account for the fact that they gave no significant relief.

CONCLUSIONS

Within the scope of this study it can be concluded that:
1. Discomfort during prolonged standing increases over time.
2. The use of the standing aids chosen for this study either were not suited to significantly alleviate the discomfort experienced or they were not used optimally.
3. Prolonged constrained standing causes discomfort predominantly in the lower back and legs.
4. Prolonged constrained standing results in increased movement
5. Subjects used the standing aids bilaterally.

REFERENCES

Grandjean, E. 1980, *Fitting the Task to the Man.* 3rd edn (Taylor and Francis, London)
Hewes, G.W. 1957, The anthropology of posture, Scientific American, **196,** 122-133.
Rys, M. and Konz, S. 1989, Standing with one foot forward, In A. Mital (ed.), *Advances in Industrial Ergonomics and Safety I,* (Taylor and Francis, London) 207-214.
Schierhout, G.H., Myers,J.E. and Bridger,R.S. 1993, Musculoskeletal pain and workplace ergonomic stressors in manufacturing industry in South Africa, International Journal of Industrial Ergonomics, **12,** 3-11.

SURVEY OF THE RECOGNITION OF ICONS ON VIDEO RECORDERS

Martin Maguire and Lindsey Butters

HUSAT Research Institute
The Elms, Elms Grove
Loughborough
Leics LE11 1RG

m.c.maguire@lut.ac.uk
l.m.butters@lut.ac.uk

This paper is concerned with the effectiveness of icons as labels for controls on electronic devices in the home. It reports on a study across which captured information about people's ability to recognise deck control symbols (or icons) on audio visual equipment. It also reports on a pilot survey investigating conventions in the use of icons in the home on a wider range of devices.

Introduction

Icons are described by Horton (1994) as small pictorial symbols used on computer menus, windows, and screens representing certain capabilities of the system. Wood and Wood (1987) take a broader view of the term icon as any symbol, image or pictograph used to represent a concept, idea or physical object. Thus icons are widely used in public buildings (train stations/airports) for visitor guidance. Their application in electronic home products is strongly encouraged for the following reasons:

• They can support of the extensive human ability of pattern recognition.
• They can offer language independence leading to easy transfer of home products between countries.
• They reduce required space for information presentation.
• They offer a certain level of aesthetic appeal.

However, the design of icons for a device has to be carried out with care if they are to be effective and not to confuse and annoy users.

Attributes of effective icons

A number of authors have presented what they consider to be important attributes for the design of effective icons. Horton (1994) for instance states that icons should be: understandable, unambiguous, informative, distinct, memorable, coherent, familiar, legible, few in number, compact, attractive and extensible.
Based on the ideas of Kato (1972) and Wood and Wood (1987), attributes of good icons include:

• Easy association with the message.
• Distinguishable from other symbols.
• Not overly complex.
• Can be combined to represent interrelated concepts.
• Suitable for different cultures and uncontroversial.
• Accords with international or accepted standards.

Easy association with the message.

An icon can represent a function or feature in a number of ways. Rogers (1989) presents four different ways by which an icon can represent its underlying concept: viz. resemblance, being an exemplar, being symbolic, or by being arbitrary. Paraphrasing Rogers (1989):

Resemblance icons present a direct or analogous image of the function or concept itself. Thus, the road sign for 'falling rocks' presents a clear resemblance of the roadside hazard. The trash can on the Macintosh desktop is another example of an icon resembling its function.

Exemplar icons provide examples to represent their meaning such as a knife and fork used in the public information sign to represent 'restaurant services'. The image shows the most basic attribute of what is done in a restaurant i.e. eating.

Symbolic icons are used to convey the underlying referent that is at a higher level of abstraction than the image itself e.g. the picture of a wine glass with a fracture to convey the concept of fragility.

Arbitrary icons bear no relationship to their intended meaning so the association must be learned e.g. the bio-hazard sign consisting of three partially overlaid circles. Note that arbitrary icons should not be regarded as poor designs, even though they must be learned. As Wood and Wood state, these symbols may be chosen to be as unique or compact as possible such as a red no entry sign with a white horizontal bar, designed to avoid dangerous misinterpretation.

Distinguishable from other symbols.

Where a range of icons are needed for related functions, similarity between icons can be useful. For example, if an up arrow is used to indicate the movement of a selection bar upwards through a menu list, then a down arrow will be taken to mean movement downwards through the list. However, if too many similar symbols are used for different functions, this can cause confusion. For example, three clock symbols are shown on a teletext handset: one to display the time, one to set a time when a page is displayed and a third to cancel that time.

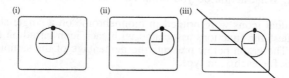

Not overly complex.

A common problem with icon design is excessive complexity. An icon may be designed on a large scale which is then hard to recognise when reproduced on a smaller scale on a key-top. In a teaching exercise to design icons for telephone functions that HUSAT occasionally carries out, delegates or students often draw complex icons that would need to be greatly simplified to be practical. As well as being distinctive, icons which are not too complex are also more easily reproduced on different displays at varying scales and resolutions.

Can be combined to represent interrelated concepts.

By making a good choice of icon elements, they provide a rich symbolic language. If one can learn the meaning of the basic elements, the combined elements become more self-explanatory. For example, using the play symbol ▶ to represent standard play speed, two symbols ▶▶ gives an indication of a faster speed.

Suitable for different cultures and uncontroversial.

Some symbols may be inappropriate for certain cultures. For example, the red cross symbol for medical facilities is not used by Moslem populations. Similarly, a 'thumbs up' symbol, representing OK, is regarded as a crude gesture in certain countries.

Accords with international or accepted standards.
    Where existing standards and conventions exist, these should be used in
preference to creating new icons unless there is reason to be believe that an icon will
not be understood. For example, the TV icons for brightness, colour and contrast
control are widely used and so alternatives are unlikely to replace them.
    Icon standards for IT are now being established by bodies such as the ISO
committee ISO/IEC JTC 1/SC/18 (ISO, 1994). The ISO standard has reached a draft
stage and presents a framework for considering icons and how they link to functions
and objects (part 1). Part 2 offers a core set of icons for the electronic desktop.
Recommended methods of testing icons are also proposed e.g. Gili et al, 1991 and
Horton, 1994.
    In addition to the various good attributes of icons, Horton (op cit.) also cites
some common fallacies about icons he considers 'common nonsense' e.g. that the
designer can use icons to totally replace words, that icons necessarily make products
easier to use, and that an icons must be perfectly obvious in order for them to be
good ones.

Study of VCR tape deck icon recognition.

    A survey was carried out, as part of the ESPRIT FACE 6994 project, on
people's preferences for and usage of a range of electronic home devices. (Maguire
et al, 1993). Part of this survey looked at people's ability to recognise tape deck icons
commonly used on a video recorder front panel or remote control unit. The survey
involved interviewing 100 people in each of four European countries (UK, France,
Germany and Italy). Each respondent was either an owner or user of a video recorder
of four years old or less. A balance was struck between male and female users and
across different age ranges.
    The particular question posed was as follows:

    "These are symbols sometimes used on the keys of TV and video recorder
    handsets. What action do you think each key would perform?"

    The individual icons were printed in monochrome on a single sheet. People
were asked to guess the functions associated with each icon based on their experience
of using VCRs. The graph below presents the percentage of the sample that made
correct responses for each icon type.

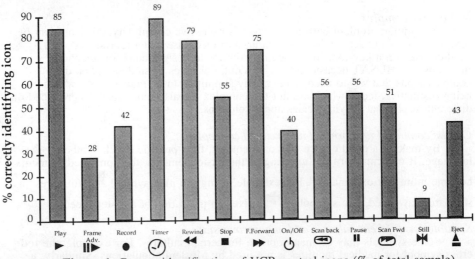

Figure 1. Correct identification of VCR control icons (% of total sample)

The symbols with the highest rate of recognition (over 75%) were:

Timer	Play	Rewind	Fast forward
89%	85%	79%	75%

Some interesting misconceptions included mistaking the filled circle 'record' symbol for 'stop'; perhaps people equated it with a traffic circular stop sign. The symbols for 'scan back' and 'scan forward' were understood by 50% and 51% of people respectively while sizeable numbers (17% and 16%) equated them with simple 'rewind' and 'fast forward'. This is understandable given the visual similarity of the two sets of symbols. The least well known symbol was 'still picture' which was recognised by only 9% of respondents. It is clear that while certain commonly-used VCR icons such as 'play', 'rewind' and 'fast forward' are reasonably well known, others such as 'pause', 'stop' and 'record' are much less familiar. The table below provides brief comments on the effectiveness of each icon in terms of the principles outlined earlier in the paper.

Table 1. Correct recognition of VCR icons (aggregate data)

SYMBOL		%	COMMENT
Timer		89	Easy association with physical object. Easily distinguished from other icons except on/off power switch
Play	▶	85	Commonly used. Good representation of action.
Rewind	◀◀	79	Commonly used. Combined element based on play symbol (reversed).
Fast forward	▶▶	75	Commonly used. Combined element based on play symbol.
Pause	‖	56	Slightly abstract.
Scan back		56	Not easily distinguished from rewind icon. Slightly complex.
Stop	■	55	Abstract symbol.
Scan forward		51	Not easily distinguished from fast forward icon. Slightly complex.
Eject	▲	43	Fairly good representation for top-loading VCRs. Less meaningful for front-loaders.
Record	●	42	Often thought of as 'stop', particularly if coloured red. Connotations of traffic lights, 'no entry' sign or full stop.
On/Off	⏻	42	Abstract symbol. More commonly understood in engineering.
Frame advance	‖▶	28	Fair representation of action although may be thought of as continuous movement (fast or slow).
Still		9	Fair representation of action. Not commonly used,

Comparison of icon recognition across countries.

The following table compares the results obtained within the four survey countries.

Table 2. Identification of VCR Control icons by country (%)

Symbol		Correct identification			
		UK	France	Ger	Italy
Play	▶	74	84	92	91
Frame Adv	‖▶	7	17	36	53
Record	●	17	52	44	59
Timer		86	89	93	87
Rewind	◀◀	78	68	86	84
Stop	■	39	57	59	69

Symbol		Correct identification			
Fast Forward	▶▶	77	65	83	77
On/off	⏻	23	41	34	63
Scan back	◀◀	23	72	54	57
Pause/still	II	49	56	51	69
Scan fast fwd	▶▶	24	73	54	56
Still	▶I	8	7	16	6
Eject	▲	21	45	36	72

Icon recognition seemed slightly poorer in the UK than in the other European countries surveyed. This may be because many VCRs sold in France, Germany and Italy often have icon labels written in English and that users may find it easier to learn the meanings of the iconic symbols rather than the English labels.

Pilot study of icons across a range of devices

Work has commenced on the use of icons across a range of household devices to investigate conventions that are used. From a small pilot across 10 households, a few initial findings are presented. There is a clear set of conventions which are used in relation to TVs with regard to picture and volume control (i), with some additional icons for sound quality (ii), but these appear less well known:

(i) ☼ ◑ ◐ ◀ ✕ (ii) ♭ ⌐ (tenor and bass)

Clearly, deck control icons on a VCR or Hi-Fi are well established amongst manufacturers: ◀◀ ▶▶ ▶ ■ II ● ▲ II▶ with I◀◀ ▶▶I being used for direct access devices (e.g. CD players) to jump to the previous and next tracks. Device 'on and off' representation varies, with symbols such as (1) being used on audio-visual devices and (2) being used on washing machines. Similarly cancel symbols on microwaves take a number of different forms e.g. (3):

1) ⏻ ⏼ ∘∘ ∘• ■ ▬   2) ⨁   3) ⊘ ✕ ⊘

A sequential set of icons are often used to represent washing and cooking intensity with different characteristics being selected for each:

Dishwasher   Cooker   Washing machine

The example dishwasher icons show a direct representation of contents, the cooker icons show the quantity of heat vapour rising and the washing machine icons show an indication of water temperature (lower number) and clothing type (upper number). Note that the use of solid colour for the first symbol indicates higher strength compared to an open character. However one model of microwave used the same symbol for intensity ◀ as the volume symbol on a television. The use of the star symbol is widely used in different contexts: e.g. to indicate (i) defrost on a microwave, (ii) a cold wash on a washing machine, (iii) lime removal on a dishwasher, and of course, special service on a telephone.

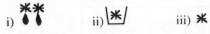

The symbol (a) was found to represent a fast spin on some washing machines and a long spin on others. Complementary symbols were (b) half length spin and (c) gentle spin.

a) 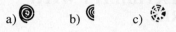 b)   c)

Some concepts seem hard to represent in iconic form and so letters are used e.g. (1) biowash-prewash and (2) economy wash. Symbols for programming a VCR are often related to a clock and the record symbol (3), while cookers may also incorporate a clock while the bell symbol is used frequently as being linked with the idea of an alarm sounding when the cooking period finishes (4). There is sometimes inconsistency between a clock symbol being used to indicate set clock and to indicate the setting of a programming period.

1) 2) 3) 4)

Conclusion

It seems that only a small subset of the commonly-used icons on electronic devices are well-understood. Icons are successful if, in the context of the device, they provide an easy association with the function. The 'play' icon is one such example. The use of combinations of the symbol to represent 'fast forward' and 'rewind' is a simple form of icon language put to good effect. However, confusion arises and mistakes are generated when a symbol such as the one used for 'record'has a strong but incorrect association ('stop'). Other common VCR symbols (such as those used for 'pause' and 'stop') are fairly abstract which restricts their ability to be memorised. These findings make the case for providing textual labels in addition to icons. The user can refer to the text label for learning purposes and the icon for more rapid use once familiarity with the product has been achieved. Icons clearly have an important role in conveying the meanings of controls or system states quickly and easily. Further work is needed to assess how effectively icons can be applied to a wide range of devices and possibly across devices in the home.

Acknowledgement: Thanks are due to T. Hewson, G. Allison and S. Hirst of HUSAT for their help with this paper.

References

Gili, J., Chacón, J.C., Menéndez, S. and Rodrígues, I., Telefónica, I+D and Clarke, A., Brown, S., Richardson, J., Hirst, S., HUSAT, 1991,*"Icons in Multipoint Videotelephony: An International Study"*, European Telecommunications Standards Institute, ETSI HF 1.1 (91), Version 2, 1991.
Telefónica Investigación y Desarrollo, Emilio Vargas, 6-28043, Madrid, Spain.
Horton, W., 1994
"The Icon Book. Visual symbols for Computer Systems and Documentation", John WIley and Sons, Inc.
ISO, 1994,*"Draft British Standard Implementation of ISO/IEC 11581-1 Information technology - User System Interfaces - Icon symbols and functions"*. Part 1: Icons - general, Part 2: Object Icons. BSI, 2 Park Street, London W1A 2BS.
Kato, S., 1972, "A New Universal Language for the New Human Environment", In *International Conference on Highway Sign Symbology*, International Road Federation and U.S. Department of Transportation, Washington, D.C. pp 13-19.
Maguire, M.C., McKnight, C. and Butters, L.M., *"European Market Survey"*, July 1993, 122 pp, CEC ESPRIT Project 6994, FACE Deliverable D8.
Rogers, Y, 1989, "Icons at the Interface: Their Usefulness". In: *Interacting with Computers*, vol 1, no 1, pp105-117.
Wood, W.T. and Wood, S.K.,1987, "Icons in Everyday Life". In: G. Salvendy, S.L. Sauter and J.J. Hurrell Jr. (ed), *Social, ergonomic and stress aspects of work with computers: proceedings of the second international conference on human-computer interaction.* Honolulu, Hawaii, August 10-14, (Elsevier Science Publishers B.V.) 1987.

FURTHER EXPLORATION OF COMPATIBILITY WITH WORDS AND PICTURES

RB Stammers

Psychology Group, Aston University
Birmingham, B4 7ET

A comparison is made between the results from a compatibility questionnaire reported on by Sidney L. Smith in 1981 and those gathered in 1994. Despite the difference in time and location of the studies, most results were supportive of Smith's findings. Where differences were found they could be explained on the basis the cultural and experience differences between the groups.

Introduction

Compatibility has been a central concept in Ergonomics for over 40 years. It is promoted as a goal of good ergonomics design without there being much explanation of what it is, or how to achieve it. It is usually couched in terms of naturalness, or population stereotypy of interface elements. Given its centrality to the discipline the concept of compatibility warrants closer examination.

One way in which this has been done is to examine how individuals respond to described and depicted situations. Smith reported on an 18 item compatibility questionnaire in 1981. He had data from three groups and was able to demonstrate strong consistency of response for some items, strong variability for others. There were also some interesting differences between groups, including gender differences.

Smith's Questionnaire and Samples Used

The questionnaire has items concerning display/control relationships and it also explores the use of language in relation to depicted scenes. It was felt interesting to compare the findings from the earlier work with a contemporary sample of subjects. However, on close examination of the original questionnaire some difficulties were seen. Some questions did not apply easily to a UK compared with a US sample, and some changes were made.

Smith's groups were:
1. 92 male engineers, age range 21-46 years, mean = 30.8
2. 80 female associates and relatives of group 1, age range 14-52 years, mean = 28.1
3. 55 human factors specialists, mostly male, age range 25-60 years, mean = 40.0.

The new sample consisted of 94 students taking psychology degrees, both single and joint honours. They were in their first term of study and had not yet had any ergonomics classes. Their details are as follows:

Overall age range 17.5-38.3, mean 20.9
72 were female, age range 17.5-38.3, mean = 20.8
22 were male, age range 18.3-33.8, mean = 21.4.

Results

The new results are presented for 11 of the 18 questions and they are compared with the earlier findings of Smith (1981). The coverage is limited for reasons of space, a fuller report is available. Smith's three groups are in the columns numbered 1-3, with the new results in column 4. In addition, the new data are also divided by gender under columns headed F (female) and M (male). Percentages are used throughout. The sub-heading numbers and titles used by Smith are repeated here.

1. Knob Turn

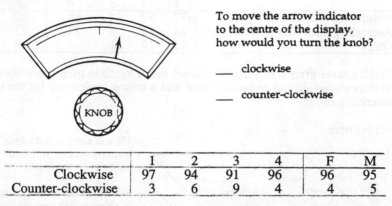

To move the arrow indicator
to the centre of the display,
how would you turn the knob?

____ clockwise

____ counter-clockwise

	1	2	3	4	F	M
Clockwise	97	94	91	96	96	95
Counter-clockwise	3	6	9	4	4	5

The original findings suggested a "strong, shared population stereotype" of the kind illustrated in many ergonomics textbooks. On the basis of very similar results this stereotype was shared by the subjects in the new sample.

2. Quadrant label

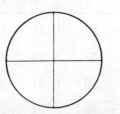

In what order would you label
the four quadrants of the circle?

Write in the letters A, B, C
and D, assigning one letter to
each quadrant.

A great deal of diversity was found in the original samples. As the data below illustrate the new sample adds to this diversity.

	1	2	3	4	F	M
Clockwise from upper right	33	26	45	26	28	18
Clockwise from upper left	19	11	5	27	23	41
Counter-clockwise from upper right	34	3	5	0	0	0
"Reading" order	14	54	43	44	46	36
Other	0	6	2	3	4	5

The current group was closer to Smith's female and human factors specialists (HFS) groups than to his engineers. The "reading" preference of "left-to-right, top-to-bottom" was popular with groups with 2 and 3 and with the new group, but the new group differed from the other two with a proportion choosing "clockwise from upper left".

4. Numbered keys

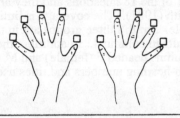

A worker is required to duplicate numbers as they appear on a screen, by pressing 10 keys, one for each finger.

Label the diagram to show how you would assign the 10 numerals to the 10 fingers.

	1	2	3	4	F	M
Ascending from left to right	70	70	84	68	69	67
Ascending outwards from thumbs	18	16	5	24	23	28
Other	12	14	11	8	8	5

Smith's three groups of subjects did not differ much in their preferences and the new data shows similar patterns. There was a strong preference for the left to right ascending order.

5. Stove burners

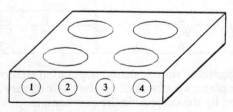

Here is a stove, with four burners on top, and four controls on the front.

Put a number on each burner to show which control should operate it.

Burner/Control Matching	1	2	3	4	F	M
1 4 / 2 3	13	5	7	17	15	23
1 3 / 2 4	20	23	33	27	29	18
2 4 / 1 3	41	53	42	20	21	16
2 3 / 1 4	26	16	16	20	22	14
Other	0	4	2	15	13	27

Smith's subjects showed a preference for one configuration, although not strongly. The new data shows a much wider spread of preferences. Why this should be so is not clear, but may reflect a much wider variety of configurations now commercially available.

6. Cross Faucets*

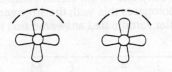

Here are two knobs on a bathroom sink, looking down at them.

Put an arrow on each dotted line, to show how you would operate these knobs to turn the water on.

Smith's subjects showed a variation in response here. The female group showed the strongest preference for the "paired clockwise turning", whilst the other two groups showed their strongest preference for the counter-clockwise pairing. The new data shows a fairly even split between these alternatives However, a closer examination of the gender differences in the new results shows a similar split to that found in the first study.

Left Faucet	Right Faucet	1	2	3	4	F	M
C	C	17	34	22	35	35	36
C	CC	23	20	13	17	18	14
CC	C	13	26	16	17	22	0
CC	CC	47	20	49	31	25	50

7. Refrigerator Door

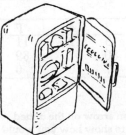

Here is a refrigerator.

Is its door

____ left-opening?

____ right-opening?

Smith found no strong pattern of interpretations here, although a slight reversal by the HFS was noted. The new group produce results that are almost at the mean of the earlier groups.

	1	2	3	4	F	M
"Left-opening'	36	41	56	46	49	36
"Right-opening	64	59	44	54	51	64

9. Digital counter

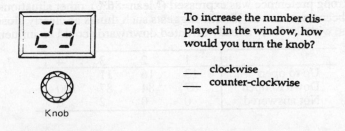

Knob

To increase the number displayed in the window, how would you turn the knob?

____ clockwise
____ counter-clockwise

---

* i.e., taps, although neither term was used in the questionnaires.

This example showed an almost universal response in the new sample with only one of the 94 subjects showing reversal. Only one of the earlier samples, the HFS, showed such a strong response. Smith compares this with the stronger pattern of responses to Item 1. A slightly smaller sample had answered this item in the first two samples.

	1	2	3	4	F	M
Clockwise	87	79	95	99	99	100
Counter-Clockwise	13	21	5	1	1	0

11. River Bank

Here is a river, flowing from East to West.

Is the church on the

___ left bank?

___ right bank?

A strong preference for the church being on the right bank was demonstrated in the earlier samples. Almost exactly the same results were found in this new sample.

	1	2	3	4	F	M
Left bank	18	16	13	12	11	14
Right bank	82	84	80	87	89	86
Not answered	0	0	7	0	0	0

14. Door handle

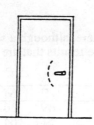

Put an arrow on the dotted line, to show how you would operate the handle to open this door.

It would seem that the universal experience of subjects in the new sample is of a down to open movement. This is in contrast to the earlier samples where, although a strong preference was expressed (Mean=86%), other situations had presumably been encountered. Smith suggests such things as clearly closed office cabinet doors, with the lever already oriented downwards could be influential.

	1	2	3	4
Up to open	13	16	11	0
Down to open	87	84	87	100
Not answered	0	0	2	0

16. Lever control

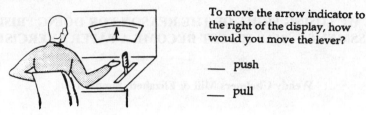

To move the arrow indicator to the right of the display, how would you move the lever?

___ push

___ pull

Traditional direction of motion stereotypes are reflected in the original data, and are supported by the new findings. A suggestion of a gender difference in the original samples is not borne out by the new data.

	1	2	3	4	F	M
Push	76	59	71	80	83	67
Pull	24	41	25	20	17	33
Not answered	0	0	4	0	0	0

17. Lever Faucets

Here are two knobs on a bathroom sink, looking down on them.

Put an arrow on each dotted line, to show how you would operate these knobs to turn the water on.

A very similar pattern of results was found in the new and the old data, with subjects expecting a "bilateral symmetry in forward or backward motion"

Left faucet	Right faucet	1	2	3	4	F	M
Back	Back	25	33	16	16	17	14
Back	Forward	3	3	5	13	8	27
Forward	Back	10	4	7	6	6	9
Forward	Forward	62	60	71	65	69	50

Discussion

It is reassuring that many of Smith's original findings are supported here, despite the passage of time and the fact that the new, younger, sample comes from a different country. Indeed, some of the "classic" stereotypes receive even stronger support than given in his samples. Some of the gender differences suggested in the two sets of results warrant closer attention. The question arises, of course, of how much paper exercises can be generalized to real equipment. Although this question remains to be explored, it is interesting that studies by the author of the ratings of computer screen icons for appropriateness showed them to be predictive of subsequent identification performance.

Reference
Smith, S.L. (1981) Exploring compatibility with words and pictures. **Human Factors, 23,** 305-315.

ARE WE LOSING SIGHT OF THE REASON FOR DOING "RISK ASSESSMENT " - HAS IT JUST BECOME A PAPER EXERCISE?

Wendy Chalmers Mill & Elizabeth Simpson

INTERACT
46 Kingsway
LONDON WC2B 6EN

Introduction

In order to analise RISK could be defined as:-
> *-THE CHANCE OF EXPOSURE TO THE ADVERSE CONSEQUENCES OF FUTURE EVENTS.*

If we then we consider **all** of the factors which contribute to these *"adverse consequences of the future"*, surely the quantification of the RISK will then never stay within fixed formulae. Therefore, when considering Health & Safety, as we are dealing with human beings, the risk will never be constant. However it is important that *"risk"* is viewed in a consistent way throughout an organisation and this consistency requires that a common perception of risk should be created and communicated to everyone within the organisation and perhaps to all professionals who carry out any Risk assessments.

Why do we do risk assessment ?

> *- TO COMPLY WITH LEGISLATION*
> *- TO PREVENT INJURIES IN THE WORKING*
> * ENVIRONMENT.*
> *- TO REDUCE ABSENTEEISM*
> *- TO IMPROVE INDIVIDUAL WORK EFFICIENCY AND*
> * IN TURN COMPANY PROFITS*

How many companies employ a consultant to provide the company with a risk assessment and then believe that they comply with 1992 D.S.E Regulations. It must be remembered that there are **guidelines** on **minimum** standards but over and above this staff must be assessed for any individual risk. *e,g,*

> *question -* *DO "our" CHAIRS COMPLY WITH THE LEGISLATION*
> *IF THEY ADJUST UP AND DOWN FROM THE SEAT*
> *AND THE SEAT BACK GOES IN AND OUT ?*
> *answer -* *DO YOU EMPLOY 5ft 5" CLONES ?*

The risk assessment is not just to see whether the company meets the minimum standards. It appears that consultants or *"risk assessors"* get engulfed by reams of documentation and loose sight of the practical application and the reason for assessments and therefore long term benefits to the company are lost

> *-DO WE IN FACT JUST FOLLOW `A` PROCEDURE*
> *TO THE EXCLUSION OF THE PURPOSE ?*

How many consultants are faced with a deluge of formulas, checklists, theories and strategies and totally loose sight of why the legislation was initially instigated and what the aims of the assessor should be :-

> *(1) TO REDUCE THE RISKS OF HEALTH HAZARDS AND INJURY FOR*
> *EMPLOYEES*
> *(11) TO REDUCE THE RISKS TO THE COMPANY RESULTING FROM ILL*
> *HEALTH WITH RESPECT TO:-*
> > *- REDUCTION IN WORK PERFORMANCE*
> > *- ILL-HEALTH ABSENCE COVER*
> > *- POTENTIAL FOR LITIGATION/COMPENSATION.*

How can we ensure that we do not loose the basic understanding that we are dealing with *individual* human beings in fixed working environments.

> *AREAS FOR CONSIDERATION*
> *(1) THE HUMAN(a variable factor)*
> *(11) THE WORK ENVIRONMENT(which can more easily be*
> *categorised within the "overused" checklist system)*

Checklists

Surely we then have to consider a means of risk assessment other than check lists or "furniture audits". Do these address all risks ? Will they in turn meet the aims as above? A check list system is a useful tool and it *may* address the work station however how often is their use inadequate ? Do they always address the employee who after all is the reason for a risk assessment.

Self assessment forms

Is the employee educated about or aware enough of potential hazards and risks to accurately complete a self assessment form ? As a means for on-going monitoring **after** an initial assessment a self assessment checklist may be of value.

However if we continue to only look at parts of the equation only half the answer, at best, is found and, in turn all the risks may not have been considered and there will be less potential benefit to industry.

Who should perform the assessment ?

We must also consider who is qualified to perform such a task. It now appears that we have a boom in *"qualified consultants"* performing **Risk** assessments. From:- Architects, Engineers, Osteopaths, Furniture suppliers, Ergonomists, Occupational health clinicians, Physiotherapists to the in-house assessors.

How do we then compile a system which addresses all the risks and then enables the risk assessor to prioritise the risks according firstly to the human being taking into consideration the 1992 Display Screen Equipment Regulations.

Points which we should consider.

- Should management have awareness training prior to the risk assessment ? Management should be aware that by reducing the sources or impact of risk, they will in turn increase their likelihood of success.
- How do we create a questionnaire or perhaps questionnaires which consider legislation requirements and *people*.
- What preparation or information may be required prior to the assessment.
- Should the staff have user training prior to the risk assessment
- Who is qualified to do the assessment.

Risk assessment must be looked at as a positive step towards optimal Human Resource Management and not an expensive waste of time and money. This attitude is constantly re-enforced unless risk assessments are carried out effectively. The following system is actively used in many companies and it continues to prove its value with regard to long term reduction and management of Risk.

How to classify the Human factors

The following system for prioritisation within Risk Assessment is focused on the prevention of ill health. Prior to assessment a questionnaire is completed by all staff to ascertain the nature of the work task and any existing musculo/skeletal or visual symptoms. The format of the questionnaire should allow an assessor to distinguish between 'normal' aches and significant symptoms which may be work related. NB (the assessor should not necessarily have any medical training). However in any Risk Assessor training these basics should be taught. Staff can then be classified on the basis of a potential for long term effects and damage, these physical factors as in Table 1, can be the causal factors in demotivating staff, resulting in a reduction in work quality standards.

STAGE 1.   Classification Of Staff

1. Pattern of symptoms
Postural aches ?
Significant symptoms ?
(Categorised according to numeric values on questionnaire 1. 2. 3. none.)

2. Intensity of D.S.E. use
Complexity of task ?
work organisation ?

INTERACT **SYMPTOM SURVEY**

NAME _____ **DATE** _____

COMPANY _____ **DEPT.** _____

AGE _____

<u>**TASK ANALYSIS**</u>   <u>**AVERAGE HOURS DAILY**</u>

Display screen
Equipment (D.S.E.)

<u>D.S.E. Intensive</u>   _____

<u>D.S.E. non Intensive</u>   _____

<u>Writing</u>   _____

<u>Telephone</u>   _____

<u>Clerical</u>   _____

<u>Subject to deadlines</u>   yes   no

<u>Breaks taken</u>   yes   no

Have you had any pain or discomfort over the last year ?   yes   no

If yes please number any areas of the drawing which cause you discomfort.

1. ache
2. pain
3. tingling or numbness

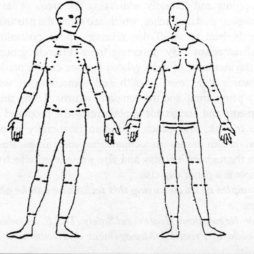

Table 1

(According to task analysis)

STAGE 2.   Priority for Assessment and time allocation
        (A). Any staff describing symptoms 3.
        (B). Any staff describing symptoms 2. or high intensity users
        (C). Staff describing symptoms 1. or no symptoms reported.
    It is advised that staff in categories A&B should be initially assessed by an outside consultant, to ensure that **all** the risks are eliminated, however all remaining staff could be assessed by in-house assessors. Following the completed initial assessment in-house assessors should monitor and maintain the programme, only using an outside consultant for specific difficulties which fall outside their remit.
Ref. A.C.O.P. Display screen equipment work,
    27 (e)" Those responsible for the assessments should   recognise their own limitations as to assessment so that further expertise can be called on if necessary. "

STAGE 3.   Priority for Corrective Action
    High risk   -Category **A**. where symptoms appear to be related to aspects
                  of the workstation or work organisation.
    Medium Risk - Category **B**. where symptoms appear to be related to aspects of
                  the workstation, work organisation , high intensity users or
                  where areas of on-compliance are identified
    Low Risk   - When areas of non compliance of the workstation or work
                  organisation are identified, which are not related to any
                  symptoms or high intensity users.

Conclusion

    This system is designed to produce an effective three stage approach to Risk assessment and implementation. Initially identifying and managing any existing work related symptoms and secondly addressing the areas of less urgency. It is designed to pick up common postural aches, which are often the precursors to work related injuries, and frequently lead to staff demotivation poor concentration and work quality and increased absenteeism. By targeting these different groups of individuals it has been found that the incidence of work related injuries can be markedly reduced, staff feel more positive towards their own   health and awareness at work thereby promoting their productivity programme and the company returns. Using this system companies can plan risk assessments and budget for implementing the proposed solutions in stages which are appropriate to risk, and which   are also financially manageable. Using this system, focusing on health issues in conjunction with fixed standards, will reduce general absenteeism throughout industry and also provide an effective, practical solution to what has now become a paper exercise.
N.B. :- Examples of studies using this technique can be obtained from INTERACT

References
Guidance on Regulations Health and Safety (D.S.E. Regulations 1992)
Approved code of Practice (Management of Health and Safety at Work Regulations 1992)

RISK FACTORS IN THE DEVELOPMENT OF UPPER LIMB DISORDERS IN KEYBOARD USERS

Craig Simmons and Elizabeth Lloyd

UserData Limited,
7-11 Kensington High Street,
London W8 5NP.
Tel: 0171 243 2925
Compuserve: 100431,2416

The role of ergonomic risk factors in the onset of various types of upper limb disorders in computer keyboard workers remains poorly understood. The study described in this paper considers the relationship between 12 ergonomic risk factors and the incidence of upper limb disorders in 18 keyboard users. The number of problems identified ranged from 3 to 10 per user. Incidence rates on the ergonomic criteria ranged from 11% (keyboard too low) to 100% (no health and safety training). A positive association was found between 'inadequate back support' and a diagnosis of 'diffuse upper limb pain'. As a result of this study further research is suggested to investigate the possible multi-factorial relationship between the ergonomic criteria and certain types of injury.

Introduction

In a recent Health and Safety Executive survey (Hodgson, Jones, Elliot and Osman, 1993), self-reports of musculoskeletal conditions far exceeded those of any other disease category. Whilst most of these were related to back complaints, a large proportion were classified as disorders of the upper limbs where work factors were thought to play a part. A similar pattern of injury is seen in the USA where over half of all worker compensation claims are attributed to 'Cumulative Trauma Disorders', a term synonymous with the HSE's preferred nomenclature of Upper Limb Disorders (or ULD's).

Such disorders are not new. As far back as the 18th Century the Italian physician Bernardini Ramazzini (1713) observed the work of scribes and notaries identifying their "certain violent and irregular motions and unnatural postures of the body" as a cause of musculoskeletal injuries. He concluded that the main factors in the onset of such conditions were a static working posture; repetitive motions and psychological stress. His observations still hold true today.

The modern 'scribe' may have replaced the quill with the keyboard but the recent upward trend in compensation claims would suggest that the risks involved in such work are similar. Epidemiological research studies in this field are scarce but have provided some evidence to link keyboard use with increased rates of ULD's (for example see English, MacLaren, Court-Brown, Hughes, Porter, Wallace, Graves, Pethick and Soutar, 1989; Hunting, Laubli and Grandjean, 1981; Rossignol, Morse, Summers and Pagnotto, 1987). Further evidence is provided by ergonomics and biomechanics studies and, despite their very different approaches to the investigation of work-related upper limb disorders, all three methodologies are now considered by the Health and Safety Executive (HSE) to be of equal validity (HSE, 1990).

Various ergonomic factors in keyboard work have been posited to put users at increased risk of upper limb disorders by their presence or absence. Most research also stresses that combining adverse conditions (poorly designed furniture with static posture, for instance) can increase the injurious effect (Hunting et al, 1981; Scalet, 1987; Mullaly and Grigg; 1988). The following broad categories of ergonomic factors have been identified (an exhaustive list is beyond the scope of this paper).

Posture

It is thought that the best posture for keyboard use is achieved with the elbows at the side of the body, the forearms approximately parallel to the floor and the wrists in a neutral position. There are those that believe this posture to be more of an ideal than achievable, due to the constraints imposed by the design of the traditional keyboard, which induces pronation of the forearms and wrist deviation in many individuals. See Pheasant (1986), Scalet (1987) and Ferguson and Duncan (1974) for further details.

Keying Rates

The general advice given is that keying rates in excess of 10,000 key depressions per hour put a user at increased risk of injury (London Hazards Centre, 1988) but the link between high rates of keying and musculoskeletal symptoms is obviously not straightforward. Pheasant (1991) quotes a study by Ong in 1984 where ergonomic improvements led to both an increase in keying rates and a reduction in discomfort.

Keying force

The reader is referred to Rose (1991) for a discussion of keyboard actuation forces.

Desk/keyboard height

The height of the keyboard surface is dependent on the height of the work surface and 'thickness' of the keyboard unit. The height relative to the user affects the degree of wrist extension/flexion and the forearm angle.

Document holders

Document, or copy, holders have been advocated as a means of decreasing neck and eye strain (see Hunting et al , 1981).

Footrests

The provision of a suitable footrest can enable a user to obtain a comfortable forearm height where the desk keyboard combination is too high, without losing lower limb and lumbar support from the chair.

Seating

Most important, from an upper limb disorder perspective, is that the seat provides good postural support and is sufficiently adjustable to facilitate the adoption of a comfortable, safe posture. However, as Pheasant (1986) points out, anthropometric data suggests that the relative adjustment between chair and keyboard is, for more than 50% of users, necessarily a compromise. General seating requirements are covered in a number of basic ergonomic texts (for example, Pheasant 1986, 1991; Grandjean, 1988) and HSE guidance (HSE, 1991).

Work duration/rest breaks

Using a computer for more than four hours a day, and in particular more than seven hours a day, has been associated with increased health risks (Rossignol et al, 1987). Rest breaks are important as they allow recovery time for fatigued muscles (Mullaly and Grigg, 1988; Aaras, 1987). Better job design, to include a mix of computer and non-computer-based tasks, has also been advocated (for example, see Linton and Kamwendo, 1989).

Psychological stress

This has been implicated in the development of musculoskeletal disorder and is thought to be particularly pertinent in computer users (Westgaard and Bjorklund, 1987; Smith).

Other factors

Factors such as training and the physical environment are also frequently considered to indirectly affect the health and well-being of the worker (for a discussion see Pheasant, 1991 and Green and Briggs, 1989 respectively).

In the course of our consultancy and litigation work we have noticed that the violation of accepted 'good ergonomic practice' is common, and felt that it would be informative to ascertain whether specific upper limb disorders could be related to these violations.

Method

18 computer keyboard users took part in this study. Each user had been diagnosed as suffering from an upper limb disorder by a registered orthopaedic surgeon. At the time of injury, their ages ranged from 22 to 51, all but one were female. In all cases their injuries were thought to be work-related.

All the users worked in different offices. With one exception, they all worked in different companies. Each used a unique combination of computer equipment and furniture from a range of manufacturers.

The users were categorised into the following groups based on the provided diagnosis (numbers in each group are shown in brackets. Note: some users suffered from more than one complaint): Tenosynovitis (9), Lateral epicondylitis (8), Carpal tunnel syndrome (3), Neck pain (5), Rotator cuff tendinitis (2) and Diffuse upper limb pain (7).

Each subject was interviewed and their re-constructed workplace examined. Anthropometric data was also recorded for each user. In some cases not all of the equipment or furniture used at the time of the injury was available for inspection. For these users 'best guess' estimates were used based on the available information.

As a result of this exercise the presence or absence of certain ergonomic risk factors was noted. Details of these follow.

The keyboard height was judged to be relatively too high if the height of the home row was 50mm or more above the seated elbow height. Similarly, the keyboard height was judged to be relatively too low if the height of the home row was 50mm or more below the seated elbow height. All figures were adjusted for postural slump. Keyboard heights within this range were judged to be acceptable. The horizontal angle of the keyboard to the front edge of the desk was measured. Where this exceeded 20° in either direction, the keyboard angle was judged to be unfavourable.

A footrest was seen as necessary where it was not provided and the seat height was adjusted in excess of the popliteal height of the user plus an allowance for footwear. When a footrest was provided or deemed unnecessary (due to the anthropometrics of the user and/or the seat height) the situation was judged to be acceptable.

A document holder was deemed necessary where it was not provided in circumstances where the user's work involved copy typing. Where the work involved audio typing or a document holder had been provided then the situation was judged to be acceptable.

Back support was judged to be poor where the chair backrest was either broken, the design or adjustment of the backrest was such that it did not give adequate lumbar support, or the user adopted a seated posture such that their back was not in contact with the backrest.

The angle of the screen surface to the front edge of the desk was measured. Where this exceeded 20° in either direction, the screen position was judged to be unfavourable.

The user was judged to have had sufficient rest breaks from computer-working if a typical day was made up of 20% or more non-sedentary work or they were provided with a break or change of activity for more than 5 minutes per hour.

Where a user recalled being in receipt of health & safety training and/or literature on computer usage prior to their injury this was noted. For the purposes of this study, no account was taken of the quality or content of the information received.

Where users reported high levels of mental stress at work in the period leading up to their injury this was noted.

Where users reported that they typed at more than 10,000 key depressions per hour (two-handed) or 5,000 per hour (one-handed) sustained for more than four hours per day, their typing load was judged to be high.

The workplace environment was judged to be poor if the user made a complaint concerning the lighting, heating, or ventilation.

Results and Analysis

The percentage of users whose workplace failed on each of the ergonomic criteria is shown in Tables 1 and 2. The number of problems found ranged from 3 to 10 per user (out of a maximum of 12).

Table 1. Percentage of users whose workplace failed the criteria (observed data)

Keyboard too low	Keyboard too high	No footrest	No copy holder	Poor back support	Keyboard at angle	Screen at angle
11%	72%	61%	72%	33%	28%	50%

Table 2. Percentage of users whose workplace failed the criteria (self-reported data)

Inadequate breaks	No H&S training	High Stress	Typing load	Poor environ.
89%	100%	83%	61%	83%

Ratings of the workplace, along with age and sex, were correlated with the diagnosis for each user using the Contingency Coefficient, a statistic suitable for nominal data. The resultant association matrix yielded one significant correlation; between 'diffuse upper limb pain' and 'inadequate back support' ($p < .005$).

Discussion

A generally high prevalence of ergonomic problems was found. Whether this is a feature of the users sampled or representative of the general population of keyboard users is, without the benefit of a control group, difficult to determine.

What is clear is that, in the case of this group, much readily available advice on the healthy and safe use of VDU's, such as that issued by the Health and Safety Executive (1983, 1986, 1991) has not been heeded. This situation may have been different had health and safety training been given.

Whilst it is not possible to establish any causal relationship within the constraints of the current study, the strong association found between the occurrence of diffuse pains and inadequate lumbar support warrants further investigation. Two possible explanations are suggested. Firstly, the absence of a back support is known to increase general upper body fatigue (Pheasant, 1991) which in turn has possible implications for the development of diffuse pain. Secondly, the tendency when presented with inadequate back support is to lean forward reducing the distance to the keyboard. To maintain contact with the keys then requires abduction of the elbows and increased ulnar deviation of the wrists. Stressing the wrists in such a way has been implicated in the development of ULD's (Rose, 1991; Duncan and Ferguson, 1974). It is known that postural factors are often of prime importance in determining the development of ULD's, and these have not been considered directly in this work. Implicit use of individual anthropometric data has been made in the development of some of these measures (keyboard height and footrest, for instance) but further work is planned which will use postural data more directly.

It is also worth highlighting some of the problems inherent in such a study. Whilst the workplaces were assessed by the same ergonomist, to ensure consistency, it was not feasible to arrange for the medical examinations to be undertaken by the same surgeon. Experience suggests that considerable variability exists in the diagnosis of ULD's.

The reader's attention is also drawn to the self-reported nature of some of the data. Whilst such a method is arguably appropriate for some measures (e.g. training, stress), quantitative data is perhaps more accurate in the case of others (e.g. keying rate, environmental conditions). Efforts were made to validate the self-reported data but in the majority of cases this was not practicable.

As acknowledged in the introduction to this paper, ULD's are multi-factorial in origin and future analyses aim to explore the more complex interactions that undoubtedly exist within the current data set.

References

Aaras, A., 1987, Postural load and the development of musculoskeletal illness, Scandinavian Journal of Rehabilitation, Supplement **18**, pp36.

Duncan, J. and Ferguson, D., 1974, Keyboard operating posture and symptoms in operating, Ergonomics, **1 7**(5), 651-662.

English, C.J., MacLaren, W.M., Court-Brown, C., Hughes. S.P.F., Porter, R.W., Wallace, W.A., Graves, E.J., Pethick, A.J., and Soutar, C.A., 1989. *Clinical epidemiological study of relations between upper limb soft tissue disorders ad repetitive motions at work.* IOM Report no. TM/88/19 UDC 616.74.

Ferguson , D., and Duncan, J., 1974, Keyboard design and operating posture, Ergonomics, **1 7**, 731-744.

Grandjean E., 1988, Fitting the task to the man (Taylor & Francis, London)

Green, R.A, and Briggs, C.A., 1989, Effect of overuse injury and the importance of training on the use of industrial workstations by keyboard operators, Journal of the American Medical Association, **3 1** (6), 557-562.

Health & Safety Executive, 1983, *Visual Display Units* ISBN 0 11 883685 (HMSO, London)

Health & Safety Executive, 1986, *Working with VDUs* IND(G) 36(L) (HMSO, London)

Health & Safety Executive, 1990, *Work Related Upper Limb Disorders - A guide to prevention* ISBN 0 11 885565 4 (HMSO, London)

Health & Safety Executive, 1991, *Seating at Work* ISBN 0 11 885431 3 (HMSO, London)

Hodgson, J.T., Jones, J.R., Elliot, R.C., and Osman, J. 1993, *Self-reported work related illness.* Health and Safety Executive; ISBN 0 7176 06074.

Hunting, W., Laubli, T., and Grandjean, E., 1981, Postural and visual loads at VDT workplaces 1: Constrained postures, Ergonomics, **2 4**(12), 917-931.

Linton, S.J. and Kamwendo, K., 1989. Risk factors in the psychosocial environment for neck and shoulder pain in secretaries, J. Occup. Medicine, **3 1**(7), 609-613.

London Hazards Centre, 1988, *Repetition Strain Injuries: Hidden harm from over-use* (London Hazards Centre, London).

Mullaly, J. and Grigg, L. RSI:Integrating major theories, Australian Journal of Psychology, **40**(1), 19-33.

Pheasant, S., 1986, *Bodyspace* (Taylor & Francis, London).

Pheasant, S., 1991, *Ergonomics, Work and Health* (Macmillan, London).

Ramazzini, B., 1713, *De Morbis Artificum.* Trans. W.C. Wright, 1940, Diseases of Workers (Chicago: University of Chicago)

Rose, M.J., 1991, Keyboard operating posture and actuation force: implications for muscle over-use, Applied Ergonomics, **2 2**(3), 198-203.

Rossignol, A.M., Morse, E.P., Summers, V.M., and Pagnotto, L.D., 1987, Video display terminal use and reported health symptoms among Massachusetts clerical workers, Journal of Occupational Medicine, **29**(2), 112-118.

Scalet, E., 1987, *VDT Health and Safety* (London: Ergosyst Associates).

Smith, M.J., Coehn, B.F. and Stammerjohn, L.W., 1981, An investigation of health complaints and job stress in video display operations, Human Factors, **2 3**,387-400.

Westgaard, R.H. and Bjorklund, R., 1987, Generation of muscle tension additional to postural load, Ergonomics, **3 0**(6), 911-923.

Author Index

Subject Index